Hybridized and Coupled Nanogenerators

Hybridized and Coupled Nanogenerators

Design, Performance, and Applications

Ya Yang

WILEY-VCH

Author

Ya Yang
Beijing Institute of Nanoenergy and Nanosystems,
Chinese Academy of Sciences
School of Nanoscience and Technology, University of Chinese Academy of Sciences
Rm. 315A, Tiangong Plaza, Tower C
30 Xueyuan Road, Haidian District
100083 Beijing
China

Cover Image
Courtesy of Ya Yang

Library of Congress Card No.:
applied for

British Library Cataloguing-in-Publication Data
A catalogue record for this book is available from the British Library.

Bibliographic information published by the Deutsche Nationalbibliothek
The Deutsche Nationalbibliothek lists this publication in the Deutsche Nationalbibliografie; detailed bibliographic data are available on the Internet at <http://dnb.d-nb.de>.

Print ISBN: 978-3-527-34634-9
ePDF ISBN: 978-3-527-82238-6
ePub ISBN: 978-3-527-82239-3
oBook ISBN: 978-3-527-34633-2

Typesetting SPi Global, Chennai, India
Printing and Binding Markono Print Media Pte Ltd., Singapore
Printed on acid-free paper

10 9 8 7 6 5 4 3 2 1

Contents

Forewords

Maxwell's equations are fundamental key equations in physics, which are the foundation of modern wireless communication, light communication, and energy science. Nanogenerators (NGs) are based on utilizing Maxwell's displacement current as the driving force to effectively convert mechanical energy into electricity, regardless of whatever used materials are nano or not. To date, NGs have attracted extensive interest due to potential applications as self-powered sensor networks, small power sources, and even large-scale blue energy harvesting. Hybridized NGs are based on integrating different NGs into a system that can simultaneously/individually scavenge the different types of energies in our living environment, so that a stable and sustainable power supply can be obtained by using whatever energy is available in the environment. Since we reported the first hybrid NGs in 2009 (Chen Xu, Xudong Wang, and Zhong Lin Wang, Nanowire structured hybrid cell for concurrently scavenging solar and mechanical energies, *JACS*, **131**, 2009, 5866–5872), a variety of NGs based on different structures have been demonstrated, including mechanical energy, solar energy, thermal energy, and even chemical energy. In recent years, hybridized NGs also include the integration of different NGs to scavenge the same type of energy but using different methods for largely enhancing the energy conversion efficiency. Part of this book will give a detailed summary about the design, performance, and applications of hybridized nanogenerators.

Zhong Lin Wang, Professor
Laureate of Albert Einstein World Award of Science
Laureate ENI award

Regents' Professor and Hightower Chair Professor
School of Materials Science and Engineering
Georgia Institute of Technology, USA

Director and Chief Scientist
Beijing Institute of Nanoenergy and Nanosystems, Chinese Academy of Sciences

Multifunctional materials have attracted more and more interest due to potential applications in energy scavenging units, sensors, and physical sciences. As one of the multifunctional materials, many ferroelectric materials have piezoelectric, pyroelectric, photovoltaic, and triboelectric effects, where the coupling among these effects is interesting. By utilizing these multifunctional materials, various one-structure-based coupled nanogenerators have been achieved to simultaneously or individually scavenge different energies from the living environment wherever and whenever any energy is available. The core of coupled nanogenerators is the multifunctional material, so that looking for the high-performance materials is very important to realize the maximized complementary scavenging of multiple energies from the ambient environment. Moreover, coupled nanogenerators have potential applications in multifunctional sensors due to the use of the same materials and electrodes, resulting in higher resolution, smaller volume, and lower cost as compared with those of various integrated sensors reported for the simultaneous detection of multi-physical signals. Some new physical effects such as thermo-phototronic effect are based on the coupled nanogenerators, where the new coupling enhancement phenomenon has been found between the thermoelectric and photovoltaic effects in some semiconducting materials. Part of this book will give a detailed summary about the design, performance, and applications of coupled nanogenerators.

Yue Zhang, Professor
Department of Materials Physics and Chemistry
School of Materials Science and Engineering
University of Science and Technology Beijing, China

Member of the Chinese Academy of Sciences

Preface

Scavenging multiple energies from our living environment is of critical importance to meet stable and sustainable energy needs. Various energies such as mechanical, thermal, and solar energies are not always available at the same time due to the weather or the working conditions. Nanogenerators are based on utilizing the Wang term $\partial \boldsymbol{P}_s/\partial t$ as the driving force to convert energies into electricity, regardless of whatever used materials are nano or not. The purpose of developing hybridized and coupled nanogenerators is to simultaneously or individually scavenge multi-mode energies from the environment; while hybridized nanogenerators are based on integrating the different nanogenerators into a system to scavenge the same or different energies for enhancing the energy conversion efficiency, and coupled nanogenerators are based on one material and one structure-based energy scavenging device with the same electrodes but different energy harvesting functionalities. Several new physical effects including pyro-phototronic effect, ferro-pyro-phototronic effect, and thermo-phototronic effect are based on the coupled nanogenerators; while both the pyro-phototronic effect and thermo-phototronic effect are focused on the semiconducting materials, the ferro-pyro-phototronic effect is focused on the ferroelectric materials. These physical effects have potential applications in solar cells, photodetectors, light emission diodes, and so on.

In this book, we present a comprehensive discussion about the design, performance, and applications of hybridized and coupled nanogenerators. The recent advancements have been summarized including wind-driven triboelectric nanogenerators, electromagnetic–triboelectric hybridized nanogenerators, photovoltaic–pyroelectric coupled nanogenerators, multi-effects coupled nanogenerators, and some new physical effects. The advantages and challenges of hybridized and coupled nanogenerators are discussed, and related perspectives on the opportunities for materials and new physical effects have been summarized.

Chapter 1 (contributor: Ya Yang) introduces the overall development of hybridized and coupled nanogenerators. The hybridized nanogenerators include hybrid energy cells, electromagnetic–triboelectric hybridized nanogenerators, and other hybridized nanogenerators. The coupled nanogenerators include pyroelectric and photovoltaic coupled nanogenerators and multi-effects coupled nanogenerators. The corresponding applications are also discussed.

Chapters 2 and 3 (contributors: Yang Wang, Ya Yang) discuss the recent development of wind-driven triboelectric nanogenerators and electromagnetic–triboelectric hybridized nanogenerators. The different working mechanisms with the different devices are introduced. The device working modes, structures, materials, performances, and applications are summarized.

Chapter 4 (contributors: Kai Song, Ya Yang) introduces the other hybridized nanogenerators, which includes hybridized photoelectric and piezoelectric nanogenerators, hybridized photoelectric and triboelectric nanogenerators, and hybridized photoelectric and pyroelectric nanogenerators.

Chapter 5 (contributors: Bangsen Ouyang, Ya Yang) summarizes the various devices by hybridizing nanogenerators and sensors. The sensors include pressure, strain, temperature, and magnetic sensors, and photodetectors. The performance parameters of these sensors include sensitivity, response speed, and stability.

Chapters 6 and 7 (contributors: Ding Zhang, Ya Yang) discuss the recent progress in the development of hybridizing nanogenerators and energy storage devices for simultaneous energy harvest and storage, and thermal energy technologies including pyroelectric and thermoelectric nanogenerators. Focus will be on materials selection, structure designs, performance optimization, and potential applications of different nanogenerators and nanogenerator-based energy harvest and storage devices. Particularly, the effects of nanomaterials with various morphologies and different structure designs on the performance of devices are compared and analyzed systematically. Finally, the opportunities, challenges, and perspectives these devices face within the energy field are discussed.

Chapters 8 and 9 (contributors: Yun Ji, Ya Yang) introduce recent advancements of multi-effects coupled nanogenerators, including photovoltaic–pyroelectric effect nanogenerator, pyro-piezoelectric effect nanogenerator, tribo-piezo-pyro-photoelectric effect nanogenerator, and so on. The two chapters emphasize the working principle, materials, device configuration, output performance, and application of the multi-effects coupled nanogenerators. Particularly, pyroelectric effect, photovoltaic effect, thermal effect, piezoelectric effect, triboelectric effect, and interactions among them are described in detail. Multifunctional ferroelectrics and semiconductors are introduced, such as $BaTiO_3$, $BiFeO_3$, ZnO, and SnS. State-of-the-art device structure and the relevant output electric signals (output current, voltage, and power) are presented. Applications of the multi-effects coupled nanogenerators in the area of energy-storage device charging, photodetection, multifunctional sensing, and image sensing are reviewed.

Chapter 10 (contributor: Ya Yang) elaborates the three new physical effects, which are based on coupled nanogenerators. Pyro-phototronic effect is based on the three-way coupling among semiconductor, pyroelectricity, and photoexcitation. The ferro-pyro-phototronic effect is based on the three-way coupling among ferroelectric materials, pyroelectricity, and photoexcitation. The thermo-phototronic effect is based on the three-way coupling among semiconductor, thermoelectricity, and photoexcitation.

The objective of writing this book is to systematically introduce hybridized and coupled nanogenerators. Understanding of the fundamental mechanism and related technological applications of the hybridized and coupled nanogenerators can be seen in this book. The potential readership includes scientists, engineers, undergraduate and graduate students in materials, physics, energy, nano-science, and other related fields from science and industry.

First, I would like to thank my postdoctoral supervisor (Prof. Zhong Lin Wang) and doctoral supervisor (Prof. Yue Zhang) for good guidance and strong support in these years. It would not have been possible to achieve this research without their guidance and support. Second, I would like to thank my students for their contributions to this book: my postdoctoral student (Yang Wang), and my doctoral students (Kai Song, Bangsen Ouyang, Ding Zhang, Yun Ji). Third, I would like to thank my current and former group members who have made outstanding contributions to the development of hybridized and coupled nanogenerators: Yingchun Wu, Shuhua Wang, Kewei Zhang, Xiandai Zhong, Xue Wang, Ting Quan, Kun Zhao, Qingbin Zhai, Xi Liu, Yang Wang, Jia Qi, Nan Ma, Yun Ji, Bangsen Ouyang, Kai Song, Tiantian Gao, Qiang Jiang, Bo Chen, Yuanming Wang, Yuting Jiang, Ding Zhang, Rudai Zhao, Xue Zhao, Weiqi Qian. Moreover, I thank my family members for the support and understanding. Finally, I want to acknowledge funding by the Beijing Institute of Nanoenergy and Nanosystems, Chinese Academy of Sciences, the National Natural Science Foundation of China (Grant No. 51472055, Grant No. 61404034), External Cooperation Program of BIC, Chinese Academy of Sciences (Grant No. 121411KYS820150028), the 2015 Annual Beijing Talents Fund (Grant No. 2015000021223ZK32), the Beijing Natural Science Foundation (Grant No. 2154059), the University of Chinese Academy of Sciences (Grant No. Y8540XX2D2), the National Key R&D Program of China (Grant No. 2016YFA0202701), and Qingdao National Laboratory for Marine Science and Technology (Grant No.2017ASKJ01).

Ya Yang, Professor
Beijing Institute of Nanoenergy and Nanosystems,
Chinese Academy of Sciences
School of Nanoscience and Technology,
University of Chinese Academy of Sciences
*E-mail: yayang@binn.cas.cn
Website: http://www.researcherid.com/rid/A-7219-2016

Beijing, China
20 January 2020

1

Overview

1.1 Introduction

Nanogenerators are based on the use of the displacement current of Maxwell as the driving force to convert environmental energies into electric signals, exhibiting various potential applications in wearable electronics, sensor systems, robotics, and other energy-related science. Prof. Z. L. Wang and coworkers invented the first piezoelectric nanogenerator in 2006 [1], and invented the first triboelectric nanogenerator (TENG) in 2012 [2]. Both these nanogenerators are based on the polarization $\boldsymbol{P}_s$ produced due to the mechanical motions induced electrostatic surface charges, where it is not the electric field produced medium polarization $\boldsymbol{P}$.

Usually, Maxwell's equations can be expressed by

$$\nabla \cdot \boldsymbol{D} = \rho \tag{1.1}$$

$$\nabla \cdot \boldsymbol{B} = 0 \tag{1.2}$$

$$\nabla \times \boldsymbol{E} = -\frac{\partial \boldsymbol{B}}{\partial t} \tag{1.3}$$

$$\nabla \times \boldsymbol{H} = \boldsymbol{J} + \frac{\partial \boldsymbol{D}}{\partial t} \tag{1.4}$$

where $\boldsymbol{D}$ is the electric displacement vector ($\boldsymbol{D} = \varepsilon_0\boldsymbol{E} + \boldsymbol{P}$). Z. L. Wang added an additional term $\boldsymbol{P}_s$ in $\boldsymbol{D}$ in 2017 [3, 4]. Thus, $\boldsymbol{D}$ can be given as

$$\boldsymbol{D} = \varepsilon_0\boldsymbol{E} + \boldsymbol{P} + \boldsymbol{P}_s \tag{1.5}$$

where the polarization vector $\boldsymbol{P}$ is associated with the appearance of the external electric field, while the additional term $\boldsymbol{P}_s$ is associated with the appearance of the surface charges that can be independent of external electric fields [5]. Maxwell's displacement current is then expressed as

$$J_D = \frac{\partial \boldsymbol{D}}{\partial t} = \varepsilon\frac{\partial \boldsymbol{E}}{\partial t} + \frac{\partial \boldsymbol{P}_s}{\partial t} \tag{1.6}$$

where the first term $\varepsilon\frac{\partial E}{\partial t}$ (electric field induced) is associated with the electromagnetic waves theory, while the $\frac{\partial \boldsymbol{P}_s}{\partial t}$ added by Wang (called **Wang term**) is due to the non-electric field-induced, strain-related polarization; this is the

*Corresponding author: yayang@binn.cas.cn

Hybridized and Coupled Nanogenerators: Design, Performance, and Applications, First Edition. Ya Yang.

practical application of Maxwell's equations in the energies scavenging field as nanogenerators [5].

Various energies such as thermal, mechanical, chemical, and solar energies exist in the living environment. However, the occurrence of these energies depends on some working conditions such as weather or some other factors. The purpose of developing hybridized nanogenerators is to scavenge the different energies at the same time by integrating the different energy scavenging units into a system, so that we can obtain stable and sustainable power supply, regardless of whatever energy is available in the environment [6]. Prof. Z. L. Wang and coworkers invented the first hybrid energy cell in 2009 [7]. The first electromagnetic–triboelectric hybridized nanogenerator has been reported to scavenge one vibration energy by two different energy scavenging units in 2015 [8], which can largely enhance the efficiency of conversion from mechanical energy to electric energy. The existing hybridized nanogenerators are based on effectively stacking individual nanogenerators together in parallel or in series, where the individual nanogenerator has independent device structures and output electrodes. This is not suitable for miniaturization of the device dimension and for massive production. It is highly desirable to utilize multifunctional materials to obtain multi-effects coupled nanogenerators with the same structure, material, and electrodes. By using piezo–tribo–pyro–photoelectric effects, Prof. Ya Yang and coworkers invented the first coupled nanogenerators in 2015 [9], which have the same materials, the same electrodes, and simultaneous different energies scavenging abilities. This book will give a detailed summary about the design, performance, and applications of the hybridized and coupled nanogenerators.

1.2 Hybridized Nanogenerators

Hybridized nanogenerators are based on integrating the different energies scavenging units into a system for realizing simultaneous multiple energies scavenging, which has two advantages as compared with the reported individual energy scavenging techniques: (i) increasing the total output electric performances; (ii) providing a more stable and sustainable small power source. By integrating the electromagnetic generators (EMGs) and TENGs, the research group of Prof. Ya Yang reported various electromagnetic–triboelectric hybridized nanogenerators to scavenge different types of mechanical energies [8, 10–19], where all the hybridized nanogenerators have the same concept of integrating different nanogenerators to scavenge one mechanical energy for realizing obvious enhancement of energy conversion efficiency. Moreover, there have been more reports about hybridized nanogenerators in other countries such as Korea and Singapore in recent years [20–23].

1.2.1 Hybrid Energy Cells

In 2009, mechanical and solar energies have been reported to be scavenged with a hybrid device by using a piezoelectric nanogenerator at the bottom surface and a dye-sensitized solar energy harvester at the top surface of the device [7]. The output voltage of the solar energy harvester can be about 0.591 V and the output current density can be about 6.9 μA/cm^2. When the piezoelectric nanogenerator is working, the increased output voltage can be about 0.6 V and the output current density still remains at a fixed value, where the enhancement of the voltage is due to the piezoelectric nanogenerator [7]. By integrating a piezoelectric nanogenerator, the used dye-sensitized solar energy harvester has been also utilized to create a hybrid energy cell for concurrently scavenging solar and mechanical energies [24]. The corresponding output voltage of the solar cell was found to increase from 0.415 to 0.433 V due to the piezoelectric nanogenerator.

From 2011 to 2014, various hybrid energy cells were reported to scavenge multi-types of energies at the same time for realizing the enhancement of the conversion efficiency of the related energy devices, exhibiting practical applications in sensor networks and electrochemical reactions. In 2013, Yang et al. reported a hybrid energy cell consisting of a TENG and a micropyramid Si solar cell, which can be utilized to scavenge mechanical and solar energies at the same time [25]. As displayed in Figure 1.1a, the working of the TENG is due to the periodical contact and separation between the indium tin oxide (ITO) film and the polydimethylsiloxane (PDMS) nanowires. The fabricated solar cell consists

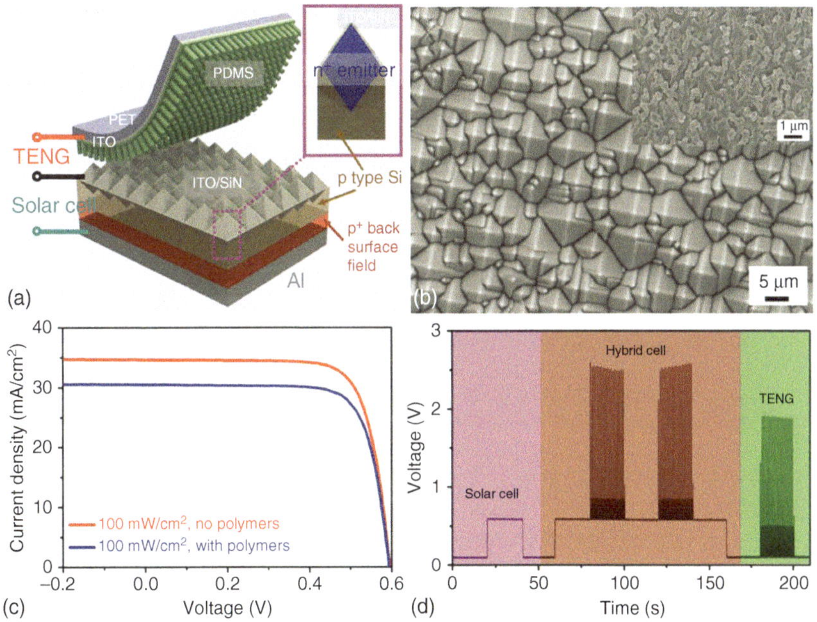

Figure 1.1 Hybrid energy cell for scavenging solar and mechanical energies. (a) Schematic diagram of the designed device. (b) Scanning electron microscopy (SEM) image of the Si pyramid structures. The inset displays an SEM image of the PDMS nanowire array. (c) *J*–*V* curves of the solar cell covered with and without PTFE(polytetrafluoroethylene)-PDMS polymers under a light illumination intensity of 100 mW/cm^2. (d) *V*–*t* curves of the solar cell, hybrid energy cell, and TENG. Source: Reproduced with permission from Yang et al. [25]. Copyright 2013, American Chemical Society.

of Si-based micropyramid p–n junctions, where the used Si pyramids have sizes ranging from 1 to 10 μm, as presented in Figure 1.1b. The inset of the Figure 1.1b shows the used PDMS nanowires. Figure 1.1c presents the *J–V* curves of the used solar energy harvester with and without the transparent PDMS nanowires on the surface of the device, where the output voltage of the device can be about 0.6 V and the output current density can be about 35 mA/cm^2 under simulated sunlight illumination, respectively. After using the PDMS nanowire array on the solar cell, the corresponding conversion efficiency of the fabricated device was decreased from 16% to 14%. Figure 1.1d nanomaterial displays the *V–t* curves of the hybrid energy cell, indicating that the rectified output voltage signals of TENGs exhibit DC peak characteristics.

Yang et al. also reported many other hybrid energy cells [26–29], including TENGs, piezoelectric nanogenerators, solar cells, thermoelectric nanogenerators, and pyroelectric nanogenerators. All the hybrid energy cells are based on effectively integrating the multimode energy scavenging units into a system to obtain sustainable power supply. The purpose of developing hybrid energy cells is to maximize the energies obtained from our living environment. How to effectively integrate the different energy scavenging units is still a challenge in practical devices. Moreover, the ratios among different energy scavenging abilities also need to be considered to make sure that this integration is useful in the system.

1.2.2 Electromagnetic–Triboelectric Hybridized Nanogenerators

The electromagnetic effect is due to the electromagnetic induction in Faraday's law, where the magnet and the coil have relative movements to induce the voltage/current signals. The working of a TENG is based on the coupling effect between the triboelectrification effect and the electrostatic induction in the periodical mechanical motion process [2]. In 2015, Prof. Ya Yang and coworkers first developed a method to integrate an EMG and a TENG in one device [8], so that the same mechanical motions can generate more electric energy due to the use of two energy scavenging devices. The corresponding energy conversion efficiency from mechanical energy to electric energy can be largely increased. This method has been rapidly extended to scavenge many kinds of other mechanical energies, such as wind energy [30], biomechanical energy [31], rotational energy [11], and so on.

To harvest vibration energy, Wu et al. demonstrated a spring-based hybridized nanogenerator including an EMG and a TENG [8], as illustrated in Figure 1.2a. In the vibration process, the electromagnetic and triboelectric effects can coexist in the same mechanical motions, which can effectively enhance the energy conversion efficiency. Figure 1.2b displays a photograph of the device, where the TENG can work under the periodical mechanical motions. Moreover, the working of an EMG is based on the relative mechanical motions between the magnet and the coils. Figure 1.2c illustrates the photograph of the vibration part. The pyramid microstructures with average sizes smaller than 10 μm can be found on the

Figure 1.2 Electromagnetic–triboelectric hybridized nanogenerator for scavenging vibration energy. (a) Schematic diagram of the electromagnetic–triboelectric hybridized nanogenerator. (b) Photograph of the fabricated device. (c) Photograph of vibration part in the device. (d) SEM of the PDMS surface with micro/nanostructures. Source: Reproduced with permission from Wu et al. [8]. Copyright 2015, Elsevier.

surfaces of the PDMS films, as shown in Figure 1.2d. The output voltage of the TENG can be up to 600 V, and the corresponding output current can be about 3.5 μA. Under the same mechanical motions, the output voltage of the EMG can be about 3 V and the output current can be about 1 mA.

To scavenge wind energy, Wang et al. designed an electromagnetic–triboelectric hybridized nanogenerator, which is based on a tube structure [12]. As displayed in Figure 1.3a, the Kapton film can vibrate in the tube when wind goes through the tube from the right side of the device to the left side of the device [32]. By utilizing a high-speed camera, we can find that the effective contact between the two electrodes and the Kapton film for the TENG is mainly at the left side of the acrylic tube (marked with red dot line). There is no contact between the Kapton film and the two electrodes at the right most region of the tube, where most of the vibration energy was wasted due to the required effective contact/separation for the working of the TENGs. By integrating an EMG at the right region of the device in Figure 1.3b, the wind-driven vibration of the organic film can drive the working of the EMG and generate more electricity [12]. To understand the ratios of the produced energies for the two generators, we can calculate the electricity generated in the time of 10 seconds, where the TENG can give 2.6 mJ while the EMG can give 3.2 mJ. The total energy for the electromagnetic–triboelectric hybridized nanogenerator is 5.8 mJ, which

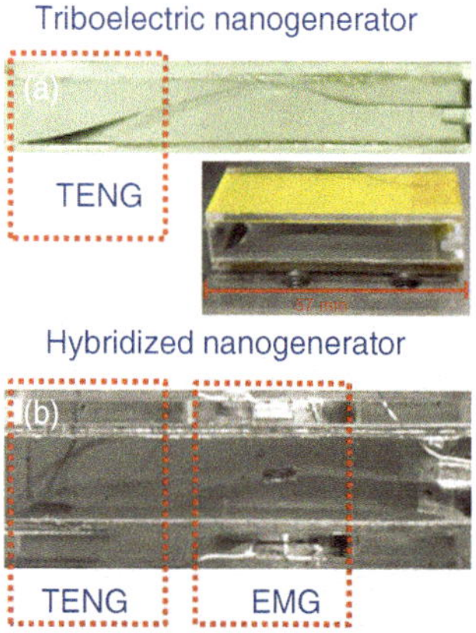

Figure 1.3 Electromagnetic–triboelectric hybridized nanogenerator for scavenging wind energy. (a) Photographs of the triboelectric nanogenerator (TENG) for harvesting wind energy. Source: Reproduced with permission from Wang et al. [32]. Copyright 2015, John Wiley and Sons. (b) Photograph of the hybridized nanogenerator including a TENG and an electromagnetic generator (EMG) for harvesting wind energy. Source: Reproduced with permission from Wang et al. [12]. Copyright 2015, American Chemical Society.

is larger than that of the individual TENG by over 130%. Moreover, Wang and Yang reported a hybridized nanogenerator, where the two ends of the Kapton vibration film have been fixed to harvest wind energy [19]. Although many kinds of electromagnetic–triboelectric hybridized nanogenerators have been reported, all these devices are based on the same idea of integrating the two kinds of energy scavenging units into one system to scavenge the same energies for enhancing the energy conversion efficiency.

1.2.3 Other Hybridized Nanogenerators

Smart cities can be realized by integrating wind and solar energies scavenging units with city buildings to power wireless sensor networks. However, it is very difficult to integrate conventional wind harvesters with Si-based solar cells, as illustrated in Figure 1.4a. The largest issue is the required rotation motion for the conventional wind harvesters, which is not suitable for integrating it with planar solar cells, as displayed in Figure 1.4b. In 2016, Prof. Ya Yang first designed a new hybridized nanogenerator by integrating silicon solar cells with a wind-driven TENG in one device for harvesting wind–solar energies at the same time [30]. As illustrated in Figure 1.4c, these hybridized nanogenerators can be effectively integrated on the top surface of houses because of the non-rotating part of the TENGs. Figure 1.4d presents an optical image of the hybridized nanogenerator, where the solar cells can be fixed on the surfaces of the wind-driven TENG. Under a device size of 12 × 2.2 cm, the total output power of the solar cells is 8 mW, but the corresponding output power for the wind-driven TENG can be about 26 mW. If the hybridized nanogenerator array can be arranged on the top surfaces of city buildings to simultaneously scavenge wind and solar energies, self-powered sensor networks in smart cities can be realized in future.

For this technique, the solar cells can be easily attached on the wind-driven TENGs. The most important issue is how to realize the large size of the device. To solve this issue, Prof. Ya Yang first designed the meter-scale TENG to harvest wind energy and applied the corresponding patents, as shown in Figure 1.5a, where corresponding output current signals of single device can be larger than 100 μA, as depicted in Figure 1.5b. Figure 1.5c illustrates that the corresponding

Figure 1.4 Hybridized nanogenerator for scavenging solar and wind energies. (a) Photograph of conventional wind turbine generator. (b) Photograph of the solar cells on the top of a building. (c) Schematic diagram of the hybridized nanogenerators on the roof of a house model. (d) Photograph of the fabricated hybridized nanogenerator. Source: Reproduced with permission from Wang et al. [30]. Copyright 2016, American Chemical Society.

output power of the fabricated TENG is larger than 220 mW for the single device. Figure 1.5d displays that TENGs with different sizes can be fabricated, which is very important for meeting different environmental needs. This technique has demonstrated the ability of designing different devices with sizes ranging from centimeters to hundreds of meters to simultaneously scavenge wind and solar energies, as illustrated in Figure 1.5e [33].

Figure 1.5d illustrates a schematic diagram of a power station including thousands of large-scale TENGs. Before 15 March 2018, Prof. Ya Yang first designed the large-scale "electric wall" including tens of TENGs as a wall for scavenging wind energy and applied the corresponding patents, as displayed in Figure 1.6a. Each wind wall can provide the electric power outputs ranged from 1 W to 10 W, which can easily light up tens of lamp bulbs, as presented in Figure 1.6b. One thousand of these wind walls can provide a total electric power output up to about 10 kW, which can be utilized in many windy places such as both sides of high-speed train road, desert island, sea surface, mountain area. If the solar cells were fixed on the devices, the energy produced can be much larger than 10 kW. Several megawatt or gigawatt power stations can also

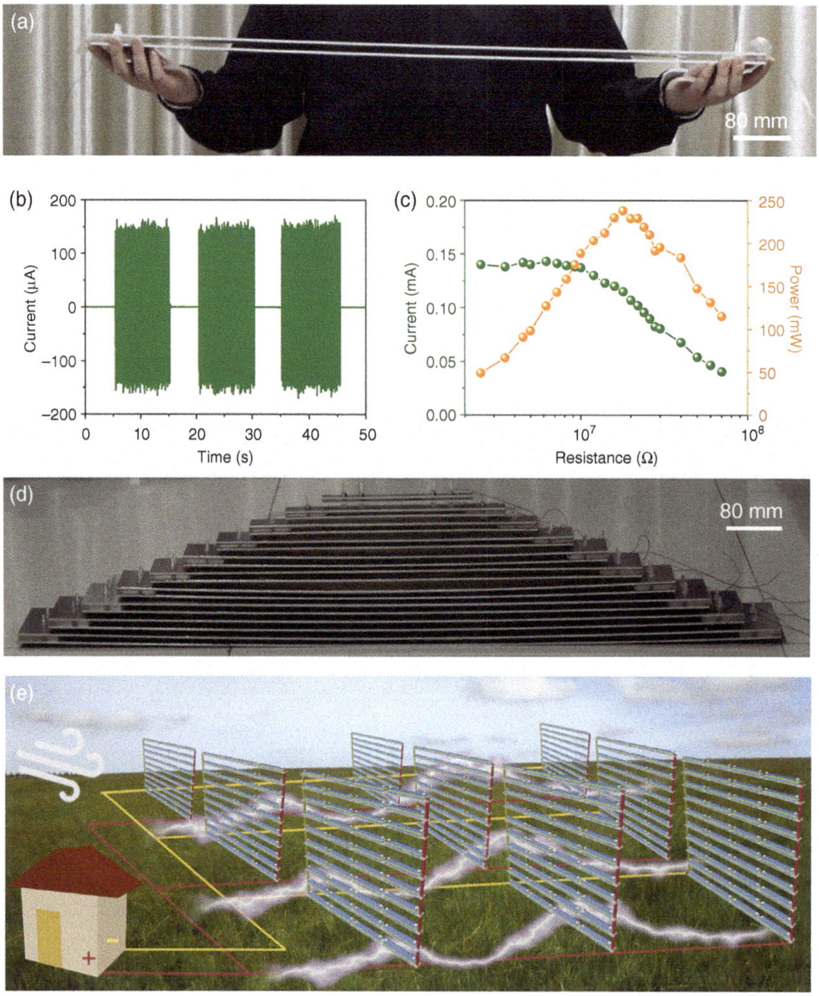

Figure 1.5 Large-scale triboelectric nanogenerators (TENGs) for scavenging wind energy. The large-scale TENG to scavenge wind energy was first designed and fabricated by Prof. Ya Yang before 15 March 2018. The corresponding patents have been also filed based on it. (a) Photograph of the designed large-scale TENG. The data were provided by Prof. Ya Yang. (b) The output current signals of the large-scale TENG. The data were provided by Prof. Ya Yang. (c) The output current signals of the large-scale TENG under different loading resistances and the corresponding output powers. The data were provided by Prof. Ya Yang. (d) Photograph of the designed different-sized TENGs. The data were provided by Prof. Ya Yang. (e) An illustrative diagram of large-scale TENGs wind walls on farm land. Source: Reproduced with permission from Chen et al. [33] Copyright 2017, John Wiley and Sons.

be designed based on this new technique. The electric power obtained can be transformed by power substations and then be transported through cables for long-distance electricity supplement. These large-scale new hybridized nanogenerators have potential applications in the wind and solar energies scavenging field.

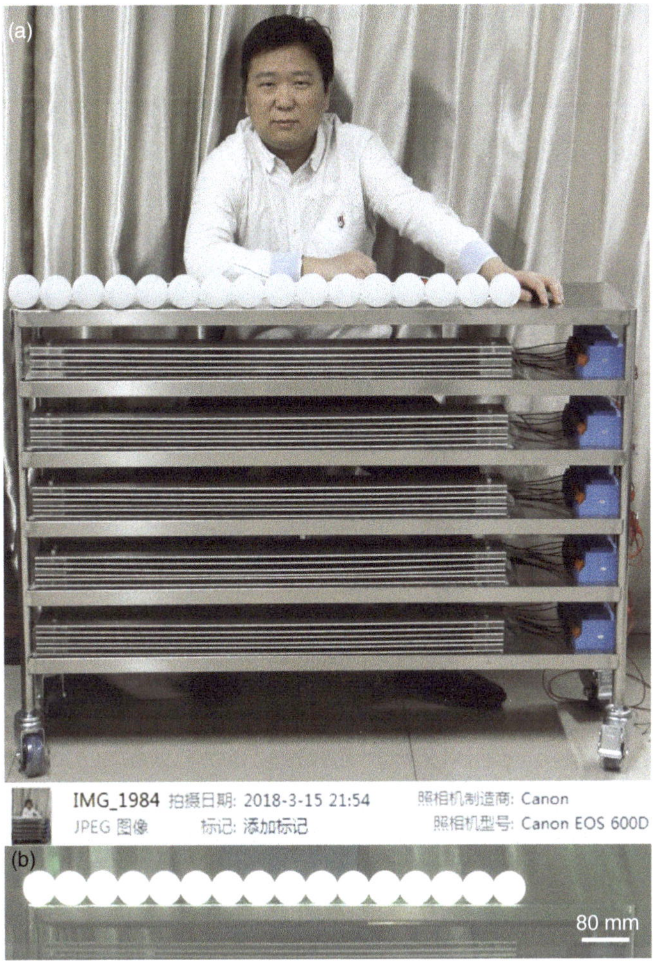

Figure 1.6 Large-scale triboelectric nanogenerators (TENGs) for scavenging wind energy. The mobile meter-scale wind wall includes tens, even hundreds, of meter-scale TENGs, which was first designed and fabricated by Prof. Ya Yang before 15 March 2018. The corresponding patents have been also filed based on it. (a) Photograph of the large-scale TENGs together with Prof. Ya Yang. The time of taking the photo is 15 March 2018. The data were provided by Prof. Ya Yang. (b) Photograph of 16 LED bulbs lighted by the TENGs to scavenge wind energy. The data were provided by Prof. Ya Yang. Source: Courtesy of Prof. Ya Yang.

1.3 Coupled Nanogenerators

Hybridized nanogenerators can simultaneously scavenge the different energies from our living environment by integrating the different nanogenerators into a system. Different nanogenerators have different electrodes and materials, which makes it difficult to realize device miniaturization and decrease the device cost. To solve this issue, in 2015, Prof. Ya Yang and coworkers first invented the coupled nanogenerators, which have the same materials and electrodes but different energy scavenging functionalities [9]. One-structure-based coupled nanogenerators can be fabricated by using the various physical properties to simultaneously/individually harvest various energies in the environment wherever/whenever energy types are available, where the core of the coupled nanogenerators is the multifunctional materials.

1.3.1 Pyroelectric and Photovoltaic Coupled Nanogenerators

Some ferroelectric materials have both pyroelectric and photovoltaic properties such as $LiNbO_3$ [34–36], $Pb(Zr,Ti)O_3$ (PZT) [37–39], $BaTiO_3$ (BTO) [40–42], and $BiFeO_3$ [43–45]. Owing to the wide bandgaps of about 2.7–4 eV, these ferroelectric materials are more suitable to be used as photodetectors instead of as solar cells. When the light is illuminated on the ferroelectric materials, the light-induced heating can generate pyroelectric current, and photovoltaic current can be also observed. The pyroelectric and photovoltaic coupled nanogenerators are based on understanding the mechanisms about how to couple the two current/voltage signals in the ferroelectric materials.

Here, we used the ferroelectric $BaTiO_3$ material as an example to describe the pyroelectric and photovoltaic coupled nanogenerators [46]. The Ag and ITO were utilized as the bottom and top electrodes of the device, respectively. As illustrated in Figure 1.7a, there will be no current signals when the $dT/dt = 0$, where T is the temperature and t is the time. The reason is that there is no change of the polarization in the $BaTiO_3$ material, so that no output current signals can be seen in the external circuit. When the temperature variation dT/dt is larger than zero, the corresponding polarization in the $BaTiO_3$ material will be decreased, indicating that the electrons in the external circuit can move from the bottom Ag electrode to the top ITO electrode to deliver a pulsed output current signal. The pyroelectric current signals are AC signals, which can only flow in the external circuits.

For the photovoltaic effect, the spontaneous polarization-induced electric field can effectively drive the light-induced electrons and holes to separate and flow to the two electrodes, as displayed in Figure 1.7b. Owing to the built-in electric filed in these ferroelectric materials, only one material can realize the appearance of photovoltaic current instead of p–n junctions. As depicted in Figure 1.7c, the pyroelectric effect induced by light illumination can modulate the charge transfer processes of the light-induced electrons and holes such as generation, separation, and recombination. As a result, the coupling enhancement "$1 + 1 > 2$" for the output current signals can be seen in this system due to the coupling effect between the pyroelectric current and the photovoltaic current signals. The new physical effect of ferro–pyro–phototronic effect can be found in this process.

1.3.2 Multi-effects Coupled Nanogenerators

Multi-effects coupled nanogenerators are focused on multifunctional materials such as ferroelectric materials. As illustrated in Figure 1.8, Prof. Ya Yang and coworkers invented the coupled nanogenerators in 2015 [9], representing a new

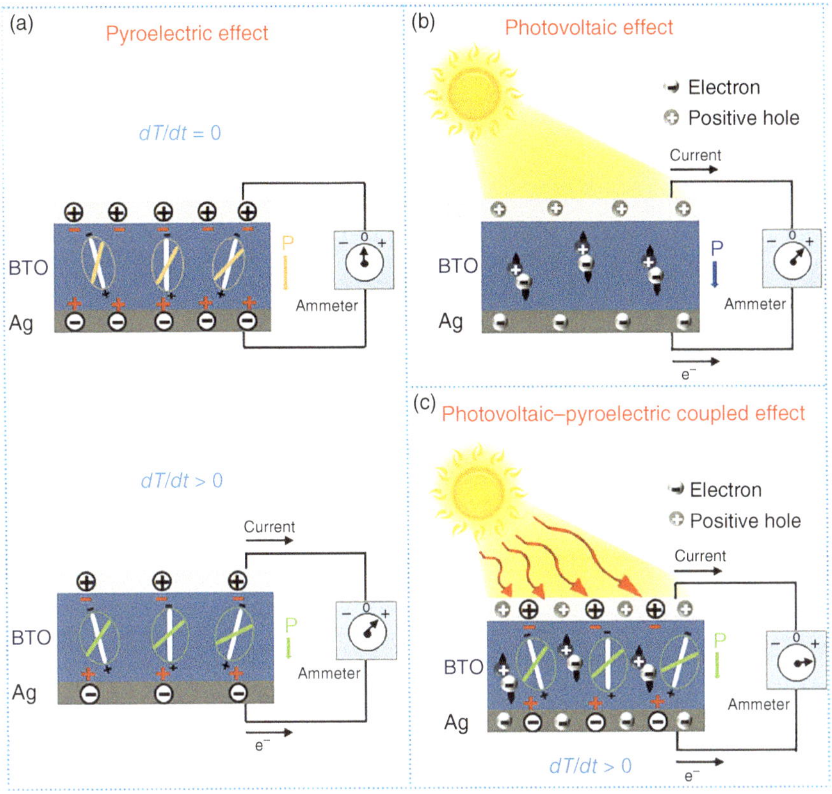

Figure 1.7 Working principle of pyroelectric and photovoltaic coupled nanogenerators. (a) Schematic diagram of the mechanism of pyroelectric effect. (b) Schematic diagram of the mechanism of photovoltaic effect. (c) Schematic diagram of the mechanism of pyroelectric and photovoltaic coupled effect. Source: Reproduced with permission from Ma et al. [46]. Copyright 2017, John Wiley and Sons.

research field in scavenging multiple energies in the environment. By utilizing the piezo–tribo–pyro–photoelectric effects, the multi-effects coupled nanogenerator with one structure and one material can be used to scavenge various energies at the same time. The core of multi-effects coupled nanogenerator is the multifunctional material, where the first material that was used for the coupled nanogenerator is the ferroelectric $PbZr_xTi_{1-x}O_3$ (PZT) material [9]. As depicted in Figure 1.9a, the coupled nanogenerator consists of a PZT material, a polyamide material, and two electrodes, where the PZT material has the piezoelectric, pyroelectric, and photovoltaic effects as the active material. The polyamide film as the triboelectric material was utilized to generate the triboelectric current/voltage signals. The top electrode consists of the ITO film together with Ag nanowires–polydimethylsiloxane film. Figure 1.9b presents the produced charge amounts in the different parts of the coupled nanogenerators. Owing to the pyroelectric effect, the produced charge amount is about 0.39 μC. The produced charge amount is about 1.37 μC due to the photovoltaic effect and that due to piezoelectric and triboelectric effects is about 5.91 μC. When the

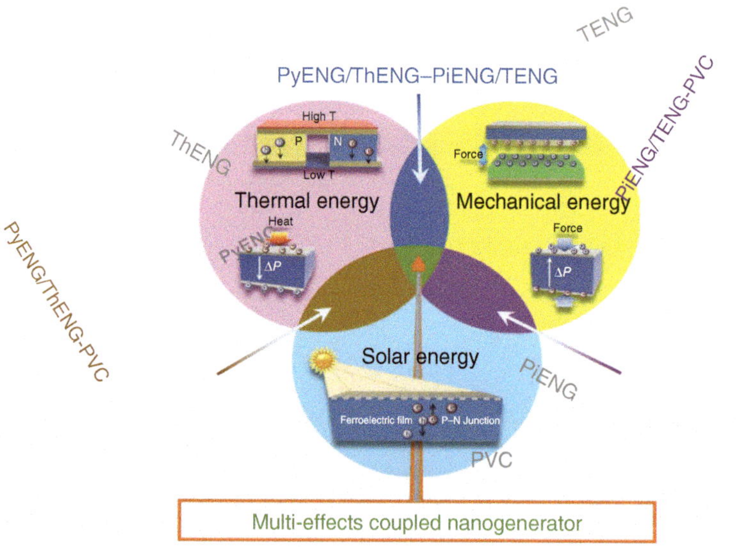

Figure 1.8 Illustration of multi-effects coupled nanogenerator toward multi-energies scavenging. Source: Reproduced with permission from Zhang et al. [9]. Copyright 2017, John Wiley and Sons.

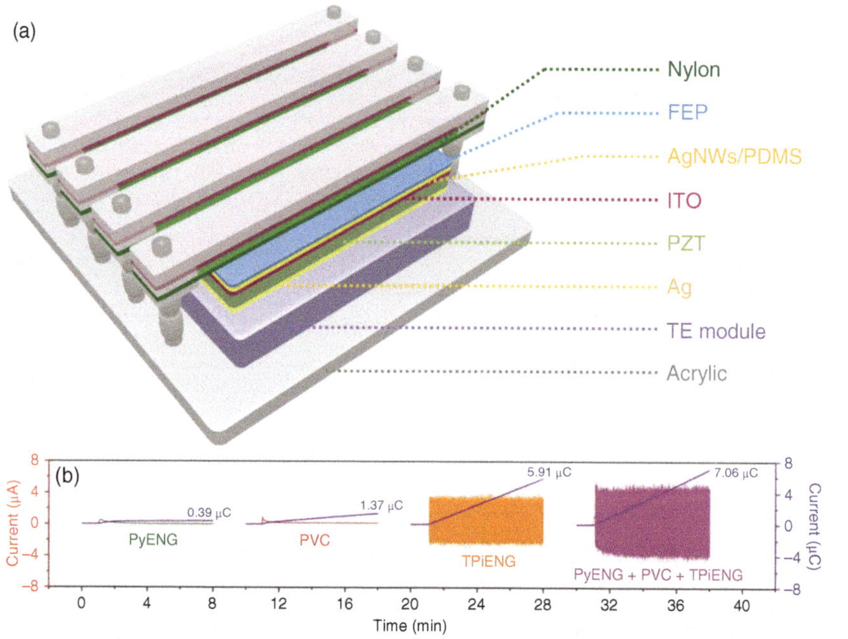

Figure 1.9 One-structure-based multi-effects coupled nanogenerator. (a) Schematic diagram of the designed one-structure-based multi-effects coupled nanogenerator. (b) Transferred charges of the different devices. Source: Reproduced with permission from Zhang et al. [9]. Copyright 2017, John Wiley and Sons.

different energies are applied on the device, the produced charge amount can be up to 7.06 μC, which is much larger than that of individual energy scavenging methods.

The ferroelectric PZT material is toxic due to the presence of Pb. It is necessary to look for other high-performance ferroelectric materials without Pb for the coupled nanogenerators. It has been reported that a polyvinylidene fluoride (PVDF)-based one-structure-based coupled nanogenerator has been utilized to scavenge mechanical and thermal energies at the same time by using tribo–piezo–pyroelectric effects [47]. By charging a capacitor, the coupled nanogenerators have the best charging performance. However, the piezoelectric constant of the PVDF is too small with about 20 pC/N. Ji et al. reported a ferroelectric BTO material-based multi-effects coupled nanogenerator [48], where the piezoelectric constant of BTO can be larger than 300 pC/N. On comparing the charging curves of the different conditions, it is seen that the coupled nanogenerator exhibited faster charging performance than the other individual effects.

As compared with the other hybridized nanogenerators with simple physical integrations, the one-structure-based multi-effects coupled nanogenerators have more advantages such as simpler structure, smaller volume, and lower cost, representing a new research trend in multifunctional materials-based all-in-one multiple energy harvesting in our living environment. Moreover, these coupled nanogenerators have potential applications in multifunctional sensor systems. Owing to the multifunctionalities achieved by one material, these new sensor systems exhibit more advantages than the conventional methods such as higher resolution, smaller device size, and so on. Obtaining high-performance multifunctional material is still a challenge, and more research work is needed along this research direction.

1.4 Applications

Hybridized nanogenerators have extensive practical applications in environmental energies scavenging technologies and self-powered sensor networks. Wind and solar energies can be simultaneously scavenged by the hybridized nanogenerators, consisting of TENGs and Si-based solar cells. Moreover, the sizes of the hybridized nanogenerators can range from centimeters to hundreds of meters for harvesting large-scale energies, pushing the potential applications of the hybridized nanogenerators instead of the conventional wind–solar energy harvesters in some environments. Moreover, the energy conversion efficiency is still a challenge for these energy scavenging devices. The stability of these hybridized nanogenerators in the natural environment still needs to be considered. Hybridized nanogenerators have been used as various self-powered sensors, where the different nanogenerators can detect the different physical signals to realize simultaneous detection. Because different nanogenerators were used for detecting the different physical signals, no signal interference can be found in the detection process.

The coupled nanogenerators have potential applications in multi-energies scavenging and multifunctional sensor systems. The core of the coupled nanogenerators is the multifunctional material; therefore, looking for a high-performance material is one of the most important challenges for the applications. Ferroelectric materials-based coupled nanogenerators have demonstrated the ability of scavenging multi-energies from the environment, indicating that the coupled nanogenerators have better performances than individual energy scavenging devices. However, energy conversion efficiency of these devices still needs to be enhanced. Ways to obtain more electricity from these devices also need to be considered. For the application of multifunctional sensor systems, coupled nanogenerators exhibit more advantages than the conventional methods of integrating the different sensors in a system, such as smaller sizes, higher resolutions, and lower cost. The coupling effect among the different physical signals is very interesting due to the possible new mechanism. However, how to change the current/voltage in one circuit by the coupling needs to be investigated to quantify these signals in the multifunctional sensing process.

The hybridized nanogenerators have more chances to be used as a practical large-scale energy scavenging technique as compared with the coupled nanogenerators. However, the coupled nanogenerators have more scientific significance than the hybridized nanogenerators, where the coupling enhancement among the different physical signals is interesting for future research.

1.5 Conclusion and Prospects

Hybridized nanogenerators integrating two or more different energy harvesting units into an energy scavenging system for simultaneously harvesting multi-types of energies in our living environment have the potential for effectively enhancing the total output performance. It is one of the most significant multiple energies scavenging technologies. The development of hybridized nanogenerators is based on maximizing the harvested energy from one type of energy by integrating the different energy scavenging units and achieving complementary harvesting of multiple energies. The large-scale hybridized nanogenerator consisting of solar cells and TENGs will be a very powerful technology such as creating new wind–solar complementary energies scavenging systems.

The core of the coupled nanogenerators is based on multifunctional materials such as ferroelectric materials. The coupled nanogenerators have the same materials and the same electrodes, which can obtain various advantages of smaller sizes, lower cost, and higher integrations than the conventional integrated different nanogenerators. The coupling enhancement effect among the different physical effects has been found in the coupled nanogenerators. Considering the highlighted advantages and the ongoing related research hotspots, more breakthroughs in hybridized and coupled nanogenerators with respect to the design, performance, and applications for multiple energies scavenging and self-powered multifunctional sensor networks will be seen in future.

References

1 Wang, Z.L. and Song, J. (2006). Piezoelectric nanogenerators based on zinc oxide nanowire arrays. *Science* 312: 242.
2 Fan, F., Tian, Z., and Wang, Z.L. (2012). Flexible triboelectric generator! *Nano Energy* 1: 328.
3 Wang, Z.L. (2017). On Maxwell's displacement current for energy and sensors: the origin of nanogenerators. *Mater. Today* 20: 74.
4 Wang, Z.L., Jiang, T., and Xu, L. (2017). Toward the blue energy dream by triboelectric nanogenerator networks. *Nano Energy* 39: 9.
5 Wang, Z.L. (2020). On the first principle theory of nanogenerators from Maxwell's equations. *Nano Energy* 68: 104272.
6 Dresselhaus, M.S. and Thomas, I.L. (2001). Alternative energy technologies. *Nature* 414: 332.
7 Xu, C., Wang, X., and Wang, Z.L. (2009). Nanowire structured hybrid cell for concurrently scavenging solar and mechanical energies. *J. Am. Chem. Soc.* 131: 5866.
8 Wu, Y., Wang, X., Yang, Y., and Wang, Z.L. (2015). Hybrid energy cell for harvesting mechanical energy from one motion using two approaches. *Nano Energy* 11: 162.
9 Zhang, K., Wang, S., and Yang, Y. (2017). A one-structure-based piezo–tribo–pyro–photoelectric effects coupled nanogenerator for simultaneously scavenging mechanical, thermal, and solar energies. *Adv. Energy Mater.* 7: 1601852.
10 Zhang, K., Wang, X., Yang, Y., and Wang, Z.L. (2015). Hybridized electromagnetic–triboelectric nanogenerator for scavenging biomechanical energy for sustainably powering wearable electronics. *ACS Nano* 9: 3521.
11 Zhong, X., Yang, Y., Wang, X., and Wang, Z.L. (2015). Rotating-disk-based hybridized electromagnetic–triboelectric nanogenerator for scavenging biomechanical energy as a mobile power source. *Nano Energy* 13: 771.
12 Wang, X., Wang, S., Yang, Y., and Wang, Z.L. (2015). Hybridized electromagnetic–triboelectric nanogenerator for scavenging air-flow energy to sustainably power temperature sensors. *ACS Nano* 9: 4553.
13 Quan, T., Wang, X., Wang, Z.L., and Yang, Y. (2015). Hybridized electromagnetic–triboelectric nanogenerator for self-powered electronic watch. *ACS Nano* 9: 12301.
14 Zhang, K. and Yang, Y. (2016). Linear-grating hybridized electromagnetic–triboelectric nanogenerator for sustainably powering portable electronics. *Nano Res.* 9: 974.
15 Wang, X., Wang, Z.L., and Yang, Y. (2016). Hybridized nanogenerator for simultaneously scavenging mechanical and thermal energies by electromagnetic–triboelectric–thermoelectric effects. *Nano Energy* 26: 164.
16 Zhang, K., Wang, Z.L., and Yang, Y. (2016). Conductive fabric based stretchable hybridized nanogenerator for scavenging biomechanical energy. *ACS Nano* 10: 4728.

17 Quan, T. and Yang, Y. (2016). Fully enclosed hybridized electromagnetic–triboelectric nanogenerator for scavenging vibration energy. *Nano Res.* 9: 2226.
18 Quan, T. and Yang, Y. (2016). A shared-electrodes-based hybridized electromagnetic–triboelectric nanogenerator. *ACS Appl. Mater. Interfaces* 8: 19573.
19 Wang, X. and Yang, Y. (2017). Effective energy storage from a hybridized electromagnetic–triboelectric nanogenerator. *Nano Energy* 32: 36.
20 Salauddin, M., Toyabur, R.M., Maharjan, P. et al. (2019). Design and experimental analysis of a low-frequency resonant hybridized nanogenerator with a wide bandwidth and high output power density. *Nano Energy* 66: 104122.
21 Rahman, M.T., Salauddin, M., Maharjan, P. et al. (2019). Natural wind-driven ultra-compact and highly efficient hybridized nanogenerator for self-sustained wireless environmental monitoring system. *Nano Energy* 57: 256.
22 Hou, C., Chen, T., Li, Y. et al. (2019). A rotational pendulum based electromagnetic/triboelectric hybrid-generator for ultra-low-frequency vibrations aiming at human motion and blue energy applications. *Nano Energy* 63: 103871.
23 Sun, C., Shi, Q., Hasan, D. et al. (2019). Self-powered multifunctional monitoring system using hybrid integrated triboelectric nanogenerators and piezoelectric microsensors. *Nano Energy* 58: 612.
24 Xu, C. and Wang, Z.L. (2011). Compact hybrid cell based on a convoluted nanowire structure for harvesting solar and mechanical energy. *Adv. Mater.* 23: 873.
25 Yang, Y., Zhang, H., Liu, Y. et al. (2013). Silicon based hybrid energy cell for self-powered electrodegradation and personal electronics. *ACS Nano* 7: 2808.
26 Yang, Y., Zhang, H., Chen, J. et al. (2013). Simultaneously harvesting mechanical and chemical energies by a hybrid cell for self-powered biosensors and personal electronics. *Energy Environ. Sci.* 6: 1744.
27 Yang, Y., Zhang, H., Lin, Z.H. et al. (2013). A hybrid energy cell for self-powered water splitting. *Energy Environ. Sci.* 6: 2429.
28 Yang, Y., Zhang, H., Lee, S. et al. (2013). Hybrid energy cell for degradation of methyl orange by self-powered electrocatalytic oxidation. *Nano Lett.* 13: 803.
29 Yang, Y., Zhang, H., Zhu, G. et al. (2013). Flexible hybrid energy cell for simultaneously harvesting thermal, mechanical, and solar energies. *ACS Nano* 7: 785.
30 Wang, S., Wang, X., Wang, Z.L., and Yang, Y. (2016). Efficient scavenging of solar and wind energies in a smart city. *ACS Nano* 10: 5696.
31 Donelan, J.M., Li, Q., Naing, V. et al. (2008). Biomechanical energy harvesting: generating electricity during walking with minimal user effort. *Science* 319: 807.
32 Wang, S., Mu, X., Yang, Y. et al. (2015). Flow-driven triboelectric generator for directly powering a wireless sensor node. *Adv. Mater.* 27: 240.
33 Chen, B., Yang, Y., and Wang, Z.L. (2017). Scavenging wind energy by triboelectric nanogenerators. *Adv. Energy Mater.* 8: 1702649.

34 Glass, A.M., Von der Linde, D., and Negran, T.J. (1974). Multiphoton photorefractive processes for optical storage in $LiNbO_3$. *Appl. Phys. Lett.* 25: 233.
35 Carnicero, J., Caballero, O., Carrascosa, M., and Cabrera, J.M. (2004). Superlinear photovoltaic currents in $LiNbO_3$: analyses under the two-center model. *Appl. Phys. B* 79: 351.
36 Arizmendi, L. (2004). Photonic applications of lithium niobate crystals. *Phys. Status Solidi A* 201: 253.
37 Zhang, J., Su, X., Shen, M. et al. (2013). Enlarging photovoltaic effect: combination of classic photoelectric and ferroelectric photovoltaic effects. *Sci. Rep.* 3: 2109.
38 Yang, X., Su, X., Shen, M. et al. (2012). Enhancement of photocurrent in ferroelectric films via the incorporation of narrow bandgap nanoparticles. *Adv. Mater.* 24: 1202.
39 Ichiki, M., Maeda, R., Morikawa, Y. et al. (2004). Photovoltaic effect of lead lanthanum zirconate titanate in a layered film structure design. *Appl. Phys. Lett.* 84: 395.
40 Koch, W.T.H., Munser, R., Ruppel, W., and Würfel, P. (1975). Bulk photovoltaic effect in $BaTiO_3$. *Solid State Commun.* 17: 847.
41 Xing, J., Jin, K.J., Lu, H. et al. (2008). Photovoltaic effects and its oxygen content dependence in $BaTiO_{3-\delta}$/Si heterojunctions. *Appl. Phys. Lett.* 92: 71113.
42 Liu, F., Fina, I., Gutiérrez, D. et al. (2015). Selecting steady and transient photocurrent response in $BaTiO_3$ films. *Adv. Electron. Mater.* 1: 1500171.
43 Yi, H.T., Choi, T., Choi, S.G. et al. (2011). Mechanism of the switchable photovoltaic effect in ferroelectric $BiFeO_3$. *Adv. Mater.* 23: 3403.
44 Bhatnagar, A., Chaudhuri, A.R., Kim, Y.H. et al. (2013). Role of domain walls in the abnormal photovoltaic effect in $BiFeO_3$. *Nat. Commun.* 4: 2835.
45 Yang, S.Y., Seidel, J., Byrnes, S.J. et al. (2010). Above-bandgap voltages from ferroelectric photovoltaic devices. *Nat. Nanotechnol.* 5: 143.
46 Ma, N., Zhang, K., and Yang, Y. (2017). Photovoltaic–pyroelectric coupled effect induced electricity for self-powered photodetector system. *Adv. Mater.* 29: 1703694.
47 Wang, S., Wang, Z.L., and Yang, Y. (2016). A one-structure-based hybridized nanogenerator for scavenging mechanical and thermal energies by triboelectric–piezoelectric–pyroelectric effects. *Adv. Mater.* 28: 2881.
48 Ji, Y., Zhang, K., and Yang, Y. (2017). A one-structure-based multi-effects coupled nanogenerator for simultaneously scavenging thermal, solar, and mechanical energies. *Adv. Sci.* 5: 1700622.

2

Wind-Driven Triboelectric Nanogenerators

2.1 Introduction

Wind, a sustainable source of clean energy, has been widely used to conquer global warming and energy crisis, and about 4.3% of global electricity came from wind power by 2015 [1]. The electromagnetic wind turbine based on the electromagnetic effect is the main type of wind energy harvester [2–10]. However, this wind harvester can hardly be applied in our daily living environment owing to its inherent drawbacks, such as the complex structure, the large volume, and the requirement for high wind speed [11–17].

In order to adequately exploit wind energy and enforce the wind harvester on small electronic devices, it is necessary to develop new techniques to effectively utilize weak wind on miniaturized devices. Triboelectric nanogenerators (TENGs) based on a combined mechanism of contact electrification and electrostatic induction have been widely used to convert multiform mechanical energy to electric energy and showed outstanding advantages including simple structure, low cost, and high power density [18–27]. The wind-driven triboelectric nanogenerator (WD-TENG), an important portion of TENG family, is an ideal harvester for utilizing wind energy in the living environment. In this chapter, we will first review conventional wind energy harvesters and their applications. Special emphasis is given to the WD-TENG including fundamental structures, materials, performance, and applications. We will conclude this chapter with a comparison between conventional wind energy harvesters and WD-TENGs.

2.2 Conventional Wind Harvester

2.2.1 Working Mechanisms and Devices Structure

Horizontal axis wind turbines (HAWTs) and vertical axis wind turbines (VAWTs) are two main types of conventional wind turbines (Figure 2.1a,b) [28, 29]. They usually consist of three main parts, including a wind turbine, a cabin, and a tower. In the wind turbine, blades are very important, which can transform wind energy into the driving force of the turbine rotation. The

*Corresponding author: yayang@binn.cas.cn

Hybridized and Coupled Nanogenerators: Design, Performance, and Applications, First Edition. Ya Yang.

Figure 2.1 Main types of conventional wind turbines. (a) Horizontal axis wind turbines. (b) Vertical axis wind turbines. (c) Applications of micro-wind turbines on homes. Source: Reproduce with permission Ayhan and Sağlam [30]. Copyright 2012, Elsevier. (d) Applications of turbines installation on urban expressways. Source: Reproduced with permission from Ishugah et al. [28]. Copyright 2014, Elsevier.

mechanical energy from the driving force can be converted to electric energy by the generator. The tower not only bolsters the cabin but also limits the vibration produced by the change in the wind speed. The wind energy is defined as

$$E_k = \frac{1}{2}mv^2 = \frac{1}{2}\rho V v^2 = \frac{1}{2}\rho A d v^2 = \frac{1}{2}\rho R^2 \pi d v^2 \tag{2.1}$$

where E_k is the wind energy, m is the wind mass, v is the wind velocity, ρ is the air density, A is the area of the wind wheel, R is the length of the blade, and d is the thickness of the wind disk. According to the theoretical derivation, wind velocity is one of the key factors that affect wind power, and the power is given by

$$P = \frac{E_k}{t} = \frac{1}{2}\rho R^2 \pi \frac{d}{t} v^2 = \frac{1}{2}\rho R^2 \pi v^3 \tag{2.2}$$

HAWTs, which have the axis of rotation of the blades in a horizontal position, are widely used in urban environment because of their high efficiency. To effectively harvest wind energy, the propeller-type rotor, which is mounted on a horizontal axis, uses a yaw motor to face the wind direction. The primal problems for these types of wind harvesters are the dangers to birds and aircraft due to their big blade sizes. The VAWTs adopt a vertical axis of rotation of blades, which is perfectly suitable for harvesting wind energy with different directions. In addition, the generator and gearbox of VAWTs can be mounted at ground level, making them easy to modulate and repair. Smaller VAWTs have been widely used in the urban environment because of their advantages of simple structures and low manufacturing cost.

2.2.2 Applications

In some ground locations or elevated locations, standalone wind turbines have been used to harvest wind energy. To realize broader applications, some researchers expect to design small-scale standalone wind power sources, and they must conquer some key questions, such as challenges of low capacity factor, high costs, and finite capacity to store electricity. HAWTs and VAWTs have been widely used in the urban environment to transform wind energy to electricity generation because their sources are close to the electrical load. Recently, there are some systems that have been integrated into the ambient environment. For example, micro-wind turbines, which can be mounted on existing homes, form one of the techniques for application in urban environment (Figure 2.1c) [30]. On urban expressways, novel turbines are used to capture the energy generated by driving cars at high speeds (Figure 2.1d) [30, 31].

2.3 Triboelectric Nanogenerators for Scavenging Wind Energy

In January 2012, the TENG was first introduced to harvest mechanical energy. Owing to its low cost and high efficiency, TENG provides a novel way for scavenging wind energy in our daily living environment. There are many types of TENGs to harvest wind energy. Among them, vibrating plate-based TENGs have been widely used because of the high output performance.

2.3.1 Fundamental Modes and Structure

Currently, TENGs mainly depend on four modes, including vertical contact-separation mode, in-plane sliding mode, single-electrode mode, and free-standing triboelectric-layer mode. In consideration of the application, vertical contact-separation mode was more suitable for harvesting wind energy.

2.3.1.1 Vibrating Plate-Based TENGs

Yang et al. proposed vibrating plate-based TENGs to harvest wind energy in our daily living environment, which consists of two aluminum (Al) foils and a fluorinated ethylene-propylene (FEP) film in a rectangular acrylic tube, as shown in Figure 2.2a [32]. The Al foils were stuck on the upper and lower ends, which can be used as both electrodes and triboelectric surfaces. To vibrate along with the wind, one side of the FEP film was clamped to the rectangular acrylic tube. The periodic contact between the FEP film and the Al foils can generate an output current across an external circuit. The side view of the fabricated plate-based TENG shows that the device is light and handy (Figure 2.2b). Figure 2.2c shows the front view of the device.

The working mechanism of the plate-based TENG is schematically represented in Figure 2.3. At the initial state, contact between the surface of the Al foil and the FEP film led to electrons transfer from Al to FEP, resulting in positive charges on the surface of the Al foil and negative charges on the surface of the FEP. There is

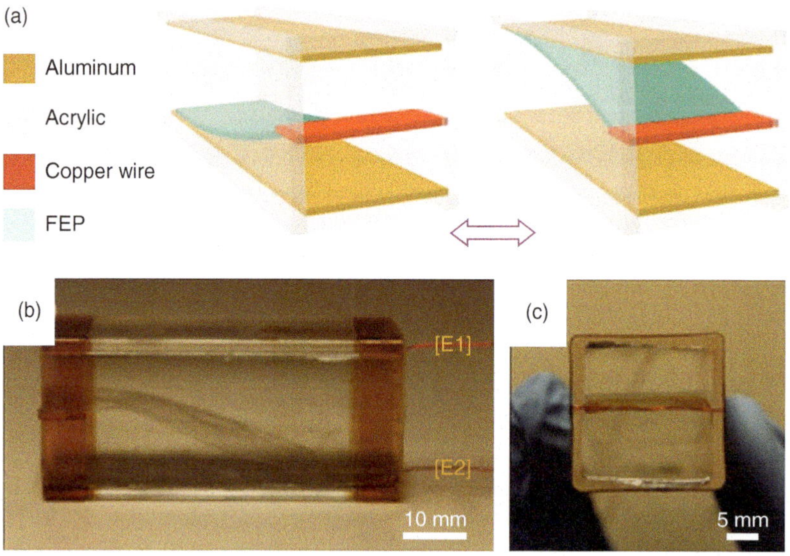

Figure 2.2 Diagram of the plate-based TENG. (a) Schematic diagram of the TENG. (b,c) The side and front views of the TENG. Source: Reproduced with permission from Yang et al. [32]. Copyright 2013, American Chemical Society.

no current in the external circuit because the positive and negative triboelectric charges balance each other (Figure 2.3a). When the FEP film segregates the bottom Al foil due to wind-induced vibration the balance is destroyed, leading to the opposite current at the two Al foils (Figure 2.3b). The electrostatic-induced current will be maintained consistently until the negative triboelectric charges on the FEP film are completely screened from the positive charges, where the current signals disappear at the foils, as shown in Figure 2.3c. Subsequently, the FEP film was close to the bottom Al foil, leading the influent and the effluent current signals to the top and the bottom Al foils, respectively, as depicted in Figure 2.3d. This process finishes once the FEP film makes contact with the bottom Al foil due to the balance formed between them, as shown in Figure 2.3e. Figure 2.3f illustrates that the FEP film is close to the top Al foil, leading to the opposite current signals compared with those in Figure 2.3d. Consequently, plate-based TENG, which can be driven between the two Al electrodes under vibration situation, can be used to harvest wind energy.

2.3.1.2 Enhanced Plate-Based TEGs

The device structures and materials can affect the output performance of triboelectric generators (TEGs). To increase the actual efficiency for harvesting wind energy, some improved structures have been invented by optimizing the structures of the devices. For example, Yang's group proposed enhanced plate-based TEG, which consists of two polytetrafluoroethylene (PTFE) films, two Al foils, and a Kapton film coated with Al electrodes on its surfaces (Figure 2.4a) [33]. To bring wind-induced vibration, one side of the Kapton film was fixed. The side view of the fabricated plate-based TEG shows that the device is small, with the

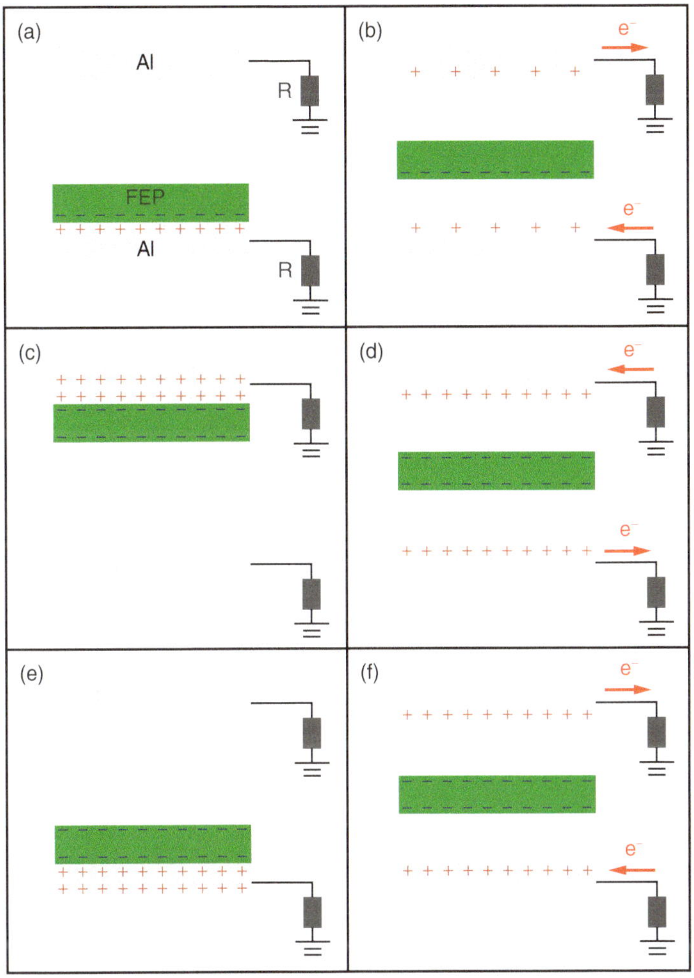

Figure 2.3 Sketches illustrating the electricity generation process (a–f) in plate-based TENG. Source: Reproduced with permission from Yang et al. [32]. Copyright 2013, American Chemical Society.

dimensions of 22 mm × 10 mm × 57 mm (Figure 2.4b). Figure 2.4c shows the front view of the device, with a thickness of about 2 mm.

The working mechanism of the enhanced plate-based TEG is schematically depicted in Figure 2.4d. At the initial state, both TEG-1 and TEG-2 are in the static condition, resulting in no output signals (Figure 2.4d(1)). When the Al electrode contacts with the top PTFE film in the TEG-1, electrons will be transferred due to different triboelectric series, and their surfaces will respectively form positive charges and negative charges. This process will generate a balance due to the nature of the insulator, leading to no output signals (Figure 2.4d(2)). When the distance between the Al electrode and the Kapton film was increased under the external force, the balance was broken, which could generate the current signal through an external load in the TEG-1, as shown in Figure 2.4d(3).

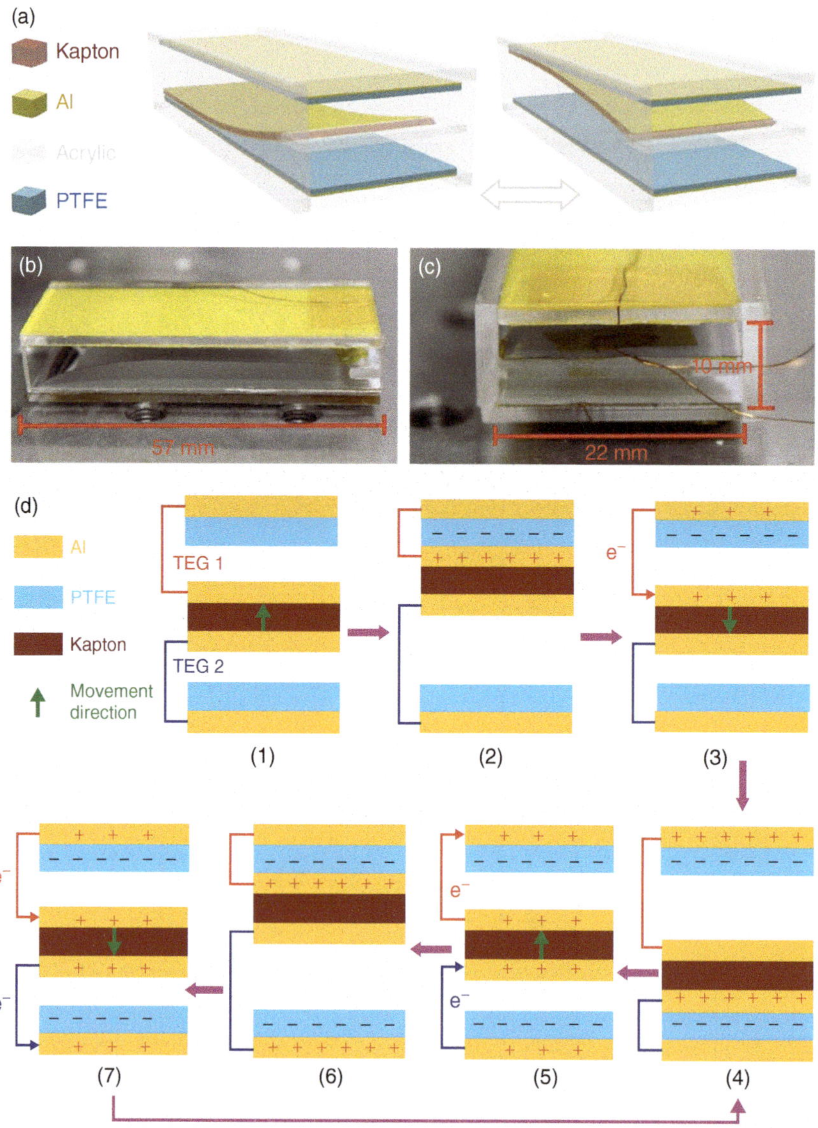

Figure 2.4 Diagram of the enhanced plate-based TEG. (a) Schematic diagram of the TENG. (b,c) The side and front views of the TEG. (d) Sketches illustrating the electricity generation process in enhanced plate-based TENG. Source: Reproduced with permission from Wang et al. [33]. Copyright 2015, John Wiley and Sons.

The electrostatic induced charges disappeared when a new balance is built, as descripted in Figure 2.4d(4). In this state, the surface of the Al electrode forms positive triboelectric charges, and the surface of the bottom PTFE film forms negative charges in TEG-2. Increasing the distance between the Al electrode and the bottom PTFE can induce two output signals in TEG-1 and TEG-2 due to electrostatic induction.

2.3.1.3 Elasto-aerodynamics-Driven TENGs

Under high wind speed, enhancing the stability of the devices is very important because of complex factors, such as buffeting, vortex shedding, galloping, and flutter. There are some natural mechanical problems in the vibrating plate-based TENGs due to a combination of bending and torsion. To increase the stability and output performances of TENGs, elasto-aerodynamics-driven TENG was developed, which consists of a Kapton film, two PTFE films, and Cu electrodes (Figure 2.5a) [34]. Figure 2.5b shows the front view of the device.

The working mechanism of the elasto-aerodynamics-driven TENG is schematically depicted in Figure 2.5c. Owing to couple triboelectrification and electrostatic induction, alternating electrons can flow between Cu electrodes. At the initial state, the TENG is in static condition, resulting in no output signals. When the Cu electrode makes contact with the top PTFE film because of wind-induced vibration, electrons are transferred due to different triboelectric series, and their surfaces will respectively form positive and negative charges. When the distance between the Cu electrode and the PTFE film increases under the external force the balance is broken, leading to the generation of the current signal through an external load. The electrostatic-induced charges disappear when a new balance is built between the Cu electrode and the PTFE film. In this state, the surface of the Cu electrode forms positive triboelectric charges, and the surface of the bottom PTFE film forms negative charges. Subsequently, the charges on the top electrodes and the bottom electrodes can flow respectively in the external circuits due to the electrostatic induction.

2.3.1.4 Others

To harvest wind energy in our living environment, Xie et al. produced a TENG based on a rotary-driven mechanical structure, which consists of a shaft, a flexible rotor blade, and two stators [35]. The triboelectrification and electrostatic induction effects in this TENG can be generated by utilizing a hybridization of the contact-sliding-separation-contact processes. Zhang et al. developed a single-electrode cylindrical TENG to harvest wind energy, which consists of a cylinder, a cylinder tube coated with PTFE [36]. The electric energy generated by this TENG can be used to power wireless sensors. Bae et al. reported a flutter-driven TENG based on a flexible flag and a rigid plate and studied three contact modes (single, double, and chaotic) to increase the performance of the TENG [37]. Ren et al. designed a coaxial rotatory freestanding TENG, which consists of a stator and a rotator. The stator was made of an acrylic column coated with polyvinylidene fluoride (PVDF) nanofibers membrane, and the rotator was made of another acrylic tube with Al tapes [38].

The TENG arrays could enhance the efficiency of harvesting wind energy. Quan et al. reported double-side-fixed TENG arrays based on the coupling of triboelectric frication and electrostatic induction. The TENG arrays were connected in parallel by copper wires. These TENGs consist of top and bottom PTFE/Ti/Al patches and a vibrational Al/Ti/Kapton/Ti/Al film [39]. To enhance the performance and reduce a typical rigid friction of common TENG for harvesting wind energy, Wang et al. reported a composited structure, which consists of an anemometer TENG based on freestanding mode and a wind vane

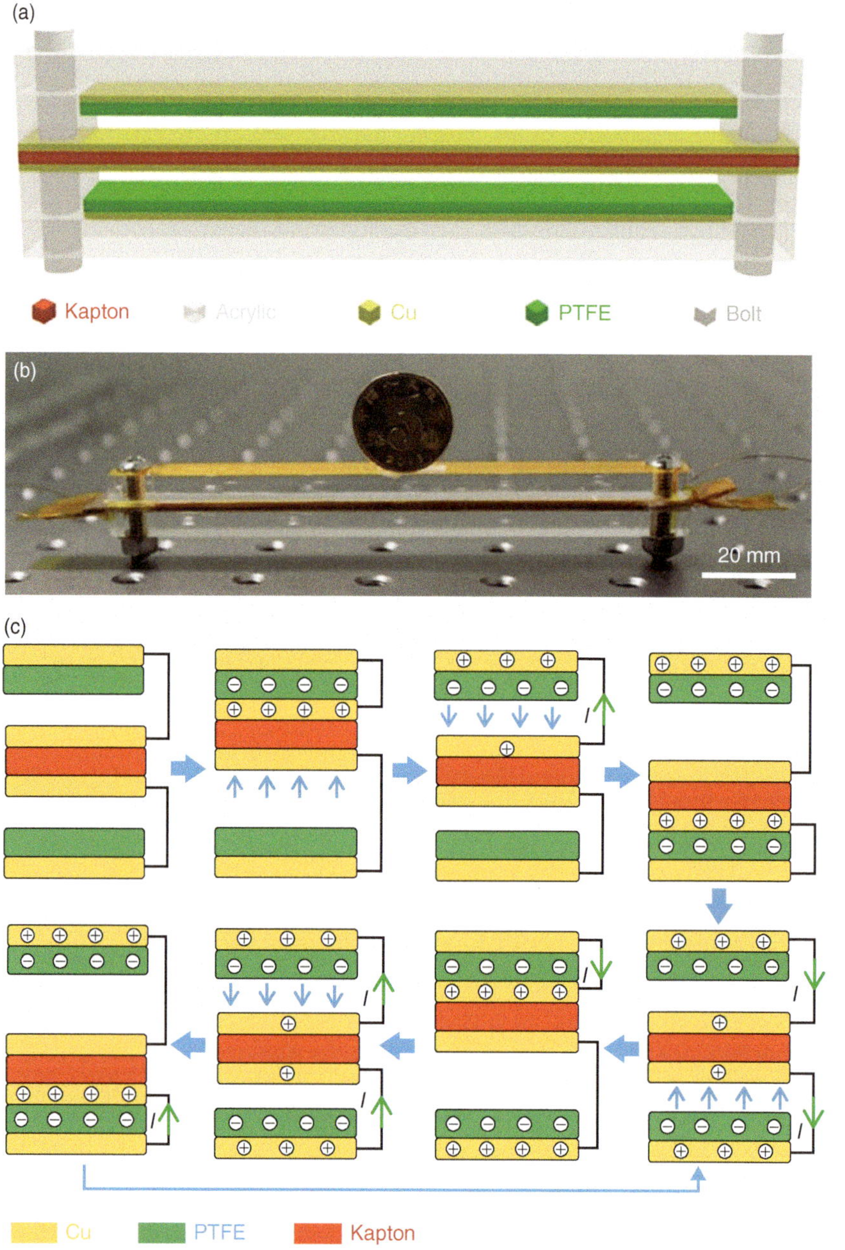

Figure 2.5 Diagram of the elasto-aerodynamics-driven TENG. (a) Schematic diagram of the TENG. (b) The front view of the TEG. (c) Sketches illustrating the electricity generation process in the elasto-aerodynamics-driven plate-based TENG. Source: Reproduced with permission from Wang et al. [34]. Copyright 2015, American Chemical Society.

TENG based on single-electrode mode [40]. The anemometer TENG was made of wind cups, rotator, and stator. Wind energy harvested through the wind cups can be used to drive the rotator. The FEP film used as a triboelectric surface was coated on the rotator. The wind vane TENG was made of a vane connected with a pie-shaped rotator and a compass coated with Cu electrodes.

Wind-driven generators normally can harvest wind in a single direction, which reduces the efficiency of wind energy utilization. To solve this problem, Zhang et al. developed a flexible and transparent TENG, which can harvest natural wind in discretional blowing directions [41]. The TENG was made of massive free-standing polymer strips similar to a forest morphology. When the wind blows, the strips comprised of indium tin oxide (ITO) with polyethylene terephthalate (PET) thin film could sway independently. Harvesting wind energy at a high altitude is difficult for normal wind-driven generators. Recently, Zhao et al. reported a woven TENG flag based on the interlaced interactions between the Kapton film and a conductive cloth, realizing collection of high-altitude wind energy [42]. The conductive belts were made of polyester textiles coated with Ni and Kapton film-sandwiched Cu belts.

2.3.2 Materials

Currently, advanced materials, such as cellulose nanocrystals (CNCs), bionic materials, and nanowires, have become the most inspiring point for TENGs, which can increase the electric or mechanical performances of the devices. To meet the exceptional properties and demands, reasonable design of various materials is indispensable.

2.3.2.1 Cellulose

Cellulose nanomaterials have many outstanding advantages including high surface area, light weight, and excellent mechanical properties, resulting in an important role in diverse applications [43–47]. Plant cell walls have a great deal of cellulose, a fibrous water-loving polymer, which is generated through the plant cellular growth process and cellulose biogenesis. Some creatures, such as fungi, invertebrates, and algae, also have cellulose. The entire process for forming cellulose needs van der Waals force and hydrogen bonds. There are two parts (crystalline and amorphous parts) in cellulose fibrils. CNCs can be generated by insulating crystalline domains, and cellulose nanofibrils (CNFs) can be produced through mechanical interventions. Both of them are nanoscale lateral dimensions, which can realize many outstanding advantages, including low density, flexibility, and large aspect ratio [48–50].

Chen et al. developed a transparent TENG, where CNCs were used as the positive triboelectric materials because of their prominent excellent mechanical properties and nano-sized slender structures [51]. CNCs in the bleached kraft pulp can be produced by acid hydrolysis method. Figure 2.6b shows the prepared CNCs suspension with high transparency. An ITO film was fixed on a PET substrate, and the suspension was dropped on the film as shown in Figure 2.6a. After the water was evaporated at room temperature, the CNC film could completely cover the ITO film. The prepared CNC/ITO film possesses great transparency, as

Figure 2.6 The CNCs/ITO film. (a) The schematic for preparing the CNCs/ITO films. (b) Photograph of the CNCs suspension. (c) The prepared CNCs/ITO electrode. (d) TEM image of the prepared CNCs. (e) AFM image of the ITO substrate. (f) AFM image of the CNCs/ITO substrate. Source: Reproduced with permission from Chen [51]. Copyright 2018, Elsevier.

illustrated in Figure 2.6c. Transmission electron microscopy (TEM) images of the CNCs show the scale of the prepared CNCs, with 50–300 nm length and 3–5 nm width (Figure 2.6d). Figure 2.6e shows the atomic force microscopy (AFM) image of the ITO film. The AFM image of CNC/ITO presents high surface roughness. When the surface of the CNC film makes contact with the FEP film under the external force, the electrons could flow from the CNC film to the surface of the FEP film because the CNC film holds a positive position in the triboelectric series.

Bacterial cellulose (BC) has been widely used due to outstanding advantages such as high elastic modulus, good degree of polymerization, and biocompatibility. Kim et al. used bacterial nanocellulose film as the triboelectric material in the TENG. The bacteria can generate the BC pellicle in a gel state. The uniform BC film was prepared by dissolving the BC pellicle at a specified concentration [52]. CNFs generated from nanosized cellulose fibrils are proverbially used in modern electronic devices due to their desirable surface roughness and electrical properties. Yao et al. produced biodegradable CNF films with nanostructured surface and applied them in the TENG [53]. The CNF hydrogel was produced into a CNF film by a treating process, where wood pulp was oxidized and then mechanically homogenized. By filtering and the drying processes, the CNF hydrogel can be transformed to transparent and flexible CNF thin films. When contacted with the FEP film, the surface of the CNF film could hold positive triboelectric charges due to its very positive position on the triboelectric series. Single cellulose films are fragile and volatile. To conquer these problems, Chandrasekhar et al. developed a cellulose/polydimethylsiloxane (PDMS) film in the TENG [54]. Composite films with irregular surface roughness were fabricated by solidifying the mixed solution of cellulose powders and PDMS on the sandpaper template. The content of cellulose could affect the output performance of the TENG. Peng et al. fabricated the composite film with PDMS and CNC flakes and found that the cellulose can increase the output performance of the TENG due to the favorable dielectric property [55].

One drawback of natural CNFs is the weak polarization, resulting in low capability to generate surface charges. Introducing different chemical groups to the CNFs could increase the capability. Yao et al. fabricated the CNF film by filtering the hydrogel and then drying, and modified the film by a nitration acid mixture of HNO_3, H_2SO_4, and water [56]. Compared to the pristine CNFs in the corresponding Fourier transform infrared spectroscopy (FTIR) curves, the nitro-CNFs possessed three new intense peaks due to asymmetric and symmetric stretching of the NO_2 group and stretching of the NO bonds. The output voltage signals of the pristine CNF–Cu pair were about 0.8 V and of the nitro-CNF–Cu pair were about 4.9 V, in the TENGs. Nanostructures on triboelectric materials can increase contacting areas, leading to increased electrostatic charges on these surfaces. Šutka et al. developed highly porous ethyl cellulose (EC) nanostructured films for the TENG [57]. The porous EC films were fabricated by phase inversion of the mixed solution (ethyl cellulose, ethanol, and toluene). The output voltage signals of the TENG based on porous EC films are higher than that of the TENG based on primary EC films. He et al. developed 1D cellulose nanofibers into the pores of cellulose microfibers skeleton, leading to the fabrication of a nanostructured paper, which can be used as template to carry other materials [58].

To boost the electron-donating tendency of cellulose, Oh et al. fabricated highly conductive ferroelectric cellulose composite papers, which consist of cellulose, silver nanowires, and $BaTiO_3$ nanoparticles [59]. The composite papers were produced through a simple filtration method, with the thickness of the prepared paper of about 70 μm, and were used in the TENG, resulting in high output performances. With the rapid development of printing technique, some researchers explored new methods to fabricate surfaces with micro/nanostructure. Qian et al. developed a biocompatible cellulose-based TENG through the all-printing method. The 3D patterned positive layer could be fabricated by printing the CNF ink onto the Ag/PET substrates [60].

2.3.2.2 Metal

Metallic materials, such as copper (Cu) foils, aluminum (Al) foils, and silver (Ag) foils, have been widely used as electrodes of TENGs, which can export the charges generated by the coupling of triboelectric friction and electrostatic induction. Al foils can be used as both a triboelectric layer and an electrode in the TENG due to the high electrical conductivity and the positive position on the triboelectric series. The surface of the Al foil would form positive charges, when it makes contact with the other film with the negative position. Superhydrophobic surfaces emerging from the cooperation of low surface energy and micro/nanostructures are widely explored in biological systems. To increase the surface roughness and environmental stability of the metallic electrodes, Zhao et al. fabricated a superhydrophobic surface on the Al substrates by using chemical etching method [61]. Al substrates were firstly polished to eliminate the dense oxidation on the Al substrates. The polished surface is smooth, as shown in Figure 2.7a,b. The contact angle on the surface is about 62°, as illustrated in Figure 2.7c. After chemical etching, nanostructures are fabricated on the Al substrates, as shown in Figure 2.7d. High magnification SEM of the etched Al substrate shows that protruding structures were formed on the substrate (Figure 2.7e). The contact angle on the etched Al substrate is about 33°, due to the cooperation of the hydrophilic property of Al and the nanostructures.

By modifying with 1*H*,1*H*,2*H*,2*H*-perfluorodecyltrichlorosilane/toluene solution, the surface energy of the etched Al substrate could be decreased, leading to the formation of a superhydrophobic surface on the Al substrate. Figure 2.7i shows that the contact angle on the prepared superhydrophobic surface is about 153°. Figure 2.7j shows the values for three different surfaces. Figure 2.7k shows a photograph of water drops with different sizes on the superhydrophobic surface. It is found that the superhydrophobic surface could form positive charges when it made contact with a polyamide film in the TENG.

2.3.2.3 Polymer

Polymers, such as PTFE, FEP, and PDMS, are ideal triboelectric materials for TENGs due to the flexibility and the excellent dielectric property. The surface roughness and environmental stability of the polymers in the TENGs can be enhanced with many advanced methods, such as plasma (ICP)-reactive ion etching, template method, and 3D printed method. PDMS possesses a great deal of advantages, such as flexibility, transparency, and high electronegativity, and can be easily produced in special shapes. To study the influence of the surface morphology for the TENG, Dudem et al. developed functional polydimethylsiloxane (NpA-PDMS) layers by soft imprint lithography [62]. The prepared layers possessed nanopillar-like architectures, which could be used to increase the output voltage of the TENG.

PTFE is a common negative triboelectric material, exhibiting strong electron-attracting ability and excellent flexibility. Guo et al. utilized inductively coupled plasma (ICP) ion etching to produce nanostructures on the PTFE film [63]. The fabricated PTFE film could form more charges when it made contact with the electrode, resulting in high output performance of the TENG. Wang et al. fabricated sponge-like porous PTFE thin films by using deionized (DI) water as the soft template [64]. Compared with that based on flat PTFE film, the output performance based on porous PTFE thin film was obviously enhanced. Zhao et al. fabricated polytetrafluoroethylene/polyethene oxide (PTFE/PEO) membranes by using electrospinning method. In order to further enhance the charge density on the composite membranes, they introduced a high amount of stable static negative charges on the surface of the membranes [65].

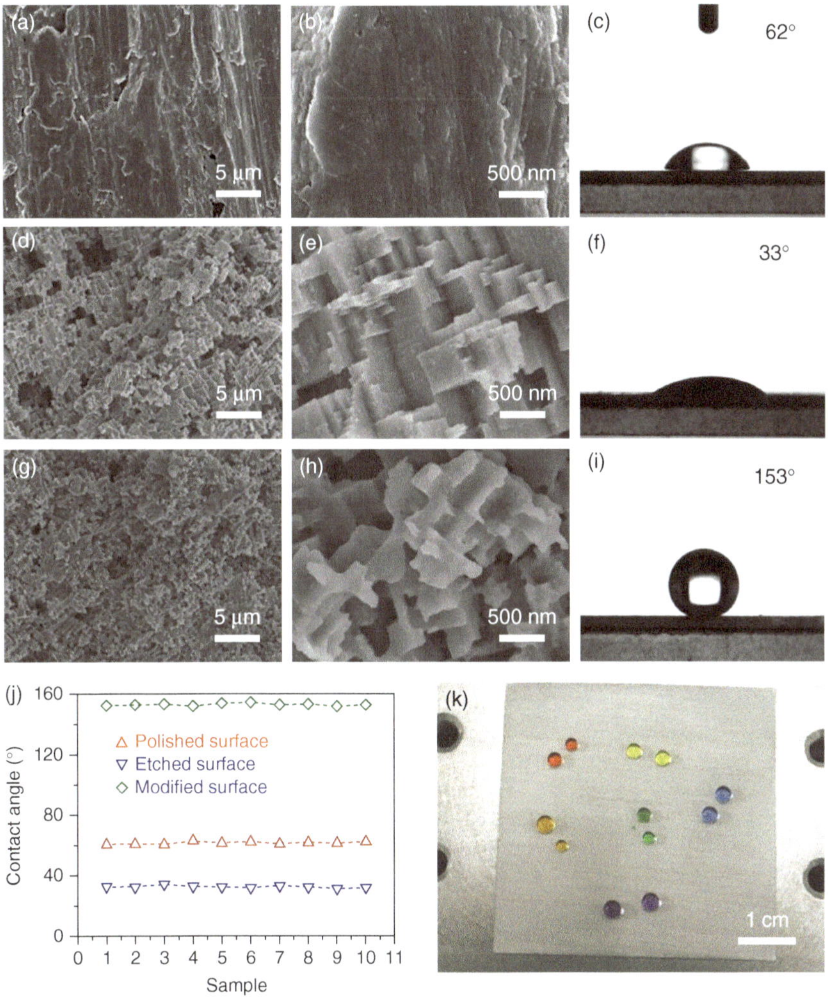

Figure 2.7 Superhydrophobic surfaces on the Al substrates. (a) SEM image of the polished Al substrate. (b) High magnification SEM image of the polished Al substrate. (c) Image of the contact angle of water on the polished Al substrate. (d) SEM image of the etched Al substrate. (e) High-magnification SEM image of the etched Al substrate. (f) Image of the contact angle of water on the etched Al substrate. (g) SEM image of the etched Al substrate modified with low surface energy material. (h) High-magnification SEM image of the modified Al substrate. (i) Image of the contact angle of water on the modified Al substrate. (j) Measured contact angles of three different Al surfaces. (k) Photograph of the water drops on the superhydrophobic surface. Source: Reproduced with permission from Zhao et al. [61]. Copyright 2016, American Chemical Society.

2.3.2.4 Nanoparticle and Nanowire

Nanoscale materials including nanoparticles and nanowires have attracted broad interest due to their unique chemical and physical characteristics. Unlike bulk materials, reducing the size of materials to the nanoscale could make them exhibit high reactivity. Nanoparticles (NPs) with high surface energy can be easily re-formed into functional materials, and nanowires (NWs) offer high surface area and restrain the mechanical degradation [66]. These nanoscale materials have been widely used in applications of energy collecting and storing devices.

NWs with high surface area that restrain the mechanical degradation could be used to optimize the performance of TENGs. Jiang et al. developed a Ag nanowire-based TENG, where Ag nanowires were used as both triboelectric layers and electrodes. Ag nanowires were fabricated through a polyol synthesis method [67]. Figure 2.8a,b shows that the diameter and length of the nanowire are about 70 nm and 10 μm, respectively. Photographic paper was used as the substrate to fix the Ag nanowire slurry, resulting in a metal electrode with excellent conductivity (Figure 2.8c). The Ag nanowires closely accumulate on the photographic paper to form the Ag nanowires membrane, as shown in Figure 2.8d. Cheon et al. fabricated PVDF–Ag NW composite by the electrospinning method. The electrostatic interactions between the Ag NWs and the

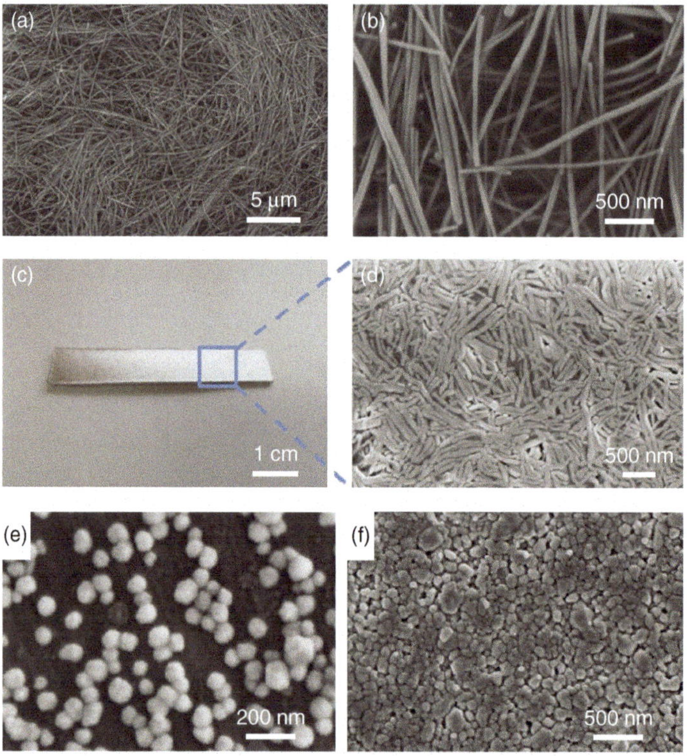

Figure 2.8 Ag nanoparticles and Ag nanowires. (a) SEM image of the Ag nanowires. (b) High-magnification SEM image of Ag nanowires. (c) A photograph of the electrode based on Ag nanowires. (d) The SEM image of the electrode. Source: Reproduced with permission from Jiang et al. [67]. Copyright 2018, American Chemical Society. (e) SEM image of the low-density Ag nanoparticles on Al foils. (f) The membrane prepared by Ag nanoparticles on a photographic paper. Source: Reproduced with permission from Jiang et al. [69]. Copyright 2017, American Chemical Society.

dipoles of the PVDF chains could promote β-phase crystal formation of PVDF, which can be further used to increase the output performance of the TENGs [68]. From a triboelectric perspective, NPs could increase the surface roughness and dielectric property of friction films, leading to enhanced performances of TENGs. Jiang et al. reported a Ag nanoparticle-based TENG, where Ag nanoparticles were used as both triboelectric layers and electrodes [69]. Ag nanoparticles with a diameter of about 50 nm were fabricated through an ice–water bath method as shown in Figure 2.8e. The Ag nanoparticles were stuck on the photographic paper to form the Ag membrane, increasing the mechanical stability of nanoparticles, as shown in Figure 2.8f. It is found that the membrane could form positive charges when it made contact with a FEP film in the TENG.

NPs can be used to increase the surface roughness of triboelectric layers. Lee et al. fabricated textile electrodes with nanostructured geometries, where Al NPs were grown by using thermal evaporation [70]. When the Al NPs were contacted with the PDMS, triboelectric charges could be generated due to their different triboelectric series. It is found that Al NPs could remarkably increase the output voltage of the TENG. Chun et al. developed an Au NP-embedded mesoporous TENG, where Au NPs were embedded into the pores of the PDMS. It is found that the contact between Au NPs and PDMS could enhance the surface potential energy, resulting in high output performance of TNEG [71]. Zhang et al. explored the effect of the output performance of TENGs based on Cu NP-embedded films and ZnO NP-embedded films, respectively. The results showed that Cu NPs can effectively increase the output performance, but ZnO NPs hardly do that [72].

2.3.3 Performance

Currently, mechanical stability and high electrical output are two main performances considered by numerous researchers for TENGs. It is found that the favorable mechanical property could increase the electrical property of the devices, especially in WD-TENGs.

2.3.3.1 Mechanical Behavior

Many flexible structures, which can be bent, folded, and twisted, were applied in the WD-TENGs. Taneda et al. explored diversiform flapping modes, such as one-node, imperfect-node, and two-node flutters. In addition, some researchers found the distinct dynamic states by using one-dimensional filaments [73]. Wang et al. developed a TEG for harvesting wind energy based on the flow-driven vibration of a Kapton film, and analyzed vibrating modes of the film [33]. It is found that an oscillating flow, which is known as Karman vortex shedding, was formed when wind flowed past a bluff body. Figure 2.9a shows that vortices are formed behind the bluff body and separated periodically from either side of the body. Thus, a flexible structure attached on the back of the body could periodically vibrate when wind flowed past this structure. Figure 2.9b shows the model of the designed structure, where a bluff body is fixed on the medium position of the acrylic tube and used to fix one side of the Kapton film. The tube could import air flow in a single direction, resulting in a preliminary adjustment of the flow. The

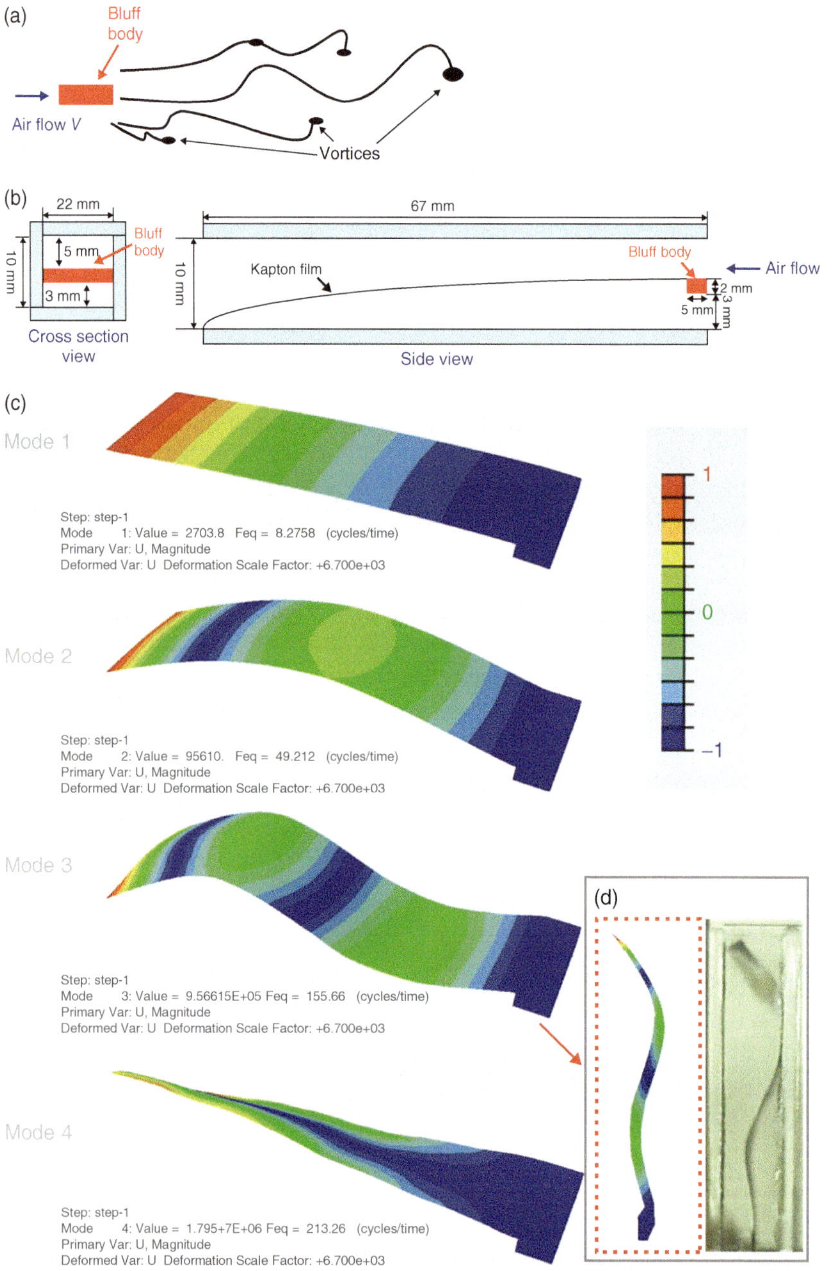

Figure 2.9 Schematic illustrations and simulation of the vibration. (a) The vortex shedding effect. (b) Schematic diagram of the device. (c) The theoretical simulation of the film. (d) The fourth vibration mode in the theoretical simulation and the real deformation of the device. Source: Reproduced with permission from Wang et al. [33]. Copyright 2015, John Wiley and Sons.

air flow could be further separated by the bluff body and vibrate the free side of the film, leading to periodic contact between the film and the walls of the acrylic tube. The frequency of Karman vortex shedding can be expressed by the Strouhal number (S_t):

$$S_t = f_s D/V \tag{2.3}$$

where V is the flow speed of the inlet port, f_s is the vortex shedding frequency, and D is the height of the perpendicular edge.

The constraint mode analysis was studied via finite element method to optimize the vibration mode of the device. Figure 2.9c shows six modes under different frequencies. The working frequency of the TEG is about 155 Hz, which coincides with the simulated fourth order mode, as shown in Figure 2.9d. Bae et al. explored dynamic characteristics of flutters in the flutter-driven TENG, where the woven flag displayed a vibrating node [37]. The fluttering amplitude was very small above the node, but the amplitude below the node increased with increasing distance. By studying different regimes of dynamic interactions, such as single-contact, double-contact, and chaotic modes, the fluttering performance could be optimized to increase the performance of the TENGs. Zhang et al. used a high-speed camera to capture the dynamic process of the contact-separation process between two strips in the lawn-structured TENG, and studied different modes via the finite element method [41]. The strips could be bent to contact together at the origin state under wind blow. The contact process could occur again when the atmospheric pressure exceeded the elastic restoration.

The wind flutter effect could affect the performance of the TENG. Wang et al. explored the interaction between the flow and the film by simulating the structure–fluid model [34]. This model consists of a fluid part and a solid mechanic part (solved with the Navier–Stokes equations). The results show that the bending mode was the main working condition for the film at different inlet wind speeds, as illustrated in Figure 2.10a–c. Figure 2.10d,e shows pressure and wind flow velocity distributions at a wind speed of 15 m/s. It was found that the vibrating film could work stably when the wind speed exceeded 10 m/s. On the other hand, the vibrating film worked on twist motion at low inlet wind speed (5 m/s), where the edge area of the thin film deformed more than the trailing edge. The twist motion would fade and fluttering mode form when wind speed approaches the critical wind velocity, which could be estimated by the frequency ratio of first order bending and twist mode and the respective damping ratio. Figure 2.10f shows the natural frequencies and the mode shapes of the mode shapes of first six order vibration mode, which coincides with the actual vibrations of the film observed via a high-speed camera.

Surface morphology could affect the mechanical performance of the TENG. Dudem et al. explored the performance of PDMS-based TENG with different nanopillar-array structures [62]. It was found that increasing the period of nanopillar architectures on the surface of PDMS could enhance the uniaxial normal stresses, but the contact area was decreased. The contact force across the PDMS and electrode was enhanced in the period of nanopillars from 100 to 200 nm, and subsequently decreased. On the other hand, the contact area also could be enhanced by increasing the diameter of nanopillar architectures.

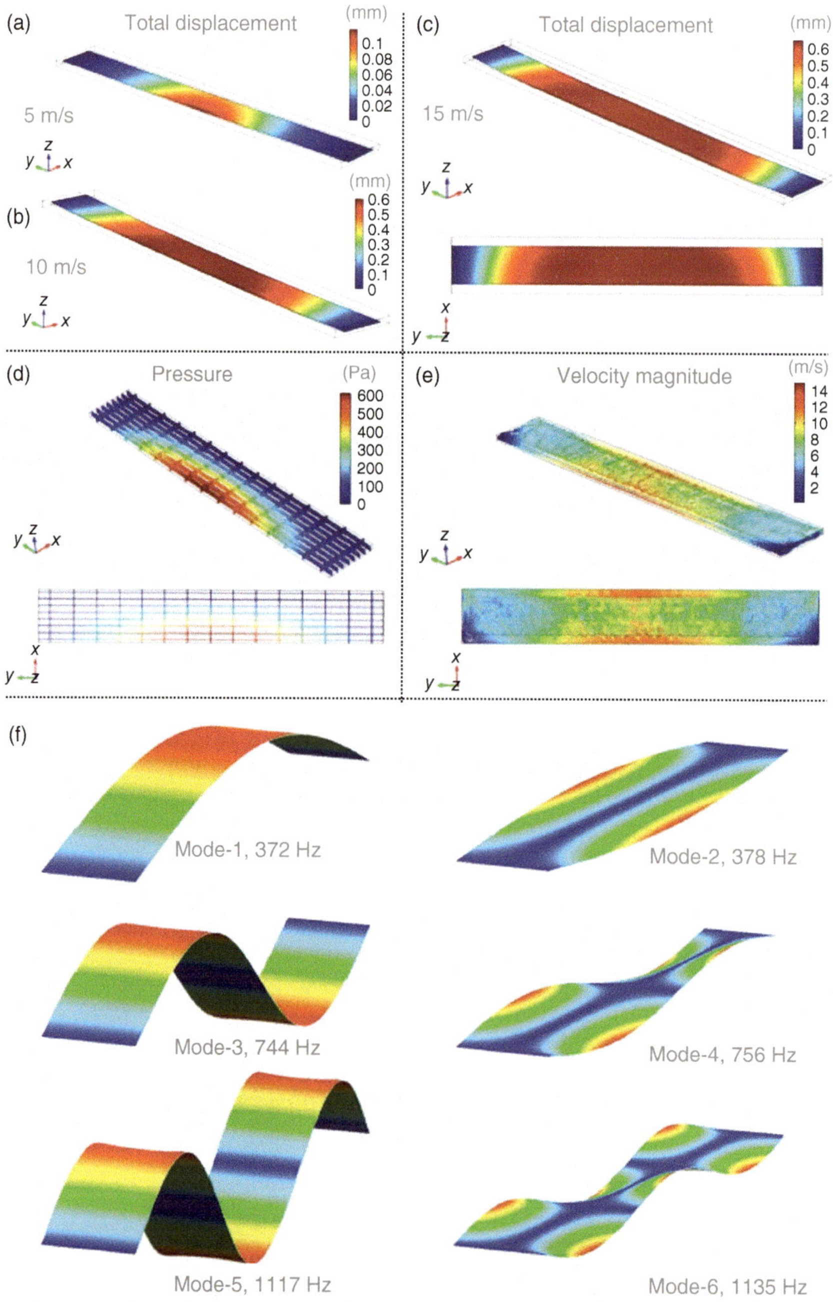

Figure 2.10 Simulation of the vibration film. (a–c) The displacement distributions of the Kapton film at different air-flow rates. (d) Pressure distribution of the film. (e) Velocity magnitude distribution of the film. (f) Six vibration modes of the film. Source: Reproduced with permission from Wang et al. [34]. Copyright 2015, American Chemical Society.

2.3.3.2 Electrical Output

The electrical output of TENGs usually includes alternative signals due to the coupling of triboelectrification and electrostatic induction. The elasto-aerodynamics-driven TENGs could deliver two part signals, where a short-circuit current of about 80 μA and an output voltage of about 240 V were generated by the top TENG, and a short-circuit current of about 60 μA and an output voltage of 220 V were generated by the bottom TENG [34]. It was found that the measured output voltage under the different loading resistances exhibited the largest power density of about 3 kW/m^3 for the top TENG and 2 kW/m^3 for the bottom TENG. It was found that the air gap could affect the output voltage, which could attain the maximum value when the distance of the gap was about 1.6 mm. On the other hand, increasing the length of the device could reduce the working frequency while increasing the power density of the TENG. The output current of the elasto-aerodynamics-driven TENG did not show any obvious change after the Kapton film vibrated to touch the PTFE film for 4.1×10^6 times, illustrating good stability.

Wind speed is an important factor to affect output performance of the wind-driven TENG. Jiang et al. found that output voltage could be enhanced with increasing the wind speed [67]. High wind speed could form more mechanical energy to generate friction between two triboelectric layers, leading to the generation of more electric energy. Xie et al. explored the output performance of the rotary TENG. The output voltage signals jumped from 0 to a maximum value of about 250 V when the TENG was blown at a wind speed of about 15 m/s [35]. The output current was small under low wind speed condition, and it could attain a saturation value of 200 μA under a wind speed of about 20 m/s. Zhao et al. explored the output performance of superhydrophobic-surface-based TENGs [61]. Figure 2.11a,b shows that the TENG can deliver an output voltage of about 218 V and an output current of about 30 μA at a wind speed of 12 m/s. The corresponding largest output power of this TENG is about 2.2 mW, as shown in Figure 2.11c. By the triboelectric charging process, the time to attain the saturated output current is about 100 seconds, as illustrated in Figure 2.11d. Increasing the wind speeds could enhance both the output performance and the corresponding working frequency, as shown in Figure 2.11e,f. On the other hand, wind direction also affects the output performance. It was found that no output signals can be measured when the angles between the wind direction and the horizontal direction of the device are under 90°. The TENG can generate a high output current when the angles are at 0° or 180°.

To realize entirely positive output signals, Wang et al. used a bridge rectification circuit to adjust the primary output signals generated by the flow-driven TEG, which contain two single TEGs [33]. The rectified output current of TEG 1 is about 50 μA, and the current of TEG 2 is about 30 μA, as shown in Figure 2.12a,b. The output current could be enhanced by connecting the two TEGs in parallel, as shown in Figure 2.12c. Figure 2.12d shows that the charging performance of a 10 μF capacitor for TEG 1//TEG 2 is better than that of TEG 1 or TEG 2. The rectified open-circuit voltage of the device is about 300 V, as shown in Figure 2.12e. The maximum output power of this device is about 2.35 mW at a loading resistance of 3 MΩ, as illustrated in Figure 2.12f.

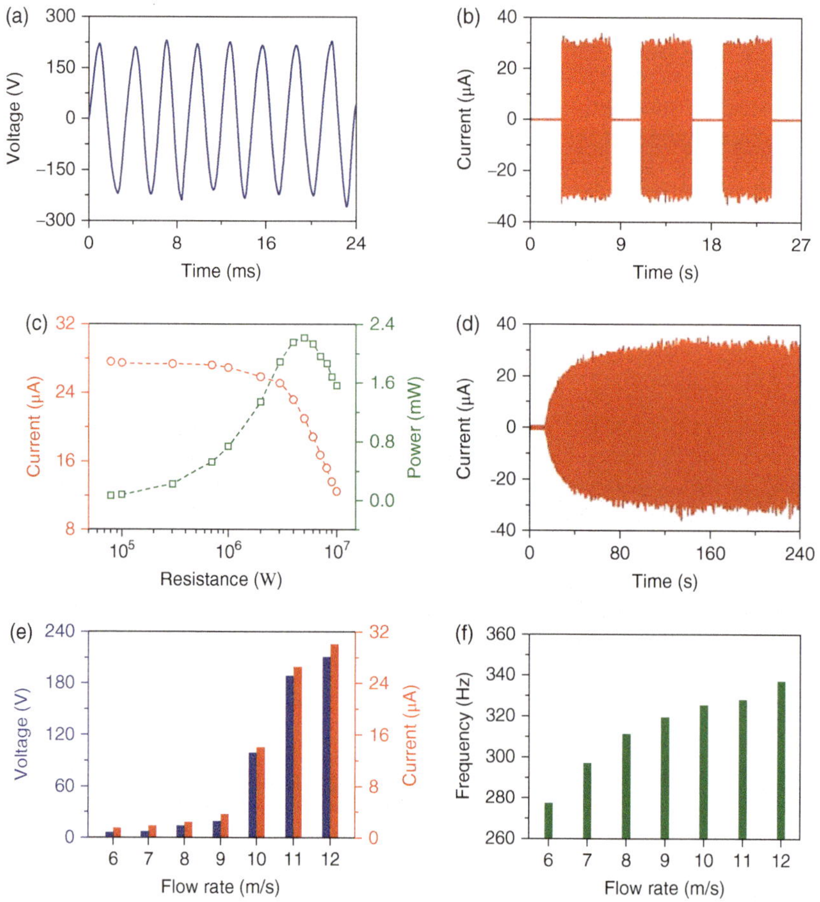

Figure 2.11 Output performance of the TENG. (a) Output voltage signals. (b) Output current signals. (c) Output current and the corresponding power of the TENG under the different loading resistances. (d) Measured original output current signals of a new TENG. (e) Output voltage/current under different wind speeds. (f) Working frequency of the TENG under different wind speeds. Source: Reproduced with permission from Zhao et al. 2016 [61]. Copyright 2016, American Chemical Society.

The height of the flow-driven TEG affects the output performance, which can attain the maximum value when the height is adjusted to 10 mm, as shown in Figure 2.13a,b. Increasing the length of the Kapton film could also enhance the output performance, as shown in Figure 2.13c,d. Then, the length of the device also can affect the output performance, as illustrated in Figure 2.13f. On the other hand, wind speeds can enhance the frequency of both TEG 1 and TEG 2, as shown in Figure 2.13g,h. Ahmed et al. explored the output performance of TENG based on the sliding mode [74]. The maximum output voltage signals and transferred charge were about 600 V and 0.15 μC when the wind flow was regulated at a rotational speed of 50 rad/s. It was found that increasing the wind speed could enhance the output current signals of the TENG. It was found that the diameter

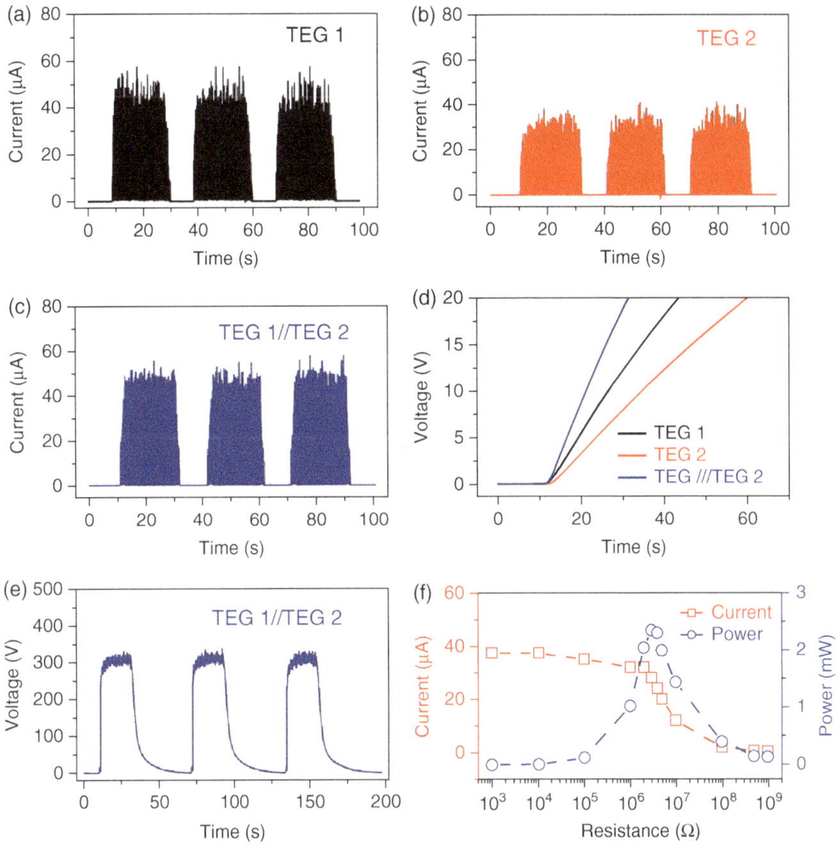

Figure 2.12 Rectified output performance of the TENG. (a) Short-circuit current of TEG 1. (b) Short-circuit current of TEG 2. (c) The current of TEG 1 and TEG 2 connected in parallel. (d) The measured voltage of a 10 μF capacitor charged by TEG 1, TEG 2, and TEG 1and TEG 2 connected in parallel. (e) Rectified open-circuit voltage of TEG 1 and TEG 2. (f) Dependence of the output current and the corresponding power. Source: Reproduced with permission from Wang et al. [33]. Copyright 2015, John Wiley and Sons.

of the rotator and grating number also could increase the performance of the TENG.

Nanostructure-like morphology on the friction layers in the TENGs could effectively improve the output performances of the devices. Dudem et al. found that nanopillar architectures distributed on the surface of PDMS could enhance the output performance of the TENG [62]. The TENG based on a flat PDMS layer generated an output voltage of about 295 V. By introducing the nanopillar architectures on the surface of PDMS layer, the output voltage can be increased to about 443 V. On the other hand, ambient humidity is a critical factor that could affect the output performances of the TENG. According to Guo et al. output current signals of the airflow-induced TENG could be decreased by increasing the ambient humidity [63]. The output current after rectification was about 3.8 μA at a relative humidity of about 20%, but it was only 0.5 μA

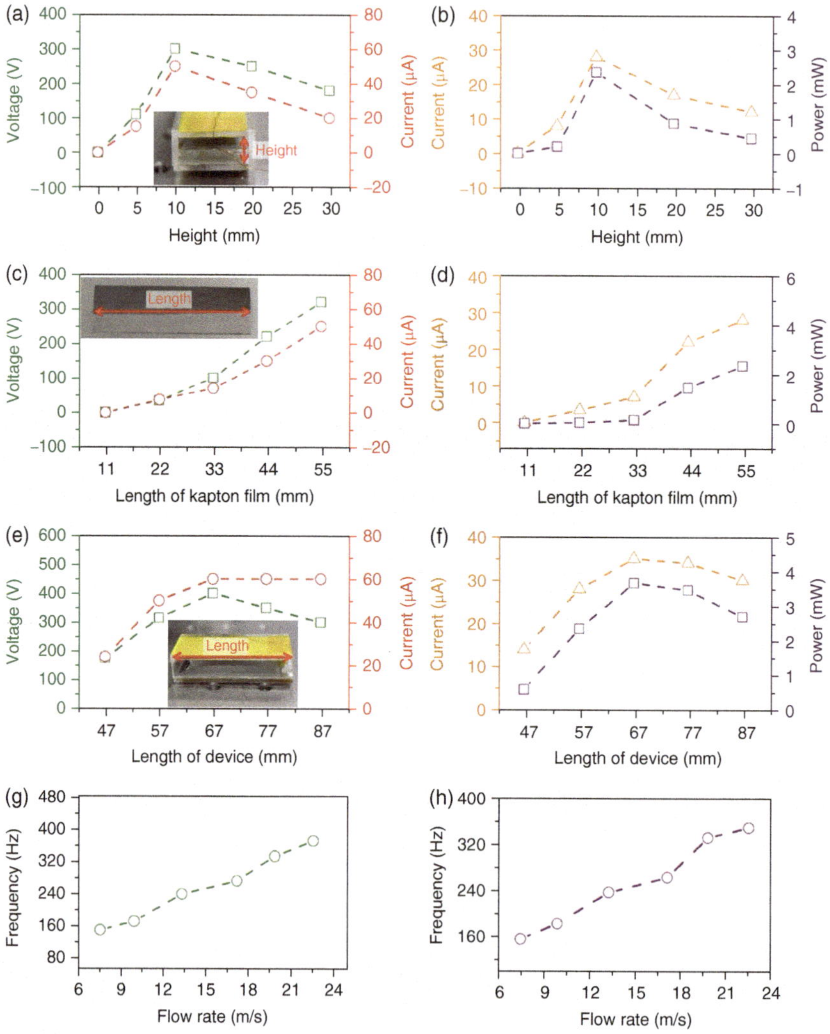

Figure 2.13 Output performance of the TENG. (a) The output voltage and the current of the device with different heights. (b) The output current and the corresponding power of the device with different heights. (c) The output voltage and the current of the device with different lengths of the Kapton film. (d) Dependence of the measured output current and the corresponding power on the length of the Kapton film. (e) Output performance of the device with different device lengths. (f) Dependence of the output current under the corresponding power on the device length. (g) The working frequency of TEG 1 under different flow rates. (h) The working frequency of TEG 2 under different flow rates. Source: Reproduced with permission from Wang et al. [33]. Copyright 2015, John Wiley and Sons.

at a relative humidity of about 100%. This could be due to the impairment of triboelectrification and electrostatic induction.

The relationship between electric output and the fluttering behavior was investigated by Bae et al. [37]. Comparing the profiles of the electrical signal and the fluttering images, they found that output current signal gradually increased when the condition of the vibrating film started from primary contact of the curved area to complete contact. A decrease in contact area then would appear at separation, leading to reduced output current signals of the TENG. In addition, the output current could be enhanced by increasing flow velocity, reaching a maximum value of about 240 V at 22 m/s. The output performance of the TENG based on the soft friction mode was explored by Wang et al. [40]. The wind flow had a light effect with the open-circuit voltage and the transferred charge. The reason could be the deformation degree of the FEP film. The electrodes were radially arrayed on the stator, where the parts of the center are small, resulting in the possibility of mistaken connection for the friction film under lower wind speed. This problem could be partly dispelled under high wind speed condition, where the buoyancy generated by the air flow will partly balance the gravity, leading to full contact between the electrode and the film. Zhang et al. developed a lawn-structured TENG and investigated output performance of the device under different gap distances between two strips [41]. The output current signals were enhanced by increasing the gap distance from 0 to 7 m. Subsequently, the electric output decreased as the gap distance increased. The maximum value of the output current of one cell comprised of two strips was about 10 μA. By connecting four cells in parallel, the value could be increased to about 40 μA.

2.3.4 Applications

Wind is one of the familiar energy sources in the surrounding environment. WD-TENGs can effectively harvest wind energy, converting mechanical energy to electrical energy. Owing to their advantages such as small scale, low cost, and easy fabrication, WD-TENGs can be used in many fields, especially for portable electronics. With the tremendous increase in low power devices, combining TENGs with other devices is vitally important because the combined systems can eliminate dependence for external energy storage by harvesting mechanical energy in the environment.

2.3.4.1 Self-Powered Printer

The printer is a common office equipment, and has been used widely in offices, schools, and supermarkets. However, immobility and high energy consumption are two main drawbacks of current printers. Therefore, Chen et al. developed a self-powered printer system, which consists of a simple handheld printer and a transparent TENG. The handheld system was assembled with a syringe and an acrylic holder [51]. The TENG could harvest wind energy, and then generate electrical energy to power the printer.

To provide unidirectional voltages to power the printer, a rectifier was used to adjust the primary output voltages generated by the TENG. Figure 2.14a shows the working mechanism of the self-powered printer system. The wind-driven TENG system can generate a DC voltage of about 130 V under a wind speed of 20 m/s, which could form an electric field between the nozzle and the copper sheet. The droplet surface could accumulate positive charges under this condition. The droplet was exposed to gravity and the downward electric field force, and could fall toward the copper sheet when the downward collective force

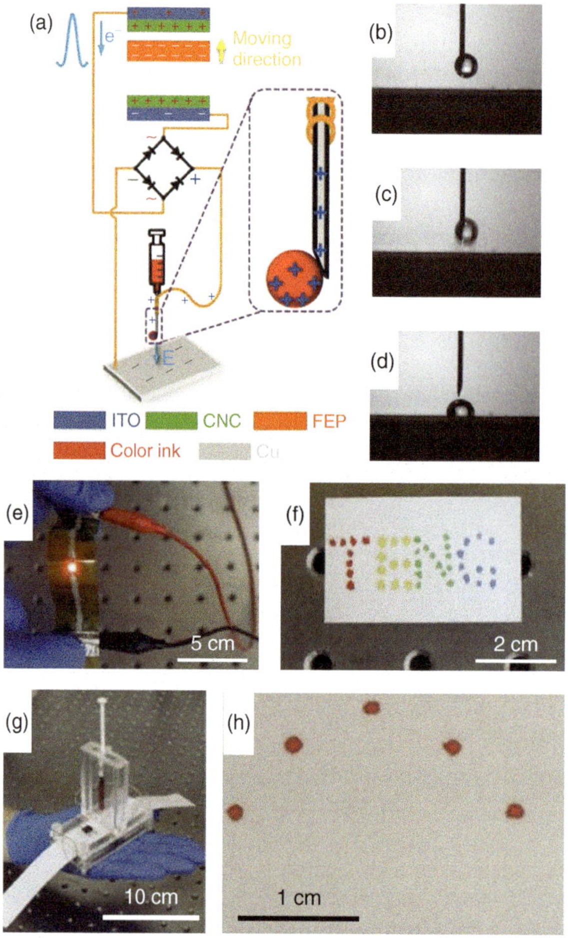

Figure 2.14 The working principle and a photograph of the fabricated self-powered printer. (a) The schematic diagram of the printer working principle. (b–d) Images of the separating process of the ink droplet. (e) The conductive circuit was printed by the printer and Ag nanowire suspension. (f) The letters were printed by the printer. (g) A photograph of the handheld printing system. (h) The curve printed by harvesting mouth-blown wind energy. Source: Reproduced with permission from Chen et al. [51]. Copyright 2018, Elsevier.

was larger than the van der Waals force between the nozzle and the color ink droplet. The separating process of the droplet was illustrated in Figure 2.14b–d. Figure 2.14e shows that the letters of "TENG" were successfully color printed by the self-powered printer system. In addition, this system can be used to fabricate a flexible conductive path comprised by printing Ag nanowire suspension, as shown in Figure 2.14f. A light-emitting diode fixed on the path can be lightened by applying a 2.5 V DC voltage. Figure 2.14g shows a handheld printing system, which can be powered by human mouth-blown air. In this condition, the value of output voltage was about 55 V. The handheld printing system can be used to print discrete points by harvesting mouth-blown wind energy, as shown in Figure 2.14h.

2.3.4.2 Wind Gauging System

A wind vector sensor system, which can detect wind speed and direction, has been widely used in weather forecasting. Traditionally, wind sensors need external power sources or conventional batteries, which limit their long-term work.

Wang et al. developed a self-powered wind sensor system based on an anemometer TENG and a wind vane TENG. This system can simultaneously detect wind speed and direction [40]. Yang et al. developed a self-powered wind vector sensor system based on four TENGs [32]. To detect wind flows from multiple directions, four independent wind-driven TENGs were fixed along four directions, as shown in Figure 2.15a. Figure 2.15b shows the external circuit diagram. The output current signals of the wind-driven TENGs could be enhanced by increasing the wind speeds. Thus, by measuring the current signals, the wind speeds could be in turn calculated, as shown in Figure 2.15c. The output voltage peaks have small fluctuations, which can be independently used to confirm the wind direction (Figure 2.15d). The output voltage signals were connected to 16 channels, forming a mapping figure. When no wind flowed along the four directions, the value of the output voltage was 0, and when the wind flowed along one direction, the wind direction can be confirmed by observing the mapping figures, as shown in Figure 2.15e–h.

2.3.4.3 Polarization of Ferroelectric Materials

Ferroelectric materials, such as $BiFeO_3$, Bi_2FeCrO_6, and $BaTiO_3$, have been widely used in photovoltaic devices due to their abnormal photovoltaic effect. Usually, these materials need a polarization process, which could induce most of the domains along a fixed direction by applying a high external electric field. Conventional polarization apparatus exhibits some drawbacks including large volume, high cost, and safety issues.

Liu et al. developed an original polarization method, which could harvest ambient wind energy through the wind-driven TENG to generate high output voltage pulse (of about 1000 V) leading to the polarization effect [75]. For the $BaTiO_3$, a large piezoelectric constant of about 150 pC/N and a pyroelectric constant of about 14 nC/(cm^2 K) were achieved by this method. Figure 2.16a illustrates the working principle of the polarization system. When the FEP film vibrated between two Cu electrodes due to the wind-induced vibration, alternate output signals were formed in the external circuit based on a combined mechanism of contact electrification and electrostatic induction. By a rectifier, the primary output violated signals can be transformed to DC voltage pulses, which were used to polarize ferroelectric material with Ag electrodes. Figure 2.16b shows a photograph of a handheld TENG-based polarization device. The piezoelectric constant could be increased with increasing the polarization time, as shown in Figure 2.16c. The piezoelectric constant of the BTO film was about 82 pC/N after a polarization time of about 20 s, as shown in Figure 2.16d.

2.3.4.4 Self-Powered Wearable Electronics

Together with the advancement of digital health, wearable electronics are attracting extensive attention. Currently, powering these electronic devices mainly depends on rechargeable batteries, limiting their further applications. To effectively extend the lifetime of the wearable electronics, a number of different research teams have begun to develop new technologies for harvesting energy from our living environment.

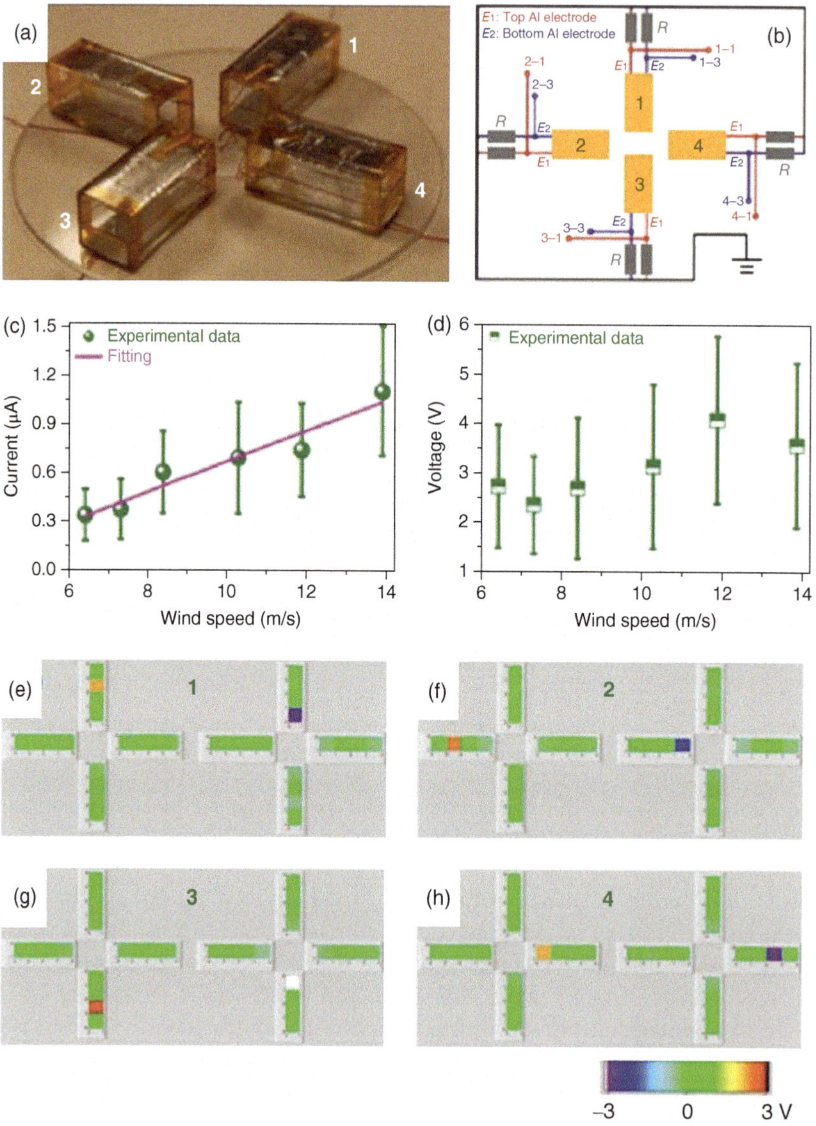

Figure 2.15 The self-powered wind vector sensor system. (a) Photograph of the four TENGs orientation. (b) Schematic diagram of the circuit. (c) Measured output current. (d) Measured output voltage. (e–h) Measured output voltage mapping figures for different wind directions. Source: Reproduced with permission from Yang et al. [32]. Copyright 2013, American Chemical Society.

Jiang et al. fabricated a smart WD-TENG, which can be used to power some wearable electronics by harvesting various human motorial energies [67–69]. By dexterous designs, the TENGs can be used as self-powered devices to monitor different human movements. Figure 2.17a shows that the TENG

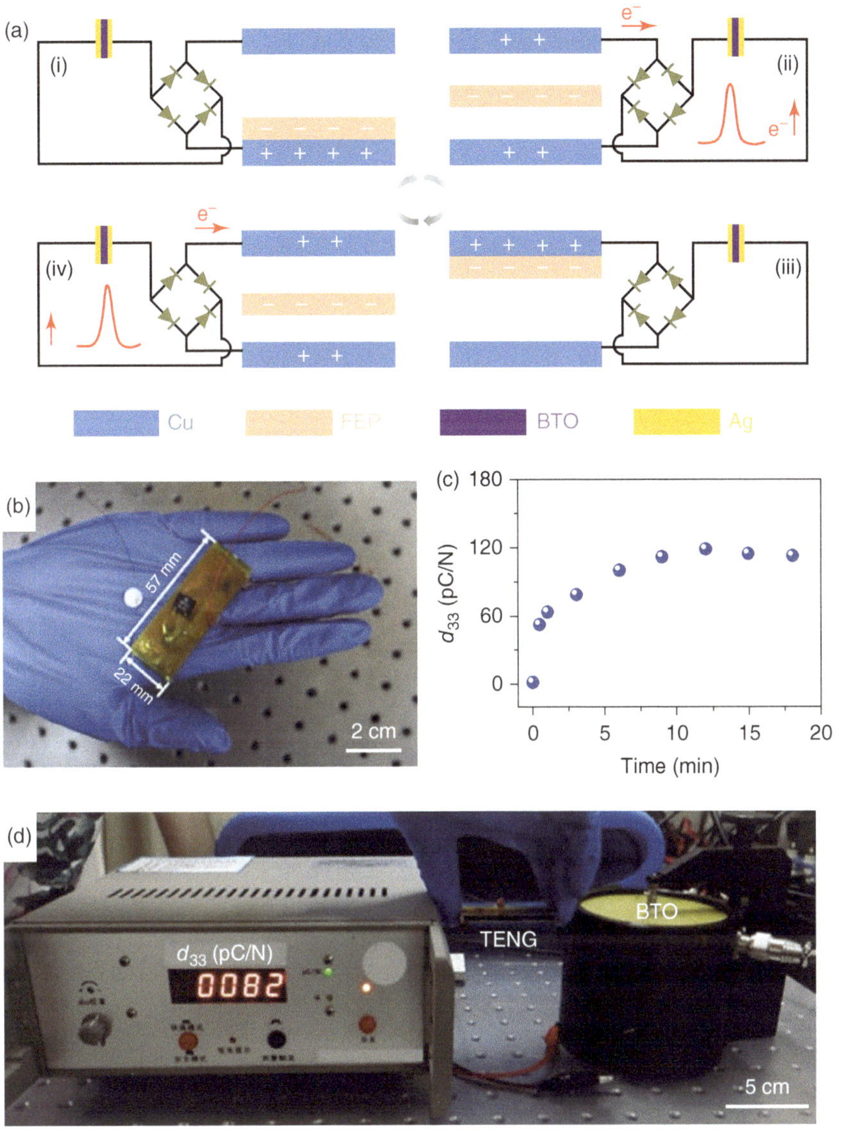

Figure 2.16 The polarization system. (a) Schematic illustration of the working principle of the polarization system. (b) Photograph of the handheld TENG-based polarization device. (c) Measured piezoelectric coefficient for different polarization times. (d) Photograph of the effective polarization result. Source: Reproduced with permission from Liu et al. [75]. Copyright 2018, Elsevier.

can be integrated into a shoe. The human motion state under the feet could be realized by measuring the output voltages under different leg movements, which respectively correspond to walking, jogging, and running. In addition, the electric energy generated by the device can power three light-emitting

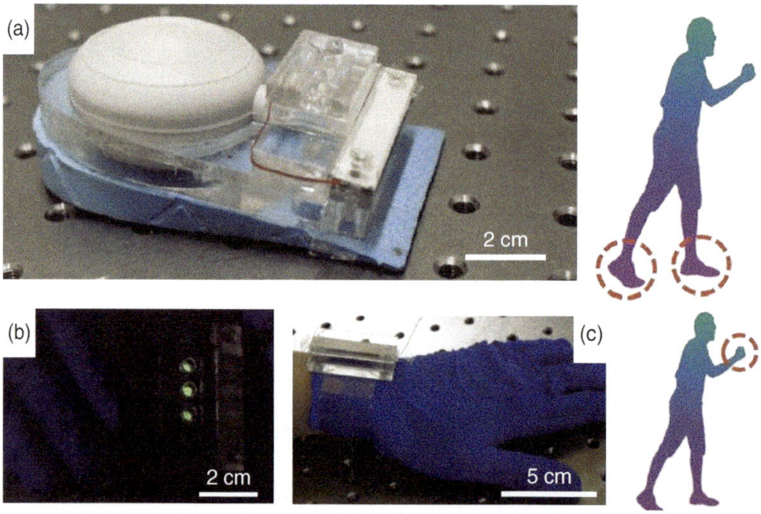

Figure 2.17 Wind-driven wearable electronics. (a) TENG-based shoe for monitoring leg movements. Source: Reproduced with permission from Jiang et al. [67]. Copyright 2018, American Chemical Society. (b) The electrical energy form the device can power LEDs. (c) The wind-driven TENG-based band for monitoring hand movements. Source: Reproduced with permission from Jiang et al. [69]. Copyright 2017, American Chemical Society.

diodes, as shown in Figure 2.17b. On the other hand, the wind-driven TENG can also be processed into a self-powered bracelet device. The mechanical energy of arm swing could be transformed to electric energy by the TENG. By measuring the output voltages, different arm movements can be monitored. Wang et al. fabricated a self-powered healthcare monitoring device based on wind-driven TENG [34]. This device can monitor human nose breath, as shown in Figure 2.18a. Figure 2.18b shows that the TENG can convert the flow from human breath to electrical energy, achieving an output voltage peak of about 30 V. By the up-flow chip, the output current of the TENG can be increased to about 350 μA, as shown in Figure 2.18c. The electrical energy generated by the TENG can be used to charge a capacitor, as illustrated in Figure 2.18d. Figure 2.18e,f shows that the temperature sensor can be powered by harvesting human nose blown air flow via the TENG.

2.3.4.5 Others

WD-TENGs, which possess the advantages of small scale, low cost, and easy fabrication, can be used as facile power sources to uninterruptedly power some small electronic devices under continuous wind flow. Yang et al. developed a small-scale wind-driven TENG, which can be used to power three light-emitting diodes and an exit sign by harvesting human mouth blowing induced wind energy, as shown in Figure 2.19a,b [32].

Wang et al. reported a TEG that can harvest flow-driven mechanical energy to achieve sufficient illumination for reading [33]. As illustrated in Figure 2.20a, 10 light bulbs could be lit up by using TEGs. Figure 2.20b shows that the

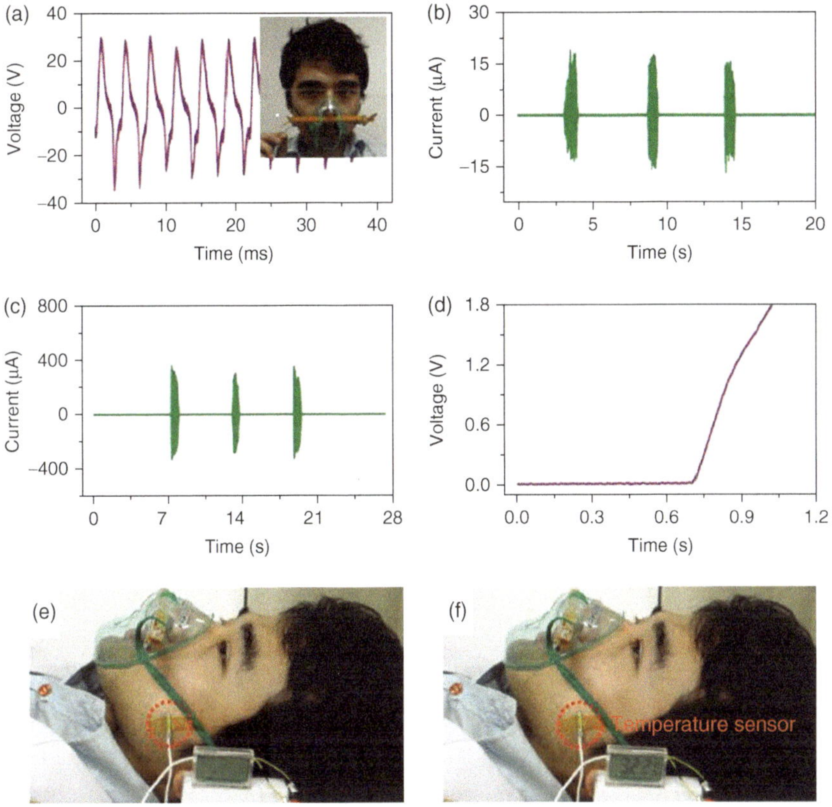

Figure 2.18 The self-powered healthcare monitoring system. (a) Output voltage signals of the TENG. (b) Output current signals of the TENG. (c) Output current signals after using the up-flow chip. (d) Measured voltage of a capacitor charged by scavenging human nose breathing via the TENG. (e) Photographs before human breath. (f) Photographs of the temperature sensor after using the TENG to scavenge human breath induced air-flow energy. Source: Reproduced with permission from Wang et al. [34]. Copyright 2015, American Chemical Society.

illumination powered by the TEGs can be used to read printed text in the dark environment. Humidity sensors are very important for food quality inspection and equipment security. Conventional humidity sensors, which are based on metal oxides, porous silicon, and polymers, would attain saturation at high humidity. Environmental humidity can affect the output performance of TENGs. The output voltage decreased with increasing the relative humidity. According to this property, Guo et al. developed a novel humidity sensor based on the airflow-induced TENG [63]. This sensor can work in a wide humidity range and exhibits several advantages, such as self-power, multifunction, low cost, and high sensitivity.

With population growth and industrial development, air pollution has developed into a main problem that threatens human endurance. Currently, air-purifying methods mainly include filtration, pulsed electron beam, and electrostatic precipitation. These approaches all face respective problems,

Figure 2.19 The wind-driven electronics light-emitting diodes. (a) Photographs of lighting light-emitting diodes via the TENG. (b) Photographs of powering an exit sign via the TENG. Source: Reproduced with permission from Yang et al. [32]. Copyright 2013, American Chemical Society.

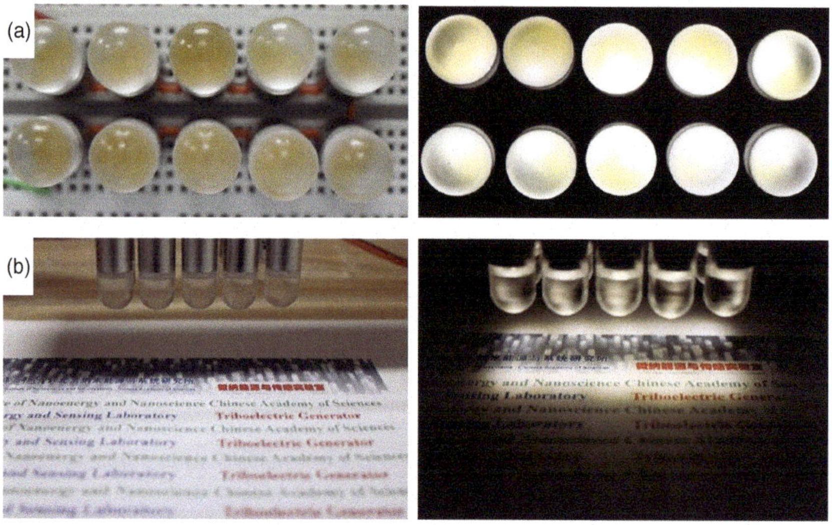

Figure 2.20 Photograph of lighting equipment power by the TEGs. (a) Ten spot lights driven by the TENGs. (b) Photograph of the printed text illuminated by the lights in the dark environment. Source: Reproduced with permission from Wang et al. [33]. Copyright 2015, John Wiley and Sons.

such as occlusion for filtration and high voltage requirement for electrostatic precipitation. Chen et al. developed a self-powered clearing system, which consists of a rotating TENG, an electric tank, and tow metal electrodes [76]. This nanogenerator can harvest wind energy and generate an output voltage of about 300 V. The voltage could not generate by-products such as ozone and NO_X compared to that in traditional electrostatic precipitation. The generated voltage signals were converted into DC electricity through a rectifying bridge, and then connected to two metal electrodes. In this system, SO_2 could be converted to H_2SO_4 by the electrochemical reaction. Hydrogen (H_2), which mainly comes from natural gas and water reduction, is one of desired energy sources because of its high energy capacity. Water splitting where water is converted to H_2 and O_2 needs an external power supply, which makes it economically inefficient. Ren et al. developed a self-powered water splitting system, which can harvest energy from our living environment for electrolytic water splitting based on a coaxial rotatory freestanding TENG [38]. The hydrogen generation rate of this system is about 6.9685 μl/min in 1 M KOH under a wind speed of about 10 m/s.

With the development of the industrial internet, wireless sensors, which can be used to apperceive the change in the surrounding environment, have attracted increasing attention in the past decade. Usually, powering wireless sensors needs an external power source such as a Li-ion battery, which is limited by the capacity of batteries. Electrical energy generated by harvesting wind energy via TEGs can be used in the wireless sensor, according to Wang et al. The output voltage signals generated by the TEG must be reduced to reasonable voltage value, which can be accomplished by the power management circuit, as illustrated in Figure 2.21a [33]. The output voltage signals were decreased to about 12 V, and the current signals were increased to about 0.9 mA, as shown in Figure 2.21b,c. The wireless system consists of a power management circuit, a wireless sensor node, and the TEGs in a transparent gas tube, as shown in Figure 2.21d. The output voltage value could be adjusted to 1.8 V via the power management circuit when the TEG continuity harvest wind energy is about 4.9 S, as shown in Figure 2.21e.

Zhao et al. fabricated a highly efficient TENG, which can harvest wind energy for sustainably powering a wireless smart temperature sensor node [61]. Figure 2.22a shows a photograph of the wireless temperature sensor. A 10 mF capacitor was used to store the energy from the nanogenerator. In the charging stage (process-1), the capacitor can be charged from 0 to 3.3 V, as shown in Figure 2.22b. The voltage of the capacitor decreased when the sensor stood at the initial stage, and then increased due to the supplement from the nanogenerator. When the TENG was shut down, no electrical energy could be generated, resulting in no temperature data on the screen of the iPhone, as illustrated in Figure 2.22c. When the TENG was working, the temperature sensor can probe the temperature of a human finger, as shown in Figure 2.22d. Lithium ion batteries, which are most effective energy storage devices, have been widely used in numerous electronic devices Jiang et al. used the Li-ion battery to store electrical energy generated by TENGs [69]. By adjusting the output voltage of the TENG via a power management circuit, the regulative electrical

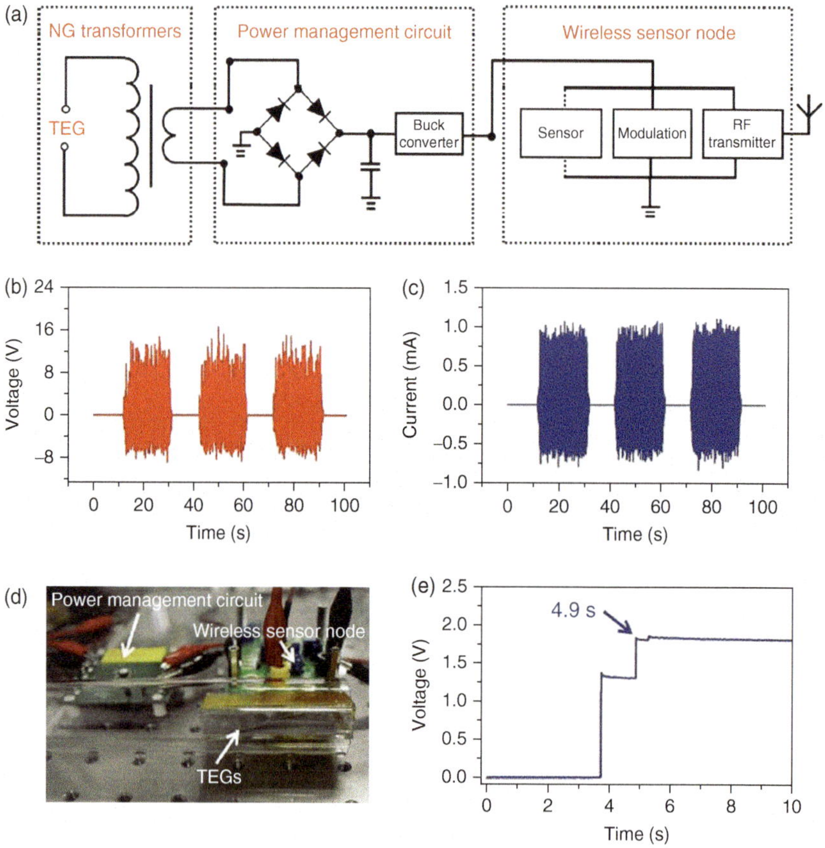

Figure 2.21 The wind-driven wireless sensor. (a) Schematic diagram of an integral powering circuit. The measured (b) regulative voltage and (c) the current at the transformer output port. (d) Photograph of the self-powered system. (e) Output voltage of the system. Source: Reproduced with permission from Wang et al. [33]. Copyright 2015, John Wiley and Sons.

energy can be used to charge the Li-ion battery, as shown in Figure 2.23a. It was found that voltages of the Li-ion battery can be increased by increasing the charging time. The voltage of the battery was about 3.6 V, when charged by the TENG for about 40 minutes. Figure 2.23b shows a photograph of the fabricated self-charging system, which consists of a TENG, a Li-ion battery, and an LED. The LED can be lit up by the Li-ion battery, which is charged by the TENG, as shown in Figure 2.23c.

On the other hand, by integrating a WD-TENG and a pressure-sensitive elastic polyimide (PI)/reduced graphene oxide (rGO) foam, Zhao et al. fabricated a self-powered pressure sensor [77]. Figure 2.24a shows a schematic diagram of the self-powered pressure system. The TENG can generate output voltage signals by harvesting the wind energy. Figure 2.24b shows photographs of the self-powered pressure system. Figure 2.24c presents four foams with different heights. The

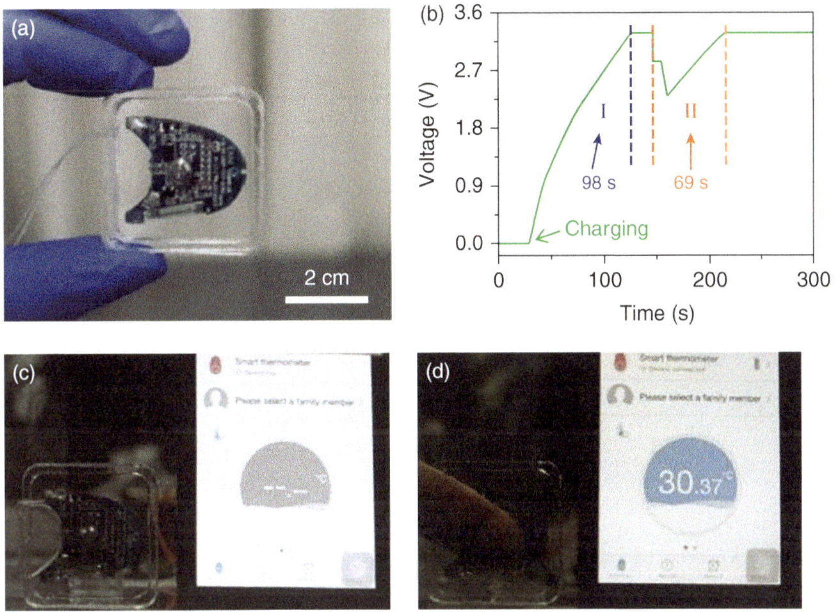

Figure 2.22 The wind-driven self-powered wireless smart temperature sensor. (a) Photographs of the temperature sensor. (b) Charging curve of a capacitor by using the TENG with a PMC. Photographs of the temperature sensor before (c) and after (d) being powered by the TENG. Source: Reproduced with permission from Zhao et al. [61]. Copyright 2016, American Chemical Society.

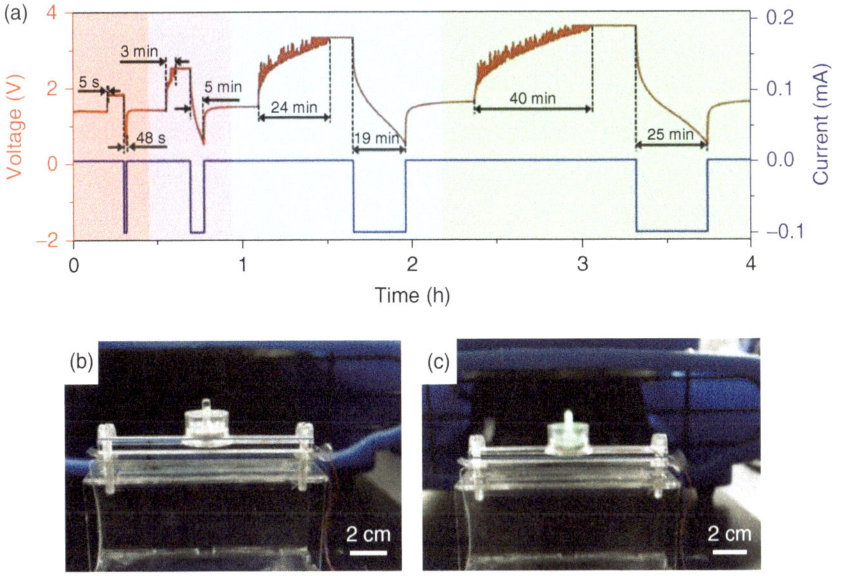

Figure 2.23 The wind-driven self-charging Li-ion battery. (a) Charging and discharging curves of the Li-ion battery via the TENG. (b) Photograph of the self-charging system. (c) Photographs of lighting up a green LED via the Li-ion battery charged by the TENG. Source: Reproduced with permission from Jiang et al. [69]. Copyright 2018, American Chemical Society.

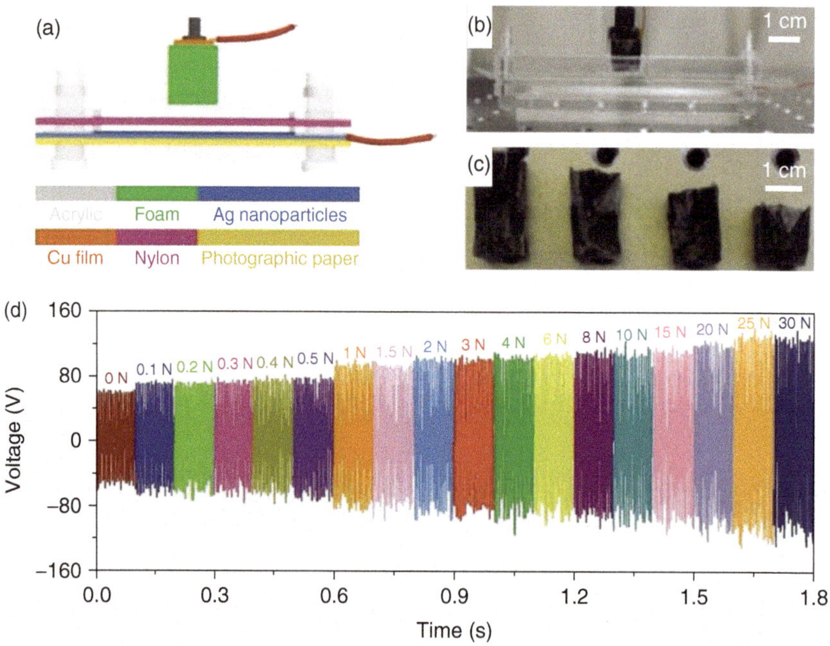

Figure 2.24 The wind-driven self-powered pressure sensor. (a) Schematic diagram of the as-fabricated self-powered pressure sensor. (b) Photograph of the prepared pressure sensor. (c) Photograph of foams with heights. (d) The output voltage of the foam connected to the TENG. Source: Reproduced with permission from Zhao et al. [77]. Copyright 2019, John Wiley and Sons.

resistances of the elastic foam that was connected to the TENG can be changed with the diversification of the external forces, resulting in changes in the measured voltage, as shown in Figure 2.24d.

2.4 Comparison

With the fast development of modern industry, energy shortage has become an important problem that restricts the further development of the human society. Conventional wind harvesters based on the electromagnetic effect have been created as one of the most important ways for harvesting wind energy. Recently, TENGs based on the cooperation of the triboelectric effect and the electrostatic induction have been used to harvest wind energy. Figure 2.25 shows the comparison between two wind harvesters, focusing on the mechanism, characteristics, and disadvantages [78]. Compared to conventional wind harvesters, TENGs can be considered as current sources with a large internal resistance, leading to characteristics of high voltage and low current. In addition, WD-TENGs possess many advantages such as low cost, high efficiency, and easy fabrication, providing them the potential to be utilized as self-powered electronics.

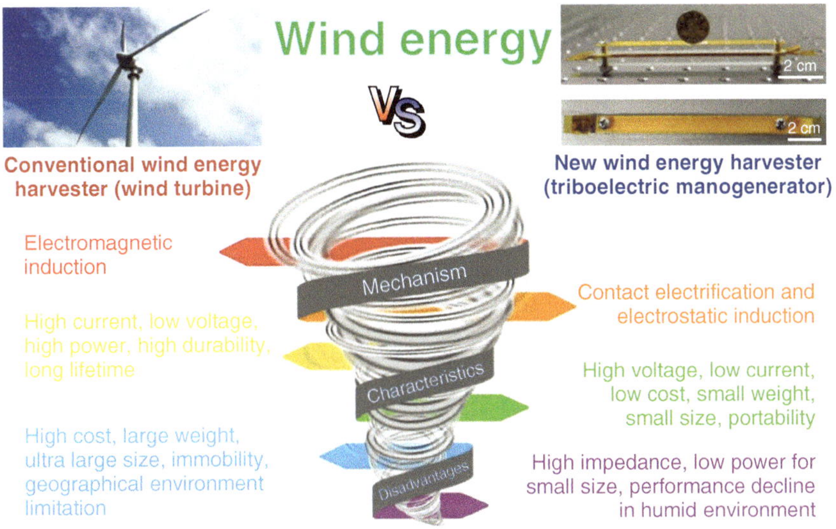

Figure 2.25 The comparison between conventional wind harvester and new wind harvester based on a triboelectric nanogenerator. Source: Reproduced with permission from Chen et al. [78]. Copyright 2018, John Wiley and Sons.

2.5 Conclusion

With the development of wearable electronics and wireless sensors, WD-TENG, which can convert wind energy to electrical energy, are researched not only to enhance output performance but also to improve compatibility with other electrical devices. It has become a new trend in the further development of self-powered devices that harvest energy from surrounding environment.

The theoretical models and working mechanisms of current WD-TENGs have been described in this chapter. The output performances of the TENGs can be improved by optimizing advanced structures and materials. Among the several structures, vibrating plate-based structure, elasto-aerodynamics-based structure, and rotary-driven mechanical structure are three popular structures. Some advanced materials, such as cellulose, superhydrophobic surfaces, and nanowires, have been used. Lastly, soma smart self-powered devices based on WD-TENGs are introduced to present wide application prospects.

References

1 Wiser, R., Jenni, K., Seel, J. et al. (2016). Expert elicitation survey on future wind energy costs. *Nat. Energy* 1: 16135.
2 Mertens, S. (2003). The energy yield of roof mounted wind turbines. *Wind Eng.* 27: 507.
3 Grant, A., Johnstone, C., and Kelly, N. (2008). Urban wind energy conversion: the potential of ducted turbines. *Renewable Energy* 33: 1157.

4 Hameed, Z., Hong, Y., Cho, Y. et al. (2009). Condition monitoring and fault detection of wind turbines and related algorithms: a review. *Renewable Sustainable Energy Rev.* 13 (1): 1.
5 Islam, M., Mekhilef, S., and Saidur, R. (2013). Progress and recent trends of wind energy technology. *Renewable Sustainable Energy Rev.* 21: 456–468.
6 Henriques, J., Da Silva, F.M., Estanqueiro, A., and Gato, L. (2009). Design of a new urban wind turbine airfoil using a pressure-load inverse method. *Renewable Energy* 34: 2728.
7 Chong, W., Fazlizan, A., Poh, S. et al. (2012). Early development of an innovative building integrated wind, solar and rain water harvester for urban high rise application. *Energy Build.* 47: 201.
8 Hoogwijk, M., de Vries, B., and Turkenburg, W. (2004). Assessment of the global and regional geographical, technical and economic potential of onshore wind energy. *Energy Econ.* 26: 889.
9 Sfetsos, A. (2000). A comparison of various forecasting techniques applied to mean hourly wind speed time series. *Renewable Energy* 21: 23.
10 Seguro, J. and Lambert, T. (2000). Modern estimation of the parameters of the Weibull wind speed distribution for wind energy analysis. *J. Wind Eng. Ind. Aerodyn.* 85: 75.
11 Al-Bahadly, I. (2009). Building a wind turbine for rural home. *Renewable Sustainable Energy Rev.* 13: 159.
12 Walker, S.L. (2011). Building mounted wind turbines and their suitability for the urban scale – a review of methods of estimating urban wind resource. *Energy Build.* 43: 1852.
13 Ricciardelli, F. and Polimeno, S. (2006). Some characteristics of the wind flow in the lower urban boundary layer. *J. Wind Eng. Ind. Aerodyn.* 94: 815.
14 Schallenberg-Rodriguez, J. (2013). A methodological review to estimate techno-economical wind energy production. *Renewable Sustainable Energy Rev.* 21: 272.
15 Herbert, G.J., Iniyan, S., Sreevalsan, E., and Rajapandian, S. (2007). A review of wind energy technologies. *Renewable Sustainable Energy Rev.* 11: 1117.
16 Safari, B. and Gasore, J. (2010). A statistical investigation of wind characteristics and wind energy potential based on the Weibull and Rayleigh models in Rwanda. *Renewable Energy* 35: 2874.
17 Mithraratne, N. (2009). Roof-top wind turbines for microgeneration in urban houses in New Zealand. *Energy Build.* 41: 1013.
18 Fan, F.-R., Tian, Z.-Q., and Wang, Z.L. (2012). Flexible triboelectric generator. *Nano Energy* 1: 328.
19 Wang, S., Lin, L., and Wang, Z.L. (2012). Nanoscale triboelectric-effect-enabled energy conversion for sustainably powering portable electronics. *Nano Lett.* 12: 6339.
20 Yang, Y., Zhang, H., Chen, J. et al. (2013). Single-electrode-based sliding triboelectric nanogenerator for self-powered displacement vector sensor system. *ACS Nano* 7: 7342.
21 Niu, S., Liu, Y., Wang, S. et al. (2013). Theory of sliding-mode triboelectric nanogenerators. *Adv. Mater.* 25: 6184.

22 Wang, S., Lin, L., Xie, Y. et al. (2013). Sliding-triboelectric nanogenerators based on in-plane charge-separation mechanism. *Nano Lett.* 13: 2226.
23 Niu, S., Wang, S., Lin, L. et al. (2013). Theoretical study of contact-mode triboelectric nanogenerators as an effective power source. *Energy Environ. Sci.* 6: 3576.
24 Liu, Y., Niu, S., and Wang, Z.L. (2015). Theory of tribotronics. *Adv. Electron. Mater.* 1: 1500124.
25 Zhu, G., Peng, B., Chen, J. et al. (2015). Triboelectric nanogenerators as a new energy technology: from fundamentals, devices, to applications. *Nano Energy* 14: 126.
26 Dharmasena, R.D.I.G., Jayawardena, K., Mills, C. et al. (2017). Triboelectric nanogenerators: providing a fundamental framework. *Energy Environ. Sci.* 10: 1801.
27 Wang, Z.L. (2015). Triboelectric nanogenerators as new energy technology and self-powered sensors–principles, problems and perspectives. *Faraday Discuss.* 176: 447.
28 Ishugah, T., Li, Y., Wang, R., and Kiplagat, J. (2014). Advances in wind energy resource exploitation in urban environment: a review. *Renewable Sustainable Energy Rev.* 37: 613–626.
29 Grant, I., Mo, M., Pan, X. et al. (2000). An experimental and numerical study of the vortex filaments in the wake of an operational, horizontal-axis, wind turbine. *J. Wind Eng. Ind. Aerodyn.* 85: 177.
30 Ayhan, D. and Sağlam, Ş. (2012). A technical review of building-mounted wind power systems and a sample simulation model. *Renewable Sustainable Energy Rev.* 16: 1040.
31 Gordeeva, L., Restuccia, G., Cacciola, G., and Aristov, Y.I. (1998). Selective water sorbents for multiple applications, 5. LiBr confined in mesopores of silica gel: sorption properties. *Kinet. Catal.* 63: 81–88.
32 Yang, Y., Zhu, G., Zhang, H. et al. (2013). Triboelectric nanogenerator for harvesting wind energy and as self-powered wind vector sensor system. *ACS Nano* 7: 9461.
33 Wang, S., Mu, X., Yang, Y. et al. (2015). Flow-driven triboelectric generator for directly powering a wireless sensor node. *Adv. Mater.* 27 (2): 240.
34 Wang, S., Mu, X., Wang, X. et al. (2015). Elasto-aerodynamics-driven triboelectric nanogenerator for scavenging air-flow energy. *ACS Nano* 9: 9554.
35 Xie, Y., Wang, S., Lin, L. et al. (2013). Rotary triboelectric nanogenerator based on a hybridized mechanism for harvesting wind energy. *ACS Nano* 7: 7119.
36 Zhang, H., Wang, J., Xie, Y. et al. (2016). Self-powered, wireless, remote meteorologic monitoring based on triboelectric nanogenerator operated by scavenging wind energy. *ACS Appl. Mater. Interfaces* 8: 32649.
37 Bae, J., Lee, J., Kim, S. et al. (2014). Flutter-driven triboelectrification for harvesting wind energy. *Nat. Commun.* 5: 4929.
38 Ren, X., Fan, H., Wang, C. et al. (2018). Wind energy harvester based on coaxial rotatory freestanding triboelectric nanogenerators for self-powered water splitting. *Nano Energy* 50: 562.

39 Quan, Z., Han, C.B., Jiang, T., and Wang, Z.L. (2016). Robust thin films-based triboelectric nanogenerator arrays for harvesting bidirectional wind energy. *Adv. Energy Mater.* 6: 1501799.
40 Wang, J., Ding, W., Pan, L. et al. (2018). Self-powered wind sensor system for detecting wind speed and direction based on a triboelectric nanogenerator. *ACS Nano* 12: 3954.
41 Zhang, L., Zhang, B., Chen, J. et al. (2016). Lawn structured triboelectric nanogenerators for scavenging sweeping wind energy on rooftops. *Adv. Mater.* 28: 1650.
42 Zhao, Z., Pu, X., Du, C. et al. (2016). Freestanding flag-type triboelectric nanogenerator for harvesting high-altitude wind energy from arbitrary directions. *ACS Nano* 10: 1780.
43 Klemm, D., Heublein, B., Fink, H.P., and Bohn, A. (2005). Cellulose: fascinating biopolymer and sustainable raw material. *Angew. Chem. Int. Ed.* 44: 3358.
44 Kalia, S., Dufresne, A., Cherian, B.M. et al. (2011). Cellulose-based bio-and nanocomposites: a review. *Int. J. Polym. Sci.* 2011: 837875.
45 Moon, R.J., Martini, A., Nairn, J. et al. (2011). Cellulose nanomaterials review: structure, properties and nanocomposites. *Chem. Soc. Rev.* 40: 3941.
46 Siró, I. and Plackett, D. (2010). Microfibrillated cellulose and new nanocomposite materials: a review. *Cellulose* 17: 459.
47 Klemm, D., Kramer, F., Moritz, S. et al. (2011). Nanocelluloses: a new family of nature-based materials. *Angew. Chem. Int. Ed.* 50: 5438.
48 Eichhorn, S.J., Dufresne, A., Aranguren, M. et al. (2010). Current international research into cellulose nanofibres and nanocomposites. *J. Mater. Sci.* 45: 1.
49 Hubbe, M.A., Rojas, O.J., and Lucia, L.A. (2015). Green modification of surface characteristics of cellulosic materials at the molecular or nano scale: a review. *BioResources* 10: 6095.
50 Hubbe, M.A. (2006). Bonding between cellulosic fibers in the absence and presence of dry-strength agents – a review. *BioResources* 1: 281.
51 Chen, B., Yang, N., Jiang, Q. et al. (2018). Transparent triboelectric nanogenerator-induced high voltage pulsed electric field for a self-powered handheld printer. *Nano Energy* 44: 468.
52 Kim, H.-J., Yim, E.-C., Kim, J.-H. et al. (2017). Bacterial nano-cellulose triboelectric nanogenerator. *Nano Energy* 33: 130.
53 Yao, C., Hernandez, A., Yu, Y. et al. (2016). Triboelectric nanogenerators and power-boards from cellulose nanofibrils and recycled materials. *Nano Energy* 30: 103.
54 Chandrasekhar, A., Alluri, N.R., Saravanakumar, B. et al. (2017). A microcrystalline cellulose ingrained polydimethylsiloxane triboelectric nanogenerator as a self-powered locomotion detector. *J. Mater. Chem. C* 5: 1810.
55 Peng, J., Zhang, H., Zheng, Q. et al. (2017). A composite generator film impregnated with cellulose nanocrystals for enhanced triboelectric performance. *Nanoscale* 9: 1428.
56 Yao, C., Yin, X., Yu, Y. et al. (2017). Chemically functionalized natural cellulose materials for effective triboelectric nanogenerator development. *Adv. Funct. Mater.* 27: 1700794.

57 Šutka, A., Ruža, J., Järvekülg, M. et al. (2018). Triboelectric nanogenerator based on immersion precipitation derived highly porous ethyl cellulose. *J. Electrostat.* 92: 1.

58 He, X., Zou, H., Geng, Z. et al. (2018). A hierarchically nanostructured cellulose fiber-based triboelectric nanogenerator for self-powered healthcare products. *Adv. Funct. Mater.* 28: 1805540.

59 Oh, H., Kwak, S.S., Kim, B. et al. (2019). Highly conductive ferroelectric cellulose composite papers for efficient triboelectric nanogenerators. *Adv. Funct. Mater.*

60 Qian, C., Li, L., Gao, M. et al. (2019). All-printed 3D hierarchically structured cellulose aerogel based triboelectric nanogenerator for multi-functional sensors. *Nano Energy* 63: 103885.

61 Zhao, K., Wang, Z.L., and Yang, Y. (2016). Self-powered wireless smart sensor node enabled by an ultrastable, highly efficient, and superhydrophobic-surface-based triboelectric nanogenerator. *ACS Nano* 10: 9044.

62 Dudem, B., Huynh, N.D., Kim, W. et al. (2017). Nanopillar-array architectured PDMS-based triboelectric nanogenerator integrated with a windmill model for effective wind energy harvesting. *Nano Energy* 42: 269.

63 Guo, H., Chen, J., Tian, L. et al. (2014). Airflow-induced triboelectric nanogenerator as a self-powered sensor for detecting humidity and airflow rate. *ACS Appl. Mater. Interfaces* 6: 17184.

64 Wang, M., Zhang, N., Tang, Y. et al. (2017). Single-electrode triboelectric nanogenerators based on sponge-like porous PTFE thin films for mechanical energy harvesting and self-powered electronics. *J. Mater. Chem. A* 5: 12252.

65 Zhao, P., Soin, N., Prashanthi, K. et al. (2018). Emulsion electrospinning of polytetrafluoroethylene (PTFE) nanofibrous membranes for high-performance triboelectric nanogenerators. *ACS Appl. Mater. Interfaces* 10: 5880.

66 Kang, H., Buchman, J.T., Rodriguez, R.S. et al. (2018). Stabilization of silver and gold nanoparticles: preservation and improvement of plasmonic functionalities. *Chem. Rev.* 119: 664.

67 Jiang, Q., Chen, B., and Yang, Y. (2018). Wind-driven triboelectric nanogenerators for scavenging biomechanical energy. *ACS Appl. Energy Mater.* 1: 4269.

68 Cheon, S., Kang, H., Kim, H. et al. (2018). High-performance triboelectric nanogenerators based on electrospun polyvinylidene fluoride–silver nanowire composite nanofibers. *Adv. Funct. Mater.* 28: 1703778.

69 Jiang, Q., Chen, B., Zhang, K., and Yang, Y. (2017). Ag nanoparticle-based triboelectric nanogenerator to scavenge wind energy for a self-charging power unit. *ACS Appl. Mater. Interfaces* 9: 43716.

70 Lee, S., Ko, W., Oh, Y. et al. (2015). Triboelectric energy harvester based on wearable textile platforms employing various surface morphologies. *Nano Energy* 12: 410.

71 Chun, J., Kim, J.W., Jung, W.-s. et al. (2015). Mesoporous pores impregnated with Au nanoparticles as effective dielectrics for enhancing triboelectric nanogenerator performance in harsh environments. *Energy Environ. Sci.* 8: 3006.

72 Zhang, Z., Chen, Y., Debeli, D.K., and Guo, J.S. (2018). Facile method and novel dielectric material using a nanoparticle-doped thermoplastic elastomer composite fabric for triboelectric nanogenerator applications. *ACS Appl. Mater. Interfaces* 10: 13082.

73 Taneda, S. (1968). Waving motions of flags. *J. Phys. Soc. Jpn.* 24: 392.

74 Ahmed, A., Hassan, I., Hedaya, M. et al. (2017). Farms of triboelectric nanogenerators for harvesting wind energy: a potential approach towards green energy. *Nano Energy* 36: 21.

75 Liu, X., Zhao, K., and Yang, Y. (2018). Effective polarization of ferroelectric materials by using a triboelectric nanogenerator to scavenge wind energy. *Nano Energy* 53: 622.

76 Chen, S., Gao, C., Tang, W. et al. (2015). Self-powered cleaning of air pollution by wind driven triboelectric nanogenerator. *Nano Energy* 14: 217.

77 Zhao, X., Chen, B., Wei, G. et al. (2019). Polyimide/graphene nanocomposite foam-based wind-driven triboelectric nanogenerator for self-powered pressure sensor. *Adv. Mater. Technol.* 4: 1800723.

78 Chen, B., Yang, Y., and Wang, Z.L. (2018). Scavenging wind energy by triboelectric nanogenerators. *Adv. Energy Mater.* 8: 1702649.

3

Electromagnetic–Triboelectric Hybridized Nanogenerators

3.1 Introduction

In the past decades, the rapid development of smart sensors and small electronic devices promoted a new invention, the self-powered device, which can generate electrical energy by harvesting the energy in the environment [1–5]. Various technologies have been devised for harvesting energies from different sources, such as solar, wind energy, and human kinetic energy [6–9]. Among them, harvesting mechanical energy from the ambient environment, such as mechanical vibration, human movement, and wind, has generated great interest because mechanical energy exists in a variety of environments and can be effectively collected by the conventional electromagnetic generator (EMG) [10–15]. Recently, triboelectric nanogenerator (TENG) based on a combined mechanism of contact electrification and electrostatic induction has provided an effective approach to produce electricity by harvesting mechanical energy [16–25]. Despite significant advances in both the EMG and the TENG in harvesting the energy, the output power is still limited by the size of the generators.

Integrating electromagnetic and triboelectric effects in one device, named as the electromagnetic–triboelectric hybridized nanogenerator, provides a promising strategy to largely increase the output power without increasing the integral size of the device. In this chapter, we will summarize some developments of the electromagnetic–triboelectric hybridized nanogenerator in the recent years. Theoretical comparison of EMG and TENG is first discussed. Then various device architectures, materials, and potential applications of the hybridized nanogenerators are demonstrated. chapter 7 concludes the chapter concludes the chapter and provides the prospects for the hybridized generators.

3.2 Working Mechanisms

The basic principle of the electromagnetic induction phenomenon is illustrated in Figure 3.1a. Moving a conductor rod connected with an external electric circuit to cut the magnetic induction lines leads to induced electrodynamic potential

*Corresponding author: yayang@binn.cas.cn

Hybridized and Coupled Nanogenerators: Design, Performance, and Applications,
First Edition. Ya Yang.

across the conductor stick [26]:

$$E = BLv \tag{3.1}$$

where B represents the magnetic flux density, L represents the length of the conductor rod, and v is the cutting velocity of the conductor.

Unlike the EMG, the TENG converts mechanical energy into electricity based on a combined mechanism of the triboelectric effect and electrostatic induction [27–29]. The triboelectric effect can lead the materials to form charges on the surface when two different materials contact each other. The sign of the charges for the two triboelectric materials depends on their opposite positions in the triboelectric series. The electrostatic induction introduces the redistribution of electrical charges on the surfaces to balance the electric field generated by the change of the distance between the two triboelectric materials. Recently, four basic operation modes of the TENG were established, including vertical contact-separation mode, in-plane sliding mode, single-electrode mode, and freestanding-triboelectric-layer mode. Among these modes, the in-plane sliding mode can more effectively generate the static charges due to the close contact between two materials. Figure 3.1b illustrates the basic principle of this mode. In the initials stage, the whole surfaces of the top metal material and the polymer interlayer intimately contact with each other. Because the ability of the metal to attract electrons is weaker than that of the polymer, net positive charges form on the surface of the metal and net negative charges with equal density transfer to the surface of the polymer, based on the triboelectric effect. Moving the metal to slide outward can decrease the contact surface area between the two materials, leading to charge separation. The separated charges will produce an electric field, which drives a current flow in the external circuit due to the electrostatic induction. By neglecting the edge effect and the vertical distance between the metal and the polymeric surface, the generated current flow can be simplified as

$$I = \frac{\sigma \Delta S}{\Delta T} = \sigma l v \tag{3.2}$$

where σ presents the triboelectric charge density, l presents the width of the friction surface, and v presents the sliding velocity of the top metal.

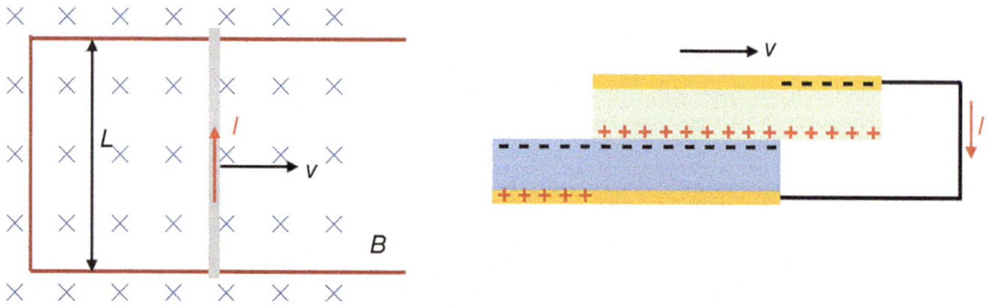

Figure 3.1 Theoretical comparison of EMG and TENG. (a) Schematic fundamental principles of the electromagnetic generator. (b) Schematic fundamental principles of the triboelectric nanogenerator.

3.3 Hybridized Devices Structure and Working Mechanisms

Structural design is very important for output performance of the generator, especially for hybridized devices [30–35]. To realize an ideal output performance, it is important to balance the different effects of generators. The working frequency is an essential factor that affects the output performance of both EMG and TENG. Thus, the structure of an electromagnetic–triboelectric hybridized nanogenerator should satisfy the demand of high frequency. In addition, the structure can fulfill the basic principles of electromagnetic induction and triboelectric effect and electrostatic induction, which can effectively utilize mechanical energy from a mechanical motion.

3.3.1 Shared-Electrode-Based EMG–TENG

Both EMG and TENG have electrodes, and optimizing the structure of electrodes can effectively economize the materials and size of the devices. Quan et al. developed a hybridized nanogenerator, which combined an EMG with a TENG by sharing the same set of electrodes [36]. As shown in Figure 3.2a, EMG consists of the Cu coils and a movable magnet, and TENG consist of an fluorinated ethylene-propylene (FEP) film, a protection layer, and a movable nylon film. Two Cu coils coated on a printed circuit board were sharing electrodes of both the EMG and the TENG. The hybridized nanogenerator can generate electric energy by harvesting the mechanical energy generated by the relative sliding movement between the fixed part and the moving part. During this process, electrons can be moved into the surface of the FEP film based on the triboelectric effect between the FEP and the nylon film at the initial stage, where Cu coils possess the same quantity of the triboelectric charges based on the electrostatic induction effect. After sliding between the FEP and the nylon film, the electrons flow from the left Cu coil to the right Cu coil through an external circuit, leading to a pulsed output current, as illustrated in Figure 3.2b. At the same time, the EMG can generate an alternating current (AC) output current signal because of Faraday electromagnetic induction between the Cu coils and the magnet.

3.3.2 Rotating-Disk-Based EMG–TENG

The rotating structure is the main structure of the conventional EMG because it can achieve an extremely high working frequency in the specific space. Based on this structure, Zhong et al. fabricated a rotating-disk-based hybridized nanogenerator by combining an EMG with a triboelectric generator (TEG), as shown in Figure 3.3a [37]. The EMG includes six magnets that are uniformly distributed in a circular acrylic substrate of the rotator and six groups of coils that are connected in series on the bottom substrate of the stator. TENG consists of radially arrayed sectors separated by equal-degree intervals with a central angle of 1°, a polyamide film, and a layer of complementary patterned electrode networks that are separated by trenches in between. Figure 3.3b illustrates the corresponding photographs of each layer.

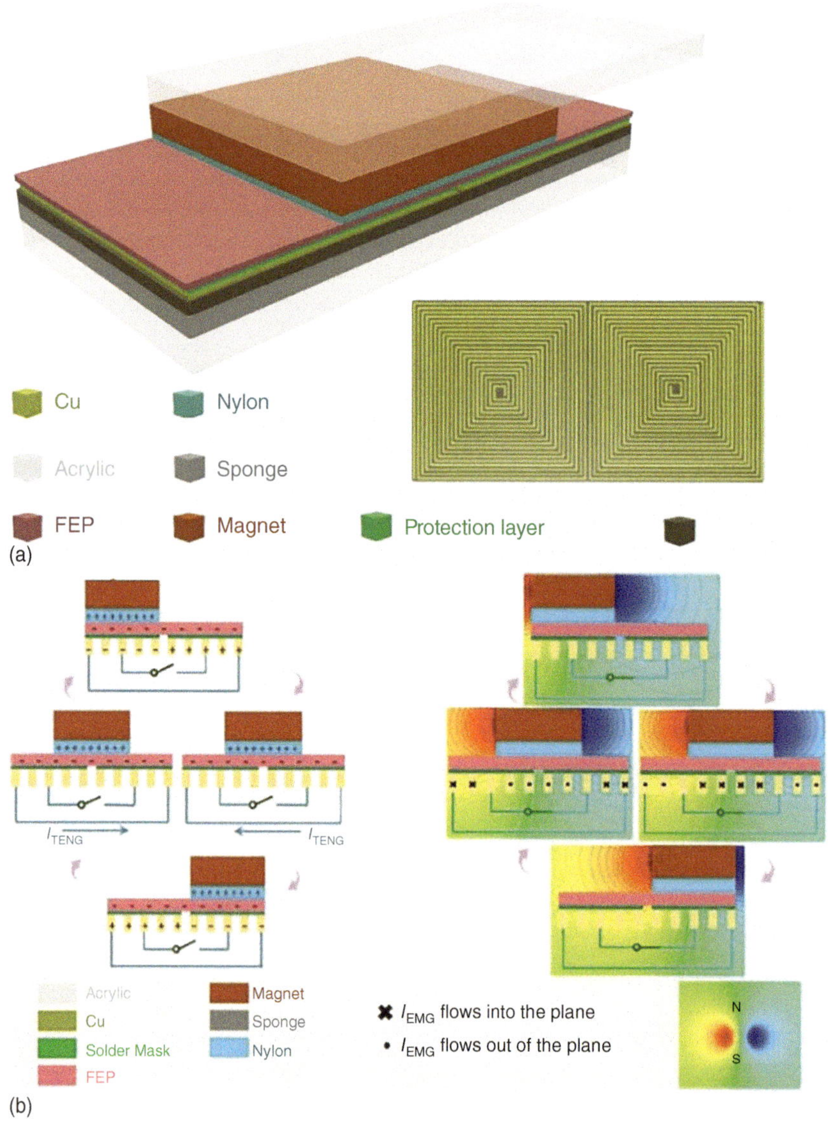

Figure 3.2 The shared-electrode-based hybridized nanogenerator. (a) Schematic diagram of the designed hybridized nanogenerator. (b) Schematic diagram of the working principle of the TENG and EMG. Source: Reproduced with permission from Quan et al. [36]. Copyright 2016, American Chemical Society.

The hybridized nanogenerator can generate electrical signals when the rotator is rotated in correspondence to the stator. For the EMG, the electricity is generated due to Faraday electromagnetic induction. There is no current in the coils at the state where the magnetic field through the coils is a constant. The magnetic flux through the coils will decrease when the rotator spins counterclockwise through 30° to reach an intermediate state, leading to the generation of an induced positive current in the coils (process 1), as shown in Figure 3.4a.

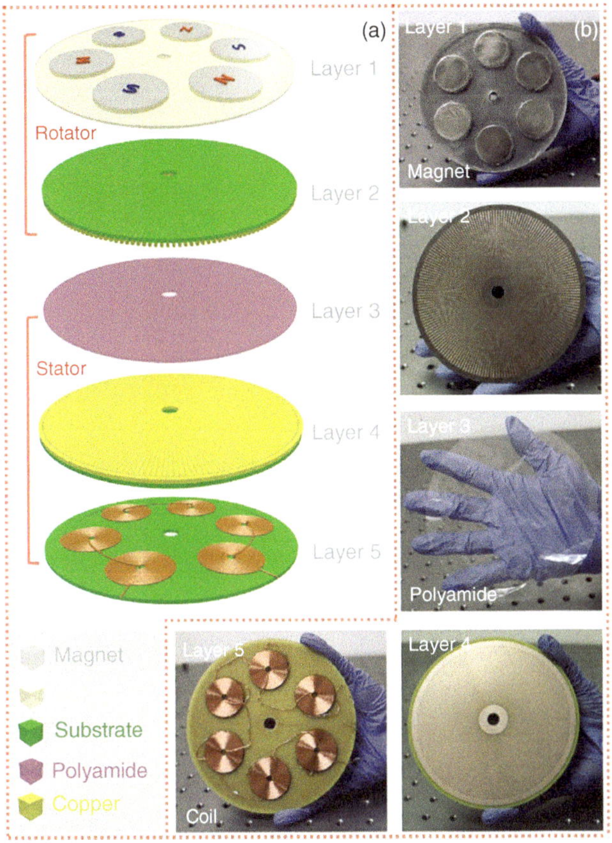

Figure 3.3 The rotating-disk-based hybridized nanogenerator. (a) Schematic diagram of the fabricated hybridized nanogenerator. (b) Photographs of the corresponding layers of the nanogenerator. Source: Reproduced with permission from Zhong et al. [37]. Copyright 2015, Elsevier.

After that, the rotator spins through another 30°, where a negative magnetic field appears in the coils (process 2). When the rotator spins through 30° again, a negative current is generated in the coils due to the decrease in the magnetic flux through the coils (process 3). Finally, with the spin of the rotator, a new state is formed, which is equivalent to the initial state (process 4). For the TENG, electron generation shifts from the polyamide (PA) film to Cu due to the triboelectric effect, when the top Cu grating slides along the surface of the PA film, as illustrated in Figure 3.4b. The same quantity of positive and negative charges exists on the two adjacent electrodes at the initial state. When the top Cu grating slides a through distance along the PA film, the electrons flow from electrode 1 to electrode 2 due to the electrostatic induction, forming a positive current signal (process 1). The current signal can remain until the Cu grating slides on the top of the electrode 1 (process 2). When the Cu grating-1 continues to slide, the electrons on electrode 2 flow to electrode 1, leading to the generation of a negative signal (process 3). Finally, the Cu grating-2 fully moves on the top of electrode 2, which is equivalent to the initial state (process 4).

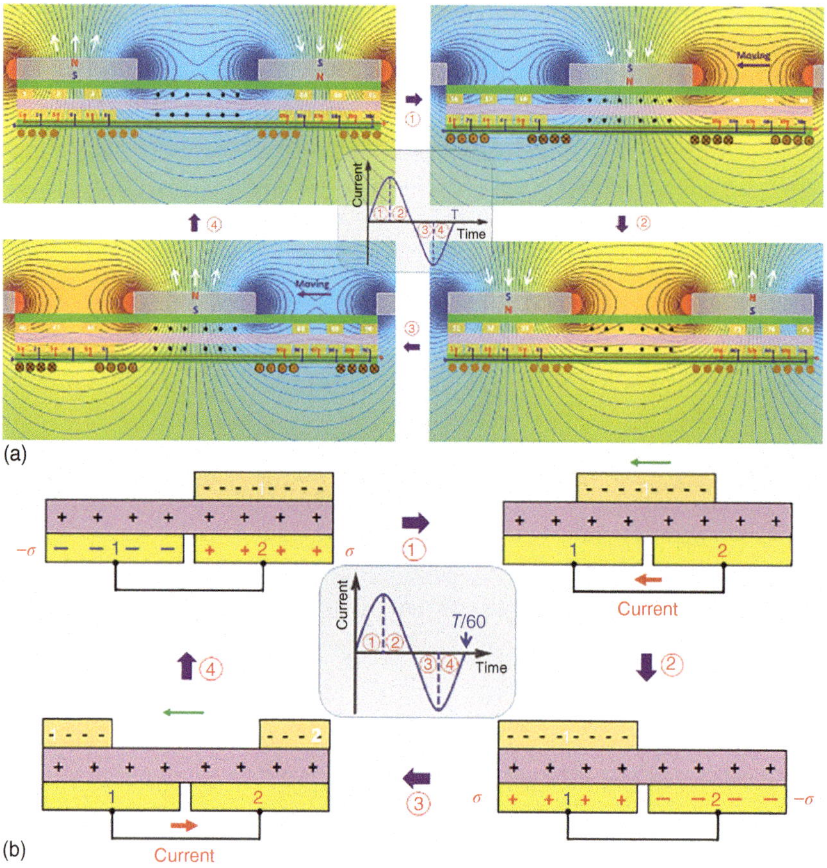

Figure 3.4 Schematic diagram of the working principle of hybridized EMG–ENG. (a) Schematic diagram of the working principle of EMG under the different states. (b) Schematic diagram of the working principle of TENG under the different states. Source: Reproduced with permission from Zhong et al. [37]. Copyright 2015, Elsevier.

3.3.3 Spring-Based EMG–TENG

With the development of wearable electronics, the structure of generators should adapt to the human body motion. Zhang et al. developed a spring-based hybridized electromagnetic–triboelectric nanogenerator, which can be used to harvest the biomechanical energy [38]. The EMG consists of a magnet and a set of coil, and the TENG consists of an Al foil and a polydimethylsiloxane (PDMS) film, as shown in Figure 3.5a. Double-layered acrylic substrates are used as the framework, which are fixed by four springs at the corners face to face. Figure 3.5b illustrates the working principle of the hybridized nanogenerator. Pressing the top substrate leads to full contact between the top PDMS film and the bottom Al film, forming triboelectric charges due to the triboelectric effect between them. The positive charges accumulate on the Al side and negative charges on the PDMS side because electrons engender the shift from Al into the

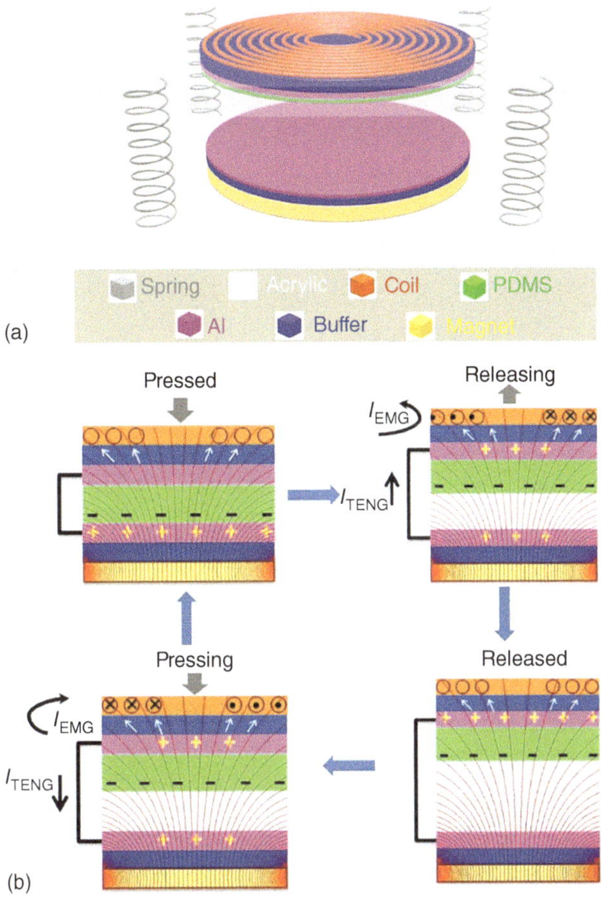

Figure 3.5 The spring-based hybridized nanogenerator. (a) Schematic diagram of the fabricated hybridized nanogenerator. (b) Schematic diagram of the working principle of TENG and EMG under the different states. Source: Reproduced with permission from Zhang et al. [38]. Copyright 2015, American Chemical Society.

PDMS. On releasing the external force, the PDMS film is separated from the Al film under the elastic force of springs, forming the inner electric field. The field will drive the electrons to flow from the top electrode to the bottom electrode in the external circuit. On the other hand, EMG can generate current signals due to a decrease in the magnetic flux across the coil when the PDMS film is separated from the Al film. With increase in the distance of separation, the positive triboelectric charges on the bottom Al film are gradually neutralized, and an equal amount of inductive charges are formed on the top electrode. Subsequently, the Al film approaches the PDMS film again under the external force, resulting in a reversed inner electric field and change of magnetic flux. This process will generate a reversed direction of current for both the triboelectric part and the electromagnetic part of the hybridized nanogenerator.

3.3.4 Stretchable EMG–TENG

Creating a stretchable generator is important for some electric devices [39–43]. Zhang et al. developed a stretchable hybridized nanogenerator based on a sandwiched structure, as shown in Figure 3.6a [44]. The top part consists

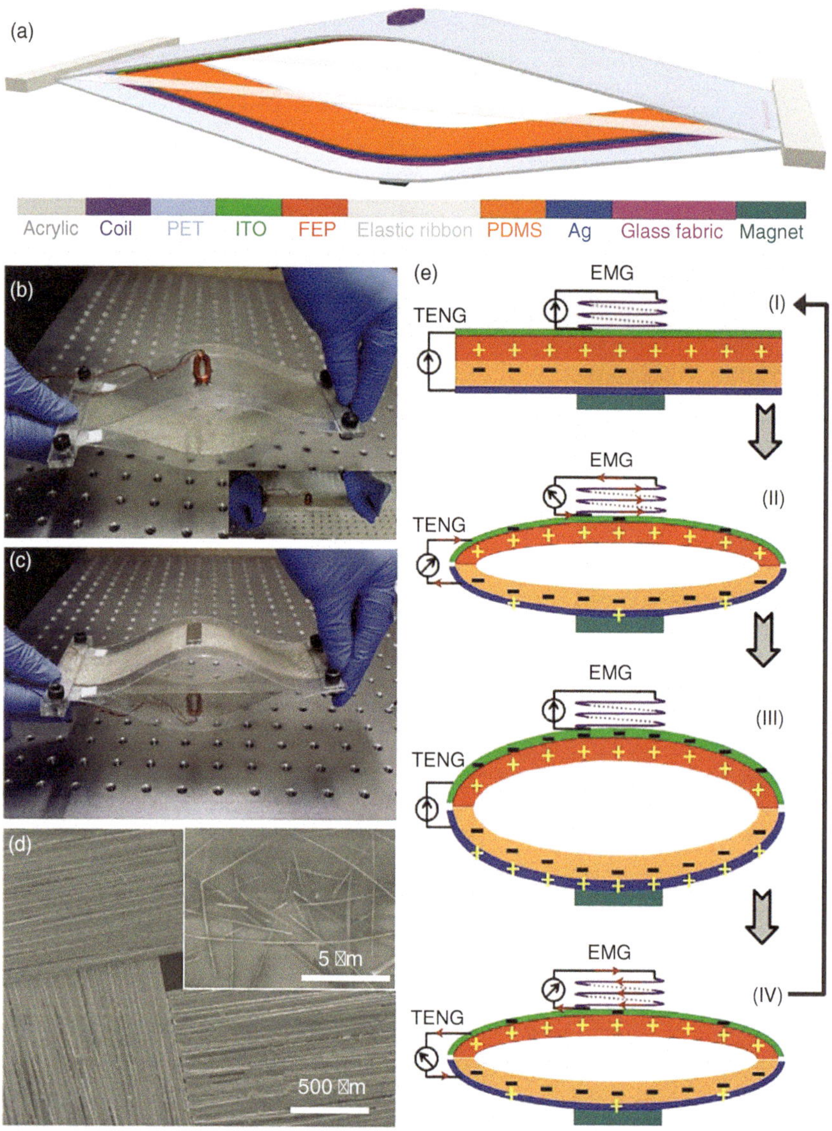

Figure 3.6 The stretchable hybridized nanogenerator. (a) Schematic diagram of the fabricated hybridized nanogenerator. (b,c) Photographs of the hybridized nanogenerator. (d) SEM image of the triboelectric layer. (e) Schematic diagram of the working principle of TENG and EMG under the different states. Source: Reproduced with permission from Zhang et al. [44]. Copyright 2016, American Chemical Society.

of a fluorinated ethylene propylene (FEP) thin film and a thick polyethylene terephthalate (PET) coated with the indium tin oxide (ITO) layer. The bottom part consists of a conductive fabric covered with a thin PDMS film (Figure 3.6d) and a PET sheet. Two parts were assembled face to face through two elastic bands, forming an arched structure, as shown in Figure 3.6b,c. A rectangular Cu coil and an NdFeB magnet were fixed at the centrality of two frameworks, respectively. Figure 3.6e illustrates the working principle of the stretchable hybridized nanogenerator. At the initial stage, the top FEP film makes contact with the bottom PDMS film absolutely under the external pulling force, leading to generation of positive charges on the FEP and equivalent negative charges on PDMS due to triboelectrification (step I). There are no current signals because of electrostatic equilibrium at this step. On removing the external force, the distance between the FEP film and the PDMS film gradually increases, resulting in an electric potential difference between them. Under this potential difference, the electrons will flow from the ITO electrode to the AgNW electrode through the external circuit, forming current signals in the TENG (step II). For the EMG, current signals are also generated in this step because of the change of the magnetic flux across the coil. The electrostatic equilibrium between the FEP and the PDMS is generated as the nanogenerator recovers the original arched structure, where no current signals are measured (step III). Subsequently, the FEP film approaches the PDMS film again under the external force, leading to the formation of a reversed inner electric field and change of magnetic flux. The reversed direction of current for both the triboelectric part and the electromagnetic part are generated until the FEP film and the PDMS film are in full contact again (step IV).

3.3.5 Others

To offset the low current output of the TENG, Zhang et al. fabricated a rotating-disk-based hybridized nanogenerator of TENG and EMG [45]. The hybridized nanogenerator consists of two parts: a rotator that is mainly composed of four acrylic intercrossing blades, polytetrafluoroethylene (PTFE) films, and four bar magnets anchored on the rotator plane, and a stator that is mainly composed of four coils and the Al foil. When the rotator spins, the friction between the PTFE film and the Al foil can generate triboelectric charges, which can further form current signals of the TENG due to the electrostatic induction. At the same time, current signals are also generated in the EMG because the magnetic flux across the coil continuously changes with the spin of the rotator. Jin et al. developed a hybridized nanogenerator, which integrates a TENG and an EMG. The hybridized nanogenerator possesses a multilayered structure, where the wave-like Al–Kapton–Al films are sandwiched by Al back-coated PTFE thin film [30]. The periodic contact between the Al film and the PTFE film will generate AC output current. On the other hand, for the EMG this vibrant process will vary the magnetic flux in the coil, leading to generation of electricity.

A large amount of vibration energy, such as in human motions, running vehicles, and even ocean waves, exists at low frequencies, where the performance of

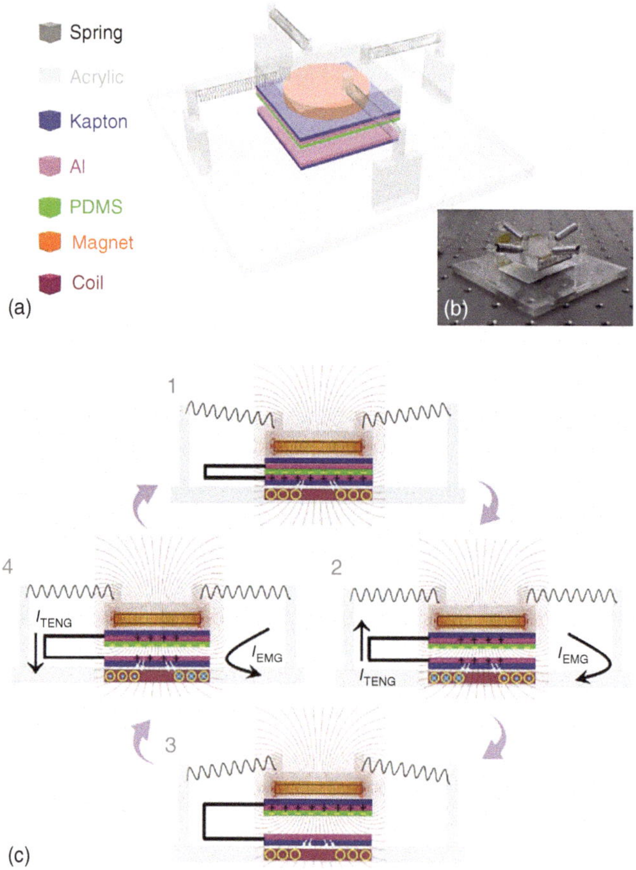

Figure 3.7 The hybridized nanogenerator. (a) Schematic diagram of the fabricated hybridized nanogenerator. (b) Photographs of the hybridized nanogenerator. (c) Schematic diagram of the working principle of TENG and EMG under the different states. Source: Reproduced with permission from Quan et al. [46]. Copyright 2015, Springer Nature.

conventional EMG is low. To solve this problem, Quan et al. reported a hybrid nanogenerator that integrated a TENG with an EMG. The TENG consists of a PDMS film and two Al electrodes, as shown in Figure 3.7a [46]. The EMG includes a magnet and a coil, which are fixed at the device face to face. Figure 3.7b illustrates a photograph of the fabricated hybrid nanogenerator. Figure 3.7c illustrates the working principle of the hybridized nanogenerator. In the initials stage, the PDMS film makes contact with the Al foil under the external force, resulting in opposite triboelectric charges on the surface of PDMS because of their different triboelectric polarities. Owing to the electrostatic balance and constancy of the magnetic flux across the coil, there is no electron flow in the TENG and the EMG (step 1). When the external force fades, the springs will induce the PDMS film to separate from the Al foil, resulting in the electron flow between the two Al electrodes because of electrostatic induction. Meanwhile, the EMG can also generate an output current in the coil due to change of the magnetic flux across the coil (step 2). The output current will reduce to 0 when the distance between the PDMS film and the Al foil reaches the maximum value (step 3). When the PDMS film approaches the Al foil, the reversed current for both the TENG and EMG can be measured (step 4).

Harvesting water kinetic energy can effectively relieve the energy crisis, especially for maritime countries. Shao et al. developed a hybrid generator, which integrated contact-separation-mode TENGs with rotary freestanding-mode EMGs [47]. By packaging the TENGs with an enclosed cylinder, the hybrid generator can steadily work in the water. Two triboelectric layers of the TENG can contact and separate periodically based on the change of attractive force generated by the magnets, which can induce current signals in the TENG. Meanwhile, the EMG can also generate an output current in the coil due to change of the magnetic flux across the coil. This cylinder-like structure is more suitable for harvesting water kinetic energy compared with other structures. To reduce the physical rubbing that exists in common TENG, Seol et al. developed a ferrofluid-based triboelectric–electromagnetic hybrid generator [48]. The TENG consists of Al foils as the electrode and a hollow cylindrical tube made of PTFE, which is used to package the ferrofluid. The EMG includes a wire coil wound around the outer wall of the tube and the magnet. Under external sway, the ferrofluid will make contact with the side walls of the cylindrical tube, resulting in current signals in the external circuit of the TENG. Meanwhile, for the EMG, the change of the magnetic flux across the coil during the process of sway induces current in the outer coil according to Lenz's law.

In order to adequately exploit wind energy, Wang et al. reported a hybridized nanogenerator based on the vibrating plate-based structure [49]. By integrating electromagnetic and triboelectric effects in this structure, the hybridized nanogenerator can engender more electricity compared with the single generator. The TENG includes the PTFE film, Cu electrodes, and the Kapton film. The EMG includes a magnet that is fixed on the Kapton film coated with Cu film, and a group of coils that is fixed on the acrylic substrate, as illustrated in Figure 3.8a,b. At the initial stage, the Cu film makes contact with the top PTFE film absolutely under the external force, leading to the generation of positive charges on the Cu film and equivalent negative charges on PTFE due to triboelectrification, as shown in Figure 3.8c. There are no current signals because of electrostatic equilibrium at this step. The distance between two films gradually increases, resulting in an electric potential difference between them. Under this potential difference, the electrons will flow from the top electrode to the Cu electrode on the PTFE film through the external circuit, forming current signals in TENG 1 (step 2). For EMG 1, current signals are also generated in this step because of the change of the magnetic flux across the coil. The output current will reduce to 0 when the distance between the PTFE film and the Cu film reaches a maximum value (step 4) and positive charges are induced on the Cu film and equivalent negative charges on the bottom PTFE. Subsequently, the Kapton film approaches the top PTFE film again under the external force, leading to the formation of a reversed inner electric field and change of magnetic flux. Reversal of direction of current for both the triboelectric part and the electromagnetic part occurs (step 6).

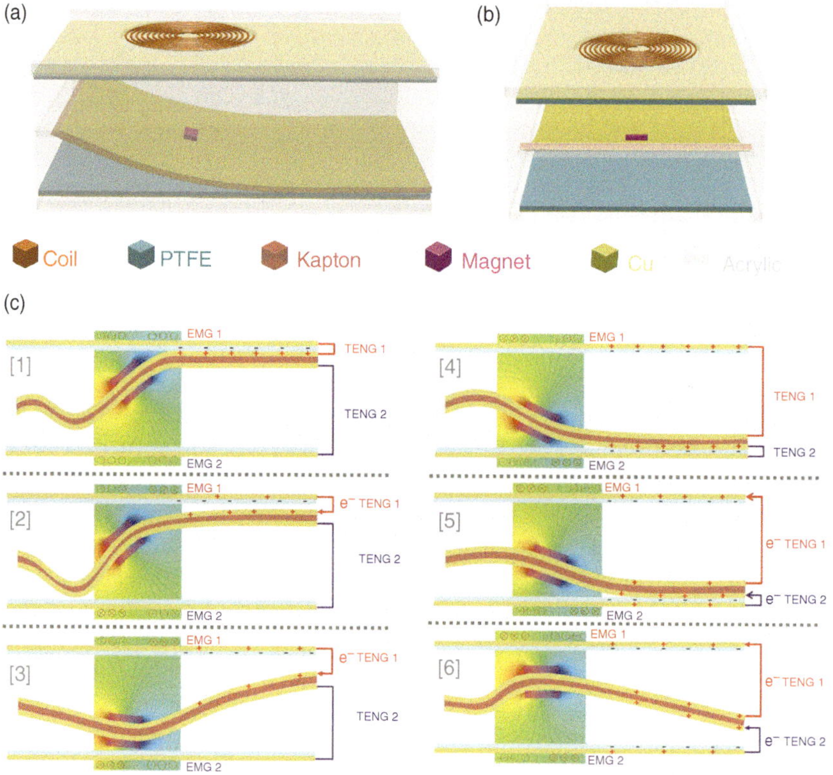

Figure 3.8 The hybridized nanogenerator based on vibrating plate-based structure. (a,b) Schematic diagram of the fabricated hybridized nanogenerator. (c) Schematic diagram of the working principle of TENG and EMG under the different states. Source: Reproduced with permission from Wang et al. [49]. Copyright 2015, American Chemical Society.

To effectively harvest wind energy, Wang and Yang developed a hybridized nanogenerator, which integrated an EMG with a TENG [50]. The structure design of the TENG is based on elasto-aerodynamics-driven structure, where a Cu electrode is attached on the acrylic substrate, and an FEP film is fixed on another Cu electrode, as shown in Figure 3.9a. The EMG includes a magnet fixed on the Kapton film that could vibrate with the wind flow and a coil fixed on the acrylic substrate. Figure 3.9b illustrates a photograph of the fabricated hybrid nanogenerator. At the initial state, the contact between the FEP film and the top Cu electrode can induce positive charges on the FEP and equivalent negative charges on PDMS due to triboelectrification, as illustrated in Figure 3.9c. There are no current signals because of electrostatic equilibrium at this step. With vibration of the FEP film, the distance between the FEP film and top Cu electrode gradually increases, resulting in an electric potential difference between them. Under this potential difference, the electrons will flow from the Cu electrode connected to the FEP film‘ to the top Cu electrode through the external circuit, forming current signals in TENG 1. For EMG 1, current signals are also measured in

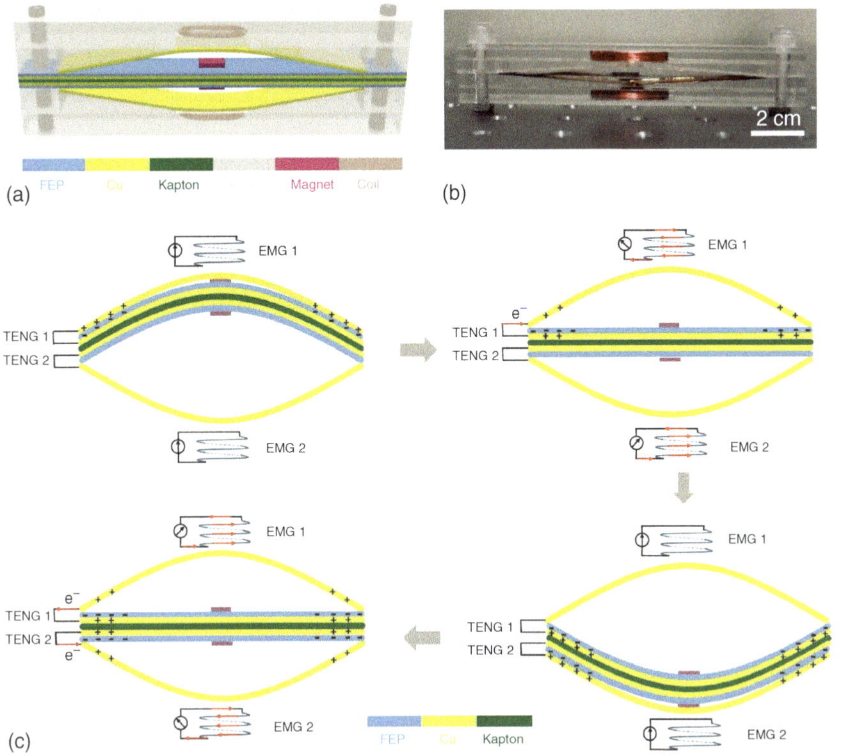

Figure 3.9 The hybridized nanogenerator based on elasto-aerodynamics-driven structure. (a) Schematic diagram of the fabricated hybridized nanogenerator. (b) Photographs of the hybridized nanogenerator. (c) Schematic diagram of the working principle of TENG and EMG under the different states. Source: Reproduced with permission from Wang et al. [50]. Copyright 2017, Elsevier.

this step because of the change of the magnetic flux across the coil. The electrostatic equilibrium between the FEP and the PDMS is formed again when the FEP film makes contact with the bottom Cu electrode, where no current signals are measured. Subsequently, the FEP film is removed from the bottom Cu electrode again, resulting in a reversed inner electric field and change of magnetic flux. The reversed direction of current for both the triboelectric part and the electromagnetic part are generated until the FEP film and the top Cu electrode are in full contact again.

The stability of the generators is very important for some electric devices in the complex environment. Quan and Yang fabricated a fully enclosed hybrid electromagnetic–triboelectric nanogenerator, which can be used in the wild environment due to its enclosed structure [51]. Figure 3.10a illustrates a schematic diagram of the hybrid electromagnetic–triboelectric nanogenerator. The TENGs and EMGs were fully packaged with an acrylic box. Figure 3.10b shows a photograph of the fabricated hybrid nanogenerator. The TENGs, which mainly consist of the Cu foil and the FEP film, are fixed on five cantilevers. EMG

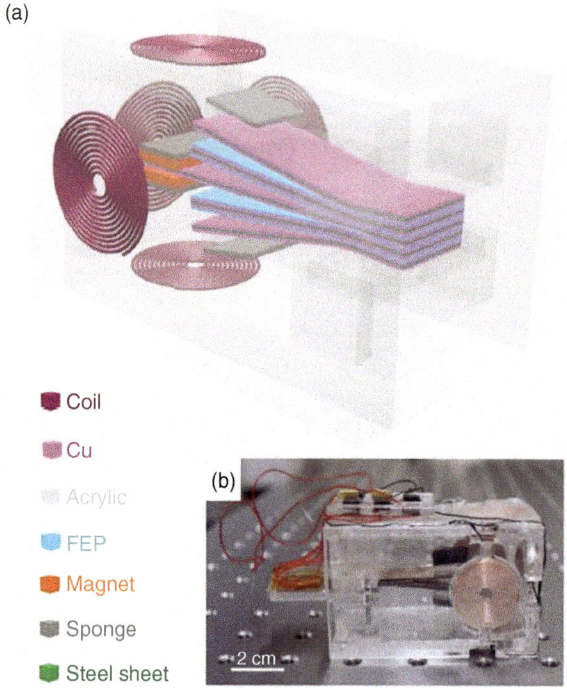

Figure 3.10 The hybridized nanogenerator based on fully enclosed structure. (a) Schematic diagram of the fabricated hybridized nanogenerator. (b) Photographs of the hybridized nanogenerator. Source: Reproduced with permission from Quan and Yang [51]. Copyright 2016, Springer Nature.

includes a magnet that is fixed on the FEP film coated with Cu film, and a group of coils that is fixed on the walls of the acrylic box.

Harvesting sliding motion, which is a familiar mechanical movement, is an effective strategy to relieve the source crisis. Zhang and Yang fabricated a linear-grating hybridized electromagnetic–triboelectric nanogenerator [52]. It mainly includes two parts: a sliding component comprised of magnets, sponge, and grated Cu electrodes, and a stator comprised of coils, ultramicroelectrode array, and nylon film, as illustrated in Figure 3.11a,c. Figure 3.11b shows a photograph of the manufactured hybrid nanogenerator. During the sliding process, the TENG can generate the output current based on the combination of the triboelectric effect and the electrostatic induction effect. At the same time, the EMG can generate an AC output current signal because of Faraday electromagnetic induction between the Cu coils and the magnet.

On the other hand, Wu et al. developed a hybrid generator by integrating a TENG and an EMG, which can be used to harvest human mechanical energy [53]. The TENG consists of a nylon film, the Cu electrode, and a polyvinylbutyral (PVB)/PDMS composite film. The EMG includes a magnetic ball packaged with the acrylic box and six coils fixed on the side wall of the box. The magnetic ball moving under external force will lead to friction between the nylon film and the PVB/PDMS composite film, generating triboelectric charges. Owing to the electrostatic induction, these charges can further form current signals of the TENG. At the same time, current signals are also generated in the EMG because the magnetic flux across the coil continuously changes with the trundle of the magnetic ball.

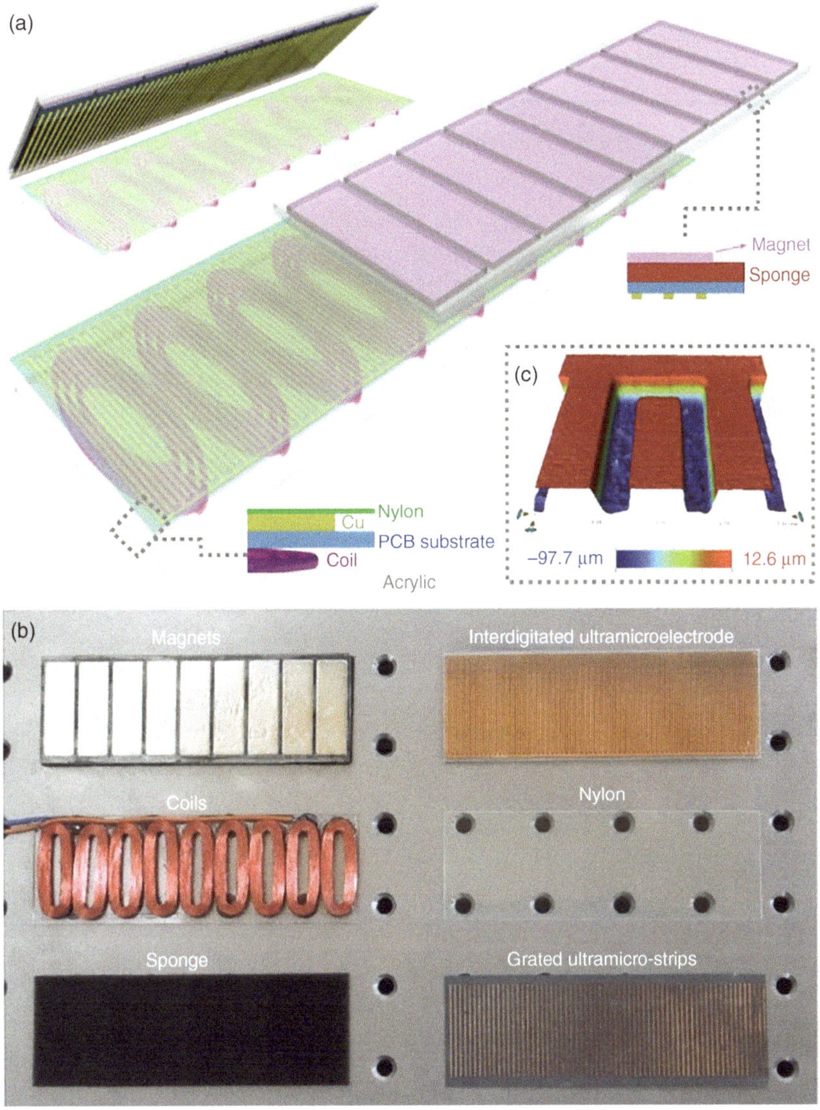

Figure 3.11 The hybridized nanogenerator based on the sliding structure. (a) Schematic diagram of the manufactured hybridized nanogenerator. (b) Photographs of the hybridized nanogenerator. (c) 3D morphology of the ultramicroelectrode array. Source: Reproduced with permission from Zhang and Yang [52]. Copyright 2016, Springer Nature.

3.4 Materials

The output performance of the electromagnetic–triboelectric hybridized nanogenerators can be increased by optimizing triboelectric materials, excepting the structure of the generators [54–60]. Changing the surface topography of materials by physical techniques will increase the contact area, resulting in more triboelectric charges. On the other hand, composite materials generated by embedding nanoparticles and nanotubes into polymer matrix can changes the surface electrification and permittivity of the materials, leading to optimization of the electrostatic induction.

3.4.1 Glass Fibers/Silver Nanowires

Metal nanowires, such as silver nanowires (AgNWs), have been widely used in flexible devices due to their favorable conductivity and extensibility. Zhang et al. fabricated a composite material composed of glass fibers/silver nanowires/PDMS, which is used as the triboelectric material, as illustrated in Figure 3.12a [44]. The prepared AgNWs, with an average diameter of ~100 nm and a length in the range of 10–30 μm, have a favorable uniformity, as shown in Figure 3.12b. Figure 3.12c shows a scanning electron microscopic (SEM) image of the glass fibers coated with AgNWs. The prepared glass fibers exhibit admirable mechanical bending characteristic and stable conductivity when bent to different lengths, as illustrated in Figure 3.12d,e. Owing to these characteristics, illumination intensity of a light-emitting diode (LED) connected by the conductive glass fibers was nearly invariant during the whole bending process, as illustrated in Figure 3.12f.

3.4.2 Composite Materials

Composite materials, which are prepared by embedding nanoparticles and nanotubes into a polymer matrix, can be used as excellent triboelectric materials to increase electrostatic induction [61–63]. Quan et al. developed the PVB nanowire/PDMS composite film, which was used as the triboelectric material [64]. Figure 3.13a displays a SEM image of PVB nanowires manufactured via an electrospinning method, with diameters of about 200 nm. By dropping a PDMS solution on the surface of the PVB nanowires, the composite film was formed after the curing process, as shown in Figure 3.13b.

3.4.3 Materials with Rough Structures

The surface topography of materials affects the contact area between two triboelectric materials, which will increase the output performance of the hybrid generators. Zhang et al. fabricated a PDMS film with pyramid-patterned structures by using an etched Si micropyramid template, as shown in Figure 3.14a [38]. In addition, Quan et al. used anodic aluminum oxide as the template, as shown in Figure 3.14b [46]. Figure 3.14c shows an SEM image of the surface of the as-prepared PDMS film, with raised structures of about 500 nm.

On the other hand, Zhang et al. fabricated nanowires on the surface of the PTFE film by reactive ion etching [45]. The nanowires can increase the contact area compared with the flat PTFE film for the TENG. By designing nanowires on the FEP film, the surface charge density was enhanced, according to Wen et al. [65]. To enhance triboelectric charge density, Seol et al. fabricated nano-grass structures on the surface of the Al substrates [66].

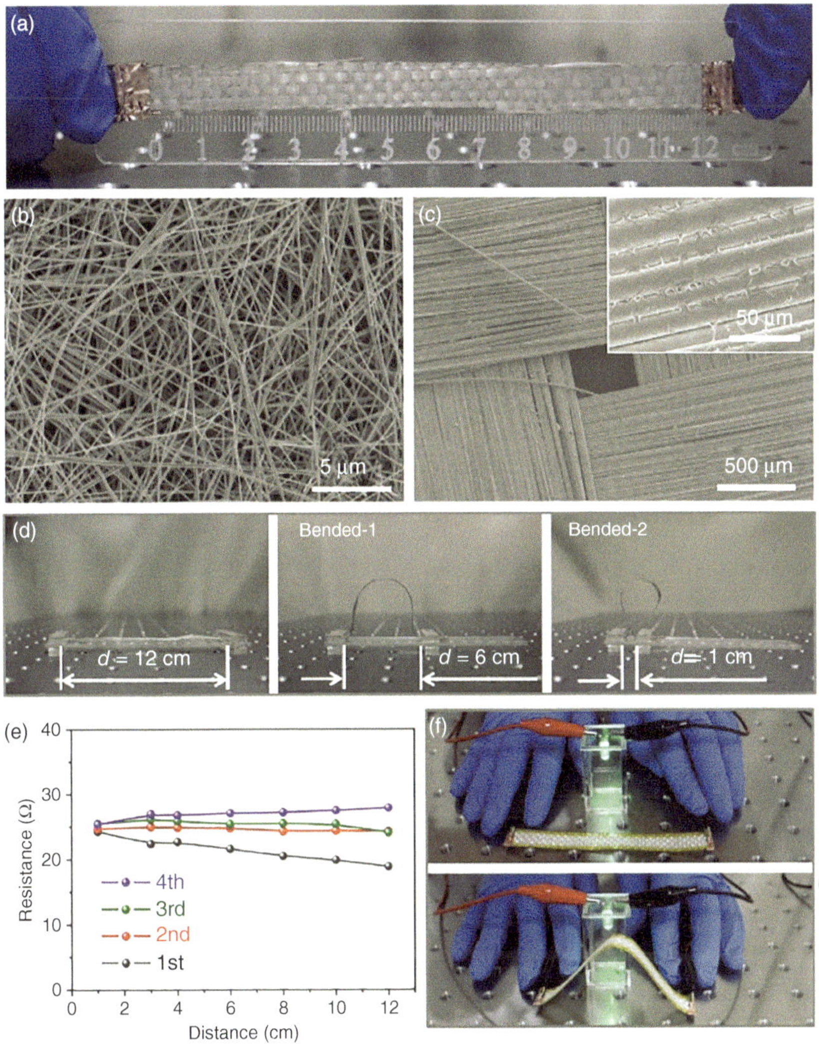

Figure 3.12 The glass fibers/silver nanowires. (a) Photograph of a conductive fabric. (b) SEM image of the AgNWs. (c) SEM image of the conductive fabric coated with AgNWs. (d) Photographs of bending conductive fabric. (e) The change of resistance with the bending distance. (f) Photographs of illumination intensity of a light-emitting diode. Source: Reproduced with permission from Zhang et al. [44]. Copyright 2016, American Chemical Society.

3.4.4 Others

Fe_3O_4 nanoparticles, which possess admirable magnetism, have been widely used in various fields. Ren et al. integrated Fe_3O_4 nanoparticles with polyvinylidene fluoride (PVDF) fibers membrane via electrospinning process [67]. The fabricated Fe_3O_4/PVDF fiber membrane not only was used as the triboelectric layer, but also was capable of responding to the external magnetic field. The

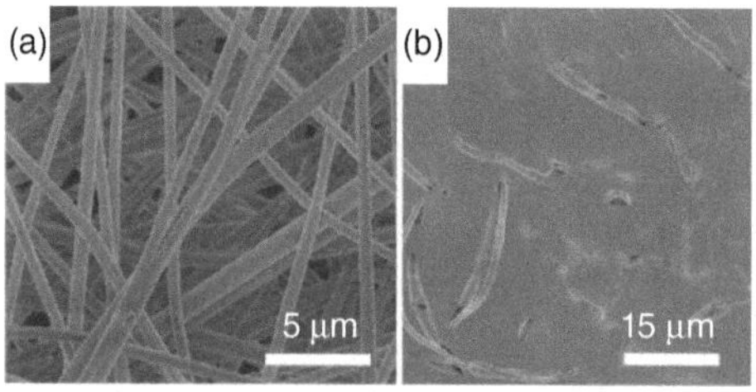

Figure 3.13 The PVB nanowire/PDMS composite film. (a) SEM image of the PVB nanowires. (b) SEM image of the surface of the PVB nanowire/PDMS composite film. Source: Reproduced with permission from Quan et al. [64]. Copyright 2015, American Chemical Society.

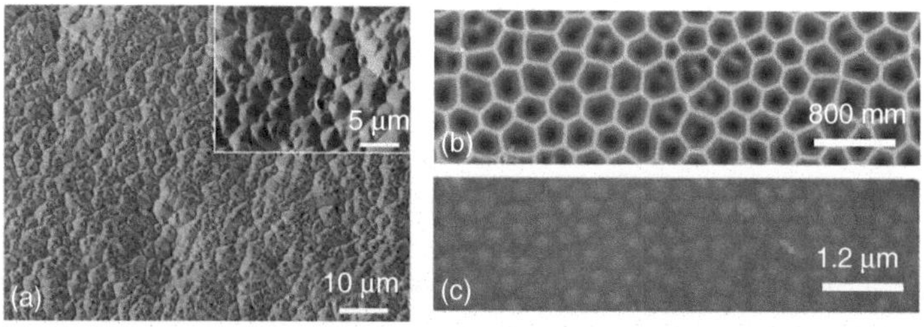

Figure 3.14 The rough structures on the surfaces. (a) SEM image of the prepared PDMS film with pyramid-patterned structures. Source: Reproduced with permission from Zhang et al. [38]. Copyright 2015, American Chemical Society. (b) SEM image of the anodic aluminum oxide. (c) SEM image of the prepared PDMS film with raised structures. Source: Reproduced with permission from Quan et al. [46]. Copyright 2015, Springer Nature.

application of composite membrane can increase the agility of the hybridized electromagnetic–triboelectric nanogenerator.

In addition, the contact friction in the electromagnetic–triboelectric hybridized nanogenerator will damage the materials, leading to a decline of output performance. To reduce the friction, Seol et al. fabricated ferrofluid-based triboelectric–electromagnetic hybrid generator, where the ferrofluids consisted of Fe_3O_4 nanoparticles, and water was used to make contact with PTFE layer [48]. The friction between the fluid and the solid will reduce the material abrasion and extend the lifetime of the generator.

3.5 Performance

Compared to individual TENGs that suffer from a low current output, hybridization of the TENG and the EMG provides a solution to utilize their advantages, resulting in the enhancement of the output performance. For example, Zhang et al. developed rotating-disk-based hybridized electromagnetic–triboelectric nanogenerator, which can deliver an output power of 17.5 mW [45].

3.5.1 Conjunction Manipulation

Zhang et al. compared the output performance of the rotating EMG and TENG. It was found that the output voltage increased with increasing resistance while the output current indicated the reverse trend [26]. Under a resistance of about

10 kΩ, both the current and the voltage inclined to saturate. The maximum output power was about 102.6 μW at the resistance of 12.3 Ω. The small internal resistance is one of the main features for the EMG due to the conductive coils. Thus the EMG can be used as a voltage source when the external resistance is large enough. For the TENG, the output voltage increased with increasing resistance while the output current indicated the reverse trend. The change of both current and voltage was slight at a resistance between 0 Ω and 1 MΩ. The maximum output power was about 140.4 μW at a resistance of 13.8 MΩ. The big internal resistance is one of main features of the TENG. Thus, the TENG can be used as a current source when the external resistance is small enough. It was found that the maximum output power of the EMG is slightly smaller than that of the TENG. In consideration of unit volume, the power of the TENG was obviously bigger than that of the EMG because the structure of the TENG could be designed with smaller volume and lower weight.

To realize the desired effect of the voltage source and the current source, the internal resistance should be close to zero and infinite, respectively. When an ideal voltage source and an ideal current source were connected in parallel, the output performance was mainly affected by the voltage source. On the contrary, if they were connected in series, the output performance was mainly affected by the current source.

3.5.2 Output Characteristics

There has been recent focus on three main parameters of output performance of the generators including the output current, the output voltage, and the maximum output power. Zhang et al. developed a conductive fabric-based stretchable hybridized nanogenerator, realizing high output performance [44]. The peak values of the measured output voltage and the output current are about 125 V and 48 μA, respectively, as illustrated in Figure 3.15a,b. It is found that a slight oscillation appears in measured curves of both voltage and current due to the destabilization during the stretchable movements. Figure 3.15c shows that the output current decreases with increasing the loading resistance, and output power attains the maximum value of about 1.9 mW at a loading resistance of 8 MΩ. For the EMG, the peak values of the measured output voltage and the output current are about 6.3 V and 8.4 mA, respectively, as illustrated in Figure 3.15d,e. Figure 3.15f shows that the output current decreases with increasing the loading resistance, and output power attains the maximum value of about 10.5 mW at a loading resistance of 800 Ω. The high output voltage of the TENG is beyond the actual powering voltage of electronic devices. To combine the TENG with the EMG, the output signals of the TENG were adjusted via a commercial transformer, realizing a low output voltage of about 5.7 V and impedance. When the TENG, the EMG, and the hybridized nanogenerator were used to charge a 47 μF capacitor, respectively, the charging performance of the hybridized nanogenerator was much better than that of others.

Quan et al. developed a hybridized electromagnetic–triboelectric nanogenerator [46]. It was found that vibration frequencies affected the output

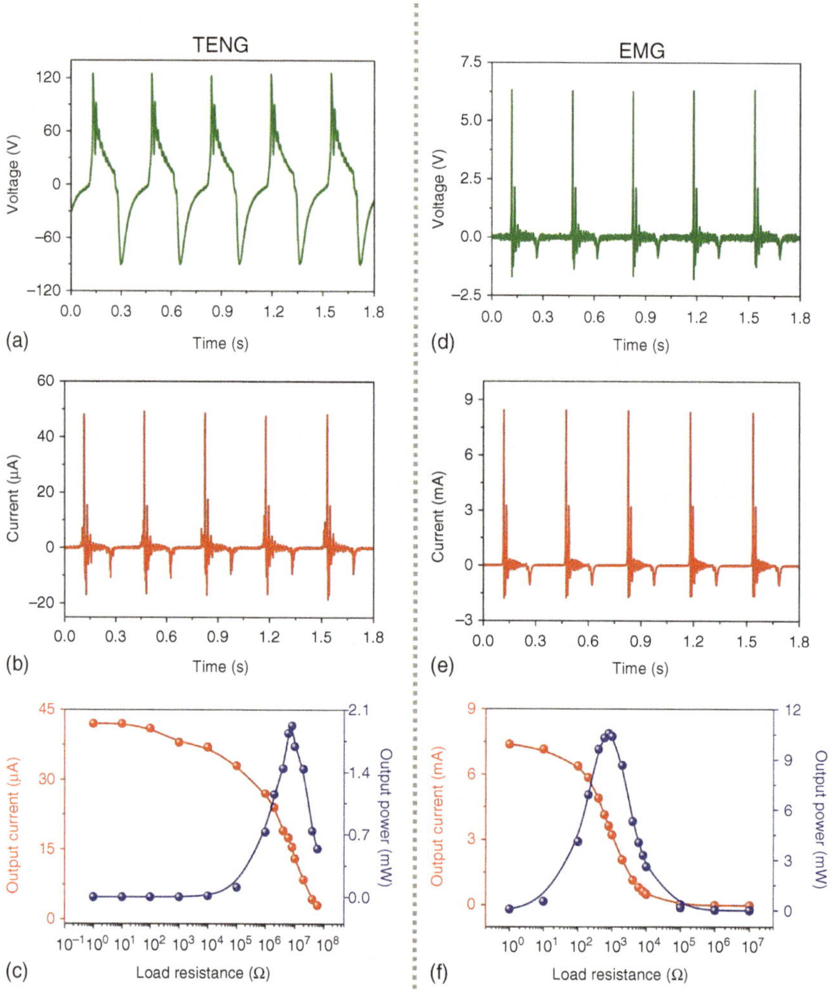

Figure 3.15 Output performance of the conductive fabric-based stretchable hybridized nanogenerator. (a) Voltage of the TENG. (b) Output current of the TENG. (c) Dependence of the output current and output power of the TENG on external loading resistance. (d) Output voltage of the EMG. (e) Output current of the EMG. (f) Dependence of output current and output power of the EMG on external loading resistance. Source: Reproduced with permission from Zhang et al. [44]. Copyright 2016, American Chemical Society.

performance of both the TENG and the EMG and the largest output signals were at the vibration frequency of about 14 Hz. Their output current decreased with increasing the loading resistance. By decreasing the impedance of the TENG via the transformer, the EMG can be effectively connected with the TENG. Figure 3.16a,b shows the output current and voltage of TENG adjusted by the transformer, where its current and voltage are 0.7 mA and 2.9 V, respectively. The corresponding output power attains the maximum value of about 0.4 mW, as shown in Figure 3.16c. Figure 3.16d shows that the output current of a

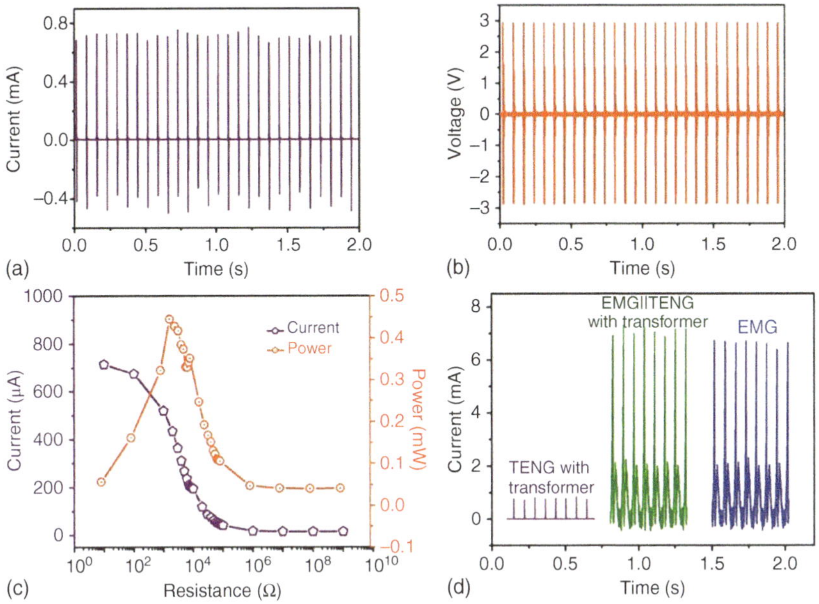

Figure 3.16 Output performance of the hybridized generator. (a) Voltage of the TENG with a transformer. (b) Output current of the TENG with a transformer. (c) Dependence of the output current and output power of the TENG on external loading resistance. (d) Comparison of output current for different generators. Source: Reproduced with permission from Quan et al. [46]. Copyright 2015, Springer Nature.

hybrid nanogenerator is larger than that of other individual energy harvesting units.

The rotating speeds affected the rotating-disk-based hybridized electromagnetic–triboelectric nanogenerator, according to Zhang et al. It was found that both the output voltage and current of the TENG would increase with increasing rotating speeds until the speed reached 600 rpm, where the output signals attained the saturated values [45]. For the EMG, both the output voltage and current increased with increasing rotating speeds in a wide range from 10 to 1000 rpm. In addition, the AC of the TENG and the EMG can be converted to direct current (DC) via two bridge rectification circuits. When the hybrid nanogenerator was used to harvest ocean energy, the frequency of the wave affected the output performances, according to Wen et al. It was found that the average power densities of both the TENG and EMG were increased with increasing motion frequency of the wave [65].

The material used as the triboelectric layer could affect the output performances of the generators. Ren et al. developed a magnetic force driven noncontact electromagnetic–triboelectric hybrid nanogenerator, where the output current could be enhanced by increasing the content of Fe_3O_4 nanoparticles-embedded PVDF fibers membrane [67]. In addition, increasing the distance between the device and magnet can enhance the output voltage and current of both the TENG and the EMG. On the other hand, Cao et al. developed a rotating-sleeve-based triboelectric–electromagnetic hybrid nanogenerator, which can be used to harvest mechanical energy [68]. The output voltage and current of the EMG can be enhanced by increasing the rotation speed of the rotator while the output voltage of the TENG remained stable at about 580 V when the rotation changed between 100 and 400 rpm.

Zhang and Yang developed a linear-grating hybridized electromagnetic–triboelectric nanogenerator, realizing the scavenging of mechanical energy from our living environment [52]. The maximum power of 102.8 mW can be attained by the hybridized nanogenerator at a sliding acceleration of 20 m/s^2. The peak voltage and current of the TENG were about 118.4 V and 0.9 mA, respectively. The peak voltage and current of the EMG were about 55.7 V and 7.7 mA, respectively. The output performances and impedances of the generators could be adjusted via two commercial transformers, as shown in Figure 3.17a,d. For the TENG, the output voltage was decreased to 8.7 V, and the current was enhanced to 8.1 mA. For the EMG, the output voltage was decreased to 4.8 V, and the current was enhanced to 22 mA. The maximum output power of the TENG and EMG is 16.8 and 24.6 mW, respectively, as shown in Figure 3.17b,e. Figure 3.17c shows that electrical energy generated by the TENG is about 2 mJ during the time interval of 1.8 seconds. Figure 3.17f shows that electrical energy generated by the EMG is about 2.8 mJ during the time interval of 1.8 seconds. In addition to the commercial transformer, a power management circuit, including a transformer, a rectifier, a voltage regulator, and capacitors, was used to adjust the output signals of the hybridized nanogenerator. It took about one seconds to provide the DC output voltage of about 5 V for the TENG. For the EMG, the managed time was about 0.4 seconds.

Based on the rotating-disk-based hybridized nanogenerator that integrated an EMG and a TENG, Zhong et al. realized the collection of the mechanical energy [37]. The output voltage and current of the TENG were about 75 V and 0.33 mA, respectively, at the rotation rate of about 200 rpm, where the largest instantaneous power was about 8.6 mW. For the EMG, the output voltage and current of the TENG were about 0.62 V and 57.8 mA, at the rotation rate of about 200 rpm, where the largest instantaneous power was about 8.4 mW. The electrical energies generated by the TENG and EMG were the same (0.4 mJ), during 100 seconds. The output performances and impedances of the generators could be adjusted via two commercial transformers. For the TENG, the output voltage was decreased to 5.5 V, and the current was enhanced to 3.87 mA, as shown in Figure 3.18a,b. For the EMG, the output voltage was decreased to 4.3 V, and the current was enhanced to 4.3 mA, as shown in Figure 3.18c,d. The output AC signals of the generators can be converted to the DC signals via two bridge rectification circuits, as shown in Figure 3.18e. It was found that the output current signals of the hybridized nanogenerator were obviously higher than that of the individual generator. Figure 3.18f shows the charged voltages of the 6600 μF capacitor with different generators, where the performance of the hybridized nanogenerator is obviously better than that of the individual generator.

To transform more mechanical energy from one motion, Wu et al. developed a hybrid generator, which integrated a single-electrode-based TENG and an

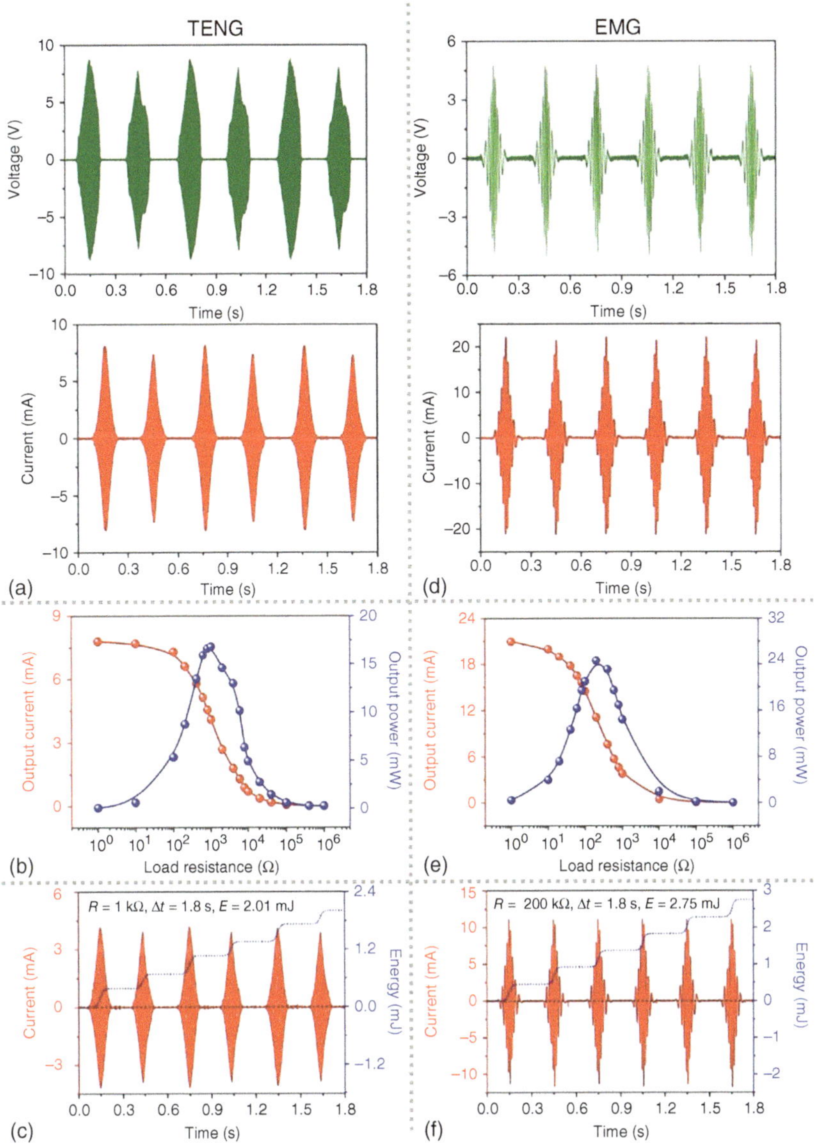

Figure 3.17 Output performance of the linear-grating hybridized generator via a transformer. (a) The output voltage and current of the TENG. (b) Output current and the corresponding power of the TENG. (c) The corresponding electricity in the TENG during the 1.8 seconds. (d) The output voltage and current of the EMG. (e) Output current and the corresponding power of the EMG. (f) The corresponding electricity in the TENG during the 1.8 seconds. Source: Reproduced with permission from Zhang and Yang [52]. Copyright 2016, Springer Nature.

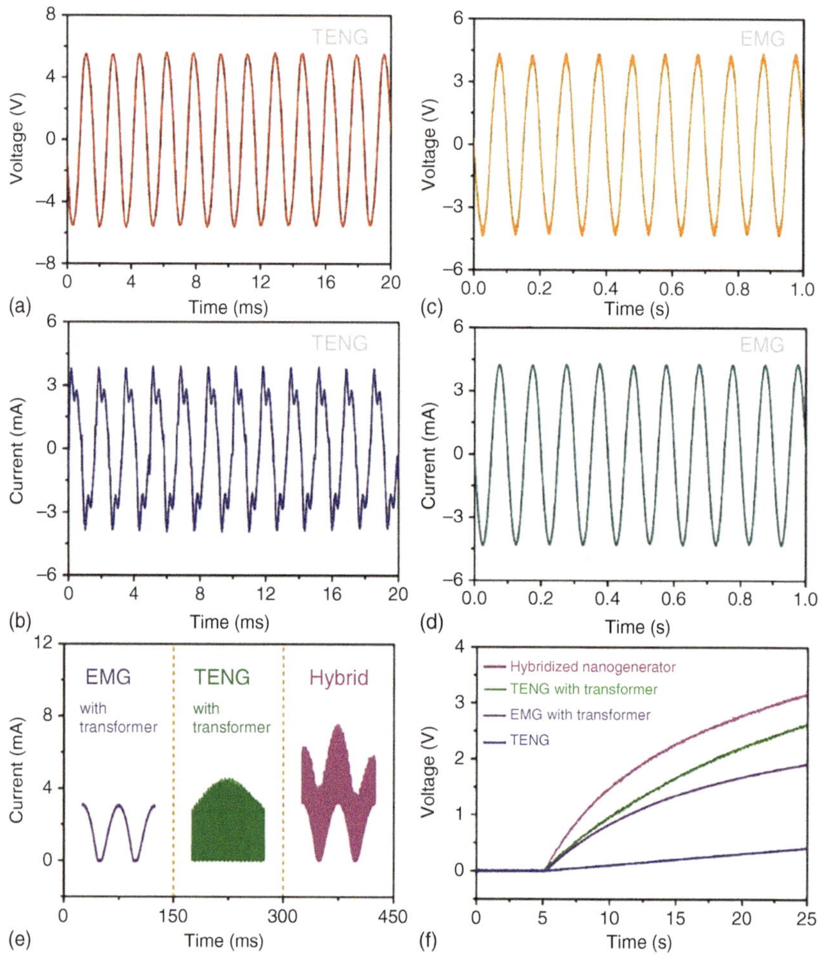

Figure 3.18 Output performance of the rotating-disk-based hybridized generator via a transformer. (a) The output voltage and current of the TENG. (b) Output current of the TENG. (c) The output voltage and current of the EMG. (d) Output current of the EMG. (e) The comparison of output current for different generators. (f) The charged voltages of a 6600 μF capacitor with different generators. Source: Reproduced with permission from Zhong et al. [37]. Copyright 2015, Elsevier.

EMG [53]. The output voltage and current of the TENG were about 600 V and 3.5 μA, respectively, at the vibration frequency of about 22 Hz, where the largest instantaneous power was about 0.25 mW at a loading resistance of 100 MΩ. For the EMG, the output voltage and current of the TENG were about 3 V and 1 mA, at the same frequency, where the largest instantaneous power was about 0.58 mW at a loading resistance of about 10 kΩ. The dropping height of the mass in the device affected the output performance of the TENG and EMG. Figure 3.19a shows that both the output voltage and current signals of the EMG enhance with increasing the dropping heights of the mass. For the

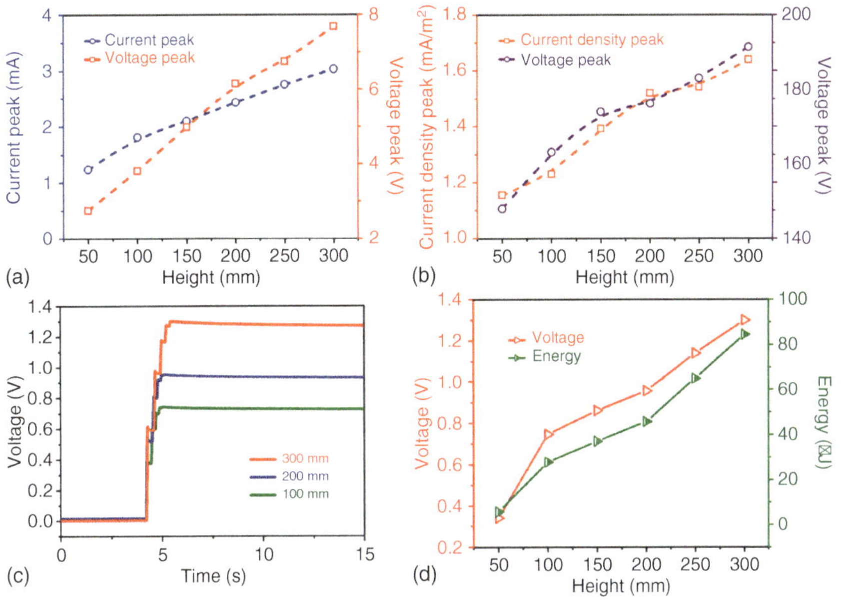

Figure 3.19 Output performance of the hybridized generator. (a) The output voltage and current of the EMG under different dropping heights of the mass. (b) Output performances of the TENG under different dropping heights of the mass. (c) The charged voltages of a 100 μF capacitor with different dropping heights of the mass. (d) Dependence of the charged voltage and the stored electrical energy in the capacitor on the dropping height of the mass. Source: Reproduced with permission from Wu et al. [53]. Copyright 2015, Elsevier.

TENG, increasing the heights also can enhance the output performance, as shown in Figure 3.19b. Figure 3.19c shows the charged voltages of the 100 μF capacitor by using the hybridized nanogenerator with dropping heights, where the performance of the hybridized nanogenerator under the dropping height of 300 mm is obviously better than that of the other heights. With the increase in the dropping height, the charged voltage of the capacitor can be enhanced, as illustrated in Figure 3.19d.

Zhang et al. developed a hybridized electromagnetic 1 triboelectric nanogenerator, which can be used to harvest the biomechanical energy [38]. The positive peak voltage of the TENG is about 61 μA and the negative peak voltage is about 32 μA during the fast contact/separation process, as shown in Figure 3.20a. The output voltage of the TENG is about 268 V, as illustrated in Figure 3.20b. For the EMG, the maximum output current and voltage were about 3.6 mA and 4.9 V, respectively, as shown in Figure 3.20c,d. By using a commercial transformer, the output signals and impedance of the TENG were matched to that of the EMG, obtaining the output current of about 1.3 mA and the output voltage of about 4.3 V. Figure 3.20e shows that the output current of hybrid nanogenerator is larger than that of other individual energy harvesting units. Figure 3.20f shows the charged voltages of the 47 μF capacitor with different generators, where the performance of the hybridized nanogenerator is obviously better than that of the individual generator.

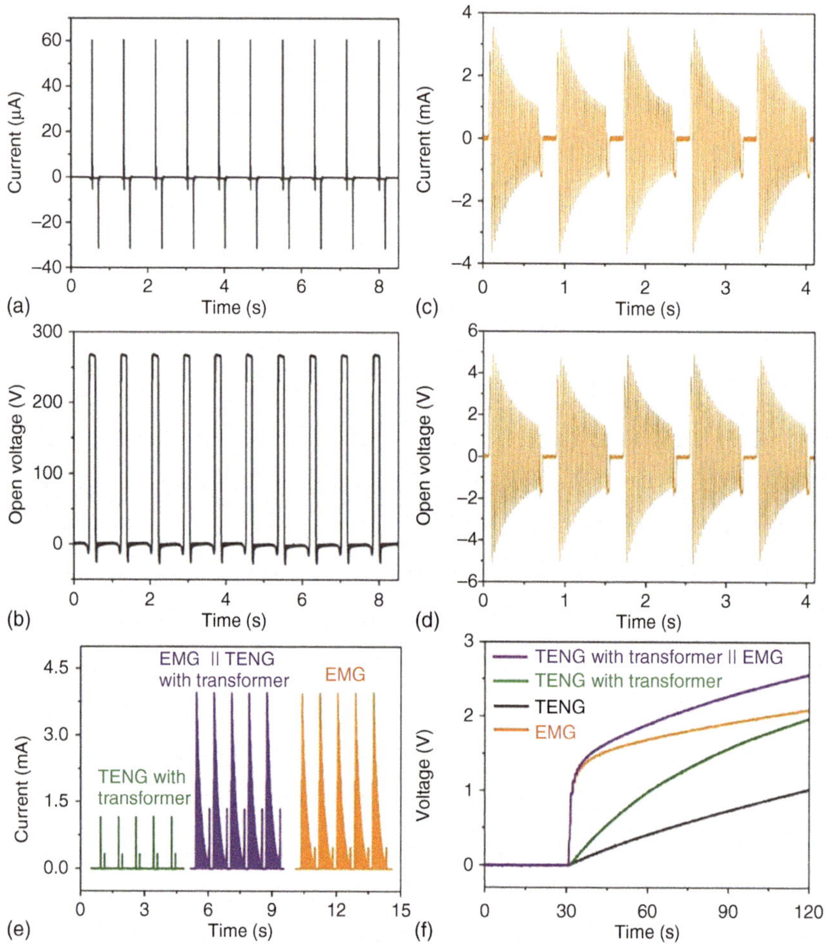

Figure 3.20 Output performance of spring-based EMG–TENG. (a) The output current of the TENG. (b) The output voltage of the TENG. (c) The output current of the EMG. (d) The output voltage of the EMG. (e) Comparison of output current for different generators. (f) The charged voltages of a 647 μF capacitor with different generators. Source: Reproduced with permission from Zhang et al. [38]. Copyright 2015, American Chemical Society.

3.6 Applications

With the rapid growth of industrial electronics, exploring new generators to solve the energy shortage has become more and more important. The electromagnetic–triboelectric hybridized nanogenerators, which integrate two generators in one device, present extensive advantages, such as small scale, low cost, and high efficiency of harvesting mechanical energies, as well as high output power. Currently, more and more efforts have been impelled to expand the applications of electromagnetic–triboelectric hybridized nanogenerators.

3.6.1 Powering Electronic Devices

At present, powering small electronics via Li-ion batteries or capacitors faces a major drawback that the lifetime of these energy storage units is limited [69–72]. The electromagnetic–triboelectric hybridized nanogenerators can effectively convert ubiquitous mechanical energies to electric energy, which can be used to charge these storage units or power the electronics directly. For example, Wang et al. developed a hybridized electromagnetic–triboelectric nanogenerator [49]. It can be used to power the temperature sensor by harvesting the wind energy. The generated electrical energy was stored in the capacitors. As shown in Figure 3.21a, four temperature sensors can be powered by the electrical energy generated by the air-flow-driven hybridized nanogenerator, displaying a room temperature of about 24 °C. The inset of Figure 3.21a shows the temperature of the finger. Wu et al. used the hybrid electromagnetic–triboelectric nanogenerator as a power source, with the nanogenerator harvesting the mechanical energy from one mechanical motion [53]. The hybrid nanogenerator was connected to commercial LEDs, where the EMG was used to power the letters "HYBRID" consisting of 57 red LEDs via a bridge rectification circuit, and the TENG was

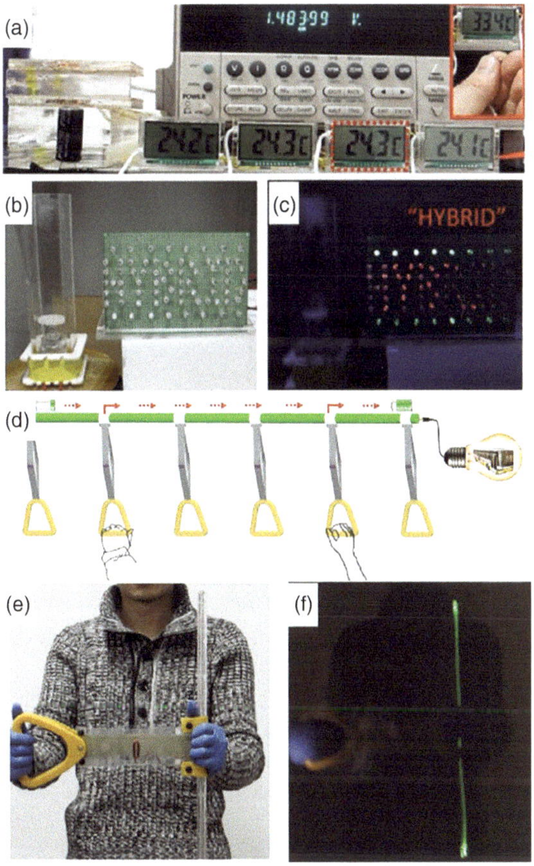

Figure 3.21 Powering electronic devices via the hybridized nanogenerator. (a) Photograph of the temperature sensor system powered by the hybridized nanogenerator. Source: Reproduced with permission from Wang et al. [49]. Copyright 2015, American Chemical Society. (b) Photograph of the LEDs connected with the hybridized nanogenerator. (c) Photograph of the LED lit up by the hybridized nanogenerator. Source: (c) Reproduced with permission from Wu et al. [53]. Copyright 2015, Elsevier. (d) Schematic diagram of the hybridized nanogenerator-based bus grips. (e,f) Photograph of the optical fiber lit up by the hybridized nanogenerator. Source: (e,f) Reproduced with permission from Zhang et al. [44]. Copyright 2016, American Chemical Society.

used to power another 16 green LEDs. When the hybrid nanogenerator started to work, all the LEDs could be lit up, as illustrated in Figure 3.21b,c.

Compared with the rigid hybridized electromagnetic–triboelectric nanogenerator, stretchable hybridized nanogenerators are more suitable for harvesting the wasted biomechanical energy. Zhang et al. integrated a fabric-based stretchable hybridized nanogenerator with a bus grip, realizing the utilization of human locomotion, as shown in Figure 3.21d [44]. Figure 3.21e shows the homemade bus grip, where the stretchable hybridized nanogenerator and an LED-based optical fiber were integrated. When the bus grip is stretching, the hybridized nanogenerator can generate electrical energy, lighting the LEDs at the two ends of the optical fiber, as shown in Figure 3.21f.

3.6.2 Self-Powered Wearable Electronics

With the rapid growth of wearable electronics, conventional power sources, such as the lithium battery, emerge with obvious drawbacks, including the limited capacity and environmental pollution produced by spent batteries. The combination of generators and wearable electronics provides a strategy to reduce the demand for lithium battery, realizing the self-powered wearable electronics that can generate electricity by themselves [73, 74]. Compared with other generators, such as the EMG and the piezoelectric generator, the electromagnetic–triboelectric hybridized nanogenerators can harvest more mechanical energy from one mechanical motion because of their advanced structures for both the electromagnetic and the triboelectric effect. Currently, more and more efforts are being made to develop self-powered wearable electronics based on the electromagnetic–triboelectric hybridized nanogenerators.

Quan et al. integrated the hybridized nanogenerator with an electronic watch in one device, as shown in Figure 3.22a [64]. The fabricated device was worn on the wearer's wrist, and it could generate electricity via the part of the hybridized nanogenerator when the wearer's wrist swayed, as illustrated in Figure 3.22b. Figure 3.22c,d shows the output current of the hybridized nanogenerator connected with the rectification circuits, corresponding to different motive states of the wrist. To steadily power the watch, the generated electricity was stored in the capacitor, where the voltage can be charged to 2 V during about 104 seconds. As the watch was powered, the charged electricity was consumed, resulting in the debasement of voltage of the capacitor, as illustrated in Figure 3.22e,f.

On the other hand, Zhang et al. fabricated a hybridized nanogenerator-based system, which can convert the harvested motive energy of walking to electrical energy through the hybridized nanogenerator, with the electricity being used to power some small electronic devices [38]. Figure 3.23a shows the fabricated system, which consists of a shoe with the LED-based lamp belt and two hybridized nanogenerators installed in the heel of the shoe. The nanogenerator can convert the mechanical energy from the movement of the person to electricity, which was used to light the lamp belt, as illustrated in Figure 3.23b. In addition, a self-powered counting system was developed by integrating the hybridized nanogenerator with the pedometer, as shown in Figure 3.23c. Figure 3.23d,e shows that a capacitor of 1000 μF charged by the hybridized nanogenerator

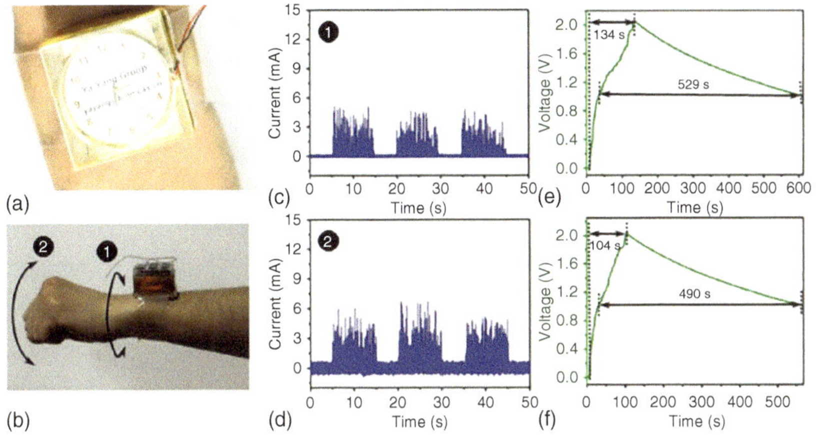

Figure 3.22 Self-powered wearable electronic watch. (a) Photograph of the fabricated device. (b) Photograph of the sway motion. (c) The output current signals corresponding to motion-1. (d) The output current signals corresponding to motion-2. (e,f) The corresponding curves of a 100 μF capacitor. Source: (e,f) Reproduced with permission from Quan et al. [64]. Copyright 2015, American Chemical Society.

can power the pedometer, displaying the correlative motorial parameters of the person in several minutes. The counting system can be optimized by using the wireless pedometer, as shown in Figure 3.23f,g. When it was beaten by the hand, the hybridized nanogenerator can generate electricity to power the wireless pedometer, obtaining the motorial parameters collected by the iPhone, as illustrated in Figure 3.23h,i. In addition, Ren et al. developed an electromagnetic–triboelectric hybrid nanogenerator to harvest biomechanical energy, where the TENG and EMG can generate electricity during the change of the magnetic force from an external magnetic field [67]. The fabricated nanogenerator and a magnet were fixed on the clothes' pocket and gusset, respectively. The relative movement between the arm and body generated the varied magnetic force, triggering the TENG and the EMG.

3.6.3 Others

Harvesting wave energy of ocean for conventional generators, such as EMG, is difficult because the frequency of wave is low and the form has irregular amplitude. To enhance the performance of the EMG, Wen et al. fabricated a triboelectric–electromagnetic hybrid nanogenerator by integrating a spiral-interdigitated electrode TENG and a wrap-around EMG in one device, where the TENG was commendably protected through packaging, avoiding the interference to generating triboelectric charges from the external environment [65]. In addition, it was capable of meeting the wide frequency range of the water wave based on the complementary harvesting strategy, where the TENG was mainly used to harvest the wave energy at low rolling speeds (<100 rpm) and fluctuant frequencies (<2 Hz), and the EMG was used to harvest the wave energy at high frequencies (>10 Hz).

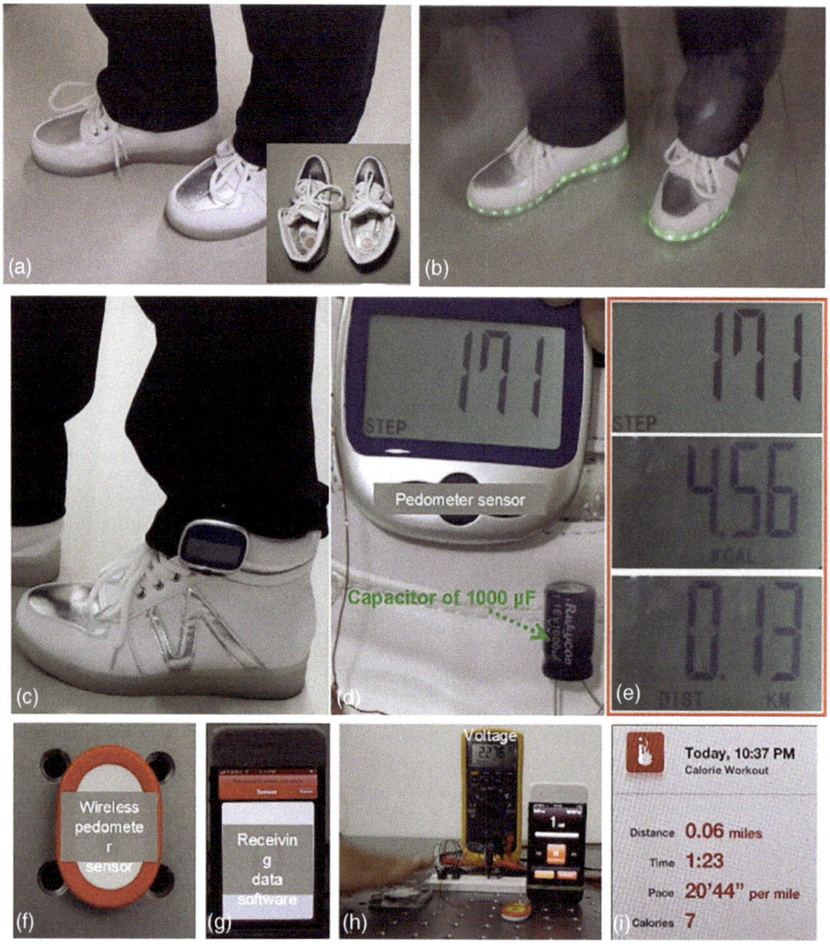

Figure 3.23 Self-powered wearable devices. (a) Photograph of the fabricated self-powered shoe. (b) Photograph of lighting a lamp belt by the hybridized nanogenerator. (c) Photograph of the shoe with a pedometer. (d) Photograph of the pedometer powered by the hybridized nanogenerator. (e) The motorial parameters collected by the pedometer. (f) Photograph of a wireless pedometer. (g) Photograph of an iPhone to collect the data. (h) Photograph of the wireless pedometer driven by the nanogenerator. (i) Collected data from a wireless pedometer. Source: Reproduced with permission from Zhang et al. [38]. Copyright 2015, American Chemical Society.

With the rapid growth of various sensors, such as temperature sensors, humidity sensors, and pressure sensors, the strategies to reduce power consumption or realize the self-driven function have received great interest. Wang et al. developed a self-driven sensor for detecting the magnitude of both the external touching force and touching velocity, which consisted of a contact-separation mode TENG and an EMG [75]. It was found that the output voltage of the TENG was related to the magnitude of the external touching force, and the output current of the EMG was related to the maximum velocity of the magnet. Zhong et al. developed

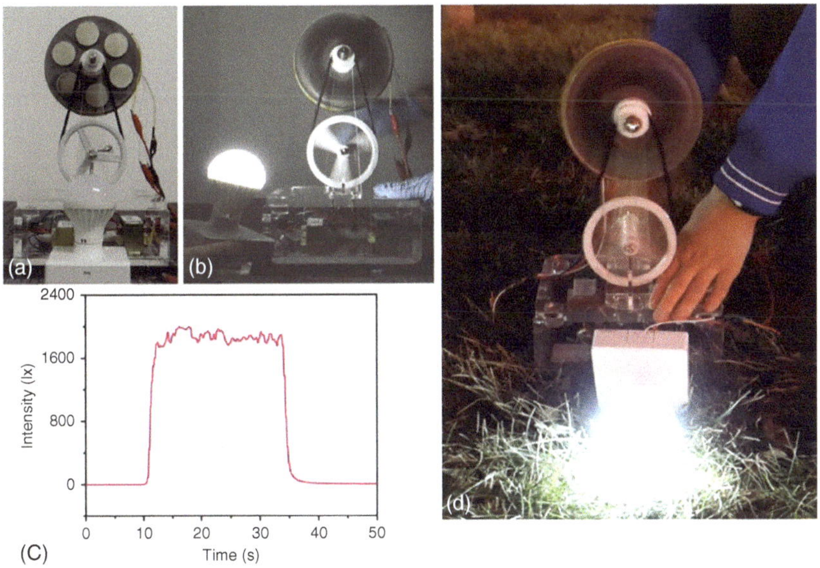

Figure 3.24 Powering a white globe lamp via the hybridized nanogenerator. (a) Photograph of the hybridized nanogenerator connected to the light. (b) Photograph of the lighting lamp. (c) The illumination intensity of the lighting lamp. (d) Photograph of the lighting lamp in the wild. Source: Reproduced with permission from Zhong et al. [37]. Copyright 2015, American Chemical Society.

a rotating-disk-based hybridized electromagnetic–triboelectric nanogenerator, which can harvest the rolling mechanical energy [37]. Owing to its advanced structures, the hybridized nanogenerator with high output power can be used as a mobile power source to drive some electronics. Figure 3.24a shows that the hybridized nanogenerator was used to light a white globe lamp through a capacitor of 6600 μF used to store the electricity from the nanogenerator. The light with the intensity of illumination of about 1700 lx can illuminate the dark surrounding environment, as shown in Figure 3.24b,c. It was capable of meeting the demand in the wild, as illustrated in Figure 3.24d.

On the other hand, it is important to explore the strategy for integrating the generators with energy storage units, especially for the electromagnetic–triboelectric hybridized nanogenerator. Among various storage units, Li-ion batteries with high output performance are capable of meeting the demand to store the electricity generated by the hybridized nanogenerator. To explore the performance of Li-ion batteries, Quan et al. used a hybrid nanogenerator to charge a self-made Li-ion battery [46]. The battery consists of cases, a buffer layer, an anode, a monolayer, and a cathode, as illustrated in Figure 3.25a. The first charge–discharge curve of the battery connected to the hybrid nanogenerator is shown in Figure 3.25b. It was found that the electricity generated by the hybrid nanogenerator could effectively be stored in the battery during continuous charge–discharge process, as shown in Figure 3.25c.

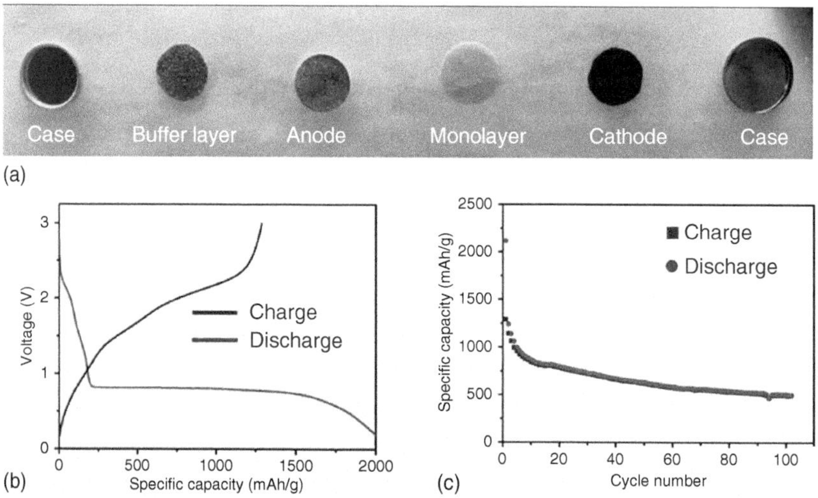

Figure 3.25 Charging a self-made Li-ion battery via the hybridized nanogenerator. (a) Photograph of the self-made Li-ion battery. (b) The first charge and discharge curves of the Li-ion battery. (c) Long-term stability test of the Li-ion battery. Source: Reproduced with permission from Quan et al. [46]. Copyright 2015, Springer Nature.

3.7 Summary and Perspectives

In summary, this chapter presents the current developments of electromagnetic–triboelectric hybridized nanogenerators. Firstly, the basic principles of the TENG and EMG are introduced to exhibit the respective advantages of the two generators, where the EMG is based on electromagnetic induction, and TENG is based on a combined mechanism of the triboelectric effect and electrostatic induction. Combining the two principles, the hybridized generators integrated the TENG with the EMG; such hybridized generators are basically introduced, including structural design, material selection, output performance and applications in this chapter. The structural design is very important for output performance of the hybridized devices. In this chapter, various structures are exhibited for different applications. The output performance of the hybridized generators can be improved by optimizing advanced structures and materials. Finally, some smart self-powered devices based on the hybridized nanogenerators are presented to show wide applicative prospects. In recent years, electromagnetic–triboelectric hybridized nanogenerators have shown a great deal of accomplishments, especially in harvesting various mechanical energies. As for promoting the developments of the hybridized nanogenerators, several problems and strategies need to be considered.

Firstly, it is essential to further explore the fundamental principle and advanced structures of the electromagnetic–triboelectric hybridized nanogenerators, especially for the part of the TENG. There are some models to illustrate the charge generation process in the TENG. For example, Wang, in 2019, minutely expounded the theoretical origin, Maxwell's equations of current, which was used as the driving force to convert mechanical energy into electricity [76]. This research also pointed out the future direction to enhance the output performance of the TENG. Based on the basic theoretical analysis of both the TENG and EMG, some advanced structures should be proposed to satisfy the demand in the coming epoch – the massive need for small-sized and flexible electronics.

Secondly, it is essential to improve the durability and output stability of the TENG. Owing to the mechanization of generating charges in the TENG, friction is inevitable and will damage the materials, resulting in the decline of output performance. Exploring hard-wearing materials would solve these problems partly. In addition, by optimizing the structures and adopting the solid–liquid friction, the durability would be enhanced.

Finally, integrating the electromagnetic–triboelectric hybridized nanogenerators with other electronics has been the further trend. Hence, how to store or utilize the electricity generated by the hybridized nanogenerators should be considered compared with conventional electronics. Recently, Li-ion batteries have been shown to be capable of storing the electricity from the hybridized nanogenerators, and some small electronics can be powered by the nanogenerators. These processes need a power management circuit to reduce the high output voltage of the TENG [77]. Thus, it is essential to develop up-to-date power management circuits to meet the special demand of hybridized nanogenerators.

References

1 He, X., Zi, Y., Guo, H. et al. (2017). A highly stretchable fiber-based triboelectric nanogenerator for self-powered wearable electronics. *Adv. Funct. Mater.* 27: 1604378.
2 Chen, J. and Wang, Z.L. (2017). Reviving vibration energy harvesting and self-powered sensing by a triboelectric nanogenerator. *Joule* 1: 480.
3 Zhao, K., Yang, Y., Liu, X., and Wang, Z.L. (2017). Triboelectrification-enabled self-charging lithium-ion batteries. *Adv. Energy Mater.* 7: 1700103.
4 Pu, X., Guo, H., Chen, J. et al. (2017). Eye motion triggered self-powered mechanosensational communication system using triboelectric nanogenerator. *Sci. Adv.* 3: e1700694.
5 Wang, Y., Jiang, Y., Wu, H., and Yang, Y. (2019). Floating robotic insects to obtain electric energy from water surface for realizing some self-powered functions. *Nano Energy* 63: 103810.
6 Tian, B., Zheng, X., Kempa, T.J. et al. (2007). Coaxial silicon nanowires as solar cells and nanoelectronic power sources. *Nature* 449: 885.
7 Dresselhaus, M.S., Chen, G., Tang, M.Y. et al. (2007). New directions for low-dimensional thermoelectric materials. *Adv. Mater.* 19: 1043.
8 O'Regan, B. and Grätzel, M. (1991). A low-cost, high-efficiency solar cell based on dye-sensitized colloidal TiO_2 films. *Nature* 353: 737.
9 Xu, S., Qin, Y., Xu, C. et al. (2010). Self-powered nanowire devices. *Nat. Nanotechnol.* 5: 366.

10 Saha, C.R., O'Donnell, T., Loder, H. et al. (2006). Optimization of an electromagnetic energy harvesting device. *IEEE Trans. Magn.* 42: 3509.
11 Arnold, D.P. (2007). Review of microscale magnetic power generation. *IEEE Trans. Magn.* 43 (11): 3940.
12 Glynne-Jones, P., Tudor, M.J., Beeby, S.P., and White, N.M. (2004). An electromagnetic, vibration-powered generator for intelligent sensor systems. *Sens. Actuators, A* 110: 344.
13 Sari, I., Balkan, T., and Kulah, H. (2008). An electromagnetic micro power generator for wideband environmental vibrations. *Sens. Actuators, A* 145: 405.
14 Bouendeu, E., Greiner, A., Smith, P.J., and Korvink, J.G. (2010). A low-cost electromagnetic generator for vibration energy harvesting. *IEEE Sens. J.* 11: 107.
15 Zhu, D., Roberts, S., Tudor, M.J., and Beeby, S.P. (2010). Design and experimental characterization of a tunable vibration-based electromagnetic micro-generator. *Sens. Actuators, A* 158: 284.
16 Seung, W., Gupta, M.K., Lee, K.Y. et al. (2015). Nanopatterned textile-based wearable triboelectric nanogenerator. *ACS Nano* 9: 3501.
17 Kim, S., Gupta, M.K., Lee, K.Y. et al. (2014). Transparent flexible graphene triboelectric nanogenerators. *Adv. Mater.* 26: 3918.
18 Lee, K.Y., Chun, J., Lee, J.H. et al. (2014). Hydrophobic sponge structure-based triboelectric nanogenerator. *Adv. Mater.* 26: 5037.
19 Wang, Z.L., Chen, J., and Lin, L. (2015). Progress in triboelectric nanogenerators as a new energy technology and self-powered sensors. *Energy Environ. Sci.* 8: 2250.
20 Chen, J., Zhu, G., Yang, W. et al. (2013). Harmonic-resonator-based triboelectric nanogenerator as a sustainable power source and a self-powered active vibration sensor. *Adv. Mater.* 25: 6094.
21 Zheng, Q., Shi, B., Fan, F. et al. (2014). In vivo powering of pacemaker by breathing-driven implanted triboelectric nanogenerator. *Adv. Mater.* 26: 5851.
22 Wang, S., Xie, Y., Niu, S. et al. (2014). Freestanding triboelectric-layer-based nanogenerators for harvesting energy from a moving object or human motion in contact and non-contact modes. *Adv. Mater.* 26: 2818.
23 Khan, U. and Kim, S.-W. (2016). Triboelectric nanogenerators for blue energy harvesting. *ACS Nano* 10: 6429.
24 Zhang, X.-S., Han, M.-D., Meng, B., and Zhang, H.-X. (2015). High performance triboelectric nanogenerators based on large-scale mass-fabrication technologies. *Nano Energy* 11: 304.
25 Yi, F., Lin, L., Niu, S. et al. (2015). Stretchable-rubber-based triboelectric nanogenerator and its application as self-powered body motion sensors. *Adv. Funct. Mater.* 25: 3688.
26 Zhang, C., Tang, W., Han, C. et al. (2014). Theoretical comparison, equivalent transformation, and conjunction operations of electromagnetic induction generator and triboelectric nanogenerator for harvesting mechanical energy. *Adv. Mater.* 26: 3580.
27 Niu, S., Wang, S., Lin, L. et al. (2013). Theoretical study of contact-mode triboelectric nanogenerators as an effective power source. *Energy Environ. Sci.* 6: 3576.

28 Niu, S., Liu, Y., Wang, S. et al. (2013). Theory of sliding-mode triboelectric nanogenerators. *Adv. Mater.* 25: 6184.

29 Niu, S., Wang, S., Liu, Y. et al. (2014). A theoretical study of grating structured triboelectric nanogenerators. *Energy Environ. Sci.* 7: 2339.

30 Jin, L., Chen, J., Zhang, B. et al. (2016). Self-powered safety helmet based on hybridized nanogenerator for emergency. *ACS Nano* 10: 7874.

31 Zhou, C., Gu, A., Liang, G., and Xia, Y. (2015). Tough silica-hybridized epoxy resin/anhydride system with good corona resistance and thermal stability for permanent magnet synchronous wind-driven generators through vacuum pressure impregnation. *Ind. Eng. Chem. Res.* 54: 7102.

32 Li, Z., Saadatnia, Z., Yang, Z., and Naguib, H. (2018). A hybrid piezoelectric-triboelectric generator for low-frequency and broad-bandwidth energy harvesting. *Energy Convers. Manage.* 174: 188.

33 Wang, S., Wang, Z.L., and Yang, Y. (2016). A one-structure-based hybridized nanogenerator for scavenging mechanical and thermal energies by triboelectric–piezoelectric–pyroelectric effects. *Adv. Mater.* 28: 2881.

34 Cheng, G., Lin, Z.-H., Du, Z.-L., and Wang, Z.L. (2014). Simultaneously harvesting electrostatic and mechanical energies from flowing water by a hybridized triboelectric nanogenerator. *ACS Nano* 8: 1932.

35 Adhikari, S., Halgamuge, S.K., and Watson, H.C. (2010). An online power-balancing strategy for a parallel hybrid electric vehicle assisted by an integrated starter generator. *IEEE Trans. Veh. Technol.* 59: 2689.

36 Quan, T., Wang, Z.L., and Yang, Y. (2016). A shared-electrode-based hybridized electromagnetic-triboelectric nanogenerator. *ACS Appl. Mater. Interfaces* 8: 19573.

37 Zhong, X., Yang, Y., Wang, X., and Wang, Z.L. (2015). Rotating-disk-based hybridized electromagnetic-triboelectric nanogenerator for scavenging biomechanical energy as a mobile power source. *Nano Energy* 13: 771.

38 Zhang, K., Wang, X., Yang, Y., and Wang, Z.L. (2015). Hybridized electromagnetic–triboelectric nanogenerator for scavenging biomechanical energy for sustainably powering wearable electronics. *ACS Nano* 9: 3521.

39 Duan, Y., Huang, Y., Yin, Z. et al. (2014). Non-wrinkled, highly stretchable piezoelectric devices by electrohydrodynamic direct-writing. *Nanoscale* 6: 3289.

40 Larmagnac, A., Eggenberger, S., Janossy, H., and Vörös, J. (2014). Stretchable electronics based on Ag-PDMS composites. *Sci. Rep.* 4: 7254.

41 Jeong, C.K., Lee, J., Han, S. et al. (2015). A hyper-stretchable elastic-composite energy harvester. *Adv. Mater.* 27: 2866.

42 Zhong, J., Zhong, Q., Hu, Q. et al. (2015). Stretchable self-powered fiber-based strain sensor. *Adv. Funct. Mater.* 25: 1798.

43 Sim, H.J., Choi, C., Lee, C.J. et al. (2015). Flexible, stretchable and weavable piezoelectric fiber. *Adv. Energy Mater.* 17: 1270.

44 Zhang, K., Wang, Z.L., and Yang, Y. (2016). Conductive fabric-based stretchable hybridized nanogenerator for scavenging biomechanical energy. *ACS Nano* 10: 4728.

45 Zhang, B., Chen, J., Jin, L. et al. (2016). Rotating-disk-based hybridized electromagnetic–triboelectric nanogenerator for sustainably powering wireless traffic volume sensors. *ACS Nano* 10: 6241.

46 Quan, T., Wu, Y., and Yang, Y. (2015). Hybrid electromagnetic–triboelectric nanogenerator for harvesting vibration energy. *Nano Res.* 8: 3272.

47 Shao, H., Cheng, P., Chen, R. et al. (2018). Triboelectric–electromagnetic hybrid generator for harvesting blue energy. *Nano-Micro Lett.* 10: 54.

48 Seol, M.-L., Jeon, S.-B., Han, J.-W., and Choi, Y.-K. (2017). Ferrofluid-based triboelectric-electromagnetic hybrid generator for sensitive and sustainable vibration energy harvesting. *Nano Energy* 31: 233.

49 Wang, X., Wang, S., Yang, Y., and Wang, Z.L. (2015). Hybridized electromagnetic–triboelectric nanogenerator for scavenging air-flow energy to sustainably power temperature sensors. *ACS Nano* 9: 4553.

50 Wang, X. and Yang, Y. (2017). Effective energy storage from a hybridized electromagnetic-triboelectric nanogenerator. *Nano Energy* 32: 36.

51 Quan, T. and Yang, Y. (2016). Fully enclosed hybrid electromagnetic–triboelectric nanogenerator to scavenge vibrational energy. *Nano Res.* 9: 2226.

52 Zhang, K. and Yang, Y. (2016). Linear-grating hybridized electromagnetic-triboelectric nanogenerator for sustainably powering portable electronics. *Nano Res.* 9: 974.

53 Wu, Y., Wang, X., Yang, Y., and Wang, Z.L. (2015). Hybrid energy cell for harvesting mechanical energy from one motion using two approaches. *Nano Energy* 11: 162.

54 Peng, J., Kang, S.D., and Snyder, G.J. (2017). Optimization principles and the figure of merit for triboelectric generators. *Sci. Adv.* 3: eaap8576.

55 Chen, J., Guo, H., He, X. et al. (2015). Enhancing performance of triboelectric nanogenerator by filling high dielectric nanoparticles into sponge PDMS film. *ACS Appl. Mater. Interfaces* 8: 736.

56 Dharmasena, R. and Silva, S. (2019). Towards optimized triboelectric nanogenerators. *Nano Energy* 62: 530–549.

57 Ko, Y.H., Nagaraju, G., Lee, S.H., and Yu, J.S. (2014). PDMS-based triboelectric and transparent nanogenerators with ZnO nanorod arrays. *ACS Appl. Mater. Interfaces* 6: 6631.

58 Ko, Y.H., Lee, S.H., Leem, J.W., and Yu, J.S. (2014). High transparency and triboelectric charge generation properties of nano-patterned PDMS. *RSC Adv.* 4: 10216.

59 Dudem, B., Huynh, N.D., Kim, W. et al. (2017). Nanopillar-array architectured PDMS-based triboelectric nanogenerator integrated with a windmill model for effective wind energy harvesting. *Nano Energy* 42: 269.

60 Zhao, P., Soin, N., Prashanthi, K. et al. (2018). Emulsion electrospinning of polytetrafluoroethylene (PTFE) nanofibrous membranes for high-performance triboelectric nanogenerators. *ACS Appl. Mater. Interfaces* 10: 5880.

61 Cheon, S., Kang, H., Kim, H. et al. (2018). High-performance triboelectric nanogenerators based on electrospun polyvinylidene fluoride–silver nanowire composite nanofibers. *Adv. Funct. Mater.* 28: 1703778.

62 Yang, J., Liu, P., Wei, X. et al. (2017). Surface engineering of graphene composite transparent electrodes for high-performance flexible triboelectric nanogenerators and self-powered sensors. *ACS Appl. Mater. Interfaces* 9: 36017.
63 Mi, H.-Y., Jing, X., Cai, Z. et al. (2018). Highly porous composite aerogel based triboelectric nanogenerators for high performance energy generation and versatile self-powered sensing. *Nanoscale* 10: 23131.
64 Quan, T., Wang, X., Wang, Z.L., and Yang, Y. (2015). Hybridized electromagnetic–triboelectric nanogenerator for a self-powered electronic watch. *ACS Nano* 9: 12301.
65 Wen, Z., Guo, H., Zi, Y. et al. (2016). Harvesting broad frequency band blue energy by a triboelectric–electromagnetic hybrid nanogenerator. *ACS Nano* 10: 6526.
66 Seol, M.-L., Han, J.-W., Park, S.-J. et al. (2016). Hybrid energy harvester with simultaneous triboelectric and electromagnetic generation from an embedded floating oscillator in a single package. *Nano Energy* 23: 50.
67 Ren, X., Fan, H., Wang, C. et al. (2017). Magnetic force driven noncontact electromagnetic-triboelectric hybrid nanogenerator for scavenging biomechanical energy. *Nano Energy* 35: 233.
68 Cao, R., Zhou, T., Wang, B. et al. (2017). Rotating-sleeve triboelectric–electromagnetic hybrid nanogenerator for high efficiency of harvesting mechanical energy. *ACS Nano* 11: 8370.
69 Hu, X., Li, S., and Peng, H. (2012). A comparative study of equivalent circuit models for Li-ion batteries. *J. Power Sources* 198: 359.
70 Endo, M., Kim, C., Nishimura, K. et al. (2000). Recent development of carbon materials for Li ion batteries. *Carbon* 38: 183.
71 Wang, J.-G., Jin, D., Zhou, R. et al. (2016). Highly flexible graphene/Mn_3O_4 nanocomposite membrane as advanced anodes for Li-ion batteries. *ACS Nano* 10: 6227.
72 Shi, J.L., Xiao, D.D., Ge, M. et al. (2018). High-capacity cathode material with high voltage for Li-ion batteries. *Adv. Mater.* 30: 1705575.
73 Deng, J., Kuang, X., Liu, R. et al. (2018). Vitrimer elastomer-based Jigsaw puzzle-like healable triboelectric nanogenerator for self-powered wearable electronics. *Adv. Mater.* 30: 1705918.
74 Ning, C., Tian, L., Zhao, X. et al. (2018). Washable textile-structured single-electrode triboelectric nanogenerator for self-powered wearable electronics. *J. Mater. Chem. A* 6: 19143.
75 Wang, P., Liu, R., Ding, W. et al. (2018). Complementary electromagnetic–triboelectric active sensor for detecting multiple mechanical triggering. *Adv. Funct. Mater.* 28: 1705808.
76 Wang, Z.L. (2019). On the first principle theory of nanogenerators from Maxwell's equations. *Nano Energy* 104272.
77 Xi, F., Pang, Y., Li, W. et al. (2017). Universal power management strategy for triboelectric nanogenerator. *Nano Energy* 37: 168.

4

Other Hybridized Nanogenerators

4.1 Introduction

In nature, there is abundant surrounding energy, such as solar energy, electrochemical energy, mechanical energy, and thermal energy. Utilizing and scavenging these renewable energies is crucial for solving the energy crisis and sustainable development [1]. However, most of the researchers developed approaches that were targeted at one single energy form while the rest of the energy forms were wasted. The hybridized energy cells have the potential to utilize solar energy or electrochemical energy in the environment individually. Meanwhile, other forms of energy can also be used when they are applied to the hybrid energy unit [2].

4.1.1 Nanogenerators for Harvesting Solar Energy

As the power from the sun is accessible and renewable, there is a dramatic increase in the research on its collection and utilization. There are various application forms of utilization that scientists have explored, such as solar cells, photodetectors, and nanogenerators [3–10]. The direct bandgap and binding energy of zinc oxide (ZnO) are 3.37 eV and 60 meV, respectively [11, 12]. There are four oxygen atoms surrounding each zinc atom tetrahedrally so that the d-electrons of zinc are hybridized with the 2p-electrons of oxygen. With its structure and properties, ZnO emerges as a potential excellent material in photovoltaic applications, including field-effect transistors, photodetectors, and other semiconductor devices for the UV spectral range. The ZnO material has many exciting applications as it not only shows optoelectric effect but also has a piezoelectric effect, which has great value for comprehensive applications.

In the SEM pictures of Figure 4.1a,b, Ning Ye et al. [13] fabricated pure ZnO and Sn element-doped ZnO nanoparticles of 250 nm by hydrothermal method. Next, these nanoparticles were coated with a thickness of 100 μm onto two kinds of substrates, the FTO substrate and the bare FTO substrate with a compact ZnO layer, which were the functional layers of solar cells. Figure 4.1c shows the optical image of a P3HT and ZnO micro/nanowire solar cell. Pan et al. described the high-temperature thermal evaporation method of synthesizing the ZnO

*Corresponding author: yayang@binn.cas.cn

Hybridized and Coupled Nanogenerators: Design, Performance, and Applications,
First Edition. Ya Yang.

micro/nanowires [14]. A thin poly(dimethylsiloxane) (PDMS) layer covered the whole device to seal it on a PS substrate, which can be used as a flexible energy system [15]. By applying stress to the flexible solar cell, the induced electric signals were interpreted on the basis of the piezotronics mechanism. As the schematic diagrams in Figure 4.1d,e shows, ZnO grows along the [0001] and [000$\bar{1}$] crystal directions connected to the Ag electrode and the P3HT end. The P3HT and ZnO wire are connected, which can produce the p–n heterojunction. The ZnO nanowires in Figure 4.1f,g are fabricated by the simple method of chemical vapor deposition [16]. Optical images of the ZnO wires with two kinds of morphology – straight and bent on a glass substrate – were used as the ultraviolet photoresponse [17]. Yang et al. first presented Sb-doped ZnO nanobelts with Schottky contact side and Ohmic contact side, which can work directly as a photodetector without voltage bias [18]. In Figure 4.1g, the doped ZnO nanobelt was connected with Ag and Au. The ZnO–Ag contact form should be Ohmic contact due to the ZnO electron affinity, and the Ag work functions are 4.5 eV and 4.26 eV [19]. As the work function in the Au electrode is 5.1 eV [20], the barrier height is 0.6 eV while the ZnO–Au contact form should be a Schottky contact.

As illustrated in Figure 4.2a, the piezopotential distribution in ZnO nanowires along the growth direction of [0001]/[000$\bar{1}$] was calculated by the Lippman theory [21]. There is a polarization of anions and cations created on the *c*-axis direction, resulting in piezoelectric changes from the voltage negative side to the voltage positive side on the ZnO wire. As the band diagram in Figure 4.2b shows, the P3HT/ZnO heterojunction formed at the boundary was utilized to explain the modulation action of the strain on the V_{OC} of the P3HT/ZnO structure. The V_{OC} on the basis of p–n heterojunction was calculated by Shockley equation [22].

$$eV_{OC} = \frac{n}{r}\Delta E + nkT\ln\frac{I_{SC}}{I_{00}} \tag{4.1}$$

When tensile strain or compressive strain is applied to a device of [0001] or [000$\bar{1}$] type, the P3HT side contacts the negative piezopotential. The piezoelectricity-induced negative charges at the boundary force the ZnO band level and this phenomenon can increase the ΔE and V_{OC} according to Eq. (4.1). The schematic diagram in Figure 4.2c illustrates the pure ZnO energy band on the ZnO/FTO substrate while Figure 4.2d illustrates the energy band of element-doped ZnO. The rougher ZnO structure doping of Sn with higher surface areas has a higher absorption ability to attach to more dye molecules, which can enhance the J_{SC}. The Sn-doping increases the electrons transport capacity to enhance electron concentration at V_{OC}. Besides, Sn-doping is a typical n-type doping, which raises the Fermi levels, resulting in an increase of V_{OC}. Therefore, the photocurrent conversion efficiency of Sn-doping ZnO is increased obviously. The compact ZnO layer separates the FTO layer from the electrolyte layer, which restrains the electrons transferring from the FTO to the electrolyte. As depicted in Figure 4.2d, by doping the Sn element, the corresponding energy bandgap gets narrower than the pure ZnO energy bandgap. Hence, the element-doped ZnO conduction band becomes lower than the pure ZnO structure. The induced electrons in the compact layer flow from the Sn-doped ZnO. There is no photocurrent generated as the electron energies between the Sn-doped ZnO and ZnO compact layer conduction band

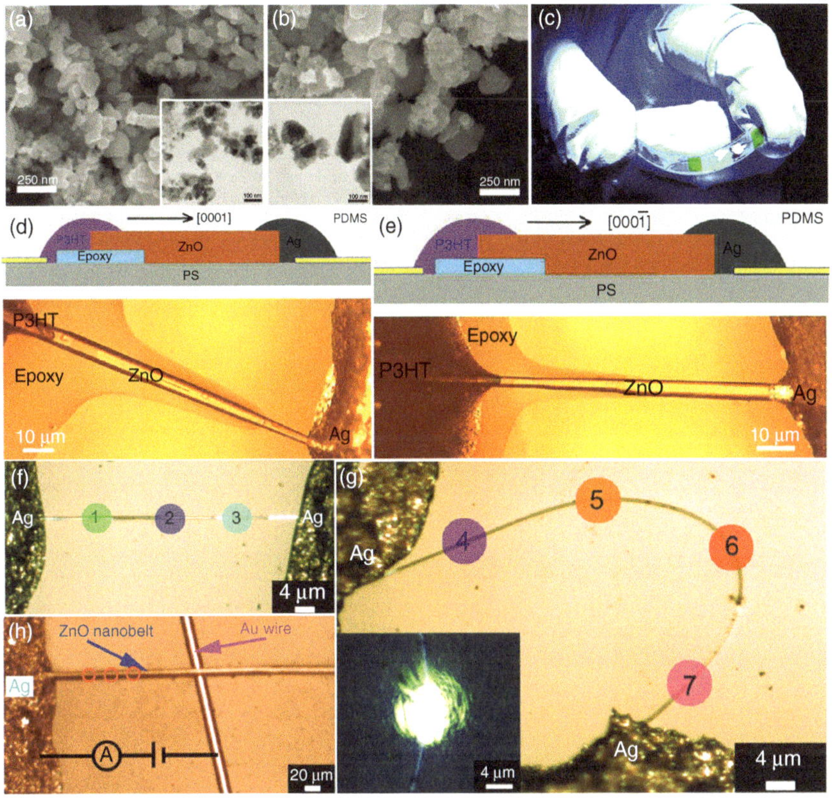

Figure 4.1 The images of different morphologic ZnO nanostructures. (a,b) SEM images of pure and Sn-doped ZnO nanoparticles. Source: Reproduced with permission from Ye et al. [13]. Copyright 2010, Elsevier. (c) The photograph of a solar cell fabricated by P3HT and ZnO nanowires. (d,e) Schematic and optical images of a fabricated device of the [0001] and [000$\bar{1}$] type. Source: Reproduced with permission from Yang et al. [15]. Copyright 2011, American Chemical Society. (f,g) Optical images of a single straight and a single bent ZnO wire between two Ag electrodes. Source: Reproduced with permission from Guo et al. [17]. Copyright 2010, American Institute of Physics. (h) Optical image of the fabricated single Sb-doped ZnO nanobelt device. Source: Reproduced with permission from Yang et al. [18]. Copyright 2010, American Institute of Physics.

are rebound by the oxidation–reduction potential. This explains the decreasing J_{SC} of Sn-doped ZnO-based solar cells compared to the undoped devices. To better understand the Sb-doped ZnO photodetector, the photoconductance principles in the designed structure were investigated in Figure 4.2e. When the 325 nm light hit the ZnO nanobelt, photoexcited electron–hole pairs were generated while there was no photocurrent observed without bias because the electron–hole recombination rate is high [23]. When the Au–ZnO Schottky contact was illuminated by the UV light, the recombination rate of this area was decreased. The Schottky barrier-induced localized electric field makes the photoexcited electrons and holes flow into two opposite directions rapidly. Then, the large photocurrent in the circuit can be measured.

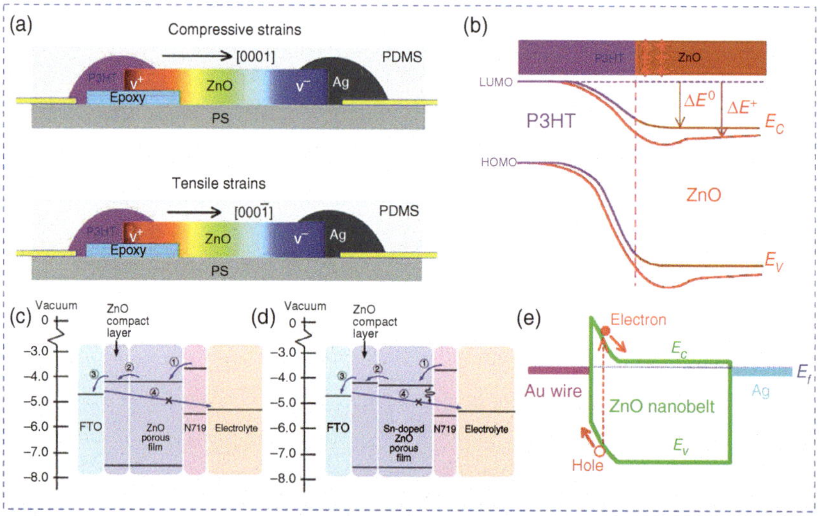

Figure 4.2 Schematic diagrams of ZnO-based devices. (a) The piezopotential distributions in the stretched device of [0001] type and compressed device of [000$\bar{1}$] type. (b) Schematic energy band diagram of P3HT/ZnO with the presence of positive piezoelectric charges. Source: Reproduced with permission from Ye et al. [13]. Copyright 2010, Elsevier. (c,d) Illustrations of the band diagrams for the solar cells with undoped and Sn-doped ZnO on the ZnO compact layer/FTO substrate. Source: Reproduced with permission from Yang et al. [15]. Copyright 2011, American Chemical Society. (e) Schematic band diagram of the Au–ZnO nanobelt Schottky contact under UV light illumination. Source: Reproduced with permission from Yang et al. [18]. Copyright 2010, American Institute of Physics.

In Figure 4.3a,c, the I–V curves of the P3HT/ZnO micro/nanowire heterojunction in [0001] and [000$\bar{1}$] directions were measured by applying different strains. The corresponding output circuit voltage increased with increasing compressive strains and showed opposite changing trend with the tensile strains. The I_{SC} and V_{OC} with increasing rate of the applied strains are also shown in Figure 4.3b,d. From these results, the performances of the solar cells produced by light can be modulated by the piezoelectric effect. The photoluminescence (PL) spectra of ZnO and Sb-doping ZnO nanobelt in Figure 4.3e show that the near band edge (NBE) peaks of pure ZnO nanobelt and the Sb-doped ZnO nanobelt are 3.27 and 3.22 eV, respectively. The PL measurement results indicated that the doping Sb element can reduce the bandgap of the ZnO nanobelt slightly. From Figure 4.3f, with the increase of Sb-doping concentrations, the output current signals of the device obviously increased, demonstrating that Sb element doping can improve the photovoltaic performances of ZnO-based UV light detectors. The PL spectra of ZnO and element doping ZnO nanoparticles in Figure 4.3g show that NBE peaks of the pure ZnO nanoparticles and ZnO nanoparticles of Sn doping are 3.22 and 3.18 eV, indicating that Sn doping can also slightly reduce the bandgap of the ZnO nanoparticles. As depicted in Figure 4.3h, the J–V curves with four kinds of structures are compared, which are an undoped solar cell without ZnO compact layer, an undoped solar cell with compact layer, a 5.5 wt%-doped solar cell without compact layer, and a 5.5wt%-doped solar cell with compact layer. The increasing charge carrier density and the rising

Fermi levels enhanced the electron transport capacity, which can increase the photocurrent due to the n-type doping. In Figure 4.3i,j, the UV photoresponse under 325 nm light illumination of the straight and bent ZnO nanowires are depicted. The current signals of the straight ZnO wires when the light hits on the three regions, "1," "2," and "3," are almost the same in Figure 4.3i. For the bent ZnO wire in Figure 4.3j, the shapes of the signals are quite different from the straight wire, and the signals of the four illuminated regions, "4," "5," "6," and "7," are also different. The photoresponse sensitivity of the bent part of the ZnO wire is remarkably higher than that of the straight part, which supports the idea that the two ZnO wires combine the optoelectronic effect and piezoelectric effect.

4.1.2 Nanogenerators for Harvesting Electrochemical Energy

Water energy is one of the most sustainable energies in nature, which has the potential to be converted into different energy forms by developing new techniques [24–28]. Water energy could be available in the ocean, flowing rivers, and raindrops, which all have great potential to be converted into usable electricity. Lin et al. [29] designed and fabricated a self-powered nanogenerator device based on superhydrophobic surfaces that can harvest electrostatic energy from the water drop. Jun Yin et al. [30] showed that moving a small amount of seawater or other solutions on monolayer graphene can generate electricity, which can drive sensors and electrical appliances. Herein, the redox-induced electricity with saline water participation attracts more attention as it can be easily produced by the difference in chemical potential between two different materials.

As seen from the SEM image of the commercial graphene powders in Figure 4.4a, graphene with thin layers easily wrinkles together to minimize the graphene surface energy. [31] High surface energy is disadvantageous for the formation of the required uniform graphene film. The lower graphene surface energy can be obtained by adding cetyltrimethylammonium bromide (CTAB) surface-active agent. Then the ecoflex as a kind of environmental and flexible elastomer polymer covers the surface of graphene to form a membrane, as shown in Figure 4.4b. As the SEM images in Figure 4.4c demonstrate, the diameter of the superhydrophobic Cu wire is 0.3 mm [32]. The thickness and width of the fabricated leaf-like CuO nanosheets on the surface of the Cu wire are 50 and 300 nm as shown in Figure 4.4d. As the image of the CuO nanosheets observed on TEM presented in Figure 4.4e shows, the nanostructures of the CuO increase the surface areas at the interface of the electrode wires and water droplets to improve the electrochemical performances. The high-resolution HRTEM image in the red square in Figure 4.4e displays the lattice fringe of the CuO nanosheets in Figure 4.4f. Yang Wang et al. designed and fabricated a biomimetic water strider-shaped microrobot to obtain electrical energy from the water surface [33]. The key reason for the water strider sliding on the water surface is illustrated in Figure 4.4g, which shows that the superhydrophobic legs of the water strider are covered by numerous oriented tiny hairs. These legs existing in nature can provide an excellent solution to fabricate the artificial device structures to skate on the surface of the water. The magnesium wires and copper wires are the basic components of the biomimetic water strider legs. As the SEM image in Figure 4.4h shows, the magnesium wires were coated by a

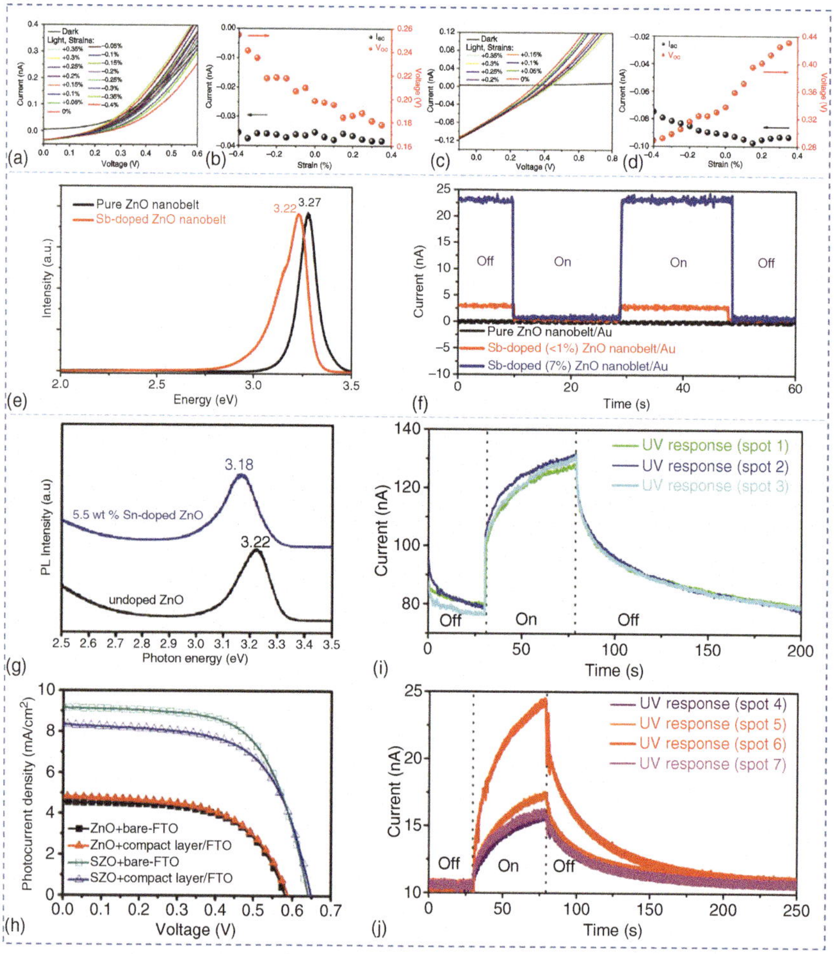

Figure 4.3 The performances of the ZnO-based devices. (a,c) *I–V* curves of the device of [0001] type and [000$\bar{1}$] type under different strains. (b,d) Dependence of I_{SC} and V_{OC} of the device of [0001] type and [000$\bar{1}$] type on applied strains. Source: Reproduced with permission from Ye et al. [13]. Copyright 2010, Elsevier. (e) PL spectra of single pure and Sb-doped ZnO nanobelts on a glass substrate. (f) The current response curves of the devices based on single ZnO nanobelts with different Sb-doping concentrations. Source: Reproduced with permission from Yang et al. [18]. Copyright 2010, American Institute of Physics. (g) PL spectra from undoped ZnO and Sn-doped ZnO. Enlarged part of the current response curve under UV light illumination. (h) The *J–V* curves with and without ZnO compact layer with undoped and Sn-doped ZnO porous film. Source: Reproduced with permission from Yang et al. [15]. Copyright 2011, American Chemical Society. (i,j) The UV photoresponses from the straight and bent ZnO wires at the different regions. Source: Reproduced with permission from Guo et al. [17]. Copyright 2010, American Institute of Physics.

layer of superhydrophobic TiO_2, which was fluorinated by fluoroalkyl silane. The nanorod structures on the surface of the copper wires as legs are depicted in Figure 4.4i, which were fabricated by a simple solution-immersion process.

The photograph in Figure 4.5a illustrates the structure of a flexible graphene-based strain device, which consisted of a graphene/ecoflex strip

Figure 4.4 The SEM and TEM images of the devices based on electrochemical devices. (a) SEM image of graphene powder. (b) Top view of the fabricated graphene–polymer nanocomposite. Source: Reproduced with permission from Wang et al. [31]. Copyright 2018, John Wiley and Sons. (c,d) SEM images of the superhydrophobic Cu electrode. Source: Reproduced with permission from Wang and Yang [32]. Copyright 2019, Elsevier. (e,f) TEM image and HRTEM image of the superhydrophobic Cu electrode. Insets is the SADE pattern. (g) SEM image of a water strider's leg showing numerous setae. (h) SEM image of a magnesium wire showing nanoparticle. (i) SEM image of a copper wire showing nanorod arrays. Source: Reproduced with permission from Wang et al. [33]. Copyright 2019, Elsevier.

and meandering zinc wires sealed together in a stretchable ecoflex model. In the model chamber, NaCl solution was injected using a syringe so that the electrochemical signals obtained by the redox-induced reactions can be utilized to detect the change of the strain. From the photographs in Figure 4.5b–d, the self-powered strain detector shows that it is easy to be stretched, bent, and deformed. As demonstrated in Figure 4.5c, the whole device simply consisted of superhydrophobic zinc and copper electrodes with electrolyte solution droplets. A green-light LED was lighted successfully by the fabricated compact devices while three droplets of sodium chloride (3 wt%) were dropped in the middle of the two metal electrodes. The photograph in Figure 4.5d illustrates the structure of an artificial biomimetic water strider, which can be floated on the water as the superhydrophobic wire with the contact angle over 150° and light weight.

The typical electrochemical principles and mechanism of the graphene-based device are illustrated in Figure 4.6a. With the NaCl solution as the electrolyte, the electrons can be generated on the surfaces of both graphene and zinc as their chemical potentials are different. Once the switch of the circuit is open, there is an immediate reduction reaction taking place on the surface of graphene. Meanwhile, each oxygen molecule dissolved in the water can get four

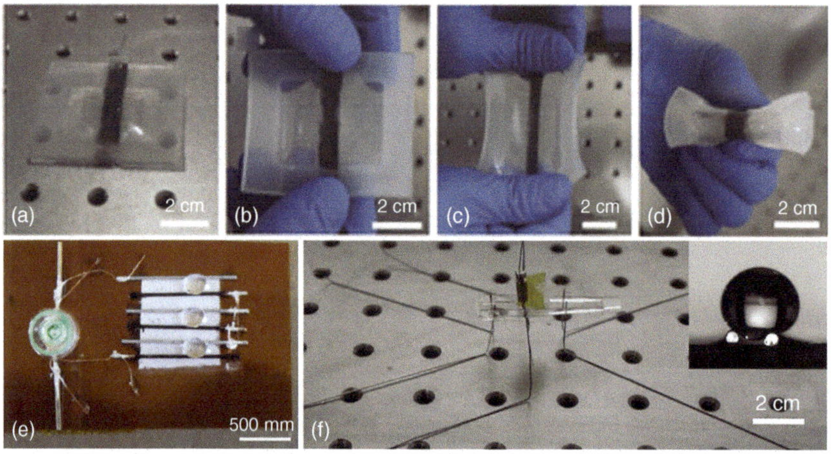

Figure 4.5 The photographs of the electrochemical devices. (a–d) Photographs of the strain sensor (a,b) before and (c,d) after being stretched as well as rolling up. Source: Reproduced with permission from Wang et al. [31]. Copyright 2018, John Wiley and Sons. (e) Photographs of a green LED after being lit up by the integrated devices consisting of three droplets of NaCl solution. Source: Reproduced with permission from Wang et al. [32]. Copyright 2019, Elsevier. (f) A photograph of fabricated water strider microrobot. The legs are coated with superhydrophobic coating, resulting in a high contact angle (inset image). Source: Reproduced with permission from Wang et al. [33]. Copyright 2019, Elsevier.

electrons and convert into hydroxyl ions. The chemical reaction formula is as follows: $O_2 + 2H_2O + 4e \rightarrow 4OH^-$. The surface of zinc is oxidized and the zinc atoms lose electrons, and then bond with hydroxide into zinc hydroxide. The chemical reaction formula is as follows: $2Zn - 4e + 4OH^- \rightarrow 2Zn(OH)_2$. The electrochemical reactions occurring between the Cu and Zn wire electrodes are depicted in Figure 4.6b. When a droplet of electrolyte was dropped on the two metal wires, an immediate reduction reaction takes place on the copper wire surface as the cathode ($O_2 + 2H_2O + 4e \rightarrow 4OH^-$) and the oxidation reaction occurs on the zinc wire surface as the anode ($2Zn - 4e \rightarrow 2Zn^{2+}$). The directional electron flow produces electrochemical signals when the electrodes are connected. As demonstrated in Figure 4.6c, a three-phase contact point (TPCP) was formed on an air–liquid–solid phase reaction environment. The construction of electrode architecture at TPCP as the electrode surface area and wetting can significantly affect the overall performances [34]. Sufficient oxygen was trapped at the TPCP by superhydrophobic electrodes to benefit the gas diffusion, which can show electrochemical performances [35]. Moreover, the TPCP was quite different under different wetting states such as the Wenzel state (W-state), Wenzel–Cassie coexistent state (WC-state), and Cassie state (C-state), as shown in Figure 4.6d. According to the research, the droplet wetting process was progressive. At the beginning of the reaction, the wetting state was at C-state as the superhydrophobicity on the surface of the electrodes was high. The electrolyte could immerse into the electrode surfaces, which formed the WC-state as the reactions progress. Finally, the liquid could completely wet the solid electrode at the W-state. The principles and mechanism of the biomimetic water strider are illustrated in Figure 4.6e. On the copper surface, the reduction reaction occurs at the cathode ($O_2 + 2H_2O + 4e \rightarrow 4OH^-$). On the surface of the magnesium legs, the oxidation reaction occurs at the anode ($2Mg - 4e \rightarrow 2Mg^{2+}$). The reactions are enhanced due to the enhanced gas diffusion when the electrodes were under WC-state on the three-phase contact line (TPCL) [36]. The concentration and the immersion time of the fluoroalkyl silane reagent have an effect on the wetting state to enhance electrochemical performances. Thus, the TPCL on the superhydrophobic metal wires as shown in Figure 4.6f can exploit these enhancement features. Figure 4.6g demonstrates a real water strider walking on the water surface by moving its middle legs to produce a forward thrust. As in Figure 4.6h, the superhydrophobic metal legs driven by a miniature motor on the top simulated the movement process the same as a real water strider. In Figure 4.6i, the metal legs were forced by the motor backward to prop up the air space under the water on the basis of COMSOL simulation data. This leads to the difference between water resistance (F_{hr}) and surface tension (F_{hs}) as shown in Figure 4.6k. Meanwhile, the legs undergo viscous drags from water in Figure 4.6j, which combine the generated torque to swing backward. Based on the above factors, the biomimetic water strider could skate on the water like the real ones.

As the photographs depict in Figure 4.7a, the graphene/ecoflex film with a width of 8 mm and effective length of 10 mm was stretched from the initial state of 0% strain to 200% strain. The measured resistance can be repeatable as shown in the resistance cycle curves in Figure 4.7b. The corresponding resistance variations ($\Delta R/R_0$) in the sensing property of the flexible graphene/ecoflex film were measured from the strain range of 0–200% as in Figure 4.7c, which showed the resistance variations increasing from 0% to 3474% with increasing strains. The graphene/ecoflex strip and meandering zinc wires were assembled into a complete electrochemical device with the NaCl solution filling into the chamber space. When the device was accessed into the circuit by a boat-shaped switch, which can be controlled periodically on/off, the output current and voltage were measured in Figure 4.7d,e. The positive current value and voltage value from the redox-induced reaction were 75 μA and 0.83 V, respectively. As illustrated in Figure 4.7f, unlike the individual graphene/ecoflex strip, the entire compact device can be stretched from 0% to 150%, which exhibited excellent elasticity. In order to investigate the influence of the applied strains on the current output, the current signals were measured as in Figure 4.7g with a 25% strain change as the applied gradient on the strain sensor. Corresponding to each 25% strain increment, the current signals decreased as a step shaped with increasing strains and increased with decreasing strains, which suggests a good response to the applied strain changes. Moreover, with the excellent deformation and flexibility characteristics, the strain sensor can be utilized as a wearable device. A sensor fixed on the human knee by two lines can detect the configuration and motions. From the photographs in Figure 4.7h, the bending angles of the human knee are measured by distinguishing the change detected in the signals while the current increases as the bending angles enlarge. Besides, the different body motions can be discriminated through the curved shapes of the measured current signals as in Figure 4.7i, such as walking, jumping, and running. From the measured

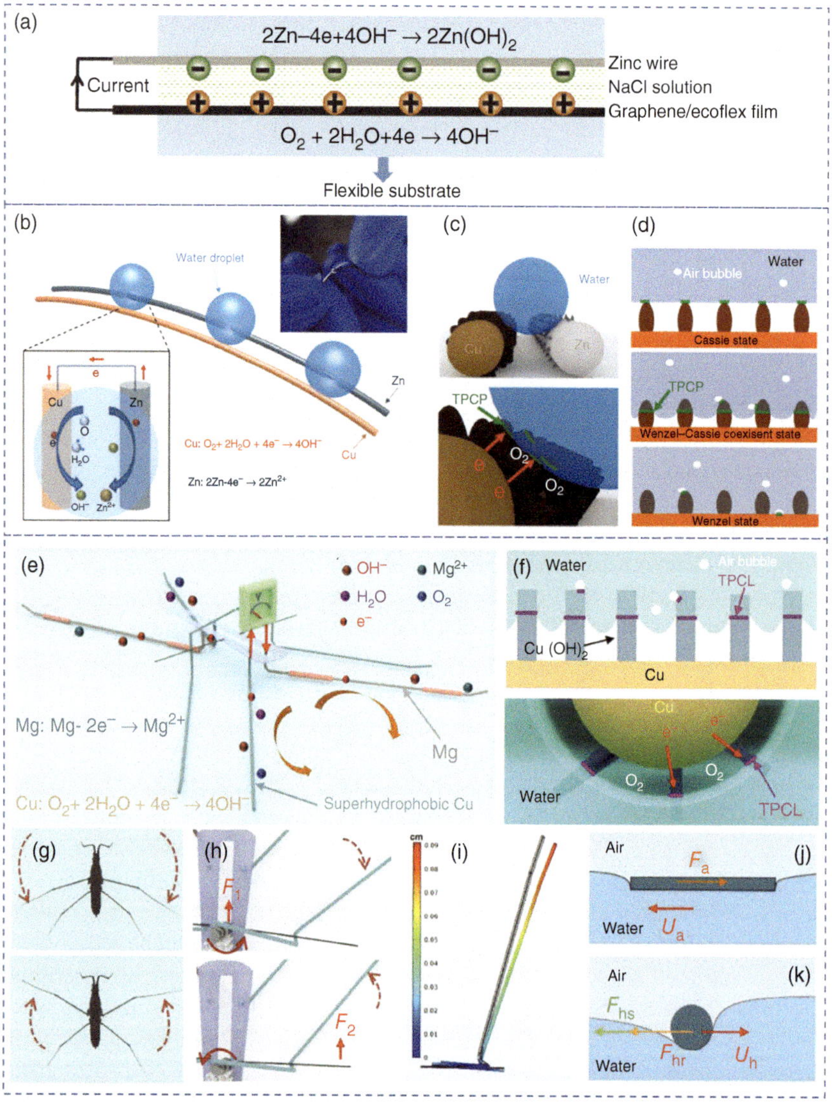

Figure 4.6 The mechanisms of the devices based on redox reaction. (a) The reaction formula of generating current. Source: Reproduced with permission from Wang et al. [31]. Copyright 2018, John Wiley and Sons. (b) Schematic illustration of the chemical reaction on superhydrophobic Cu/Zn electrodes. (c) Gas diffusion on three-phase contact point (TPCP). (d) Three types of wetting state for the hydrophobic surface. Source: Reproduced with permission from Wang et al. [32]. Copyright 2019, Elsevier. (e) Illustration of the microrobot floating on the water and mechanism of voltage generation in legs. (f) Gas diffusion on the three-phase contact line (TPCL). (g) Skating behavior of a water strider. (h) Bioinspired robot skating kinematics feature. (i) The COMSOL simulation data of the legs. (j) Illustration of the component force of the leg in the axial direction. Ua is velocity in the axial direction. (k) Two-dimensional models of the horizontal component force. Uh is the paddling velocity. Source: Reproduced with permission from Wang et al. [33]. Copyright 2019, Elsevier.

signals of three typical motion types, running produced smaller current changes than walking while the frequency of the running signals was larger, which can be regarded as a typical feature to identify. Similar to the jump signal in Figure 4.7j, the squat motion before landing produces a small characteristic peak in the opposite direction. These test results can be analyzed to provide a visible prospect for wearable applications.

As displayed in Figure 4.8a, the stabilization and measured output signals of the electrochemical structure (NaCl solution 3 wt%) are affected by the wetting behaviors of electrodes due to redox reaction kinetics. For the smooth copper–zinc (Cu-Zn) electrodes, the current signals go up to 0.15 mA and then slowly decay to around 0.04 mA after 1800 seconds. The Cu electrode has oxidized leaf-like CuO nanostructure on the surface and the Zn electrode has etched nanoparticles on the surface. The device with two electrode (Nano-CuO-Zn(Hydrophile)) appears the hydrophilic performances. The current signal of this structure (Nano-CuO-Zn(H)) can be enhanced to 0.27 mA due to the increased interface surface between water and the electrodes. Furthermore, the signal of the nano-CuO-Zn-superhydrophobicity structure after superhydrophobic treatment is 0.21 mA. The current signals in Figure 4.8b increased from 0.25 mA to 1.2 mA with one individual device increasing to six parallel devices. The increasing numbers of parallel devices can enhance the output current. The power produced by the integrated device can drive a green LED light and a smart temperature sensor attached to the human hand as in Figure 4.8c,d. As the schematic diagram depicts in Figure 4.8e, the watch was designed to be driven by an integrated electrochemical device. The watch belt made with a PDMS transparent film and the integrated device under the dial plate as the battery were put together to fabricate the watch. In the photographs in Figure 4.8f, after three NaCl solution droplets were dropped to the surface, this watch could tick forward sustainably due to a voltage generation of 2.2 V. As displayed in Figure 4.8g,h, when the microrobot walks on the surface of water, the LED on the top, which is connected with a power management circuit (PMC), was lighted up by the electrochemical energy generated by the redox reaction between the metal legs and the electrolyte surface. Moreover, the artificial microrobot can directly drive a temperature sensor as in Figure 4.8h. Moreover, in order to store the generated electricity, supercapacitors with different capacities are illustrated in Figure 4.8i to collect the electricity obtained from the redox reaction by the biomimetic microrobot. It takes more time to charge a larger capacitor of 5 F and more time to recharge it. According to the charging and discharging curves demonstrated in Figure 4.8j, the microrobot with stable output electric performances can charge the supercapacitors sustainably.

From all the research above, there is great potential to utilize the electrochemical energy that comes from water energy. In the future, devices that combine electrochemical energy with other kinds of energy will show a visible prospect.

4.2 Hybridized Photoelectric and Piezoelectric Nanogenerator

In the surroundings, there are diverse energy forms, such as solar energy, electrochemical energy, and mechanical energy. Since focusing on developing only one

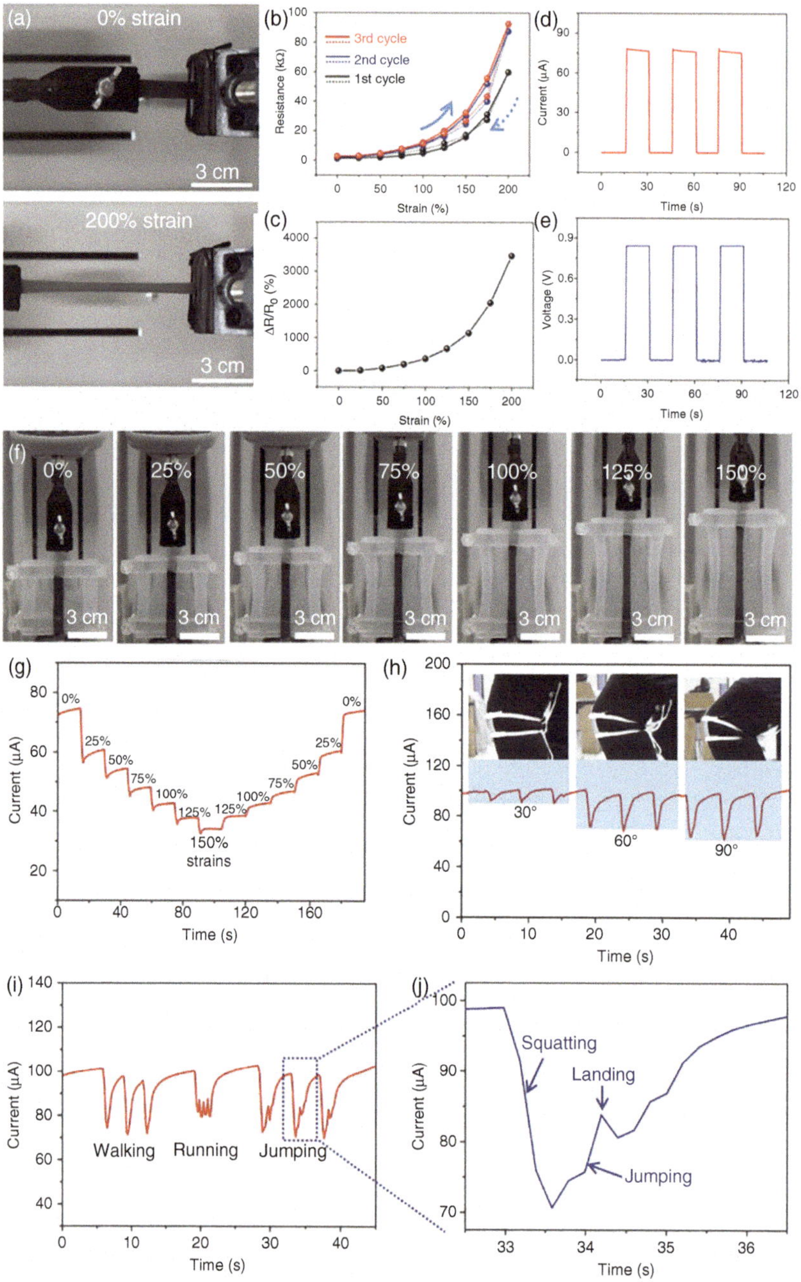

Figure 4.7 Performances of the flexible devices based on graphene/ecoflex. (a) Photographs of graphene/ecoflex film before and after being stretched by 200%. (b) Repeated resistance cycle curves of the stretching and releasing process. (c) The relative change in resistance versus strain curves from 0% to 200% strains. (d,e) Measured output current signals and output voltage signals under positive connection conditions. (f) Photographs of the sensor stretched from 0% to 150%. (g) Current variation of the strain sensor under consecutive stretching and releasing step strain process (0–150% strains). (h) Application as wearable sensor of knee motions monitoring including response signals of different bending angles. (i,j) Body motions of walking, running, and jumping, as well as the contained motions of the dashed box. Source: Reproduced with permission from Wang et al. [31]. Copyright 2018, John Wiley and Sons.

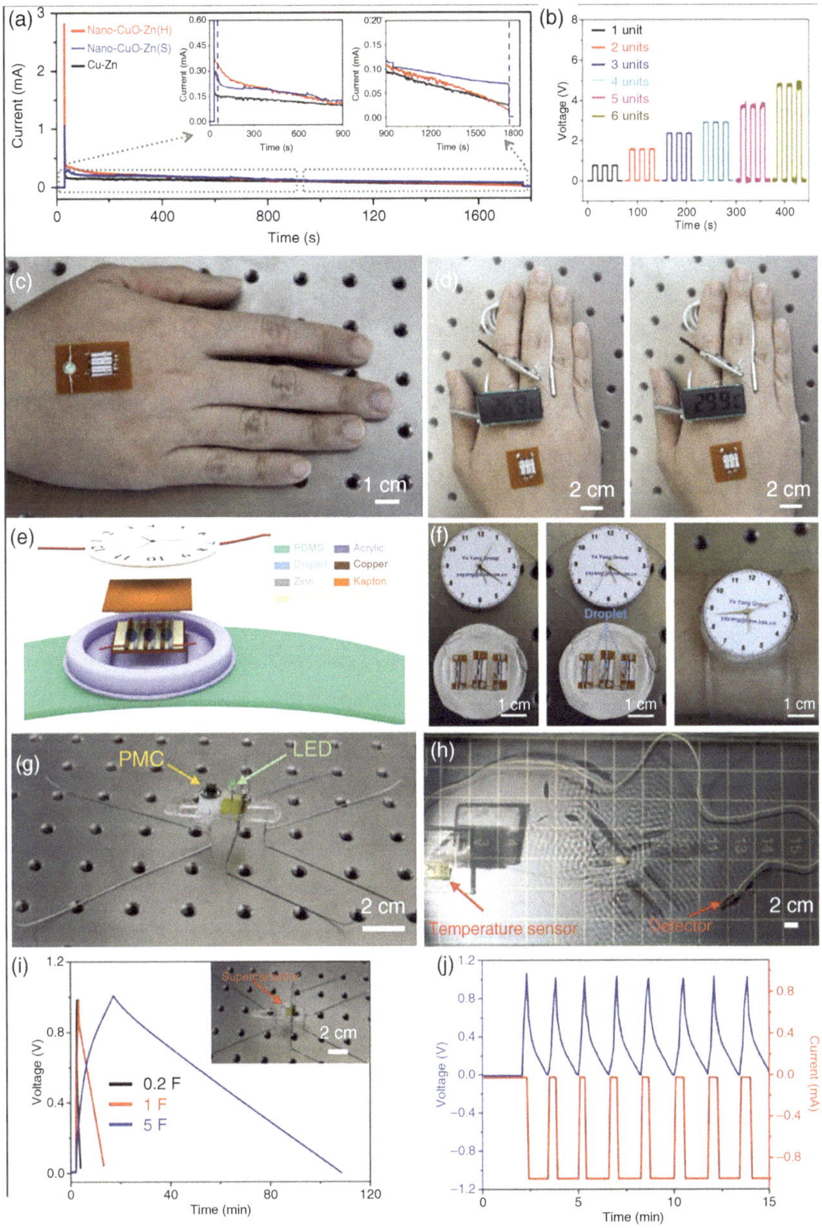

Figure 4.8 Performances of the devices based on redox reaction. (a) Stability testing of three different devices. (b) Measured output voltage signals of the integrated superhydrophobic devices in series. (c,d) Photographs of a green LED and a temperature sensor before and after being lit up by the integrated devices consisting of three droplets of NaCl solution. (e,f) Schematic structure and photograph of the superhydrophobic device connected to the watch.Source: Reproduced with permission from Wang et al. [32]. Copyright 2019, Elsevier. (g,h) The integrated microrobot consists of an LED and a PMC, and the microrobot can directly power a smart temperature sensor when it moves on the surface of water. (i) Charging curves of the different capacitors by using the microrobot directly. (j) Measured charging/discharging voltage–time curves of a 0.2 F capacitor. Source: Reproduced with permission from Wang et al. [33]. Copyright 2019, Elsevier.

individual form of energy could cause wastage of energy, it is critical to investigate methods for harvesting as many forms of energy as possible to satisfy energy needs in the future and sustainable development on earth [37, 38]. After all these years, multiple methods and innovative materials are updated and developed to transform solar energy and mechanical energy into available electrical energy. Photovoltaic conversion was designed by utilizing p–n junctions [39], organic or inorganic thin films [40], and other photoelectric devices [6, 41]. The nanogenerators based on piezoelectric effect were fabricated to utilize and harvest the mechanical energy that is easily available around us [42]. The problem is that most of the existing technologies and entities can only harvest one individual energy form. Herein, there are multiple energy sources coexisting that have the potential to be fully utilized by integrated structures whenever and wherever one or all of them are available.

4.2.1 Typical Materials and Structure Design

As the SEM photographs in Figure 4.9a,b demonstrate, the integrated device consisted of a photovoltaic unit and a piezoelectric unit on the basis of the piezoelectric effect [43]. The functional ZnO nanowire arrays grow on the ITO substrate by a solution-based growth technique. Then the ZnO nanowires were coated with dye molecules as the main functional material to separate the counter electrode from the electrolyte in solar cells [44]. Meanwhile, the ZnO nanowires array on a polystyrene-coated substrate as the top zigzag electrode periodically makes contact with and separates from the other ZnO nanowires electrode to create the piezoelectric signals. The SEM images in Figure 4.9c,d shows that the ZnO NWs were fabricated to a "teeth to teeth" configuration [45, 46]. Figure 4.9e,f shows that the ZnO nanowires array as the functional material grows on a flexible PET substrate with an ITO layer on it [47]. In Figure 4.9g, the cross-sectional SEM picture shows the ZnO nanosheets of 0.5 μm height acting as the piezoelectric active layer in the nanogenerator. There exists a layered double hydroxide (LDH) layer above the Al layer attached to the polyethylene terephthalate substrate. The thickness of the Zn:Al LDH layer is 20 nm. The Al layer on the bottom side serves as an electrode and a catalytic layer to formate ZnO nanosheets through the assembled LDH layer. As demonstrated in Figure 4.9h, the layer structure of solar cell consists of an Ag layer of 80 nm as an anode, MoO_3 of 20 nm as an electrode blocking layer, P3HT: PC60BM of 200 nm as an active layer, ZnO of 50 nm as an electron transport layer, and ITO of 200 nm as the cathode [48]. In Figure 4.9i, the cross-sectional SEM image shows multiple layers of the hybrid cell [49]. The introduced materials constitute the assemblies of hybridized generators to scavenge solar energy and mechanical energy.

4.2.2 Principles

A typical hybrid cell has been designed by integrating solar cells and piezoelectric nanogenerators. As demonstrated in Figure 4.10a, the Pt electrode on the Si substrate acting as the positive electrode, the ITO conductive glass as the negative electrode, and the aligned ZnO NWs coated with dye molecules in the electrolyte fit together to be a whole entity. For the piezoelectric nanogenerator unit, the

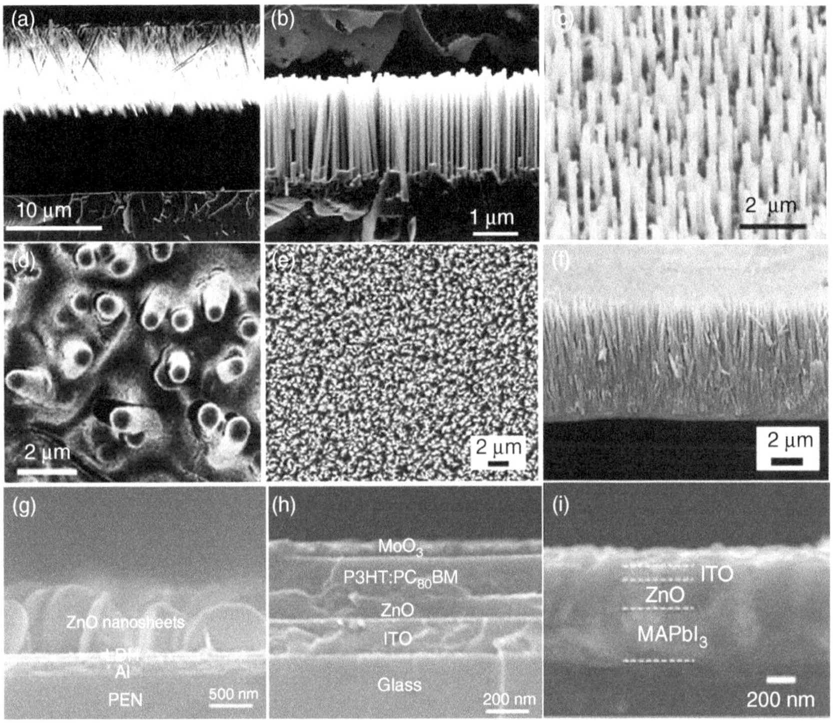

Figure 4.9 The SEM images of the materials for the hybridized solar and piezoelectric nanogenerator. (a,b) SEM image of the SC unit and the NG unit. Source: Reproduced with permission from Xu et al. [43]. Copyright 2009, American Chemical Society. (c) SEM image of the ZnO nanowire array grown for fabricating the NG. (d) Top view SEM image of the solar cell. Source: Reproduced with permission from Xu and Wang [45]. Copyright 2011, John Wiley and Sons. (e) SEM image of the ZnO nanowire array. (f) Photograph of the bent ZnO nanowire array grown on an ITO/PET substrate. Source: Reproduced with permission from Yang et al. [47]. Copyright 2013, American Chemical Society. (g) Cross-sectional SEM image of ZnO nanosheets grown on an Al film. (h) Cross-sectional SEM image of the inverted structure. Source: Reproduced with permission from Yoon et al. [48]. Copyright 2015, Elsevier. (i) SEM image of the fabricated device on glass substrate. Source: Reproduced with permission from Jella et al. [49]. Copyright 2018, Elsevier.

aligned ZnO NWs on the Pt coating substrate acting as anode and the Pt coating etching zigzag silicone substrate acting as cathodes can form a whole device. These two parts can be seen parallel as they share one anode and are connected to the cathode by wires. The schematic diagrams in Figure 4.10b demonstrate that the energy band theory was used to explain the device principles. The red dashed square on the right part of Figure 4.10b shows that the output voltage of an individual piezoelectric nanogenerator was determined by the transient change in ZnO NWs Fermi level ($E_{F,\text{ZnO-NG}}$) and Pt electrode ($E_{F,\text{Pt-NG}}$) while the Fermi level was raised as the electrons were driven in due to the piezoelectric potential from two electrodes [19]. The band energy diagram of a nanogenerator–solar cell parallel device is illustrated in the green square in Figure 4.10b. The extra electrons moved from the integrated Pt electrode side to the ITO electrode side because the Fermi level of the solar cell unit is higher. As there is a Schottky barrier on the boundary of the Pt electrode and ZnO NWs, the extra electrons cannot cross the nanogenerator. The electrons show the tendency to accumulate on the Schottky barrier side, which can raise the Fermi level ($E'_{F,\,\mathrm{ZnO\text{-}NG}}$). Then, the $E'_{F,\,\mathrm{ZnO\text{-}NG}}$ keeps consistent with the $E_{F,\,\mathrm{Pt\text{-}NG}}$ [50]. The Fermi levels are consistent as the anodes are connected in parallel configuration. Above all, the output voltage can be shown by the equation: $V_{\mathrm{p\text{-}HC}} = V_{\mathrm{SC}} = V'_{\mathrm{NG}} = E'_{F,\,\mathrm{Pt\text{-}NG}}\text{-}E_{F,\,\mathrm{ZnO\text{-}NG}}$. When the sunlight illuminates the solar cell unit, the photoexcited electrons are induced to a higher level. Then the induced electrons move to the ZnO conductive band and flow toward the cathode side. Therefore, hybrid device integrated piezoelectric nanogenerator and dye-sensitized solar cell can effectively enhance the output voltage of an individual NG. This provides an innovative design approach to drive an integrated structure to scavenge hybrid energy such as solar energy and mechanical energy in the surroundings. The structure of a compact integrated device in Figure 4.10c has been updated and improved. The ITO layer on the top of the solar cell was utilized as the cathode and the Ag pasted on the GaN substrate was utilized as the anode [51]. These two ZnO nanowire array layers were distributed as a relative configuration to work as two different functions. The electron energy band diagram in Figure 4.10d can explain the working mechanisms of hybrid behaviors. The diagram indicates that the piezoelectric and photovoltaic potentials can drive the electron to move. The voltage signals can be achieved by the Fermi-level difference between ZnO NWs in SC ($E_{F,\,\mathrm{ZnO\text{-}SC}}$) and ZnO NWs in NG ($E_{F,\,\mathrm{ZnO\text{-}NG}}$) [43]. The piezoelectric potential on the Schottky contact can drive electrons to flow through the Au–ZnO junction. Then, the charge carriers continue to move through the solid-state electrolyte to the solar cell side [52]. The available maximum photovoltaic signals are determined by the electrochemical potential in the ZnO Fermi level and the electrolyte. The output voltage curves in Figure 4.10e illustrate that the output signals can be enhanced significantly by hybridization. The energy band diagram in Figure 4.10f–h can be used to explain the phenomenon. When the SC unit is connected with an NG unit, a Schottky barrier height (Φ_1) is created on the ZnO–Au interface [53]. When light is applied on the hybrid cell, voltage signals are generated by the Fermi levels of Au electrode (E_F) changing ($\Delta\Phi$). The photoinduced electrons through the SC unit cannot move into the NG part as a Schottky barrier is produced, which increases the resistance of NG and decreases the voltage in the circuit. When mechanical pressure is applied on the top of ZnO nanosheets, the generated negative charges at the ZnO nanosheet and Zn:Al LDH interface accumulate, while the positive charges are gathered at the Au side, resulting in band change of ZnO nanosheets. Consequently, when applying pressure on the device, the V_{HC} increases owing to the flow of photoexcited electrons through the piezoelectric nanogenerator significantly.

The performance of different varieties of hybrid cells is demonstrated in the diagrams in Figure 4.11. When the sunlight hits the hybrid cell without mechanical ultrasonic wave, as depicted by the blue line in Figure 4.11a, the generated photovoltaic U_{OC} and J_{SC} are 0.591 V and 6.9 μA/cm^2, respectively. As depicted by the red line in Figure 4.11a, when sunlight and ultrasonic wave are applied on the cell simultaneously, the corresponding voltage can reach 0.60 V while the J_{SC} remains at 6.9 μA\cm^2. The U_{OC} of the s-HC is improved

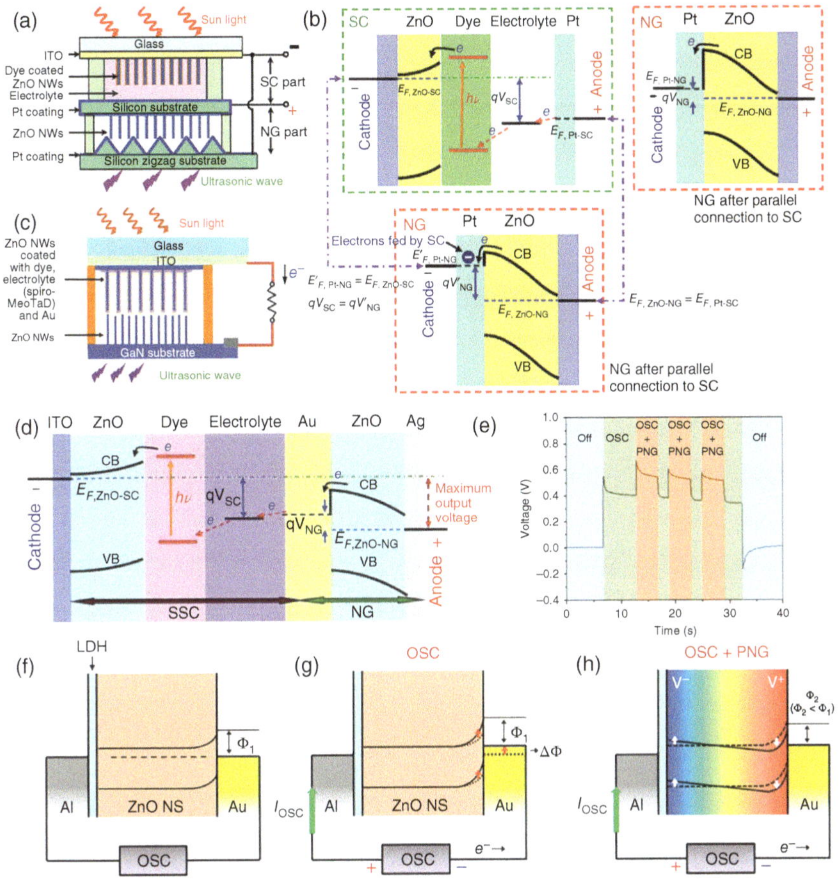

Figure 4.10 The schematic diagrams of the mechanisms for the hybrid nanogenerators. (a,b) Electron energy band diagram of the hybrid cell and its working principles. Source: Reproduced with permission from Xu et al. [43]. Copyright 2009, American Chemical Society. (c,d) Schematic structure and energy band diagram of an integrated hybrid solar cell. Source: Reproduced with permission from Xu et al. [45]. Copyright 2011, John Wiley and Sons. (e) The enhanced output voltage signal of a hybrid solar cell and a piezoelectric nanogenerator. Source: Reproduced with permission from Yang et al. [47]. Copyright 2013, American Chemical Society. (f–h) The energy band diagrams to explain the mechanisms of hybrid solar cell and piezoelectric nanogenerator. Source: Reproduced with permission from Yoon et al. [48]. Copyright 2015, Elsevier.

by 9 mV by the piezoelectric effect produced by the ultrasonic wave, which can be seen clearly. The voltage from the nanogenerator unit shown in Figure 4.11b when there is no light confirms that the U_{OC} enhancement comes from the NG part and the J–V curve of the anode and cathode of SC unit without the NG unit during the measurement. As depicted by the blue curve in Figure 4.11b, when there is only sunlight without piezoelectric source, the U_{OC} and J_{SC} exhibited by the hybrid cell are 0.415 V and 252 μA/cm^2 respectively. When the ultrasound wave and light are applied on the hybrid device simultaneously, the U_{OC} is 0.433 V and the J_{SC} remains at 252 μA/cm^2 as shown by the red line in Figure 4.11b. The HC unit voltage value had a 19 mV enhancement after applying

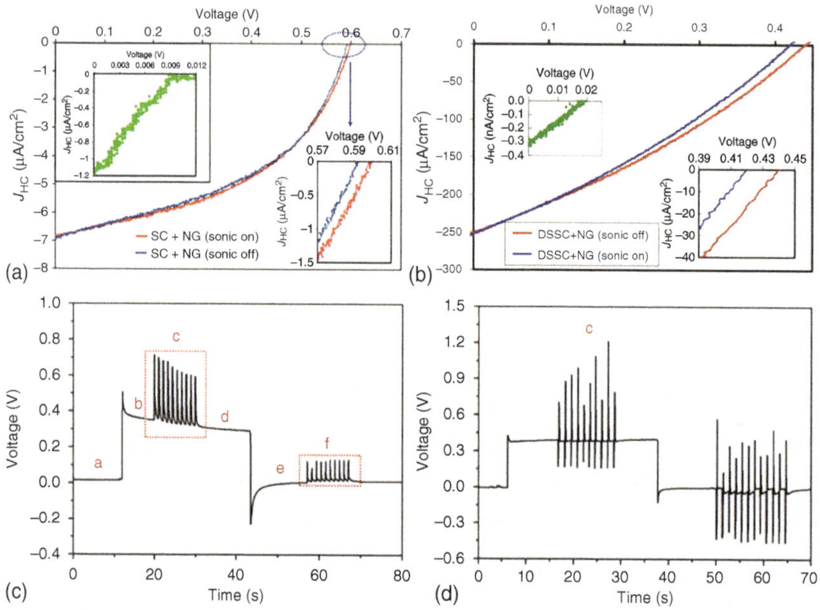

Figure 4.11 Performances of the hybrid nanogenerators. (a) A comparison of the *J–V* characteristics of a hybrid cell when illuminated by simulated sunlight with (red curve) and without (blue curve) turning on the ultrasonic wave excitation.Source: Reproduced with permission from Xu et al. [43]. Copyright 2009, American Chemical Society. (b) The *J–V* characteristics of a hybrid cell with another structure. Source: Reproduced with permission from Xu et al. [45]. Copyright 2011, John Wiley and Sons. (c) V_{HC} of the hybrid cell measured under controlled application of applied light and mechanical compressive force. (d) V_{HC} of the hybrid cell measured under vertical compressive force. Source: Reproduced with permission from Yoon et al. [48]. Copyright 2015, Elsevier.

the ultrasonic wave on it as illustrated in the inset plot of Figure 4.11b. Better than the hybrid cell introduced above, this innovative device is a solid-state device that is integrated to avoid solvent leakage and evaporation. Figure 4.11c shows the measured V_{HC} from a hybrid cell under applied light illumination and mechanical force. From the typical signal curve, there is no V_{HC} value from HC observed when the device is measured without light illumination and application of mechanical force (labeled "a"). When the light illumination is on, the photovoltaic signal V_{HC} increases to 0.4 V (labeled "b"). Then the external mechanical pressure is applied to the whole device at an interval of 1.0 seconds periodically, and V_{HC} value increases to 0.71 V (labeled "c"). Once the light is off and the mechanical pressure is removed, V_{HC} recovers to 0 V (labeled "e") as in the initial state. Besides, when applying pressure without light illumination, the V_{HC} peaks induced by piezoelectric effect can also be measured, as the device can work as a piezoelectric nanogenerator independently as well as the solar cell part (labeled "f"). In Figure 4.11d, the DC type signal is induced by photovoltaic effect under light illumination while the piezoelectric AC signal is induced by piezoelectric effect on applying compressive strain. Hence, solar energy and mechanical energy can be utilized by hybridized devices simultaneously.

4.2.3 Applications and Performances

In Figure 4.12a, the designed structure diagram of a typical hybridized energy device is illustrated. The materials fabricated for the device are flexible, which consists of two parts. The ITO film and Ag as both electrodes and ZnO NWs grown on P3HT film form the solar cell unit. The bottom piezoelectric nanogenerator part consists of a simple PVDF film and Ag electrodes. As shown in Figure 4.12b, the flexible ITO/PET substrate with ZnO nanowire arrays grown by solution-based growth technique can be bent, indicating a potential application as a wearable device. As shown by the digital photograph demonstrated in Figure 4.12c, the black area of the integrated compact structure was coated with P3HT material. When the nanogenerator was touched by hand, a compressive strain was created on the nanogenerator, which can produce piezoelectric signals. The piezoelectric nanogenerators can directly drive a liquid crystal display with the numbers "66666" as depicted in Figure 4.12d. The hybrid electric signals induced by nanogenerators and solar cells in the circuit was connected can charge a Li-ion battery to 1.5 V. Then, the charged battery was utilized to power four red-light LEDs in series as illustrated in Figure 4.12e.

Therefore, all these hybrid energy cells combining photovoltaic devices and piezoelectric devices can not only harvest solar or mechanical energy individually but also scavenge both energy forms together in one circuit. Therefore, the hybridized nanogenerators can harvest multiple energy forms whenever and wherever only one energy type exists or both energy types coexist in our environment.

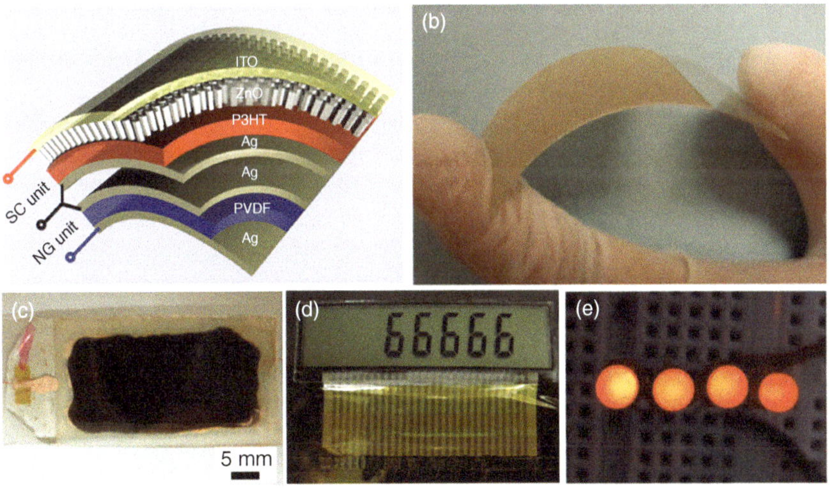

Figure 4.12 The applications of the hybrid nanogenerators. (a) Schematic diagram of the fabricated hybrid energy cell. (b) Photograph of the bent ZnO nanowire array grown on an ITO/PET substrate. (c) Photograph of the fabricated ZnO nanowire array-P3HT film hybrid cell. (d) Optical image of an LCD driven by the hybrid thermal and mechanical energies, which was induced by hand touch. (e) Optical image of four red LEDs driven by the charged Li-ion battery. Source: Reproduced with permission from Yang et al. [47]. Copyright 2013, American Chemical Society.

4.3 Hybridized Photoelectric and Triboelectric Nanogenerator

As solar energy and wind energy can easily coexist in nature, these two kinds of energy sources together with the working devices can maximize the use and conversion of natural energy. Hybrid devices based on photovoltaic effect and triboelectric effect are discussed in this section.

4.3.1 Materials and Fabrication

Ya Yang et al. first fabricated a typically hybridized energy nanogenerator, which consisted of two energy units [54], as shown in Figure 4.13d. The top PDMS layer was designed for collecting triboelectricity induced by electrostatic electrification. The PDMS nanowire arrays in Figure 4.13a were designed by anodized aluminum oxide template. The diameters and the depths of PDMS NWs on the template were around 200 and 500 nm, separately. The photoelectric component was on the bottom of the device and the main part was a micro-pyramidal Si-based solar cell. The structure included Al film and ITO/SiN film as two electrodes, a p^+ back surface field part, an n^+ emitter part, and a p-type Si layer [55]. The SEM photograph in Figure 4.13b illustrates the shape of the pyramid of the Si layer whose size was around 8 μm. The flexible transparent PDMS layer of the triboelectric nanogenerator (TENG) was designed for producing triboelectricity and acting as a protection layer of the solar energy unit. The ITO layer on the pyramidal Si works as the common electrode for the photoelectric structure and triboelectric structure. The Al film with dimensions of about 200 nm in Figure 4.13c was etched to obtain nanostructures that increased the effective surface to work [56, 57]. The structure schematic of typical devices is demonstrated in Figure 4.13e. The silicon-based solar cell units were fabricated to harvest solar energy in the environment while the TENG part, which has a simple structure of the PTFE and Al films on the acrylic tubes, was utilized for harvesting the triboelectricity. The wind enables the rotation of the outer acrylic tube to induce triboelectric charges due to the contact and separation of the charge induced layer. Therefore, the generated electrons move from the Al side to the ground side. In order to scavenge solar energy and wind energy in our living surroundings, an integrated hybrid nanogenerator was designed to be installed on top of city constructions, which comprised of a triboelectric nanogenerator and solar cells as shown in Figure 4.13f [58]. The schematic of the major components and a photograph are presented in Figure 4.13g. As the schematic shows, the structure of the TENG is simple. The Kapton film and an FEP film were fabricated as the triboelectric layer. A narrow gap was created for the air to flow between the Cu electrodes. Besides, the generated electricity was stored in a Li-ion battery. As shown in Figure 4.13h, the compact TiO_2 nanotubes on metallic titanium foil were analyzed by anodization [59].

4.3.2 Principles and Performances

As depicted in Figure 4.14a, the photovoltaic signal generated by a solar energy device is around 0.6 V. The corresponding triboelectric pulse of the fabricated

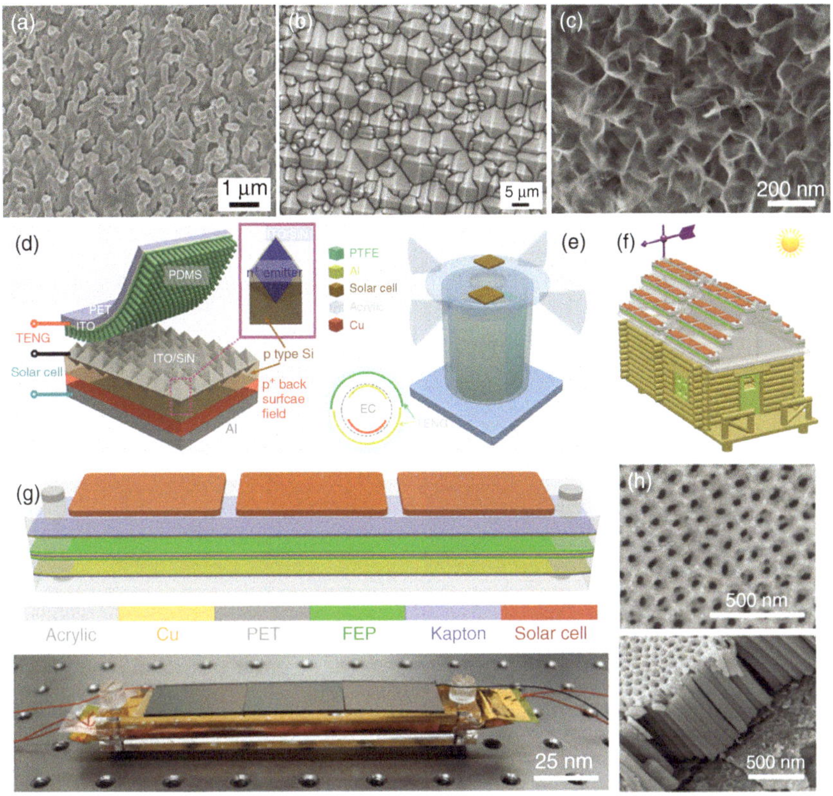

Figure 4.13 The SEM images and structure designs for hybridizing solar energy and triboelectricity. (a) SEM image of PDMS nanowire array. (b) SEM image of fabricated Si pyramids surface. Source of a, b, d: Reproduced with permission from Yang et al. [54]. Copyright 2013, American Chemical Society. (c) SEM image of Al film surface with nanostructures. Source of c, e: Reproduced with permission from Wu et al. [56]. Copyright 2014, Springer. (d) Schematic diagram of the fabricated hybrid energy cell. (e) Schematic diagram of the hybrid energy cell. The inset shows the cross-sectional schematic diagram of the device. (f) Schematic diagram of the integrated hybridized nanogenerators on the roof of a house model. (g) Schematic diagram of the fabricated hybridized nanogenerator and photograph of the fabricated hybridized nanogenerator. (h) SEM image of the as-grown TiO_2 nanotubes and cross-sectional SEM image of the TiO_2 nanotubes. Source: Reproduced with permission from Wang et al. [58]. Copyright 2016, American Chemical Society.

TENG by controlling the contact frequency between the charge sensitive layers, which consist of the ITO layer and the PDMS layer, is up to 3 V as depicted in Figure 4.14b. The electrostatic charges generated by the triboelectric effect drive electrons flowing through the circuit connected by wires in the separated charge layers [60]. The induced surface charges are transferred from the ITO electrode to the PDMS layer once the two parts make contact as the different electrons gain the ability of the two materials, which results in the observed triboelectric signals [61, 62]. The pyramidal structures on the electrode increase the friction between the triboelectric layers on the Si surface. The V_{OC} is calculated by

$$V_{OC} = \frac{\sigma_0 d_S}{\varepsilon_0} - \frac{\Delta\sigma}{\varepsilon_0}\left(d_S + \frac{d_{PDMS}}{\varepsilon_{r(PDMS)}}\right)$$

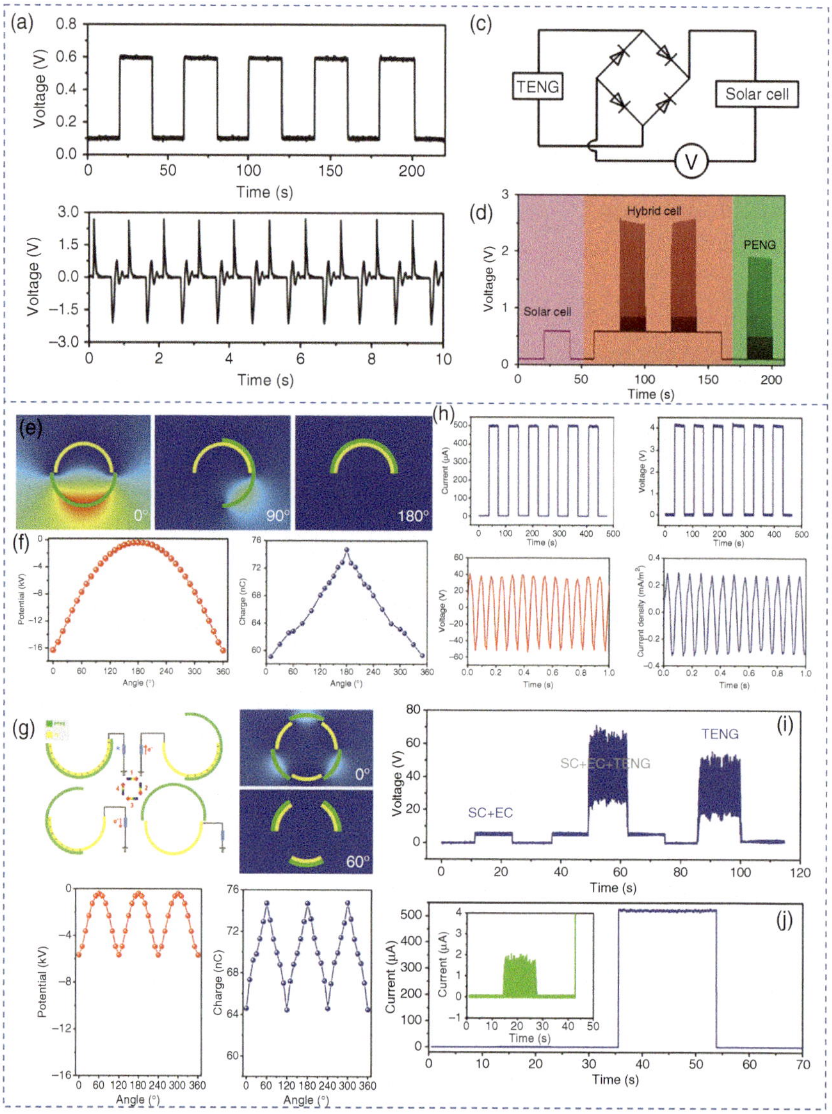

Figure 4.14 The output performances of hybrid solar energy and triboelectricity nanogenerators. (a,b) Output voltage of the fabricated Si solar cell and the fabricated TENG. Source: Reproduced with permission from Yang et al. [54]. Copyright 2013, American Chemical Society. (c) Schematic diagram of the connection between the Si solar cell and TENG. (d) Output voltage of the hybrid solar cell and TENG. (e) Electric potential distribution in the TENG with one strip unit at the rotation angles of 0°, 90°, and 180°. (f) Average electric potential on the PTFE film under the different rotation angles and the total charge amount on the Al film of the TENG. (g) Sketches that illustrate the electricity generation process in a full cycle and electric potential distribution in the TENG with three strip units, and the changes in the electric potential on the PTFE film and the charge amount on the Al film with increasing rotation angles. (h) Measured output current and voltage of the solar cell units of the electrical measurement system and measured output voltage and current density of the TENG with two strip units. (i) Output voltage of the SC + EC, TENG (after rectification), and the hybrid energy cell (SC + EC and TENG in series). (j) Output current of the hybrid energy cell. Source: Reproduced with permission from Wu et al. [56]. Copyright 2014, Springer Nature.

where σ_0 represents the charge density induced by triboelectricity, d_S represents the distance, $\Delta\sigma$ represents the transferred charges, ε_0 represents the vacuum permittivity, d_{PDMS} represents the PDMS film thickness, and $\varepsilon_{r(PDMS)}$ represents the PDMS relative permittivity. Based on the above equation, increasing the d_S can enhance the triboelectric V_{OC}. The connection method to integrate the solar cell and the rectified triboelectric nanogenerator is demonstrated in Figure 4.14c. By the series connection, the signals can always be produced no matter when and where the solar energy or mechanical energy exists. The measured hybrid signals of the device are shown in Figure 4.14d. From the three types of signals, it can be proved that the hybrid cell can work as an individual solar cell, an individual triboelectric nanogenerator, and the hybrid energy system simultaneously. Besides, the hybrid signals are enhanced significantly by the combination of photovoltaic and rectified triboelectric signals. COMSOL is a professional simulation tool that can verify the charges, behaviors, and potential distribution through finite element simulation. Figure 4.14e displays the potential distributions by COMSOL simulation when the corresponding angles of rotation are 0°, 90°, and 180° respectively. When the PTFE film and Al film are completely staggered, the potential of the PTFE film can reach the largest value of about −16 kV. When the two films are fully matched, the potential value in the PTFE film is the smallest. As shown in Figure 4.14f, the potential and charges demonstrate the changing tendency with changing rotation angles. Figure 4.14g illustrates the working process and mechanism of the TENG, which involves the whole circulation of film behaviors. The output and potential indicate that the amount of charges on the Al film depends on the strip numbers. The induced charges are decreased with increasing strip numbers [63]. Figure 4.14h displays the corresponding output signals of eight solar cells in series connection and voltage outputs and current density of the triboelectric unit, respectively. The three types of output signals are presented in Figure 4.14i, which are the signals of SC + EC units, the rectified TENG + SC + EC hybrid unit, and individual TENG, respectively. These signals prove that the hybridized energy systems can utilize both one individual energy source and the hybrid energy forms. The corresponding performances of the TENG + SC + EC hybrid units are depicted in Figure 4.14j while the smaller individual current signals generated by the TENG are illustrated in the inset graph. Therefore, the hybrid energy cells can output the enhanced hybrid signals by combining individual energy units.

4.3.3 Applications

The voltage of an individual triboelectric nanogenerator after using a transformer was about 5 V and the output voltage of an individual solar cell unit was 7 V as shown in Figure 4.15a,b. When the solar cell and TENG were combined as illustrated in Figure 4.15c, the hybridized signals reached 15 mA, which is larger than that with individual solar cells and TENG. The voltage–time curves of the metal lithium half-battery charged by hybrid energy units are presented in Figure 4.15d,e. The measured TENG power performances when the device was under the wind flow in different directions of 0–40° are shown in Figure 4.15f.

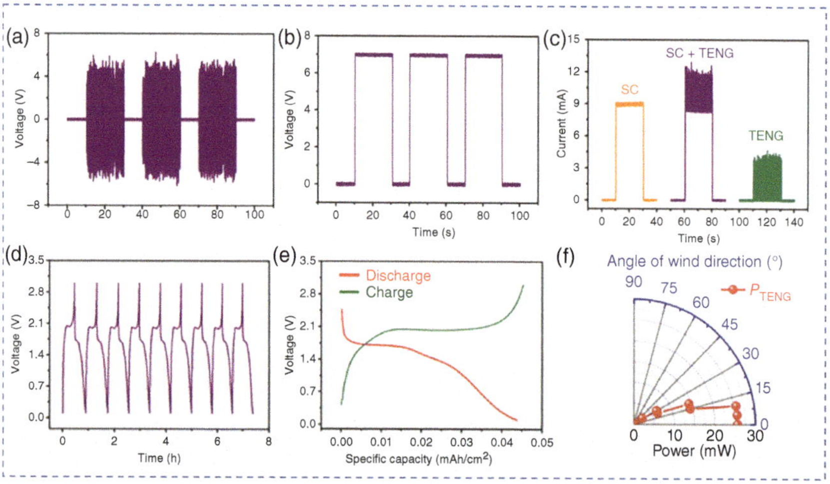

Figure 4.15 The performances of the hybrid nanogenerators and applications. (a) Measured output voltage signals of the TENG after using a transformer. (b) Measured output signals of the SC under a full-sun intensity. (c) Measured output current signals of the TENG, the SC, and the hybridized nanogenerator. (d) Charge and discharge voltage–time curves of the fabricated Li-ion battery. (e) The first charge–discharge voltage profiles of the Li-ion battery. (f) Output power of the TENG depends on the different angles between the wind input and the direction of the device length, ranging from 0° to 40°. Source: Reproduced with permission from Wang et al. [58]. Copyright 2016, American Chemical Society.

From the fan diagram of wind angles and measured powers, it is seen that the measured output powers decreased as the angles increased in the range of 0–40°.

In order to make the utmost use of the hybridized nanogenerators, a variety of applications were explored. Figure 4.16a demonstrates the absorption spectra from 400 nm to 700 nm and the photograph of the RhB solution after electrodegradation. As the electrodegradation time increases, the RhB concentration evidently decreases from 100% to 2%, which indicates that the electricity induced by the hybrid nanogenerator can degrade organic pollutants. Besides, the observed green LED light and the blue LED lights in Figure 4.16b,h indicate that the output voltage/current of fabricated energy cells can attain the operation voltage and operation current of the LEDs. A normal Li-ion battery was charged from 1.54 to 3.60 V and from 1.1 to 3.0 V then discharged by the hybrid energy cell in Figure 4.16c,g. Then a red LED light was powered by the charged battery in Figure 4.16d, which can be extensively utilized for signal systems. Figure 4.16e shows that the hybrid energy units charged the Li-ion battery to enhance the voltage from 2.60 to 3.50 V in a commercial cell phone, and the cell phone is ready for use. From Figure 4.16f, the hybridized nanogenerator has the best charging performances. As shown in Figure 4.16i, a temperature sensor and an anemometer can be driven by the charged cells. From all the explorations above, it is significant that the hybridized devices scavenging the multiple energy forms from the environment have the ability to power the lights and charge the Li-ion battery in our daily life. Figure 4.16j displays the cycling performance of the Li-ion cells, which has good cycle stability, and the hybrid energy nanogenerators can get a higher

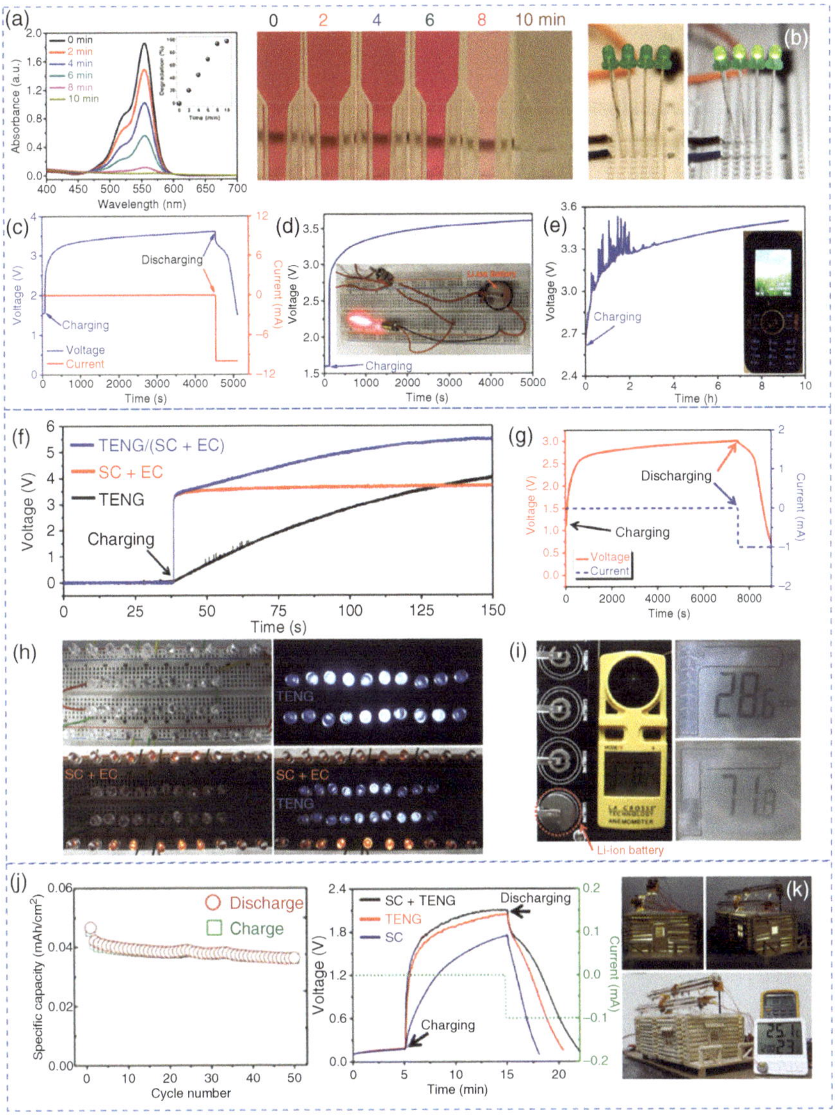

Figure 4.16 The typical applications of the hybrid nanogenerators. (a) Absorption spectra and photographs of the RhB solution under the same time intervals. (b) Optical images of four green LEDs before and after they were driven by the hybrid energy cell. (c) The hybrid energy cell-charging and the subsequent constant-current discharging curves of a Li-ion battery. (d) The charging curve of a Li-ion battery and an optical image of a red laser diode. (e) The charging curve of the Li-ion battery of a commercial cell phone. Source: Reproduced with permission from Yang et al. [54]. Copyright 2013, American Chemical Society. (f) The measured voltage of a 10 μF capacitor charged by the TENG, SC + EC, and the hybrid energy cell. (g) Voltage curve showing the entire cycle of a Li-ion battery charged by the hybrid energy cell, with the subsequent constant current discharge at 1 mA. (h) Optical images of 40 LEDs. (i) Optical image of the wind speed and temperature sensors powered by the charged Li-ion battery. Source: Reproduced with permission from Wu et al. [56]. Copyright 2014, Springer Nature. (j) Cycling performance of the Li-ion battery and the measured charging and discharging curves of the Li-ion battery by using the TENG and the hybridized nanogenerator (TENG and SC), respectively. (k) Photographs of the smart city powered by the hybrid cells. Source: Reproduced with permission from Wang et al. [58]. Copyright 2016, American Chemical Society.

charging voltage than that of individual solar cells or triboelectric nanogenerators under the same condition. As displayed in Figure 4.16j, this kind of appliance installed on the top of a smart house model can scavenge the hybrid energies in nature for powering electrical appliances, which shows a broad application in the future.

4.4 Hybridized Photoelectric and Pyroelectric Nanogenerator

In the process of energy conversion by photovoltaic devices, the induced thermal energy due to temperature change is often regarded as a by-product instead of the available electrical energy. To raise the conversion efficiency from solar energy and thermal energy, researchers committed themselves to developing innovative hybrid energy systems to utilize solar energy and other additional energy forms, especially heat energy.

4.4.1 Materials and Structure Design

Figure 4.17a,b illustrate the ZnO nanowires synthesized by hydrothermal process with diameters of 80–150 nm [64, 65]. Figure 4.17c confirms the cross-sectional device architecture observed by the SEM image. The thickness of ZnO seeds is 100 nm, which is followed by the ZnO nanowires array layer of about 3 μm thickness; then a poly(3-hexylthiophene) (P3HT) and PEDOT:PSS thin film as a photoelectric layer and Ag nanowire electrode are on the top layer. The thickness of Ag nanowires was 100 nm as shown in Figure 4.17d. From the structure surface photograph in Figure 4.17e, the compact Al nanolayer formed on the ZnO nanorods surface shows good electrical conductivity as the electrode. Figure 4.17f confirms the corresponding thickness of about 4 μm of a fabricated structure. In Figure 4.17g, the SEM picture shows the layer structure of InP/ZnO solar cells. The InP wafer with a thickness of about 350 μm can maintain a much larger temperature gradient compared to the traditional InP/ZnO junction [66]. The length of the observed ZnO nanorods layer was 700 nm, grown on a compact InP surface as in Figure 4.17h. On the ZnO surface, the diameter of the Ag nanowires present across the surface for absorbing and trapping light is 70 nm, from Figure 4.17i,j.

4.4.2 Principles and Applications

The structure of the hybridized photovoltaic system demonstrated in Figure 4.18a comprised of an ITO layer as the top electrode, ZnO NWs, P3HT film, PEDOT:PSS film, and an Ag nanowire as the bottom electrode. The optical photograph in Figure 4.18b shows that the active areas on the substrate of the device are about 1 cm × 1 cm. Figure 4.18c depicts the configuration of the InP/ZnO device based on thermo-phototronics. The integrated structure is composed of a Ag film as the bottom electrode, Zn-doped InP wafer as p-type doping, and a layer of ZnO NRs and Ag NWs as the top electrode. As the schematic diagram

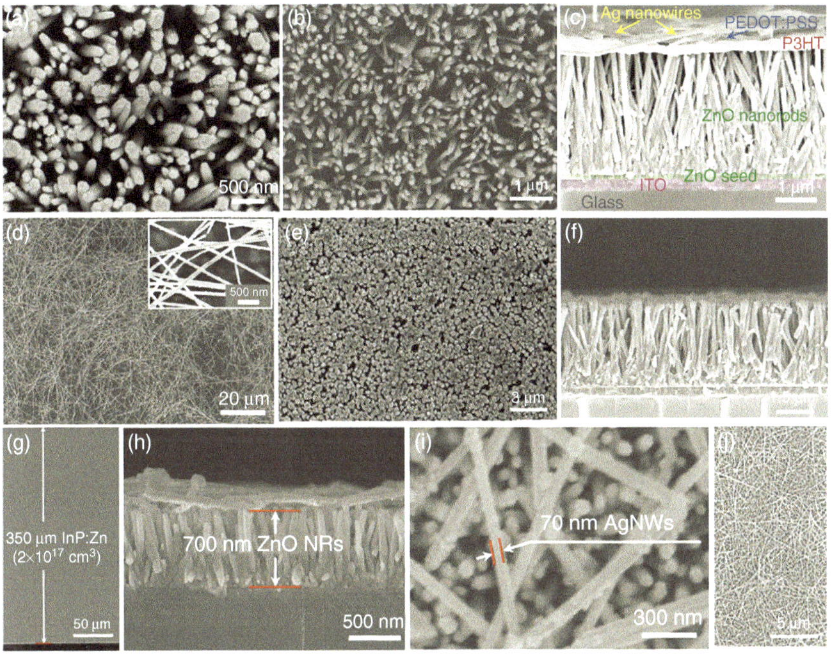

Figure 4.17 The SEM images of the typical materials for the hybrid solar energy and heat energy. (a,b) Top-view SEM image of as-grown ZnO nanowire array. Source of a, c, d: Reproduced with permission from Zhang et al. [64]. Copyright 2016, American Chemical Society. Source of b, e: Reproduced with permission from Ouyang et al. [65]. Copyright 2017, John Wiley and Sons. (c) Cross-sectional SEM view of a planar solar cell exhibiting well-separated layers. (d) Top-view SEM image of a solar cell showing cross-linked Ag nanowire networks. Inset shows a high-magnification SEM image of the Ag nanowires. (e) Top-view SEM image of the device. (f) Vertical SEM view of the as-grown ZnO nanowire array on ITO glass. (g,h) Cross-sectional SEM image of a fabricated solar cell presenting the nature of each layer. (i,j) Top-view SEM image of a fabricated solar cell showing cross-linked Ag NWs on the ZnO NRs. Source: Reproduced with permission from Zhang and Yang [66]. Copyright 2017, John Wiley and Sons.

in Figure 4.18d demonstrates, the P3HT/ZnO heterojunction photodetector structure consists of a transparent ITO film and an Al film as both electrodes, a ZnO NWs layer with P3HT, and a thin PEDOT:PSS film. Figure 4.18e shows an optical image of a fabricated multichannel device with 4 × 4 pixels. The schematic diagram of the 16 channels structure in Figure 4.18f demonstrates the structure of the device and the temperature gradient when the hybridized signals are measured.

As presented in Figure 4.19a, the schematic demonstration of the energy band in the fabricated structure relies on the energy levels reported previously [67, 68]. When the light illuminates the device, the P3HT as donor absorbs photons from light and can generate the photoexcited carriers to move to the boundary of the donor and the acceptor. The charge transfer can disassociate these excitons and cause the electron–hole pairs to be transported to different electrode sides. The directional movement of the electrons and holes can

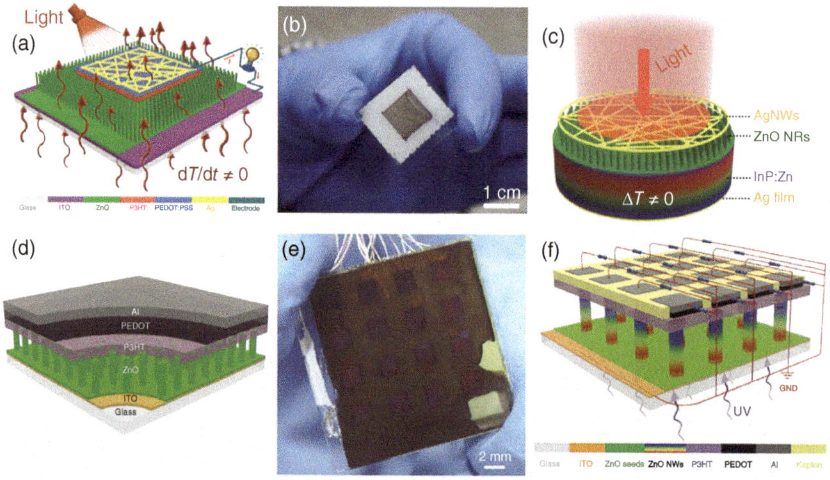

Figure 4.18 The structure designs of the hybrid devices. (a) Illustration of a P3HT/ZnO nanowire array solar cell demonstrating the pyro-phototronic effect. (b) Optical image of a fabricated device. Source: Reproduced with permission from Zhang et al. [64]. Copyright 2016, American Chemical Society. (c) Illustration of the solar cell configuration based on thermo-phototronics. Source: Reproduced with permission from Zhang et al. [66]. Copyright 2017, John Wiley and Sons. (d) Schematic of the P3HT/ZnO heterojunction photodetector. (e) Photograph of a fabricated 4 × 4 photodetector array with each pixel being 2 mm × 2 mm. (f) Schematic demonstration of the device configuration. Source: Reproduced with permission from Ouyang et al. [65]. Copyright 2017, John Wiley and Sons.

generate photovoltaic signals. The J–V curve was measured when sunlight illuminated the Ag electrode side and the ITO electrode side was attached to a thermoelectric cooler. The working mechanism of the InP/ZnO heterojunction is demonstrated by the diagram of energy band in Figure 4.19b. The corresponding electron affinities for InP and ZnO are 4.4 and 4.1 eV while the bandgaps of InP and ZnO are 1.35 and 3.2 eV separately. The InP Fermi level is 0.1 eV larger than the maximum value of the valence band [69]. The ZnO Fermi level acts the same as the InP Fermi level when the InP makes contact with the ZnO. When the light illuminates the InP device, there are photoexcited electron–holes generated. Then the photoinduced electrons swim to the Ag NWs top electrode, and at the same time the holes diffuse into the bottom Ag film electrode due to the inner field in the structure. It is worth noting that a small number of electrons are trapped in the InP heterojunction due to the combination of two main effects. One is that in the space-charge region there occurs a band-bending phenomenon, and the other is that there exists a 0.3 eV excursion of the conduction band. As seen in the diagram depicted in Figure 4.19c, the behaviors of band bending and electrons movement in the InP/ZnO hybrid device can explain the thermoelectric-enhanced performances when it is focused by light illumination. When the temperature of the InP side is reduced, a thermal potential (V_{TE}) is generated due to the temperature difference (ΔT) in the InP part and can be calculated as the values between the Fermi level difference on the two sides of the InP material. The existing V_{TE} has

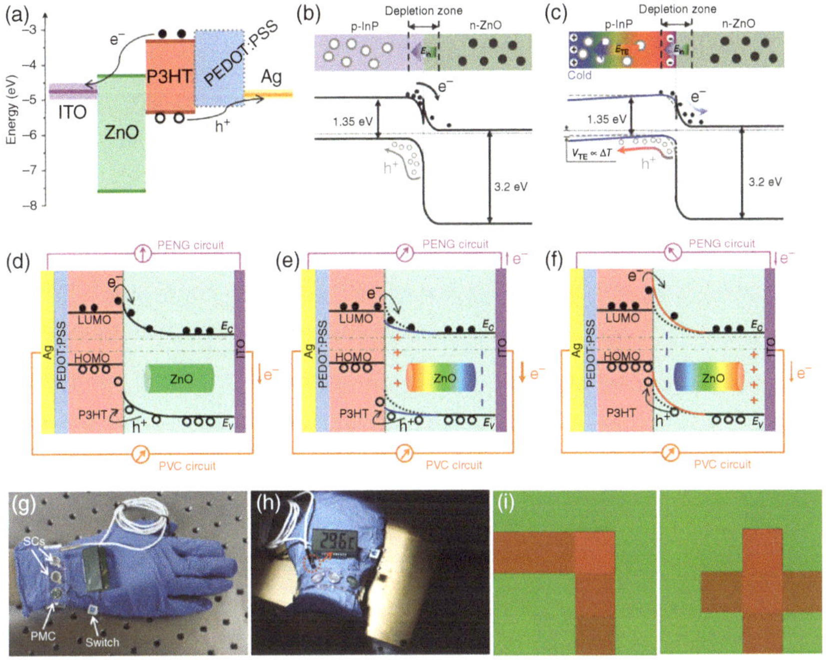

Figure 4.19 The principles and applications of hybrid devices. (a) Energy band diagram of the vertical heterostructure of ITO/ZnO/P3HT/PEDOT: PSS/Ag. Source of a, d, f: Reproduced with permission from Zhang et al. [64]. Copyright 2016, American Chemical Society. (b,c) Energy band diagram of the solar cell in the absence of and after applying temperature gradient. Source of b, c, h: Reproduced with permission from Zhang et al. [66]. Copyright 2017, John Wiley and Sons. (d–f) Schematic illustrations of the working mechanisms of the pyroelectric-enhanced solar cell. (g) Photograph of a self-powered temperature sensor. (h) Photograph of the working temperature sensor under solar simulator. (i) 7-shaped pixels and cross-shaped pixels are illuminated by UV light. Source: Reproduced with permission from Ouyang et al. [65]. Copyright 2017, John Wiley and Sons.

an accelerating effect on the photoexcited holes to swim from the InP part to the bottom side. Besides, the negative charges induced by the thermoelectricity accumulated at the boundary can increase the InP conduction band and InP valence band. The lifted valence and conduction bands can release the trapped electrons flow from InP to ZnO, which results in decreasing the recombination of the electron–holes. Hence, when the whole device is illuminated by the light and a cooling temperature is applied, the photoelectric signals can be enhanced due to the combined effects from the photons and thermoelectrics. As shown in Figure 4.19d, there is no electron flow in the entire structure due to the constant polarization when the temperature is not changing. Decreasing the temperature of ZnO results in the polarization localized positive charges across the ZnO nanowires at the P3HT/ZnO boundary in Figure 4.19e. The ZnO conduction band can be reduced by the positive charges generated on the P3HT/ZnO heterojunction interface, which results in a decrease in the barrier height. Thus, the performances of the photovoltaic device can be enhanced due to the lower barrier. Besides, the pyroelectric field induced by the temperature change applied in the device can redistribute charges in the ZnO nanowires, resulting in the movement of the electron from the ITO end to the Ag NWs end. As the electron flow directions are the same in the device, the measured signals induced by the combined principles above are increased by 18%. When the light illumination and the temperature are retained, the current signal reaches a stable plateau value. As the applied temperature is lower than the initial state, the barrier height of the P3HT/ZnO interface is changed. The photovoltaic performances of the hybrid device can be improved because band bending occurs in the P3HT/ZnO structure. Once the applied cooling operation is stopped, the temperature of the structure returns to the original state; this process can generate electron flow in the opposite direction due to the induced opposite pyroelectric field, as the current signals in the opposite direction show in Figure 4.19f. In order to utilize the InP/ZnO hybrid nanogenerator as a wearable device on the human body, the integrated device packaged on a PDMS watch band in Figure 4.19g was designed to be applied on human skin to sense the temperature. This packaged device consisted of two hybrid energy cells in series connection, a PMC chip, and a switch. When the light illuminates the energy band on the human wrist, the sensor driven by the hybrid power source can read the real-time temperature as shown in Figure 4.19h. A multichannel device can detect and scavenge the signals in different regions simultaneously without disturbances. When the 365 nm UV light illuminated the multichannel arrays in Figure 4.19i, the shapes of number-7 and cross-+ can appear as red squares that represent the voltage signals recorded by the multichannel measurement system. All these applications indicate that solar energy and heat energy can be combined, and show a prospective for utilization.

4.4.3 Performances

When the applied temperature change is depicted in Figure 4.20a, the corresponding current output signals induced in Figure 4.20b illustrate that the signals of the hybrid solar and heat energy cell are generated by a photovoltaic cell (PVC) and a pyroelectric nanogenerator (PENG). Based on the photovoltaic effect, the indoor light illumination induces free carriers to produce photocurrent from Ag to the ITO electrode (step i, Figure 4.20c). When temperature is applied to the device, the original photovoltaic signals are increased by 18% of the hybrid photoelectric and pyroelectric effects (step ii, Figure 4.20c). When the light continues and the temperature is unchanged, the signal reaches a plateau (step iii, Figure 4.20c) and the barrier height on the P3HT/ZnO boundary is changed as the temperature is lower. From the diagram in Figure 4.20d, the V_{OC} and J_{SC} of the individual solar cell unit of the integrated device are 380 mV and 0.86 mA/cm^2, respectively. The photoelectric signals increased significantly with increasing light density. Meanwhile, the voltage signals of the PENG decrease slightly as calculated in Figure 4.20e. The hybrid output signals are increased with increasing light density when periodic temperature variation is applied to the device. The enhanced output signals prove that the pyroelectric effect combined with the photovoltaic properties can improve the working properties of the P3HT/ZnO NWs hybrid cell. The temperature changes and the corresponding

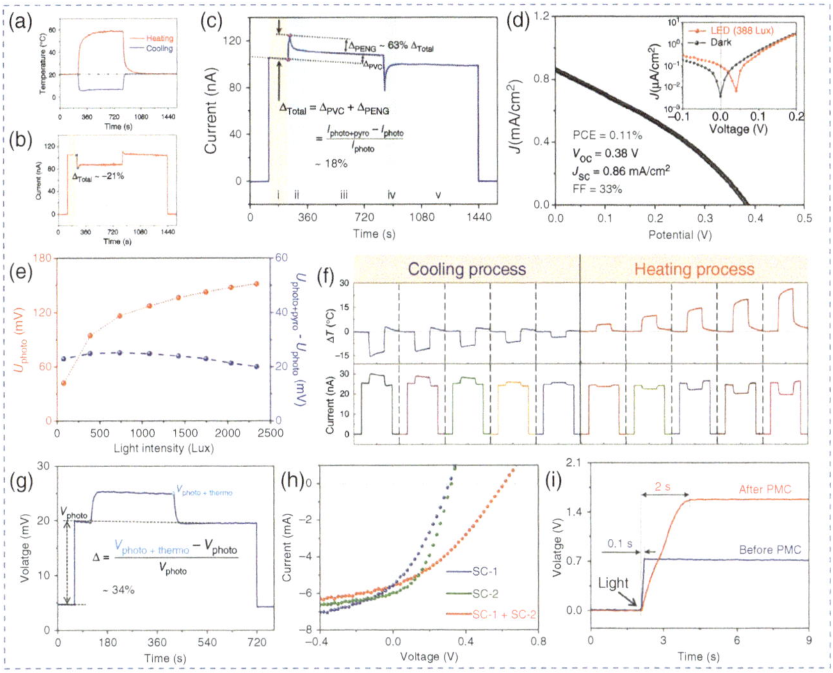

Figure 4.20 The output performances of the hybrid devices. (a) Periodic temperature changes in cooling and heating conditions. Short-circuit current curves of the pyroelectric-enhanced solar cell under cooling (b) and heating (c) conditions. (d) Representative *J–V* curve for the solar cell under 1-sun illumination. (e) The output voltage values of the PVC part and the PENG part of the device. Source: Reproduced with permission from Zhang et al. [64]. Copyright 2016, American Chemical Society. (f) Changes of temperatures and the corresponding *I–t* characteristics of the self-powered photodetector under 16.6 mW/cm^2 illumination. Source: Reproduced with permission from Ouyang et al. [65]. Copyright 2017, John Wiley and Sons. (g) Output current and voltage curves of the solar cell measured under indoor light illumination with fixed ΔT of 3.5 °C. (h) *I–V* characteristics of two InP/ZnO solar cells in series under 1 sun illumination. (i) The output performance of the two solar cells in series before and after using a power management circuit. Source: Reproduced with permission from Zhang et al. [66]. Copyright 2017, John Wiley and Sons.

current outputs under UV light illumination and different temperature variations are demonstrated in Figure 4.20f. As depicted in Figure 4.20g, it is significant that the voltage signals are increased by 34% by coupling the photovoltaic effect and pyroelectric effect for the observed which shows the voltage increment ratio is 34%. The two energy hybrid nanogenerators with almost the same output voltage values and current values show the stability and repeatability of the devices, which is crucial to the applications in Figure 4.20h. The current signals of two devices in series remain fixed and the voltage values are additive, which have no power loss. The problem is that these kinds of energy units have relatively lower voltage outputs, which has a negative influence on powering electrical appliances as an energy source. The designed PMC can regulate the performances of the power supplies by increasing the voltage and decreasing the current to match the working voltage of the electrical appliances as shown in Figure 4.20i. By using the PMC in the circuit, the output voltage signals can increase to a constant value, which provides the possibility to power most of the wearable electronics and shows a broader prospect to utilize the hybrid energy sources.

4.5 Conclusions and Prospects

After the above explanations, the fundamental mechanisms and structure designs of three typical hybridized energy systems based on solar energy combined with mechanical energy, wind energy, and heat energy are clearly introduced. The ground-breaking and innovative ideas provide the possibility to solve the energy crisis for sustainable development. Besides, the designed electrochemical devices based on the redox reaction, which can utilize ocean energy, also have the potentials to combine with other energy forms in future.

Despite these advantages that we can see, there are still several challenges and issues that need to be solved urgently. The performances of hybrid nanogenerators should be better and more stable. More hybridized nanogenerators with excellent performances will play a huge role in driving electric appliances to realize the maximum of energy utilization.

References

1 Dresselhaus, M.S. and Thomas, I.L. (2001). Alternative energy technologies. *Nature* 414: 332.

2 Xu, C., Pan, C., Liu, Y., and Wang, Z.L. (2012). Hybrid cells for simultaneously harvesting multi-type energies for self-powered micro/nanosystems. *Nano Energy* 1: 259.

3 Gräzel, M. and Hagfeldt, A. (2000). Molecular photovoltaics. *Acc. Chem. Res.* 33: 269.

4 Nazeeruddin, M.K., Kay, A., Rodicio, I. et al. (1993). Conversion of light to electricity by cis-X2bis(2,2′-bipyridyl-4,4′-dicarboxylate)ruthenium(II) charge-transfer sensitizers (X = Cl^-, Br^-, I^-, CN^-, and SCN^-) on nanocrystalline titanium dioxide electrodes. *J. Am. Chem. Soc.* 115: 6382.

5 Gräzel, M. and O'Regan, B. (1991). A low-cost, high-efficiency solar cell based on dye-sensitized colloidal TiO_2 films. *Nature* 353: 737.

6 Huynh, W.U., Dittmer, J.J., and Alivisatos, A.P. (2002). Hybrid nanorod-polymer solar cells. *Science* 295: 2425.

7 Pei, Q., Yu, G., Zhang, C. et al. (1995). Polymer light-emitting electrochemical cells. *Science* 269: 1086.

8 Yu, G., Gao, J., Hummelen, C. et al. (1995). Polymer photovoltaic cells: enhanced efficiencies via a network of internal donor–acceptor heterojunctions. *Science* 270: 1789.

9 Tian, B., Zheng, X., Kempa, T.J. et al. (2007). Coaxial silicon nanowires as solar cells and nanoelectronic power sources. *Nature* 449: 885.

10 Gur, I., Fromer, N.A., Geier, M.L., and Alivisatos, A.P. (2005). Air-stable all-inorganic nanocrystal solar cells processed from solution. *Science* 310: 462.
11 Soci, C., Zhang, A., Xiang, B. et al. (2007). ZnO nanowire UV photodetectors with high internal gain. *Nano lett.* 7: 1003.
12 Karpina, V.A., Lazorenko, V.I., Lashkarev, C.V. et al. (2004). Zinc oxide-analogue of GaN with new perspective possibilities. *Cryst. Res. Technol.* 39: 980.
13 Ye, N., Qi, J., Qi, Z. et al. (2010). Improvement of the performance of dye-sensitized solar cells using Sn-doped ZnO nanoparticles. *J. Power Sources* 195: 5806.
14 Pan, Z.W., Dai, Z.R., and Zhong, L.W. (2001). Nanobelts of semiconducting oxides. *Science* 201: 1947.
15 Yang, Y., Guo, W., Zhang, Y. et al. (2011). Piezotronic effect on the output voltage of P3HT/ZnO micro/nanowire heterojunction solar cells. *Nano lett.* 11: 4812.
16 Yang, Y., Qi, J.J., Gu, Y.S. et al. (2009). Piezotronic strain sensor based on single bridged ZnO wires. *Phys. Status Solidi RRL* 3: 269.
17 Guo, W., Yang, Y., Qi, J. et al. (2010). Localized ultraviolet photoresponse in single bent ZnO micro/nanowires. *Appl. Phys. Lett.* 97: 133112.
18 Yang, Y., Guo, W., Qi, J. et al. (2010). Self-powered ultraviolet photodetector based on a single Sb-doped ZnO nanobelt. *Appl. Phys. Lett.* 97: 223113.
19 Song, J. and Wang, Z.L. (2006). Piezoelectric nanogenerators based on zinc oxide nanowire arrays. *Science* 312: 242.
20 Park, W.I., Yi, G.-C., Kim, J.W., and Park, S.M. (2003). Schottky nanocontacts on ZnO nanorod arrays. *Appl. Phys. Lett.* 82: 4358.
21 Gao, Y. and Wang, Z.L. (2009). Equilibrium potential of free charge carriers in a bent piezoelectric semiconductive nanowire. *Nano lett.* 9: 1103.
22 Xue, J., Uchida, S., Rand, B.P., and Forrest, S.R. (2004). 4.2% efficient organic photovoltaic cells with low series resistances. *Appl. Phys. Lett.* 84: 3013.
23 Zhou, J., Gu, Y., Hu, Y. et al. (2009). Gigantic enhancement in response and reset time of ZnO UV nanosensor by utilizing Schottky contact and surface functionalization. *Appl. Phys. Lett.* 94: 191103.
24 Král, P. and Shapiro, M. (2001). Nanotube electron drag in flowing liquids. *Phys. Rev. Lett.* 86: 131.
25 Liu, K., Zhou, Y., Yuan, F. et al. (2016). Self-powered multimodal temperature and force sensor based-on a liquid droplet. *Angew. Chem. Int. Ed.* 55: 15864.
26 Krupenkin, T. and Taylor, J.A. (2012). Reverse electrowetting as a new approach to high-power energy harvesting. *Nat. Commun.* 2: 448.
27 Moon, J.K., Jeong, J., Lee, D., and Pak, H.K. (2013). Electrical power generation by mechanically modulating electrical double layers. *Nat. Commun.* 4: 1487.
28 Feng, J., Graf, M., Liu, K. et al. (2016). Single-layer MoS2 nanopores as nanopower generators. *Nature* 536: 197.
29 Lin, Z.H., Cheng, G., Lee, S. et al. (2014). Harvesting water drop energy by a sequential contact-electrification and electrostatic-induction process. *Adv. Mater.* 26: 4690.

30 Yin, J., Li, X., Yu, J. et al. (2014). Generating electricity by moving a droplet of ionic liquid along graphene. *Nat. Nanotechnol.* 9: 378.
31 Wang, Y., Wang, Y., and Yang, Y. (2018). Graphene-polymer nanocomposite-based redox-induced electricity for flexible self-powered strain sensors. *Adv. Energy Mater.* 8: 1800961.
32 Wang, Y. and Yang, Y. (2019). Superhydrophobic surfaces-based redox-induced electricity from water droplets for self-powered wearable electronics. *Nano Energy* 56: 547.
33 Wang, Y., Jiang, Y., Wu, H., and Yang, Y. (2019). Floating robotic insects to obtain electric energy from water surface for realizing some self-powered functions. *Nano Energy* 63: 103810.
34 Lu, Z., Xu, W., Ma, J. et al. (2016). Superaerophilic carbon-nanotube-array electrode for high-performance oxygen reduction reaction. *Adv. Mater.* 28: 7155.
35 Wang, C., Shen, Y., Song, Z. et al. (2017). Electro-deposition of Co-Ni sulfide nanosheet arrays on nickel foam and investigation of the pseudocapacitive performance. In: *18th International Conference on Electronic Packaging Technology*, 1250.
36 Wang, P., Hayashi, T., Meng, Q. et al. (2017). Highly boosted oxygen reduction reaction activity by tuning the underwater wetting state of the superhydrophobic electrode. *Small* 13: 1601250.
37 Yang, Y. and Wang, Z.L. (2015). Hybrid energy cells for simultaneously harvesting multi-types of energies. *Nano Energy* 14: 245.
38 Li, L., Chen, S., Wang, X. et al. (2012). Nanostructured solar cells harvesting multi-type energies. *Energ. Environ. Sci.* 5: 6040.
39 Pfann, W.G. and Roosbroeck, W. (1954). Radioactive and photoelectric p–n junction power sources. *J. Appl. Phys.* 25: 1422.
40 Halls, J.J.M., Walsh, C.A., Greenham, N.C. et al. (1995). Efficient photodiodes from interpenetrating polymer networks. *Nature* 376: 498.
41 Morii, K., Ishida, M., Takashima, T. et al. (2006). Encapsulation-free hybrid organic-inorganic light-emitting diodes. *Appl. Phys. Lett.* 89: 183510.
42 Wang, X., Zhou, J., Song, J. et al. (2006). Piezoelectric field effect transistor and nanoforce sensor based on a single ZnO nanowire. *Nano lett.* 6: 2768.
43 Xu, C., Wang, X., and Wang, Z.L. (2009). Nanowire structured hybrid cell for concurrently scavenging solar and mechanical energies. *J. Am. Chem. Soc.* 131: 5866.
44 Greene, L.E., Law, M., Goldberger, J. et al. (2003). Low-temperature wafer-scale production of ZnO nanowire arrays. *Angew. Chem. Int. Ed.* 42: 3031.
45 Xu, C. and Wang, Z.L. (2011). Compact hybrid cell based on a convoluted nanowire structure for harvesting solar and mechanical energy. *Adv. Mater.* 23: 873.
46 Xu, S., Wei, Y., Liu, J. et al. (2008). Integrated multilayer nanogenerator fabricated using paired nanotip-to-nanowire brushes. *Nano lett.* 8: 4027.
47 Yang, Y., Zhang, H., Zhu, G. et al. (2013). Flexible hybrid energy cell for simultaneously harvesting thermal, mechanical, and solar energies. *ACS Nano* 7: 785.

48 Yoon, G.C., Shin, K.-S., Gupta, M.K. et al. (2015). High-performance hybrid cell based on an organic photovoltaic device and a direct current piezoelectric nanogenerator. *Nano Energy* 12: 547.

49 Jella, V., Ippili, S., Eom, J.-H. et al. (2018). A novel approach to ambient energy (thermoelectric, piezoelectric and solar-TPS) harvesting: realization of a single structured TPS-fusion energy device using $MAPbI_3$. *Nano Energy* 52: 11.

50 Liu, J., Fer, P., Song, J. et al. (2008). Carrier density and Schottky barrier on the performance of DC nanogenerator. *Nano lett.* 8: 328.

51 Wang, X., Song, J., Summers, C.J. et al. (2006). Density-controlled growth of aligned ZnO nanowires sharing a common contact: a simple, low-cost, and mask-free technique for large-scale applications. *J. Phys. Chem. B.* 110: 7720.

52 Bach, U., Lupo, D., Comte, P. et al. (1998). Solid-state dye-sensitized mesoporous TiO_2 solar cells with high photon-to-electron conversion efficiencies. *Nature* 395: 583.

53 Kim, K.H., Kumar, B., Lee, K.Y. et al. (2013). Piezoelectric two-dimensional nanosheets/anionic layer heterojunction for efficient direct current power generation. *Sci. Rep.* 3: 2017.

54 Yang, Y., Zhang, H., Liu, Y. et al. (2013). Silicon-based hybrid energy cell for self-powered electrodegradation and personal electronics. *ACS Nano* 7: 2808.

55 Liu, Y., Das, A., Xu, S. et al. (2012). Hybridizing ZnO nanowires with micropyramid silicon wafers as superhydrophobic high-efficiency solar cells. *Adv. Energy Mater.* 2: 47.

56 Wu, Y., Zhong, X., Wang, X. et al. (2014). Hybrid energy cell for simultaneously harvesting wind, solar, and chemical energies. *Nano Research* 7: 1631.

57 Li, Z., Ding, Y., Kang, W. et al. (2015). Reduction mechanism and capacitive properties of highly electrochemically reduced TiO_2 nanotube arrays. *Electrochim. Acta* 161: 40.

58 Wang, S., Wang, X., Wang, Z.L., and Yang, Y. (2016). Efficient scavenging of solar and wind energies in a smart city. *ACS Nano* 10: 5696.

59 Zheng, D., Lv, M., Wang, S. et al. (2012). A combined TiO_2 structure with nanotubes and nanoparticles for improving photoconversion efficiency in dye-sensitized solar cells. *Electrochim. Acta* 83: 155.

60 Hu, Y., Zhang, Y., Xu, C. et al. (2011). Self-powered system with wireless data transmission. *Nano lett.* 11: 2572.

61 Yang, Y., Lin, L., Zhang, Y. et al. (2012). Self-powered magnetic sensor based on a triboelectric nanogenerator. *ACS Nano* 6: 10378.

62 Németh, E., Albrecht, V., Schubert, G., and Simon, F. (2003). Polymer tribo-electric charging: dependence on thermodynamic surface properties and relative humidity. *J. Electrost.* 58: 3.

63 Wang, Z. and Triboelectric, L. (2013). Nanogenerators as new energy technology for self-powered systems and as active mechanical and chemical sensors. *ACS Nano* 7: 9533.

64 Zhang, K., Wang, Z.L., and Yang, Y. (2016). Enhanced P3HT/ZnO nanowire array solar cells by pyro-phototronic effect. *ACS Nano* 10: 10331.

65 Ouyang, B., Zhang, K., and Yang, Y. (2017). Self-powered UV photodetector array based on P3HT/ZnO nanowire array heterojunction. *Adv. Mater. Technol.* 2: 1700208.
66 Zhang, K. and Yang, Y. (2017). Thermo-phototronic effect enhanced InP/ZnO nanorod heterojunction solar cells for self-powered wearable electronics. *Adv. Funct. Mater.* 27: 1703331.
67 Noori, K. and Giustino, F. (2012). Ideal energy-level alignment at the ZnO/P3HT photovoltaic interface. *Adv. Funct. Mater.* 22: 5089.
68 Helander, M.G., Wang, Z.B., Qiu, J. et al. (2011). Chlorinated indium tin oxide electrodes with high work function for organic device compatibility. *Science* 332: 944.
69 Nian, Q., Montgomery, K.H., Zhao, X. et al. (2015). Pulse laser deposition fabricated InP/Al-ZnO heterojunction solar cells with efficiency enhanced by an i-ZnO interlayer. *Appl. Phys. A* 121: 1219.

5

Hybridizing Nanogenerators and Sensors

5.1 Introduction

Self-powered sensors are sensors that can directly convert outer stimulations into available electrical signals or are powered by a generator that can scavenge energy from ambient environment [1, 2]. Thanks to its unique advantages such as consuming less or no energy and costing less money, it may play a key role in environmental monitoring, man–machine interaction, and the Internet of Things in the future.

The newly emerging 1D nanomaterials and 2D materials such as ZnO nanowires (NWs), carbon nanotubes (CNTs), and graphene have attracted huge research interests due to their outstanding and novel characteristics in nanoscale. The newly invented triboelectric nanogenerators (TENGs) not only broaden the way of energy harvesting but also bring new mechanism to develop novel self-powered sensors (Figure 5.1). In this chapter, sensors based on the newly discovered nanomaterials and new mechanism such as pressure sensor, strain sensor, temperature sensor, light sensor, and magnetic field sensors will be discussed from the aspect of materials, design, performance, and application.

5.2 Materials

5.2.1 Organic Materials

Figure 5.2 displays scanning electron microscopy (SEM) images of some common organic materials used for self-powered sensors. Polydimethylsiloxane (PDMS) is widely used in TENGs; PDMS with microstructure on the surface have proved to be beneficial for the enhancement of the output performance of sensors. The SEM image illustrated in Figure 5.2a hints that the PDMS has a rough pyramid structure (−5 μm) fabricated through a Si micropyramid template. The pyramid structure is further revealed by the SEM graph of the intersecting surface (see Figure 5.2b). The micropyramid structure will generate more triboelectric charges at a certain area; besides, owing to the microstructure on the rough surface, the contact angle between a water drop and the PDMS surface will be larger,

*Corresponding author: yayang@binn.cas.cn

Hybridized and Coupled Nanogenerators: Design, Performance, and Applications,
First Edition. Ya Yang.

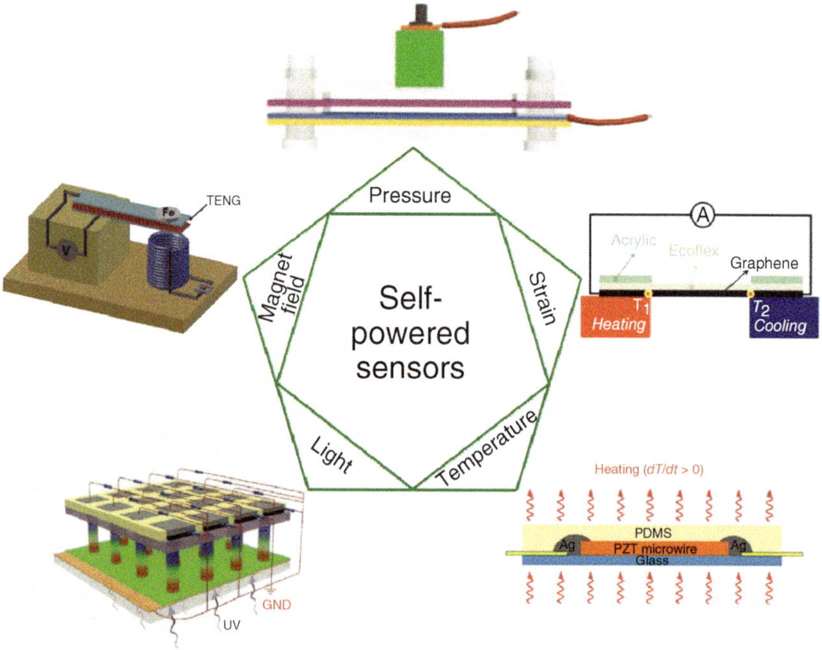

Figure 5.1 Schematic diagram of several kinds of self-powered sensors. Source: Reproduced with permission from Yang and coworkers [3–7]. Copyright 2017, Wiley; Copyright 2012, American Chemical Society; Copyright 2012, American Chemical Society; Copyright 2019, Elsevier; Copyright 2019, Wiley.

so it is harder for the water drops to stay on the surface of the PDMS, which will lead to a self-cleaning function [11]. Figure 5.2c demonstrates a PDMS microwire (−10 μm in diameter) array arranged in an orderly manner on the surface of a PDMS substrate [3].

As illustrated in Figure 5.2d,e, SEM images of the polyimides (PDs)/reduced graphene oxide composite confirm the formation of a porous structure that is caused by the freeze-drying process. Figure 5.2f demonstrates an electret film consisting of plate-like polyolefin. During the polarization process, breakdowns will be caused in the voids, and polarization charges will be generated on the surfaces of the polyolefin film [12]. Figure 5.2g is the SEM image of CNTs/silicon composite film, which indicates that the CNTs are fully packaged by the silicone [13]. Figure 5.2h illustrates a graphene–ecoflex nanocomposite; graphene is firmly attached on the flexible ecoflex; the SEM graph (the inset in Figure 5.2h) indicates that the graphene layer is about 50 μm in thickness. Poly (3-hexylthiophene) (P3HT) is an attractive high-performance p-type photoelectric material; when dissolved in chloroform P3HT can be easily spun on other nanowire materials such as zinc oxide (ZnO) nanowires and form a core–shell structure (Figure 5.2i) [14].

5.2.2 Inorganic Materials

ZnO is a semiconductor material that possesses characters of large bandgap (−3.37 eV), nontoxicity, and low cost; it has been widely used for photoelectric

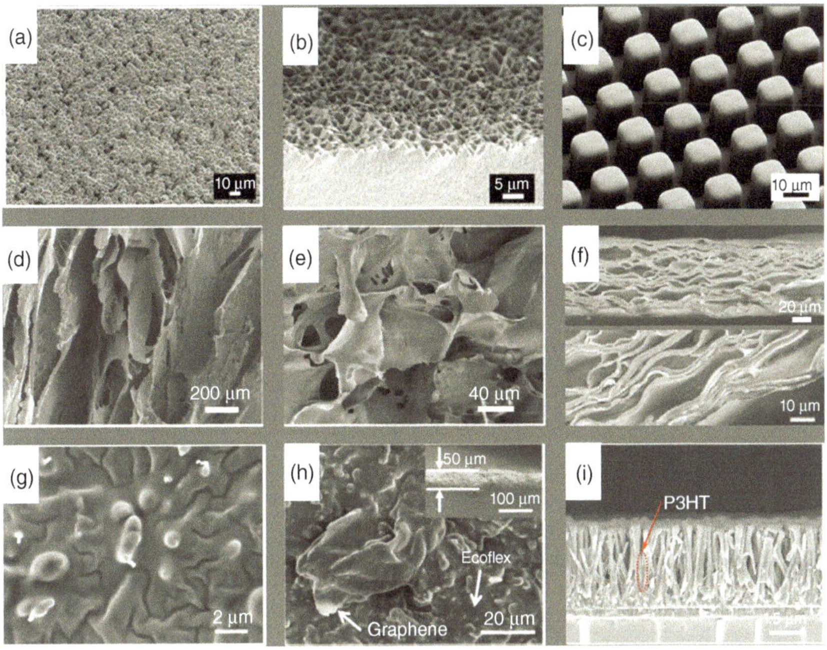

Figure 5.2 SEM images of some organic materials. SEM image of PDMS (a–c), PI/rGO (d–e), polyolefin film (f), CNTs/silicon composite film (g), graphene–ecoflex nanocomposite (h), P3HT on ZnO NWs (i). Source: Reproduced with permission from Yang and coworkers [1, 4, 7–10]. Copyright 2018, Wiley; Copyright 2018, American Chemical Society; Copyright 2013, American Chemical Society; Copyright 2014, American Chemical Society; Copyright 2019, Springer; Copyright 2019, Wiley.

device, sometimes thermoelectric device. Through Sb doping, the mechanical, electrical, electromechanical, and thermoelectric properties of ZnO can be largely improved [15]. Figure 5.3a demonstrates the photograph of a ZnO microbelt doped by Sb where the effective length of the ZnO is about 3 mm; the width of the microbelt is about 3 μm, which can be revealed by the enlarged SEM image illustrated in Figure 5.3a. Figure 5.3b shows the transmission electron microscopy (TEM) image of a ZnO microbelt doped by Sb whose effective width is nearly 300 nm. ZnO NWs grown on a glass substrate through hydrothermal method have been illustrated in Figure 5.3c,d; the NWs are densely packed on the substrate with a diameter of about 500 nm and length of about 2 μm. Lead zirconate titanate (PZT) has attracted huge research interests due to the large pyroelectric coefficient [4, 18]. Figure 5.3e demonstrates a PZT microwire fabricated by a simple mechanical method; by pressing a bulk PZT with large loading force, many PZT micro/nanowires can be obtained and then a proper sample can be selected under an optical microscope; the diameter of the PZT microwire is nearly 2 μm and the length of the PZT microwire is about 10 μm. Figure 5.3f is the SEM image of PZT nanofibers prepared by the current electrospinning method; the diameters of the nanofibers are quite uniform, which is around 200 nm. CNT is a famous 1D material due to its outstanding electronic and mechanical

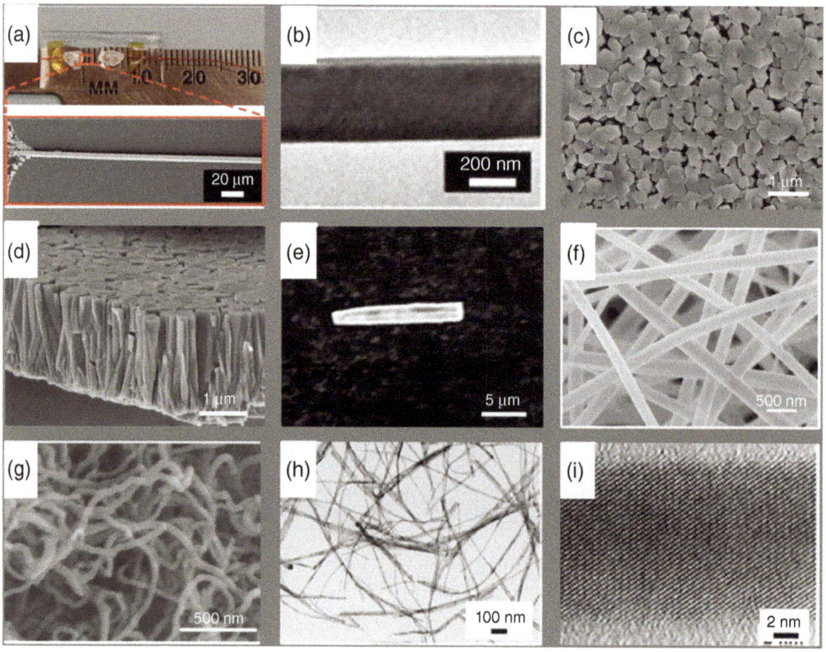

Figure 5.3 SEM and TEM images of some 1-D nanomaterials. (a,b) SEM and TEM images of Sb-doped ZnO microbelt. Source: Reproduced with permission from Yang et al. [16]. Copyright 2012, American Chemical Society. (c,d) SEM image of intrinsic ZnO nanowires grown by hydrothermal method. Source: Reproduced with permission from Ouyang et al. [12]. Copyright 2018, Elsevier. (e,f) SEM image of PZT microwire from bulk PZT and as-synthesized PZT nanofibers through electrospinning method. Source: Reproduced with permission from Yang and coworkers [5, 15]. Copyright 2017, John Wiley and Sons; Copyright 2012, American Chemical Society. (g) SEM image of CNTs, (h,i) TEM (h) and HRTEM (i) Image of Te nanowire. Source: (h,i) Reproduced with permission from Yang and coworkers [1, 17]. Copyright 2018, John Wiley and Sons; Copyright 2012, Springer Nature.

properties; Figure 5.3g is a typical SEM image of CNTs network [17]. Tellurium (Te) nanomaterials can be used for nano thermoelectric devices because of their large Seebeck coefficient (up to 408 µV/K) [2]; Figure 5.3h,i are the TEM and high resolution transmission electron microscopy (HRTEM) images, respectively; the diameter of the Te nanowires single crystal is about 12 nm.

Graphene stands out among various nanomaterials for its remarkable electrical conductivity and electron mobility. By virtue of its thermoelectric and photoelectric characteristics, graphene has found wide application in electronic devices, and temperature and light sensors [16]. Figure 5.4a is a TEM image of a row graphene sheet; the image of graphene looks semitransparent because it is only a few atomic layers thick. Figure 5.4b is the HRTEM image of graphene; the interplanar spacing obtained is about 2.1 Å, which agrees with the standard (100) face. Figure 5.4c shows the atomic force microscopic (AFM) image of graphene, which reveals that the graphene is only a few atomic layers thick [8]. The SEM image in Figure 5.4d indicates that graphene has a 3D structure like wrinkled sheets, indicating that graphene possesses a larger lateral size. TiO_2 is usually known as a wide bandgap semiconductor, which has been widely used for photoelectric devices. As shown in Figure 5.4e, the TiO_2 nanoparticles are observed by SEM, which indicates that the nanoparticles possess small diameters, of nearly 100 nm. The TEM image illustrated in Figure 5.4f further confirms that the diameter of the nanoparticles are is about 30 nm. Ferroelectric materials such as barium titanate (BTO) possess a large bandgap (−3.3 eV) [5, 9], which leads to the low absorption (less than 20%) of the solar spectrum and low power conversion (10^{-4} to 0.5%). Such a material may not be suitable for energy scavenging, but can be used for near-UV photodetection. Figure 5.4g shows the SEM image of BTO nanoparticles with a diameter of around 60 nm; Figure 5.4h is the SEM image of BTO ceramic sintering at 1200 °C – BTO has a large grain size (500–800 nm), and is densely and compactly aligned on the surface of BTO ceramic; AFM image of the as-prepared ceramic is shown in Figure 5.4i, where the length and width of the grain is 601.8 and 422.9 nm, respectively [19].

$Pb(Zr, Ti)O_3$ (PZT) has been proved to be a good pyroelectric material; usually, it can be used for detecting temperature variation or light illumination variation [4]. Figure 5.4j illustrates the SEM image of a PZT ceramic that has been etched by HF acid; besides, the diameters of the grain of the PZT is about 800 nm. Besides, among conventional ferroelectric materials, $BiFeO_3$ is attractive for it possesses the narrowest bandgap (−2.67 eV). Thus a bismuth ferrite (BFO) based photoelectric device can detect the incident light with a wavelength shorter than 465 nm. Figure 5.4k is an SEM image of BFO powders synthesized by hydrothermal method; the powder consists of layered agglomerates with sizes smaller than 1 μm [10].

Tin sulphide (SnS) is a narrow bandgap semiconductor with high absorption coefficient, whose indirect bandgap is about 1.3 eV; Moreover, SnS also has a large Seebeck coefficient (400 μV/K) and small thermal conductivity (17 mW/(cm K)) [6, 7]. Thus, SnS has been widely used for solar cells and thermoelectric generators. The SEM image of SnS thin film prepared by magnetron sputtering on an indium tin oxide (ITO) substrate is demonstrated in Figure 5.4l, from which we can find that shuttle-like SnS particles are densely packed on the ITO substrate.

5.3 Design of Self-Powered Sensors

5.3.1 Pressure Sensors

Usually, pressure sensors can be realized by utilizing the piezoelectric effect and piezoresistive effect; thus, external power sources are needed to drive those sensors. Self-powered nanotechnology based sensors such as TENG possess many inherent advantages such as consuming less or no energy and cheap price, making them superior to the traditional pressure sensors.

Figure 5.5a demonstrates a photograph of a thin PDMS sheet with micropyramid structure on the surface; subsequently, the thin sheet will be utilized for a self-powered pressure sensor. Figure 5.5b explains how the PDMS based pressure sensor works. In this device, human body itself works as a part of the functional material of the sensor; thus, signals will be produced when the human body

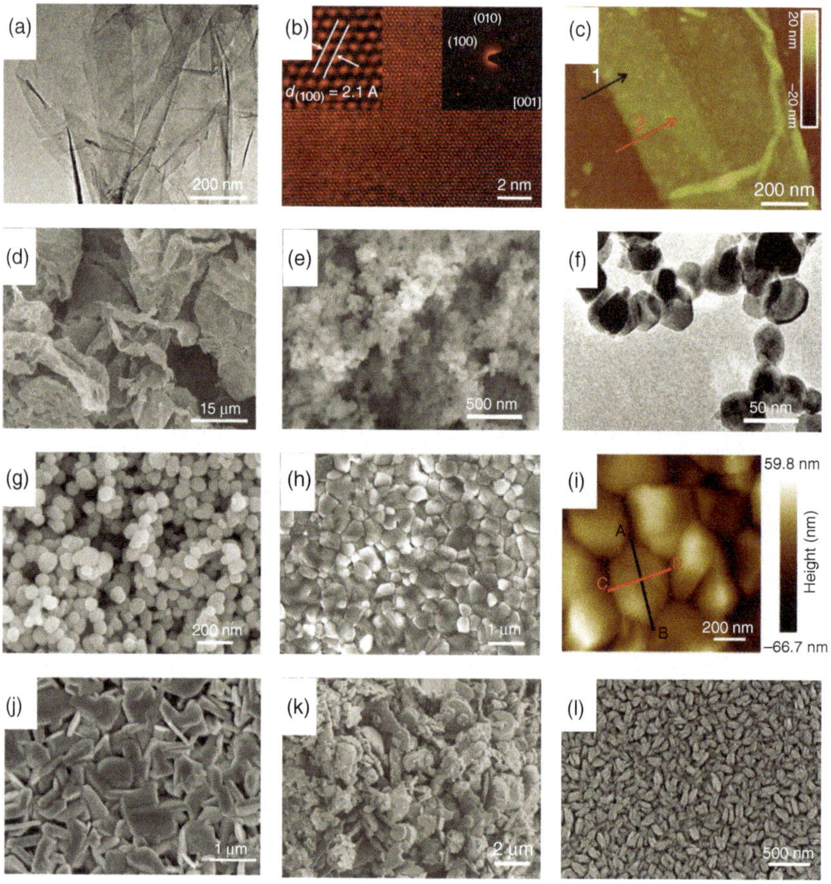

Figure 5.4 Morphology characterization of some 2-D nanomaterials or nanoparticles. (a,b) Low magnification (a) and high magnification (b) of graphene sheet. Source: Reproduced with permission from Zhang et al. [6]. Copyright 2019, Elsevier. (c) AFM image of graphene sheet. Source: Reproduced with permission from Zhang et al. [10]. Copyright 2019, Springer Nature. (d) SEM image of graphene sheet. Source: Reproduced with permission from Zhang et al. [6]. Copyright 2019, Elsevier. (e) SEM image of TiO_2 nanoparticles. Source: Reproduced with permission from Zhai et al. [19]. Copyright 2017, John Wiley and Sons. (f) TEM image of TiO_2 nanoparticles. Source: Reproduced with permission from Zhai et al. [19]. Copyright 2017, John Wiley and Sons. (g) SEM image of $BaTiO_3$ nanoparticles. Source: Reproduced with permission from Song et al. [18]. Copyright 2019, John Wiley and Sons. (h,i) SEM (h) and AFM (i) images of BTO ceramic surface. Source: Reproduced with permission from Song et al. [18]. Copyright 2019, John Wiley and Sons. (j–l) SEM image of PZT film (j), BFO nanoparticles, and SnS nanoparticles. Source: Reproduced with permission from Yang and coworkers [12, 14, 15]. Copyright 2018, Elsevier; Copyright 2018, Wiley; Copyright 2017, Wiley.

touches and separates from the PDMS film. The device consists of PDMS and ITO electrode. As shown in the first stage in Figure 5.5b, when the surface of human skin contacts the PDMS, electrons will transfer from the human skin to the PDMS because the human skin is less triboelectrically negative than PDMS; this process is usually called contact electrification. At the second stage, those

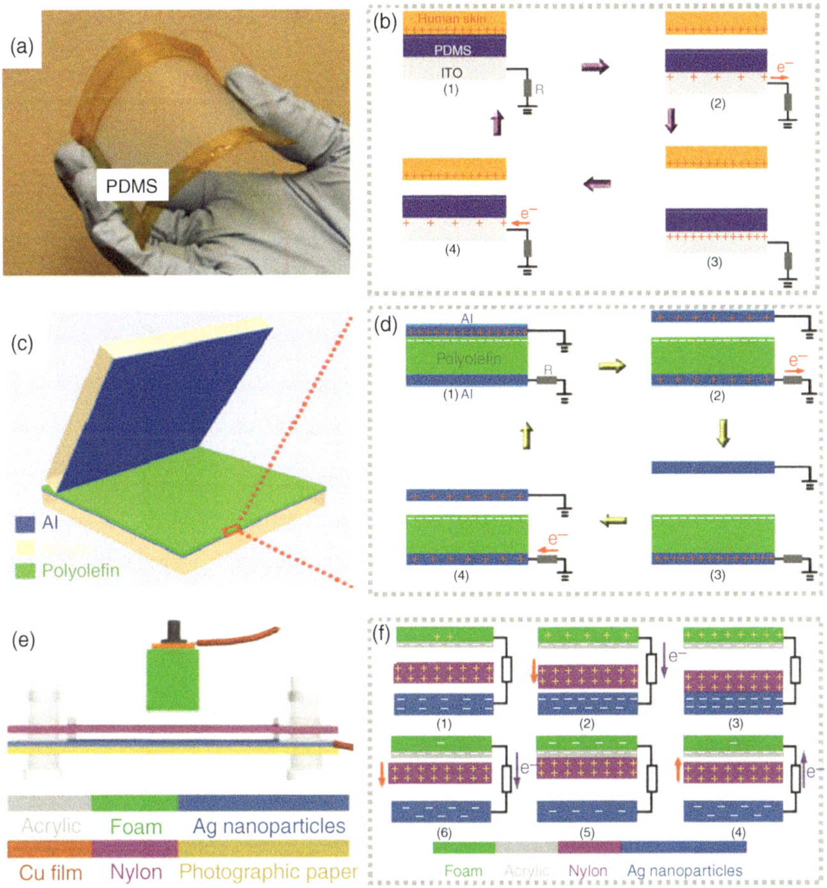

Figure 5.5 Design of pressure sensor. (a,b) Photograph (a) and working mechanism (b) of a PDMS based pressure sensor. Source: Reproduced with permission from Yang et al. [8]. Copyright 2013, American Chemical Society. (c,d) Schematic diagram (c) and working mechanism (d) of a polyolefin based pressure sensor. Source: Reproduced with permission from Yang et al. [9]. Copyright 2014, American Chemical Society. (e,f) Schematic diagram (e) and working mechanism (f) of a PI/rGO foam based pressure sensor. Source: Reproduced with permission from Zhao et al. [7]. Copyright 2019, John Wiley and Sons.

negative charges induced in the PDMS cannot be compensated when the human skin leaves the PDMS. Thus, negative charges on the PDMS will reject the electrons on the ITO electrode. At the third stage shown in Figure 5.5b, as the PDMS is far enough from the human skin, the charges on the PDMS and ITO electrode will screen each other. At the fourth stage demonstrated in Figure 5.5b, the quantity of charges of ITO will decrease when the human skin approaches the PDMS again; this process will generate a reversed output signal in the external circuit. The signal generation process of the TENG is a cycle of these four stages.

Figure 5.5c demonstrates a pressure sensor powered by a TENG constructed with polyolefin, acrylic, and Al. The Al foil attached on the acrylic not only serves as the top electrode but also works as a triboelectric surface; through corona method the other triboelectric surface polyolefin (a kind of electret) can be polarized to boost the performance of the TENG. The working mechanism of the electret film-enhanced TENG is illustrated in Figure 5.5d. At the first stage (contact electrification process), contact of the upper Al electrode with the polyolefin will lead to the transfer of electrons to the polyolefin from the Al electrode; besides, no output signal will be detected for those two kinds of charges on the Al electrode and polyolefin as they are completely screened by each other. At the second stage, once the electrode separates from the polyolefin, the unscreened negative charges on the polyolefin will reject the electrons on Al to the ground. At the third stage, equilibrium status will be realized when the top Al electrode is far away enough from the polyolefin, because of the amount of the positive charges located at the bottom Al electrode and the two kinds of charges will screen each other. At the fourth stage, when the upper electrode (Al) gets close to the polyolefin again, electrons will be injected into the bottom electrode from the ground and decrease the induced positive charges. Finally, the working status of the TENG reverts to the first stage, with the upper Al electrode fully in contact with the polyolefin film.

Figure 5.5e illustrates a self-powered pressure sensor by utilizing the pressure-sensitive property of reduced graphene oxide (rGO) composite foam and TENG as power supply unit. The working principle of the self-powered sensors is demonstrated in Figure 5.5f. At the originating state, no charge is present on the composite foam/acrylic, Ag nanoparticles, and nylon film. When the TENG is blown by an air blower, the nylon will vibrate between the top acrylic and the bottom Ag nanoparticles. Acrylic and Ag will acquire electrons due to the difference in electron affinity among acrylic, Ag, and nylon. As shown in Figure 5.5f, when the state of the TENG changes clockwise from state (3) to (5), electrons will be injected to the PI/rGO foam from the Ag nanoparticles in the external circuit, which is a consequence of the electrostatic induction of the nylon film. Subsequently, electrons will flow from the PI/rGO foam to the Ag nanoparticles when the state of the TENG varies clockwise from (5) to (3).

5.3.2 Strain Sensors

High-performance flexible strain sensors are highly desired in electronic skin, soft robotics, health monitoring, pressure sensing, and so on. Usually, the strain is obtained by measuring the resistance variation of a polymer composite. Usually, an extra voltage source is needed to reflect the change of the resistance.

Figure 5.6a shows a self-powered strain sensor by utilizing the thermoelectric effect of graphene on the flexible ecoflex film. Those commercial Peltier devices are utilized to apply a temperature difference across the graphene–ecoflex film. Figure 5.6a also demonstrates the temperature distribution across the film recorded by an IR camera; a temperature difference of about 14.5 K can be obtained; by utilizing the temperature difference and the thermoelectric property of the graphene, we can obtain an open-circuit voltage of about 0.48 mV. Figure 5.6b shows the IR images of a photothermoelectric (PTE) effect based strain sensor under different strains; besides, a 980 nm LED (36.1 mW/cm^2) is used to obtain a temperature difference across the graphene–ecoflex film (the

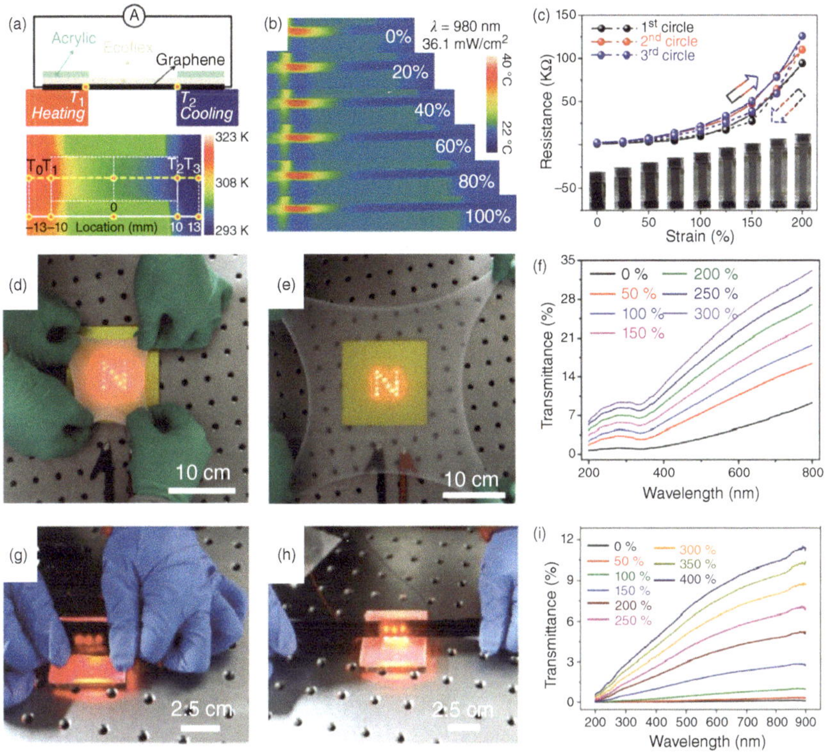

Figure 5.6 Design of strain sensors. (a) Schematic diagram and IR image of a thermoelectric device based strain senor. Source: Reproduced with permission from Zhang et al. [6]. Copyright 2019, Elsevier. (b) IR image of a PTE effect based strain sensor under different strains when illuminated by 980 nm LED. Source: Reproduced with permission from Zhang et al. [10]. Copyright 2019, Springer Nature. (c) Resistance of the graphene–ecoflex nanocomposite film as a function of strain. The inset is the photograph of the film under different strains. Source: Reproduced with permission from Zhang et al. [6]. Copyright 2019, Elsevier. (d,e) Photograph of LEDs under TiO_2 nanoparticles–silicone film without biaxial stretching and with biaxial stretching. Source: Reproduced with permission from Zhu et al. [20]. Copyright 2017, John Wiley and Sons. (f) Transmittance–wavelength curves of the TiO_2 nanoparticles–silicone film under different stretched strains. Source: Reproduced with permission from Zhu et al. [20]. Copyright 2017, John Wiley and Sons. (g,h) Photograph of different light transmittances through the CNT/silicone film without strain (g) and with a strain of 400%. Source: Reproduced with permission from Jiang et al. [1]. Copyright 2018, John Wiley and Sons. (i) Transmittance–wavelength curves of the CNT/silicone film under different stretched strains. Source: Reproduced with permission from Jiang et al. [1]. Copyright 2018, John Wiley and Sons.

high temperature region is the illumination area). The temperature at the two sides of the thin film remains unchanged under different strains (0–100%); thus, a constant PTE voltage will be obtained even if the strain sensor is under different strains. The resistance of the nanocomposite film increases monotonically with the increase of strain and shows good repeatability. The insets in Figure 5.6c demonstrate a graphene–ecoflex film with an original effective dimension of 2 cm (length) and 0.7 cm (width) under corresponding strains.

Strain detection can also be realized by measuring the transmittance change of a stretchable film. Figure 5.6d,e demonstrates a group of LEDs with "N" shape under a flexible ecoflex film with 0.5% concentration of TiO_2 nanoparticles. It is obvious that the LEDs are brighter under a larger strain due to the transmittance change caused by the strain. The strain induced transmittance change is further demonstrated in Figure 5.6f; the transmittance–wavelength curves show that the transmittance increases with increasing wavelength (400–800 nm) at a certain strain. The transmittance increases with increase of stretched strain; at a stretched strain of 300%, the transmittance reaches 30% under 880 nm incident light. Figure 5.6g,h illustrate a CNTs/silicone rubber composite film based strain sensor; the LEDs covered by the composite film are brighter when the film is under a larger strain. Transmittance of the CNTs/silicone rubber under different strains (0–400%) at each wavelength (200 to 90 nm) are depicted in Figure 5.6i, which indicates that the transmittance of the composite film can be effectively utilized for the detection of strain, because the transmittance increases monotonically with the increase of strains.

5.3.3 Temperature Sensors

A self-powered temperature sensor is mainly constructed with thermoelectric materials and pyroelectric materials. Those devices can give out an output signal when a temperature difference or temperature variation exists.

For thermoelectric materials based self-powered sensors, the output voltage is expressed as follows:

$$V = S \cdot \Delta T \tag{5.1}$$

where V denotes the open-circuit voltage, S represents the Seebeck coefficient of the material, and ΔT represents the temperature difference at the two sides of the device. Figure 5.7a is the structure diagram of a temperature sensor fabricated by ZnO microbelt doped with Sb. The effective length of the device is about 3 mm. A glass substrate is utilized to support the microbelt. Then, silver paste is used to fix the microbelt and serves as the electrode. Besides, to improve the robustness of the device, the device is sealed by a thin layer of PDMS film. Two heaters at the bottom of the glass substrate were used to establish a temperature difference across the device. Under the schematic diagram (Figure 5.7a) is the simulated temperature distribution across a glass substrate performed by COMSOL software. The Sb-doped microbelt possesses a large Seebeck coefficient, which is as high as $-350\,\mu V/K$; in addition, the power factor is about $3.2 \times 10^{-4}\,Wm/K^2$ (Figure 5.7b).

A pyroelectric effect based temperature sensor can detect time-dependent temperature fluctuation. Pyroelectric material with large pyroelectric coefficient such as PZT is suitable for the fabrication of temperature sensors. Figure 5.7c shows a schematic diagram of a PZT microwire based temperature sensor. A glass substrate with a thickness of about 0.15 mm is used to support the PZT microwire; silver paste is utilized to connect and fix the PZT microwire on the glass substrate; moreover, the device is further sealed by PDMS to protect the PZT from the atmosphere. Figure 5.7d demonstrates a $PbZr_{0.52}\,Ti_{0.48}O_3$

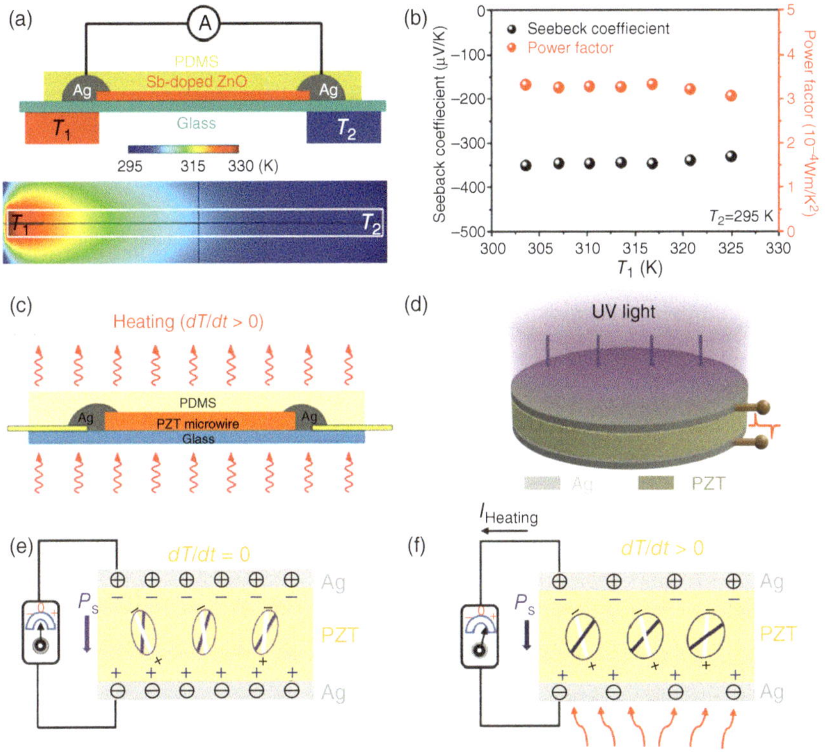

Figure 5.7 Design of temperature sensors. (a) Schematic diagram of a Sb-doped ZnO microbelt based temperature sensor and the temperature distribution across the substrate. Source: Reproduced with permission from Yang et al. [16]. Copyright 2012, American Chemical Society. (b) Seebeck coefficients and power factors of the Sb-doped ZnO microbelt under different temperatures. Source: Reproduced with permission from Yang et al. [16]. Copyright 2012, American Chemical Society. (c) Schematic diagram of a single PZT microwire based pyroelectric temperature sensor. Source: Reproduced with permission from Yang et al. [5]. Copyright 2012, American Chemical Society. (d) Schematic diagram of a PZT ceramic chip based temperature sensor. Source: Reproduced with permission from Yang et al. [5]. Copyright 2012, American Chemical Society. (e,f) Working mechanism of pyroelectric effect based temperature sensors. Source: Reproduced with permission from Song et al. [15]. Copyright 2017, John Wiley and Sons.

(PZT) ceramic chip based temperature sensor; a thin layer of Ag film deposited by magnetron sputtering works as the electrode. The diameter of the chip is about 19.8 mm. The working mechanism of the PZT ceramic is demonstrated in Figure 5.7e,f; the direction of the electric dipoles in PZT will be parallel to the direction of applied electric filed. Besides, when the applied electric field is canceled, the polarization will not disappear. As shown in Figure 5.7e, the electric dipoles oscillate within small degrees at a constant room temperature ($dT/dt = 0$); in addition, equivalent number of opposite charges will be induced on those two electrodes. When the PZT ceramic is heated ($dT/dt > 0$), the oscillation of the electric dipoles will be stronger, which will lead to a weaker effective polarization (Figure 5.7f). Then, electrons on the bottom Ag will transfer to the upper Ag electrode through the copper wire. The output current of a pyroelectric effect based self-powered sensor can be denoted by the following equation:

$$I = pA \cdot \frac{dT}{dt} \tag{5.2}$$

where p, A, and T, represent the pyroelectric coefficient, effective area of the device, and temperature, respectively.

5.3.4 Photodetectors

Figure 5.8a illustrates the schematic structure and light illumination of a self-powered photodetector based on BTO with top ITO and bottom Ag electrodes. As demonstrated in Figure 5.8b, electron–hole pairs will be generated under the excitation of incident ultraviolet light; then, due to the photovoltaic effect, electrons will drift to the Ag electrode while holes will drift toward the ITO electrode, respectively. Besides, the incident light will induce a fast temperature increase, which will produce a pyroelectric signal (Figure 5.8c).

Figure 5.8d illustrates the schematic diagram of a BTO based photodetector with a planar electrode. First, a layer of ITO is grown on one of the surfaces of the BTO ceramic chip through magnetron sputtering; then, the ITO will be sculptured by laser to fabricate an interdigital electrode. Figure 5.8e,f shows the energy band diagrams of the BTO ceramic chip based device. The work function of ITO is 4.5 eV; the work function for Ag is 4.3 eV. BTO's electron affinity is 5.3 eV and the bandgap is 3.3 eV. Figure 5.8e is the working mechanism of a top–bottom electrode device; the incident light is absorbed within a shallow region near the ITO electrode; besides, because of the long diffusion distance between the ITO electrode and the Ag electrode, the photogenerated electrons accumulate near the interface of the ITO electrode and BTO; those electrons cannot be collected by the Ag electrode with high efficiency. The working mechanism of a planar electrode BTO device is shown in Figure 5.8f; the incident light can directly reach the BTO in the gap between ITO electrodes; thus, the transportation path of carriers is shorter than the top–bottom electrode device; besides, without the absorption of ITO, the efficiency of the usage of light illumination will be higher.

Figure 5.8g shows the schematic structure diagram of a BFO ceramic chip based device; a typical hydrothermal method is used to synthesize the BFO powders; by utilizing a tablet machine, the BFO powders were compressed into wafers (the pressure is 14 MPa), whereafter those prepared wafers were sintered at 850 °C for 60 minutes; finally, ITO and Ag were separately sputtered on the two surfaces of the BFO ceramic chip. As shown in Figure 5.8h, the ITO and Ag electrodes will form Schottky contacts with BFO due to the energy band alignment. Figure 5.8i demonstrates that the polarization charges will modulate the Schottky barrier height at the interface; the negative constrained charges at the interface between ITO electrode and BFO will lift the energy band up, which will enhance the photovoltaic effect; the positive charges between the interface of Ag electrode and BFO will bring down the energy band, which will enhance the collection of carriers.

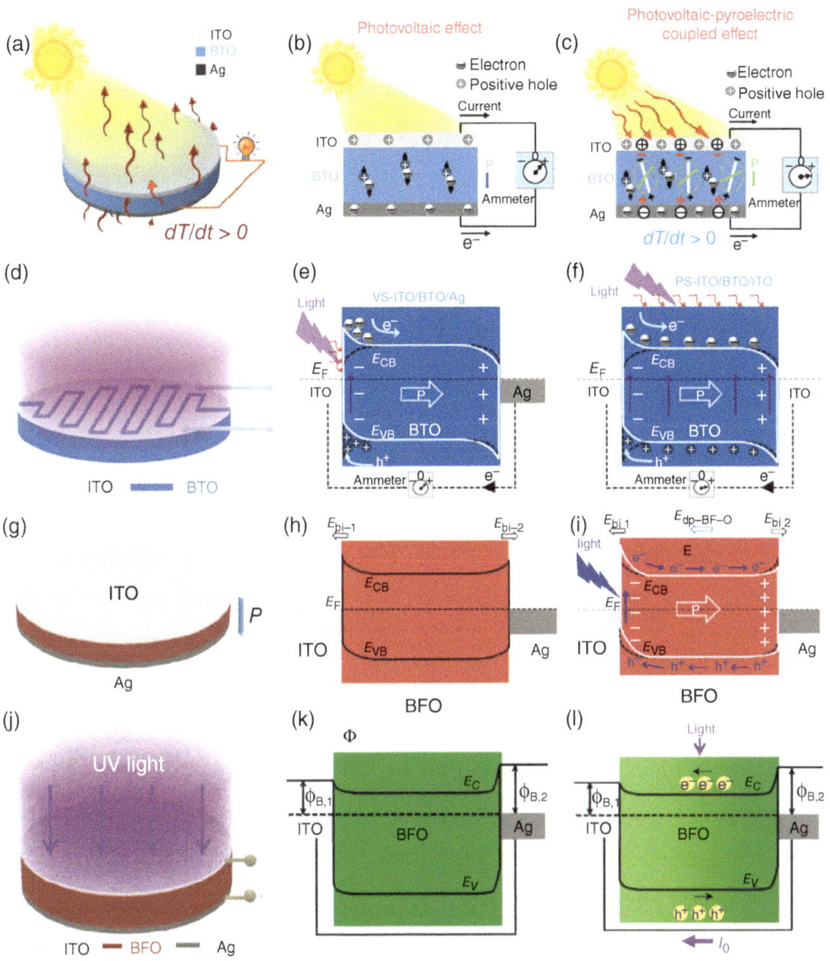

Figure 5.8 Design of ferroelectric material based photodetector. (a) Schematic diagram of a BTO based photodetector with top–bottom electrodes. Source: Reproduced with permission from Ma et al. [11]. Copyright 2017, John Wiley and Sons. (b) Photovoltaic effect in BTO ceramic chip. Source: Reproduced with permission from Ma et al. [11]. Copyright 2017, John Wiley and Sons. (c) Photovoltaic–pyroelectric coupled effect in BTO ceramic chip. Source: Reproduced with permission from Ma et al. [11]. Copyright 2017, John Wiley and Sons. (d) Schematic diagram of a BTO based photodetector with planar electrodes. Source: Reproduced with permission from Ma et al. [2]. Copyright 2018, Elsevier. (e,f) Energy band diagrams of VS-ITO/BTO/Ag (e) and PS-ITO/BFO/ITO (f) devices under light illumination. Source: Reproduced with permission from Ma et al. [2]. Copyright 2018, Elsevier. (g) Schematic diagram of BFO ceramic based self-powered photodetector. Source: Reproduced with permission from Qi et al. [14]. Copyright 2018, John Wiley and Sons. (h,i) Energy band diagram of ITO/BFO/Ag device before and after being poled. Source: Reproduced with permission from Qi et al. [14]. Copyright 2018, John Wiley and Sons. (j) Schematic diagram of a BFO UV photodetector. Source: Reproduced with permission from Qi et al. [13]. Copyright 2018, American Chemical Society. (k,l) Energy band diagram of ITO/BFO/Ag device. Source: Reproduced with permission from Qi et al. [13]. Copyright 2018, American Chemical Society.

Figure 5.8j illustrates a UV photodetector based on a BFO ceramic chip; unlike the device demonstrated in Figure 5.8g, this device will not be poled. Figure 5.8k demonstrates that Schottky junctions are formed at the interface of BFO and electrodes. Excited by the incident photons, electrons and holes will be produced in BFO ceramic and further separated near the interface; eventually, electrons will be collected by the ITO electrode while holes will be injected into the Ag electrode, respectively (Figure 5.8l) [20].

Figure 5.9a illustrates a P3HT/ZnO based heterojunction self-powered photodetector. A commercial ITO glass is ultrasonically cleaned by ethanol, and then deionized water is used to further clean the ITO glass; subsequently, a ZnO seeds layer is deposited by magnetron sputtering; the ZnO nanowires are grown by hydrothermal method; the P3HT and poly(3,4-ethylenedioxythiophene) (PEDOT):poly(styrenesulfonate) (PSS) are grown on the ZnO nanowires covered substrate by means of spin coating; finally, the top Al electrode is deposited by direct magnetron sputtering. Figure 5.9b is the energy band diagram of the P3HT/ZnO heterojunction in darkness. Under light illumination, electron–holes will be generated as the energy of the incident light exceeds the bandgap of P3HT or ZnO; then, carriers (electrons and holes) close to the interface of the p–n junction will drift toward two different directions depending on the polarity of the carriers with the help of built-in electric field; subsequently, holes will be injected into the ITO electrode and electrons will be injected into the Al electrode (Figure 5.9c).

A self-powered photodetector fabricated by inorganic materials is shown in Figure 5.9d. The substrates are commercial ITO glass purchased from a local company; a layer of absorber material (SnS) is grown on the ITO surface through magnetron sputtering; the n-type ZnO nanowires are grown by hydrothermal method; the top Ag electrode is deposited by direct magnetron sputtering. Figure 5.9e is the working mechanism of the standard photovoltaic effect of a p–n junction device; under visible light illumination, the incident light can easily reach the interface of the p–n junction; then the photogenerated electron–holes can be further separated in the depletion zone; finally holes will be injected to the ITO electrode and electrons will be collected by the Ag electrode. It is well known that ZnO nanowires exhibit excellent pyroelectric performance; the incident light will induce a transient temperature increase in the ZnO nanowires; then, bound charges will appear at the two ends of ZnO nanowires, which will bend the energy band (Figure 5.9f); finally, when the light is turned off, the bound charges will disappear and electrons on the ITO electrode will quickly drift to the Ag electrode in the external circuit (Figure 5.9i).

As illustrated in Figure 5.9g, when the 690 nm light is turned on, a pulse-like photocurrent (I_1) is obtained, which is related to the pyro-photoelectric effect (Figure 5.9f); then the current will reach a stable value (I_3), which is the outcome of photovoltaic effect (Figure 5.9e); as the light illumination is turned off, the heated ZnO nanowires will be back to room temperature, and the output current presents a negative peak due to the pyroelectric effect of ZnO nanowires (Figure 5.9i). Under ultraviolet illumination, the short-circuit current directly reaches a negative plateau (I_0) thanks to the PTE effect in the SnS thin film; as demonstrated in Figure 5.9h, there will be a temperature gradient across the

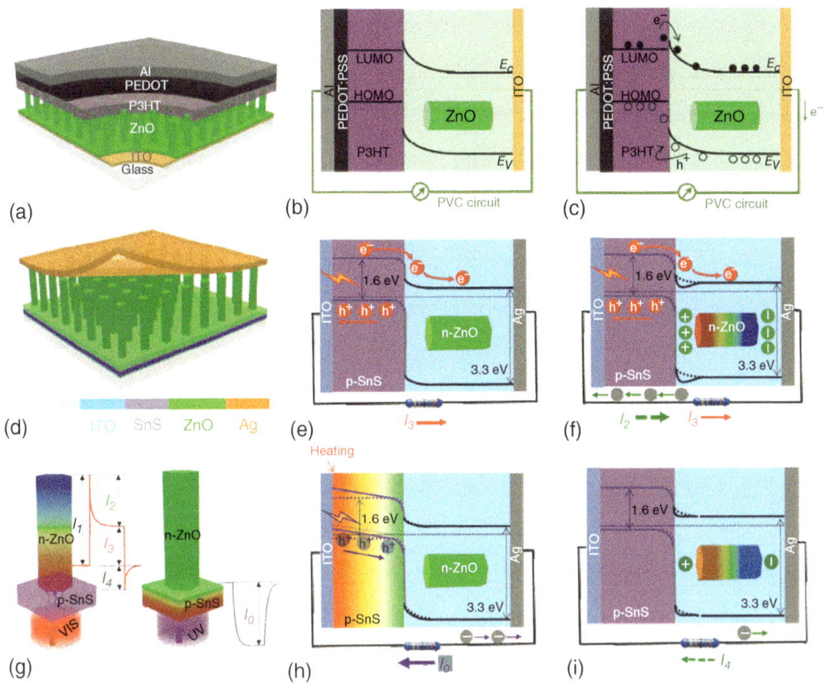

Figure 5.9 Design of semiconductor based photodetector. (a) Schematic diagram of the P3HT/ZnO heterojunction photodetector. Source: Reproduced with permission from Ouyang et al. [3]. Copyright 2017, John Wiley and Sons. (b,c) Energy band diagrams of the P3HT/ZnO heterojunction photodetector. Source: Reproduced with permission from Ouyang et al. [3]. Copyright 2017, John Wiley and Sons. (d) Schematic diagram of the SnS/ZnO heterojunction photodetector. Source: Reproduced with permission from Ouyang et al. [12]. Copyright 2018, Elsevier. (e,f) Photovoltaic effect (e) and pyro-photoelectric effect (f) in the SnS/ZnO heterojunction photodetector as the visible light is turned on. Source: Reproduced with permission from Ouyang et al. [12]. Copyright 2018, Elsevier. (g) Working mechanism of the SnS/ZnO heterojunction photodetector under visible light and ultraviolet light. Source: Reproduced with permission from Ouyang et al. [12]. Copyright 2018, Elsevier. (h) PTE effect in the SnS/ZnO heterojunction photodetector as the ultraviolet light is turned on. Source: Reproduced with permission from Ouyang et al. [12]. Copyright 2018, Elsevier. (i) Pyroelectric effect in the SnS/ZnO heterojunction photodetector as the visible light is turned off. Source: Reproduced with permission from Ouyang et al. [12]. Copyright 2018, Elsevier.

SnS thin film under ultraviolet illumination, which will induce the thermoelectric effect in SnS; the majority carriers (holes) will diffuse from the ITO side to the ZnO nanowires side, and then a thermal potential that is equal to the Femi difference at the two ends of SnS dividing the electron charges will be generated. The photocurrent's direction is from the Ag electrode to the ITO electrode side (PTE effect).

5.3.5 Magnetic Sensors

Figure 5.10a shows the schematic diagram of a magnetic sensor by utilizing a TENG as the signal generating unit; one end of the TENG was mounted on

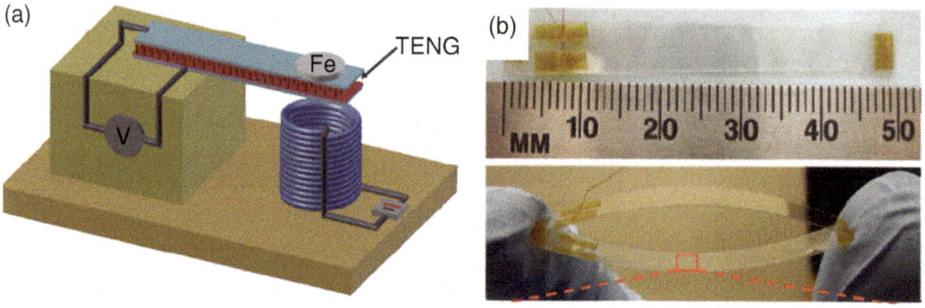

Figure 5.10 Self-powered magnetic sensor. (a) Schematic diagram of the sensor. (b) Photograph of the TENG to fabricate the sensor. Source: Reproduced with permission from Yang et al. [4]. Copyright 2012, American Chemical Society.

a substrate, while the other side of the TENG with an Fe disk was suspended. A solenoid was placed under the disk to produce the magnetic field; when the solenoid is connected to a power source, the Fe disk will be attracted by the magnetic field generated by the solenoid; then, the TENG will bend down and give out an output signal. The magnitude of the electromagnetic force (F_{mag}) will directly determine the output signal of the TENG; the detailed relationship is

$$F_{mag} = KB\frac{dB}{dz} \tag{5.3}$$

where K denotes the magnetic coefficient, z represents the distance between the solenoid and TENG, and B is the magnetic field. Besides, the deflection of the TENG is denoted by Z_m, which can be expressed as follows:

$$Z_m = \frac{F_{mag}L^3}{3YI} \tag{5.4}$$

where Y, I, and L denote the elastic modulus, momentum, and length of the TENG, respectively. According to Eqs. (5.3) and (5.4), we can conclude that the bending of the TENG will be larger under a larger magnetic field. Figure 5.10b shows the photograph of a fabricated TENG; a polyethylene terephthalate (PET) film with a thin layer of ITO electrode and a PDMS film with a bottom electrode are utilized to fabricate the TENG. The length of the TENG is about 5 cm and the width is 0.5 cm.

5.4 Performance

5.4.1 Sensitivity

Sensitivity is a key parameter of self-powered sensors, which indicates the sensing performance of the sensor. High-sensitivity self-powered sensors are highly desired in fabricating high-performance sensors. Figure 5.11a shows the output voltage of a human skin/PDMS based pressure sensor as a function of pressure. Increasing the contact area between the human hand and PDMS by increasing the pressure will induce more triboelectric charges on the PDMS; thus, the sensor's output signals increase monotonically with the increase of

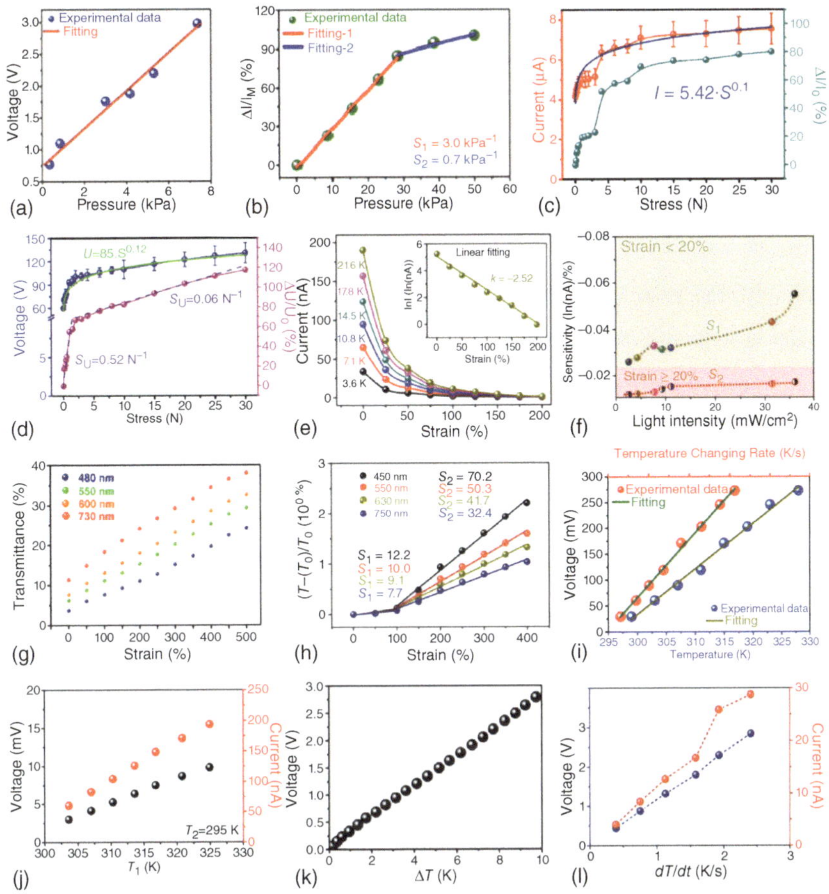

Figure 5.11 Sensitivity of self-powered sensors. (a–d) Sensitivity of some pressure sensors. Source: Reproduced with permission from Yang and coworkers [7–9]. Copyright 2013, American Chemical Society; Copyright 2014, American Chemical Society; Copyright 2019, Wiley. (e–h) Sensitivity of some strain sensors. Source: Reproduced with permission from Yang and coworkers [1, 6, 10, 19]. Copyright 2018, Wiley; Copyright 2017, Wiley; Copyright 2019, Springer; Copyright 2019, Elsevier. (i–l) Sensitivity of some temperature sensors. Source: Reproduced with permission from Yang and coworkers [5, 15–17]. Copyright 2017, Wiley; Copyright 2012, Springer; Copyright 2012, American Chemical Society; Copyright 2012, American Chemical Society.

pressure. The slope of the linearly fitted line indicates that the sensitivity is about 290 ± 20 mV/KPa.

Figure 5.11b shows the sensitivity of an electret film-enhanced TENG based pressure sensor. The TENG was used to power LEDs and the emission intensity of the LEDs will be used to indicate the pressure applied to the TENG. The sensitivity of the pressure sensor can be as per the following equation:

$$S = \frac{I_p - I_0}{I_M \cdot \sigma} \tag{5.5}$$

where I_0 is the brightness under no pressure, I_p denotes the LED's emission intensity under pressure, and I_M represents the largest brightness value of the LED. As shown in Figure 5.11b, the slope of the linearly fitted line illustrates that the sensitivities of the pressure sensor are about 3 kPa^{-1} (0–kPa) and 0.7 kPa^{-1} (30–50 kPa^{-1}).

The sensitivity of a polyimide/graphene nanocomposite foam based pressure sensor has been demonstrated in Figure 5.11c,d; the relationship between the output signals (I and V) and stress (S) is

$$I = 5.42 \times S^{0.1} \tag{5.6}$$

$$U = 85 \times S^{0.12} \tag{5.7}$$

Besides, the current and voltage variation ratios are defined as

$$\Delta I/I_0 = (I - I_0)/I_0 \times 100\% \tag{5.8}$$

$$\Delta U/U_0 = (U - U_0)/U_0 \times 100\% \tag{5.9}$$

where I_0 and U_0 are the output signals without stress. At low stress range (0–2 N), the current and voltage increase sharply with the increase of stress, which indicates the high sensitivity of the pressure sensor. At high stress range (2–30 N), the current and voltage gently increase with the increase of stress, which indicates a lower sensitivity. The definition of sensitivity is expressed as

$$S_I = \delta(\Delta I/I_0)S \tag{5.10}$$

$$S_U = \delta(\Delta U/U_0)S \tag{5.11}$$

The pressure sensor's sensitivity is about 0.52 N^{-1} in the low-stress region; in high-stress region the sensitivity is 0.06 N^{-1} (Figure 5.11d).

The sensitivity of a graphene–ecoflex nanocomposite film based strain sensor can be extracted from Figure 5.11e; the output current decreases almost exponentially as the strain increases from 0% to 200%, owing to the larger resistance under a larger strain. The inset in Figure 5.11e shows that the natural logarithm of the output current increases linearly with strain; the sensitivity of the strain sensor is 2.5 ln(nA)% when $\Delta T = 22$ K; more importantly, the sensitivity can be further improved by introducing a larger ΔT.

Figure 5.11f shows the sensitivity of a stretchable strain sensor constructed with graphene and flexible ecoflex by utilizing the photothermoelectric (PTE) effect of the graphene. The sensitivity increases gradually with increasing light illumination power density. In the small strain region (<20%), the sensitivity increases from −0.026 to −0.056 ln(nA)/% when the power density of the light illumination increases from 2.6 to 36.1 mW/cm^2; in the large strain region (>20%), the sensitivity increases from −0.012 to −0.017 ln(nA)/% as the light increases from 2.6 to 36.1 mW/cm^2.

Transmittance of a TiO_2-silicone film based strain sensor is illustrated in Figure 5.11g. As the strain varies from 0.5% to 500%, the transmittance of the composite increases monotonically. The gauge factor (sensitivity) is expressed as

$$S = (\Delta T/T_0)/\varepsilon \tag{5.12}$$

which can be obtained from the slope of the fitted line in Figure 5.11g. Under 480 nm light illumination, the sensitivity of the strain sensors is about 1.1. Besides, the sensitivity decreases when the light wavelength increases.

The sensitivity of a CNTs–ecoflex composite film strain sensor is demonstrated in Figure 5.11h. The transmittance variation ratio is expressed as $(T-T_0)/T_0$, where T_0 and T represent the transmittance of the film without and with tensile strain, respectively. The slope of the curve demonstrated in Figure 5.11h is sensitivity. Quantitatively, the sensitivity of the sensor can be expressed as

$$S = \delta((T-T_0)/T_0)/\delta\varepsilon \tag{5.13}$$

where ε is the strain. The sensitivity of the strain sensor in the 0–100% range is lower than the sensitivity in the 100–400% range. The high sensitivity of the sensor under large strain is because of the massive microcracks in the CNTs film.

Figure 5.11i shows the open-circuit voltage dependence of a zirconate titanate (PZT) microwire based self-powered temperature sensor on temperature. By utilizing the pyroelectric effect in PZT microwire, the temperature variation rate can be effectively detected. The sensitivity of the self-powered sensor can be expressed as

$$S = \Delta V/\Delta(dT/dt) \tag{5.14}$$

where V denotes the output voltage, and T denotes the temperature.

The output performance of a temperature sensor constructed with ZnO: Sb microbelt is shown in Figure 5.11j; the output signal linearly increases with the temperature at one end of the ZnO microbelt. The output signal originates from the thermoelectric effect of the Sb-doped ZnO microbelt. The sensitivity of the thermoelectric effect based temperature sensor is defined as

$$S = \Delta V/(\Delta T) \tag{5.15}$$

where ΔV represents the output voltage, and the temperature difference across the ZnO nanobelt is ΔT. The sensitivity of the Sb-doped ZnO microbelt can be obtained from Figure 5.11j through the above equation; a sensitivity of about −350 μV/K can be obtained; the sensitivity outperforms the sensitivity of an Al-doped ZnO nanostructure materials based temperature sensor (−300 μV/K). Figure 5.11k demonstrates the dependence of output voltage on the temperature difference of a flexible composite constructed with P3HT polymer and Te-nanowire. The sensitivity of the temperature sensor can be obtained through Eq. (5.13) and the output voltage–temperature data demonstrated in Figure 5.11k; the sensitivity is about 285 μV/K. The output signals of a PZT ceramic chip based self-powered temperature sensor increase monotonically with the rate of temperature change (Figure 5.11l). The definition of the sensitivity of the temperature has been demonstrated in Eq. (5.12). Thus, the sensitivity of the temperature sensor can be obtained from the slope of the line shown in Figure 5.11l.

Ferroelectric materials possess a novel photovoltaic effect that shows a great application potential in solar energy harvesting; however, the large bandgap of those materials leads to low energy conversion efficiency. Thus, those materials are usually used for ultraviolet (UV) photodetectors; besides, pyroelectric effect

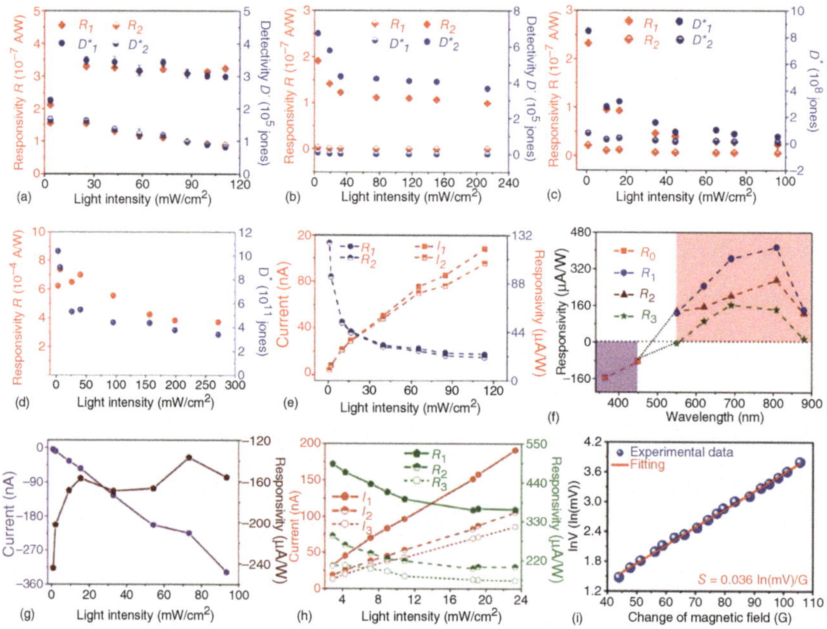

Figure 5.12 Sensitivity of self-powered sensors. (a–d) Sensitivity of some ferroelectric material based photodetectors. Source: Reproduced with permission from Yang and coworkers [2, 11, 13, 14]. Copyright 2018, Elsevier; Copyright 2017, Wiley; Copyright 2018, American Chemical Society; Copyright 2018, Wiley. (e–h) Sensitivity of some semiconductor based photodetectors. Source: Reproduced with permission from Ouyang et al. [3, 12]. Copyright 2017, Wiley; Copyright 2018, Elsevier. (i) Sensitivity of a TENG based magnetic field sensor. Source: Reproduced with permission from Yang et al. [4]. Copyright 2012, American Chemical Society.

will be induced under the incident light, which can be utilized for enhancing the sensitivity of the photodetector. The responsivity of photodetector can be expressed as

$$R = \frac{I_{ph} - I_d}{PS} \tag{5.16}$$

where I_{ph} denotes the electronic charge, I_d denotes the dark current, P represents the light's power density, and S is the effective area of the device. Figure 5.12a shows the responsivity of a top–bottom electrode BTO ceramic based photodetector; illuminated by 405 nm LED, when P increases from 3.5 to 111.1 mW/cm^2, the light induced dT/dt of the ceramic increases from 0.03 to 0.81 °C/s. Thus, pyro-photoelectric effect will be induced in the BTO based photodetector, which can be utilized to enhance the sensitivity of the photodetector; the responsivities of the photodetector under pyro-photovoltaic effect and photovoltaic effect are labeled as R_1 and R_2, respectively (Figure 5.12a). The responsivity of the photodetector can be improved by 260% when the power density is 111.1 mW/cm^2.

It is inevitable that the ITO electrode will absorb much incident light when the ITO/BTO/Ag device with a top–bottom electrode structure is used for ultraviolet detection because the ITO film will absorb much of the incident light in the ultraviolet region. Thus, the above structure will decrease the responsivity of the photodetector. As a consequence, interdigital ITO electrode with a planar structure has been invented to address the above issue. As shown in Figure 5.12b, the responsivity of the photodetector with a planar, interdigital ITO electrode under different light intensities is labeled as R_2; the responsivity of the photodetector with a top–bottom electrode is labeled as R_1. It is obvious that the responsivity of the photodetector with a planar electrode possess a higher responsivity. Among diverse ferroelectric materials, BFOs have also attracted much research interest due to their high Curie temperature. The responsivity of the photodetector can be obtained from Eq. (5.14) and the output current under different light intensities. Figure 5.12c shows the responsivity of BFO ceramic based photodetector poled at 2 kV. R_1 (under light induced pyro-photovoltaic effect) and R_2 (under photovoltaic effect) are the responsivities of the photodetector. When the power density is 0.86 mW/cm^2 (450 nm LED), R_1 is nearly ten times of R_2, which indicates that the responsivity of the BFO ceramic based photodetector can be largely improved by the light induced pyro-photovoltaic effect. As is well known, BFO is also a kind of semiconductor that can be used to fabricate p–n junction or Schottky junction devices to realize the function of photoelectric detection. Figure 5.12d shows the responsivity of a Schottky junction photodetector based on BFO. The responsivity is about 6.0×10^{-4} A/W when illuminated by a 365 nm LED (3.0 mW/cm^2), which is larger than some traditional UV photodetectors fabricated by semiconductors. Besides, the responsivity of the photodetector can also be improved by other applied conditions (electric or magnetic field) due to its multiferroic nature; thus, BFO based photodetectors possess more opportunities in practical application.

Figure 5.12e demonstrates the peak current and steady current of a P3HT/ZnO heterojunction photodetector when illuminated by 365 nm LED with different power densities. By utilizing Eq. (5.14) and the corresponding output current at different light conditions, the responsivities can be obtained. The responsivity is about 125 μA/W for both R_1 and R_2 (0.84 mW/cm^2); however, under larger light intensities R_1 is larger than R_2, which is related to the light-induced thermoelectric effect. The responsivity of an SnS/ZnO nanowires heterojunction based photodetector illuminated by LEDs with different light wavelengths is demonstrated in Figure 5.12f. The definition of the responsivity of the photodetector is shown in Eq. (5.14). Unlike the conventional photodetector, the wavelength of the light source can be utilized to modulate the polarity of the output signal, which is related to light-induced thermoelectric effect in the SnS thin film and photovoltaic effect in the heterojunction; besides, the light-induced pyroelectric signal possesses the same polarity as the photovoltaic signal. The PTE responsivity, photovoltaic responsivity, and pyroelectric responsivity are labeled as R_0, R_2, and R_3; R_1 is the sum of R_2 and R_3. In the short wavelength region (<450 nm), the responsivity (R_0) is negative due to the PTE effect. In the long wavelength region, the responsivity (R_2) is positive, which is related to the photovoltaic effect. Besides, R_3 exists only in the long wavelength region when the incident light is turned on, which possesses the same polarity as R_2. Under 365 nm light illumination, R_0 is −155 μA/W; when illuminated by 690 nm LED, R_1 and R_3 are 364 and 163 μA/W, respectively.

The responsivity of the SnS/ZnO nanowires heterojunction based photodetector under different power densities is demonstrated in Figure 5.12g,h. As shown in Figure 5.12g, under the illumination of a 365 nm LED, the output current and the calculated responsivity are always negative when the power density increases from 0.8 to 93.5 mW/cm^2, which indicates that the polarity of the outputs signal is independent of light intensity. Under 690 nm light illumination (Figure 5.12h), as the power density of the LED increases, the output current and corresponding responsivity of the photodetector are always positive, which further confirms that the polarity of the output signal is independent of light intensity (2.9–23.3 mW/cm^2). Figure 5.12i illustrates the open-circuit voltage of a magnetic field sensor under different magnetic fields. The $\ln V$–dH/dt curve exhibits a good linearity, from which the sensitivity can be acquired:

$$S = \Delta \ln V / \Delta(dH/dt) \tag{5.17}$$

where V is the output voltage, H denotes the magnetic field strength, and t represents time. The sensitivity of the magnetic field sensor can be extracted from the fitted line shown in Figure 5.12i, which is about 0.05 ln(mV)/(G/s).

5.4.2 Response Speed

Response speed is another key parameter of self-powered sensors. Usually the response speed of a sensor is defined as the time when the output signal increases (decreases) from 10% (90%) to 90% (10%) of the maximum signal. Figure 5.13a shows the pressure–time and output voltage–time curves of a TENG based pressure sensor. The rise and fall time of the TENG based pressure sensor is shorter than 0.1 seconds.

Figure 5.13b illustrates the response speed of a graphene-polymer based strain sensor, which is powered by PTE effect; the rise time of the strain sensor is shorter than 0.6 seconds and exhibits a decreasing trend as the strain varies from 0.1% to 1%; the fall time of the strain sensor also decreases monotonically with strain and the minimal recovery time is about 0.6 seconds. However, under large strain (100%), the response speed becomes a little slower than the response speed under small strain (Figure 5.13c). The response speed of a TiO_2 nanoparticles–silicone composite based strain sensor is illustrated in Figure 5.13d; the strain sensor's response time can be obtained by analyzing the enlarged force/transmittance–time curves, which is less than 1.9 seconds. Figure 5.13e,f illustrate the response speed and the recovery time of a CNTs–ecoflex composite based strain sensor; the transmittance of the CNTs–ecoflex composite synchronously increases (decreases) with the increase (decrease) of stretched force; according to the force/transmittance–time curves, the rise and recovery time are estimated to be 1.7 and 2 seconds, respectively.

The response speeds of some nanomaterials based temperature sensors are demonstrated in Figure 5.13g–l. Figure 5.13g shows the current–t curve when a human finger moves from the Sb-doped ZnO microbelt based temperature sensor; the output current decreases exponentially with time and the recovery time is about nine seconds (when current decreases to $1/e$ of the peak current). When a heat source contacts a PZT microwire based temperature sensor, sharp increase

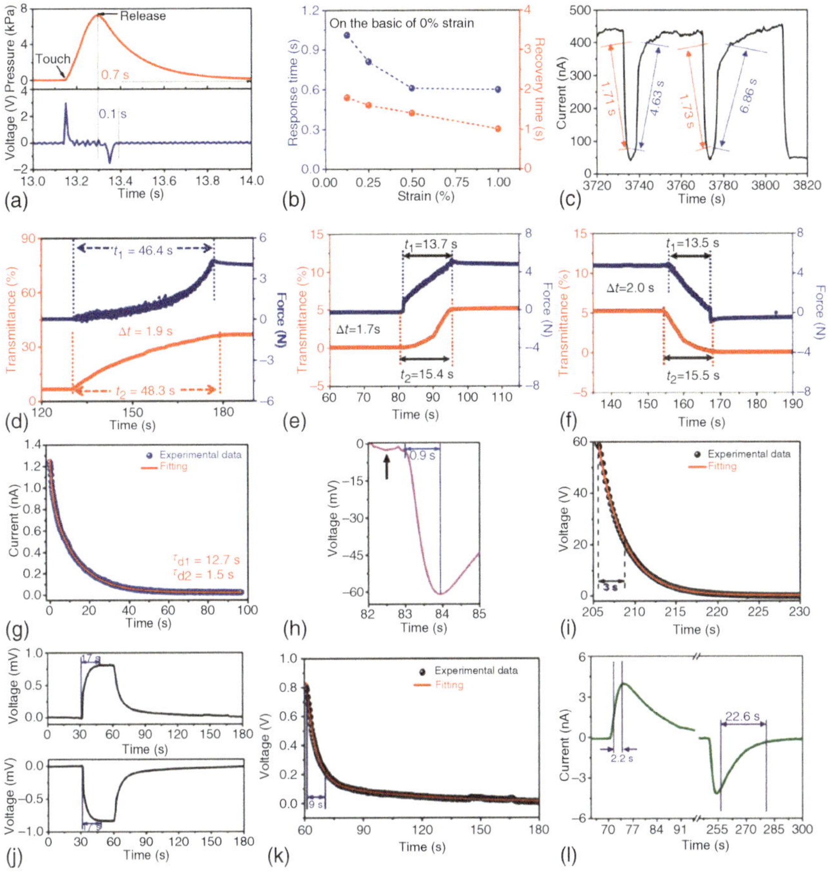

Figure 5.13 Response speed of some pressure, strain, and temperature sensors. (a) Response speed of a TENG based pressure sensor. Source: Reproduced with permission from Yang et al. [8]. Copyright 2013, American Chemical Society. (b,c) Response speed of graphene–polymer nanocomposite based strain sensor. Source: Reproduced with permission from Zhang et al. [6, 10]. Copyright 2019, Springer Nature; Copyright 2019, Elsevier. (d) Response speed of a TiO_2 nanoparticles–silicone based strain sensor. Source: Reproduced with permission from Zhai et al. [19]. Copyright 2017, John Wiley and Sons. (e,f) Response speed of a CNTs–ecoflex film based strain sensor. Source: Reproduced with permission from Jiang et al. [1]. Copyright 2018, John Wiley and Sons. (g) Response speed of an Sb-doped ZnO nanobelt based temperature sensor. Source: Reproduced with permission from Yang et al. [16]. Copyright 2012, American Chemical Society. (h,i) Response speed of a PZT microwire based temperature sensor. Source: Reproduced with permission from Yang et al. [5]. Copyright 2012, American Chemical Society. (j,k) Response speed of a Te nanowire/P3HT composite temperature sensor. Source: Reproduced with permission from Yang et al. [17]. Copyright 2012, Springer Nature. (l) Response speed of a PZT ceramic based temperature sensor. Source: Reproduced with permission from Song et al. [15]. Copyright 2017, John Wiley and Sons.

of the output voltage will be observed thanks to the PZT microwire's pyroelectric effect (Figure 5.13h); the response time is about 0.9 seconds; when the heat source gets away from the sensor, the output signal decreases exponentially with a decay time of about three seconds (the current recovery to $1/e$). Figure 5.13j,k show the output voltage–time curve when a heat source contacts and then leaves the Te-nanowire/P3HT polymer composite based temperature sensor; as shown in Figure 5.13, it takes about 17 seconds when the output voltage reaches the maximum value as the device is forward and reversely connected to the measurement system. When the heat source is moved out, nine seconds are needed for the device to recover to $1/e$ of the maximum value. Figure 5.13l demonstrated the response time of a PZT ceramic based temperature sensor; the pyroelectric signal of the PZT ceramic will be induced even when the temperature change rate is as small as 0.39 K/s; the rise and fall time of the temperature sensor is about 2.2 and 22.6 seconds, respectively. In conclusion, the temperature sensors based on pyroelectric material possess faster response speed than thermoelectric material based temperature sensors.

Photodetectors are widely used for optical communication, imaging, and flame warning. In those application scenarios, fast response speed is extremely important. Figure 5.14a,b are the rise and fall time of a BTO ceramic based photodetector; with the help of light-induced pyroelectric effect in the BTO, the photodetector shows a fast rise time of 0.4 and 1.6 seconds. Figure 5.14c,d shows that the rise and fall time of BTO ceramic based photodetector with planar electrodes are 0.6 and 0.5 seconds. The BFO ceramic based photodetector also possesses a fast response speed due to the pyroelectric effect induced by the incident light; the rise and recovery time are 0.5 and 0.8 seconds (Figure 5.14e). Figure 5.14f shows the response speed of an unpolarized BFO ceramic photodetector, which have no pyroelectric effect; obviously, the rise time (10.8 seconds) is longer over 20 times than the polarized BFO ceramic based photodetector; thus, the pyroelectric effect in the photodetector will improve the response speed of the photodetector.

Figure 5.14g shows the response speed of a P3HT/ZnO nanowires based photodetector; the response and recovery time of the photodetector are only 90 and 98 ms, respectively; the fast response speed may be associated with the 1D structure of the well-aligned ZnO nanowires. Figure 5.14h,i demonstrated the response speed of an SnS/ZnO nanowires based photodetector; the response speed of the photodetector shows an obvious difference under UV (365 nm) light illumination and visible light illumination (690 nm); the working mechanism is PTE effect under UV light illumination, which results in a long rise (49.1 seconds) and fall time (59.8 seconds). However, under visible light illumination, the photodetector shows a fast response speed (less than one second), which can be attributed to the light-induced pyro-photoelectric effect. Figure 5.15 illustrates the open-circuit voltage of a TENG based magnetic field sensor as a function of time, from which the response and decay time can be extracted. It takes 0.13 seconds for the voltage to rise from the minimal to the maximum value. As the magnetic field is turned off, a negative pulse will be induced; it takes 0.34 seconds for the voltage to recover to zero from the minimal voltage.

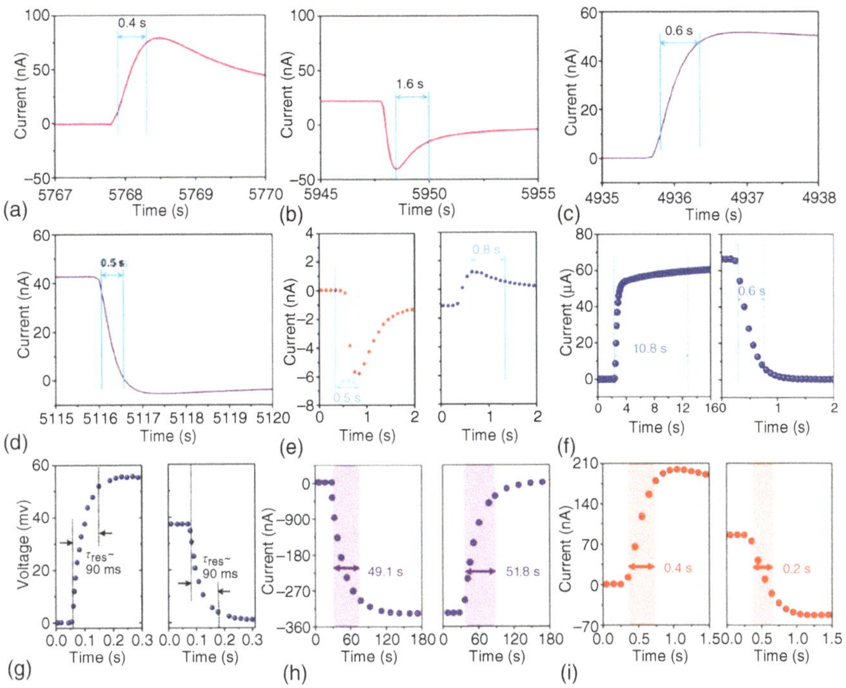

Figure 5.14 Response speed of some photodetectors. (a,b) Response speed of a BTO ceramic based photodetector with top–bottom electrodes. Source: Reproduced with permission from Ma et al. [11]. Copyright 2017, John Wiley and Sons. (c,d) Response speed of BTO ceramic based photodetector with planar electrodes. Source: Reproduced with permission from Ma et al. [2]. Copyright 2018, Elsevier. (e,f) Response speed of some BFO ceramic based photodetectors. Source: Reproduced with permission from Qi et al. [13, 14]. Copyright 2018, American Chemical Society; Copyright 2018, Wiley. (g–i) Response speed of some ZnO nanowires photodetectors. Source: (g,h) Reproduced with permission from Ouyang et al. [3, 12]. Copyright 2017, John Wiley and Sons; Copyright 2018, Elsevier.

5.4.3 Stability

High stability is an important parameter for reliable sensors, which is estimated by the repeatability of the output signals under periodical outer stimulation. Figure 5.16a illustrates the V–t and I (current density)–t curves of a human skin/PDMS based TENG when the human skin periodically contacts the PDMS; the maximum V and I reach −1000 V and 8 mA/m^2, respectively. Most importantly, the output signal shows nearly no observable difference in five circles, which indicates the pressure sensor's high stability.

Figure 5.16b illustrates the output signal of an electret film-enhanced TENG based pressure sensor under periodical outer stimulation; the output voltage shows no observable difference in seven cycles and the maximum open-circuit voltage is about −1070 V; the output current density also shows good repeatability and the maximum output current density is about 10 mA/m^2. Figure 5.16c demonstrates the output voltage of a pressure sensor fabricated by a polyimide/graphene foam (pressure sensitive unit) and powered by TENG,

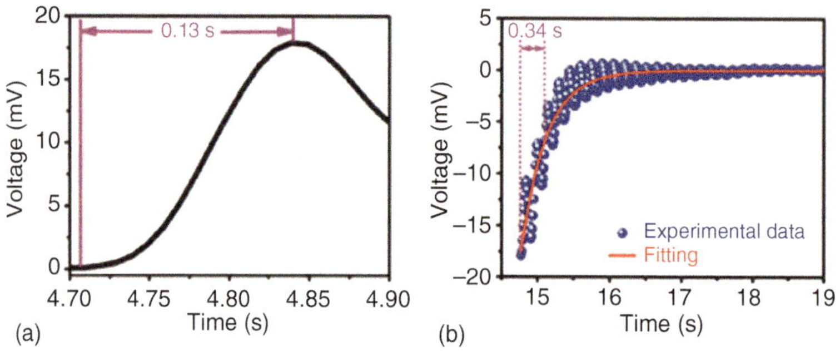

Figure 5.15 Response speed of a TENG based magnetic field sensor. Source: Reproduced with permission from Yang et al. [4]. Copyright 2012, American Chemical Society.

which exhibits good repeatability; the output voltage in cycle 1 and cycle 20 (cycle 20 means that the polyimide/graphene are periodically compressed for twenty times) are nearly the same, which indicates that the pressure sensor possesses good stability. Figure 5.16d demonstrates the short-circuit current of a graphene-polymer based strain sensor, which is powered by the thermoelectric effect of graphene; the output current shows outstanding stability in 20 hours of stretching/releasing process during which the strain varies between 0% and 100%. Figure 5.16e shows the short-circuit current of the graphene-polymer based strain sensor powered by the PTE effect of graphene; the output current of the strain sensor exhibits high repeatability in 100 cycles. Figure 5.16f,g illustrates the transmittance of TiO_2 nanoparticles based strain sensor and CNTs based strain sensor, respectively; the transmittance shows high repeatability even when over 400% strain is applied to the strain sensors for over three times. Figure 5.17a shows the output current of a Sb-doped ZnO microbelt based temperature sensor when the temperature of the left side (T_1) of the sensor periodically increases to 304 K from 295 K and the right side temperature remains at 295 K; the output current shows a high repeatability because there is no observable difference in five cycles. The output performance of Te-nanowire/P3HT polymer composite also shows good stability when the temperature difference at the two ends of the device is about 50 K despite the P3HT being a kind of organic material (Figure 5.17b); the maximum of the output current is 254 nA. When the temperature applied to the PZT microwire based temperature sensor is periodically changed, standard pulse signals will be observed (induced by pyroelectric effect) (Figure 5.17c); the output current varies synchronously with dT/dt and the peak output current reaches 0.6 nA when the dT/dt is about 1.5 K/s; besides, the current exhibits no difference in magnitude in five cycles, which indicates that the temperature sensors have outstanding stability. As shown in Figure 5.17d, the pulse-like temperature variation rate (dT/dt) is the temperature information of a PZT ceramic; the positive and negative peaks mean that the heating process starts and stops, respectively; as dT/dt increases from 0.4 to 2.4 K/s, the peak signals increase with the increase of temperature change rate; moreover, the output signal shows good stability at a certain temperature change rate.

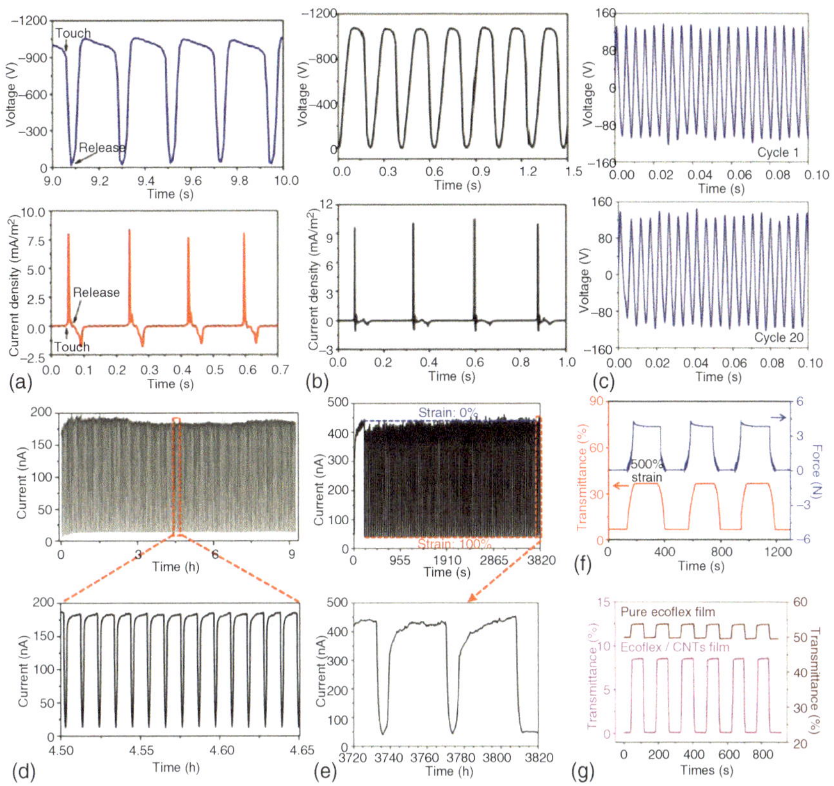

Figure 5.16 Stability of some pressure and strain sensors. (a) Open-circuit voltage and short-circuit current density of a human skin TENG based pressure sensor. Source: Reproduced with permission from Yang et al. [8]. Copyright 2013, American Chemical Society. (b) Open-circuit voltage and short-circuit current density of an electret film-enhanced TENG based pressure sensor. Source: Reproduced with permission from Yang et al. [9]. Copyright 2014, American Chemical Society. (c) Open-circuit voltage of a polyimide/graphene nanocomposite foam-based wind-driven TENG. Source: Reproduced with permission from Zhao et al. [7]. Copyright 2019, John Wiley and Sons. (d,e) Short-circuit current of graphene–ecoflex nanocomposite film based strain sensors. Source: Reproduced with permission from Zhang et al. [6, 10]. Copyright 2019, Springer Nature; Copyright 2019, Elsevier. (f) Transmittance of a TiO_2–ecoflex nanocomposite film based strain sensor under periodical force. Source: Reproduced with permission from Zhai et al. [19]. Copyright 2017, John Wiley and Sons. (g) Transmittance of a CNTs–ecoflex nanocomposite film based strain sensor under periodical force. Source: Reproduced with permission from Jiang et al. [1]. Copyright 2018, John Wiley and Sons.

Figure 5.18a shows the output current of an ITO/BTO/Ag photodetector with top–bottom electrodes when cyclically illuminated by a 405 nm LED with a power density of 25.4 mW/cm^2; both the peak current (induced by pyro-photoelectric effect) and plateau current (induced by photovoltaic effect) show high repeatability. Figure 5.18b shows the output current of ITO/BTO/ITO photodetector with planar electrodes under periodical 365 nm light (213.5 mW/cm^2); under both situations of forward and reversed connection versus the testing system, the photodetector has good stability.

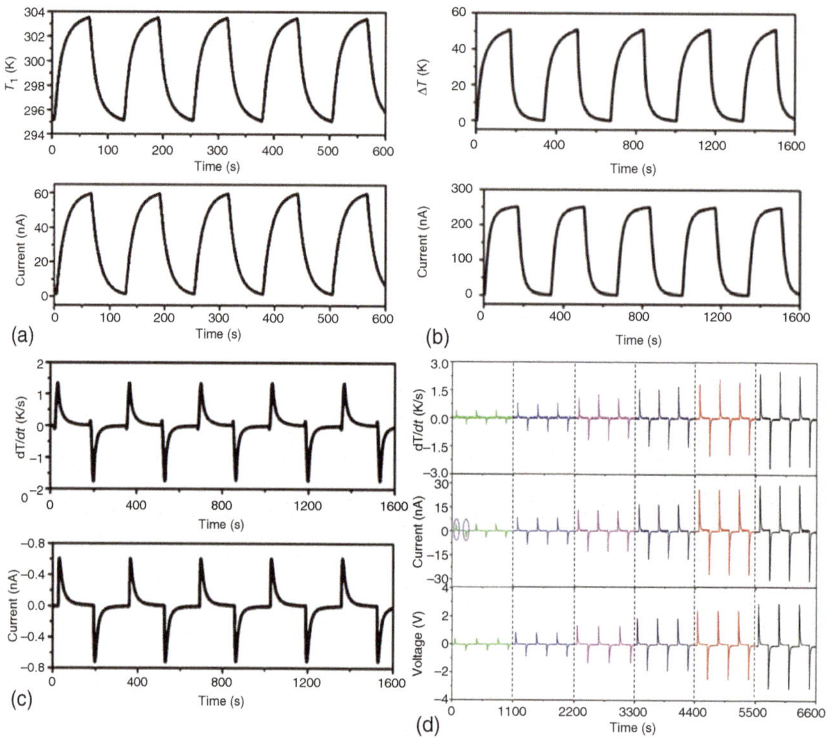

Figure 5.17 Stability of some temperature sensors. (a) Obtained short-circuit current of an Sb-doped ZnO microbelt based temperature sensor under periodical temperature variation. Source: Reproduced with permission from Yang et al. [16]. Copyright 2012, American Chemical Society. (b) Measured short-circuit of a Te-nanowire/P3HT polymer composite based temperature sensor under cyclic temperature variation. Source: Reproduced with permission from Yang et al. [17]. Copyright 2012, Springer Nature. (c) Measured short-circuit current of a PZT microwire based temperature sensor under periodical temperature variation. Source: Reproduced with permission from Yang et al. [5]. Copyright 2012, American Chemical Society. (d) Short-circuit current and open-circuit voltage of a BTO ceramic based temperature sensor when periodically heating the sensor. Source: Reproduced with permission from Song et al. [15]. Copyright 2017, John Wiley and Sons.

Figure 5.18c,d show the output signals of polarized and unpolarized BFO based ceramic chips, respectively. It is obvious that polarization of the ceramic chips will not affect the stability of the photodetector. Figure 5.18e is the output current of a P3HT/ZnO nanowires based photodetector under cyclic 365 nm light illumination (113.4 mW/cm^2); the peak and platform currents are 108 and 95 nA, respectively. Although P3HT is an organic material, the photodetector still possesses stable output performance. Figure 5.18f,g show the short-circuit current of SnS/ZnO nanowires based photodetector under 360 and 690 nm light illumination, respectively. Under periodical 365 nm light illumination, the stable output current is negative due to the PTE effect in the SnS thin film; under cyclic 690 nm light illumination, the output current shows standard positive and negative pulses, which is related to the pyroelectric effect originating from the ZnO

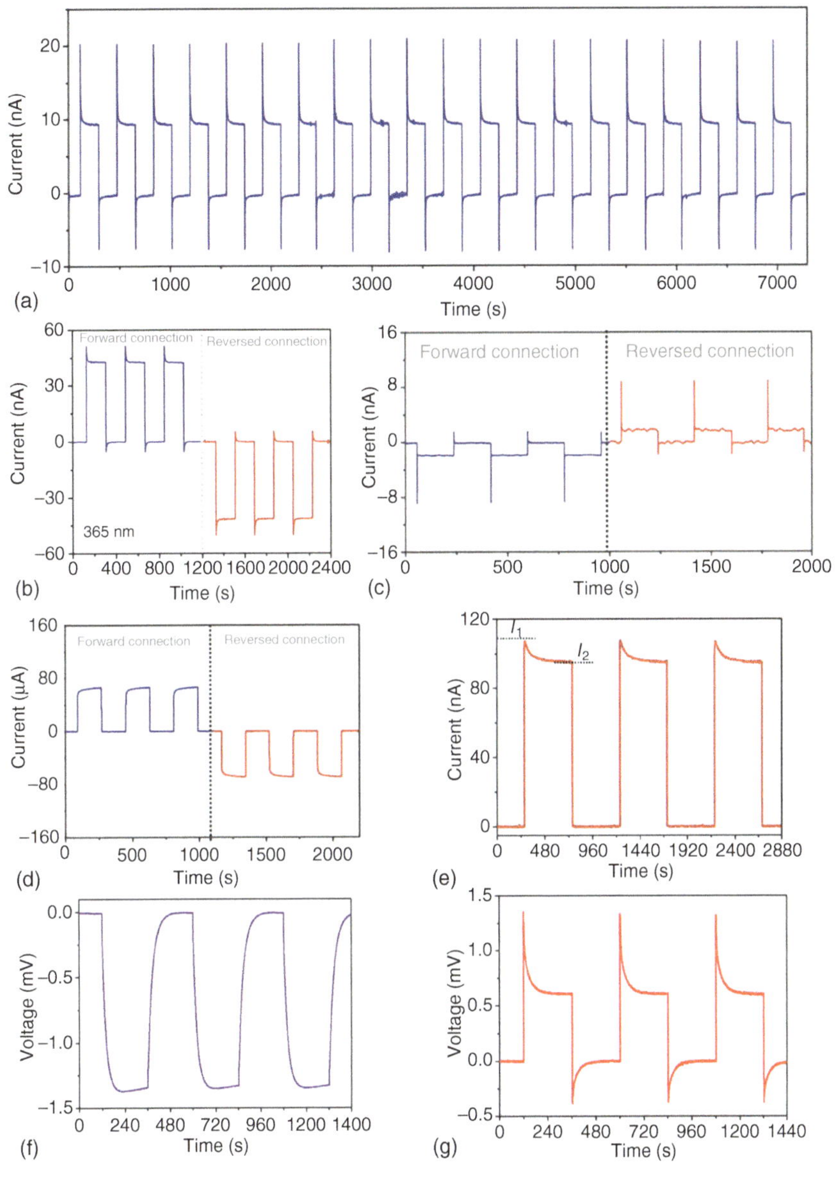

Figure 5.18 Stability of some photodetectors. (a) Output current of an ITO/BTO/Ag photodetector with a vertical structure under periodical 405 nm light illumination. Source: Reproduced with permission from Ma et al. [11]. Copyright 2017, John Wiley and Sons. (b) Output current of an ITO/BTO/ITO photodetector with a planar electrode under 405 nm light illumination. Source: Reproduced with permission from Ma et al. [2]. Copyright 2018, Elsevier. (c,d) Output current of a polarized and unpolarized ITO/BFO/Ag photodetector. Source: Reproduced with permission from Qi et al. [13, 14]. Copyright 2018, American Chemical Society; Copyright 2018, John Wiley and Sons. (e) Output current of a P3HT/ZnO heterojunction based UV photodetector. Source: Reproduced with permission from Ouyang et al. [3]. Copyright 2017, John Wiley and Sons. (f,g) Output current of an SnS/ZnO nanowires based photodetector under 360 and 690 nm light illumination. Source: Reproduced with permission from Ouyang et al. [12]. Copyright 2018, Elsevier.

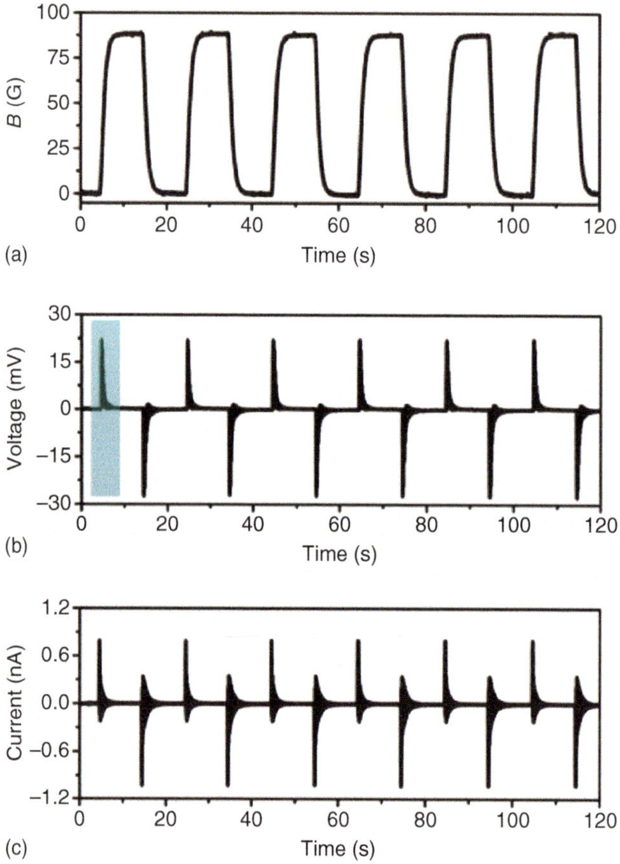

Figure 5.19 (a–c) Variation of magnetic field (a), output voltage (b), and output current (c).Output voltage and current vary with the variation of magnetic field. Source: Reproduced with permission from Yang et al. [4]. Copyright 2012, American Chemical Society.

nanowires; besides, the plateau of the current–time curve is the outcome of the photovoltaic effect of the p–n junction. Figure 5.19 demonstrates the output performance of a TENG based magnetic field sensor. The magnetic field cyclically changes between 0 and 88 G, which will generate periodical output current and voltage; as the magnetic field increases from 0 to 88 G, a positive voltage and current pulse of 22 mV and 0.8 nA will be generated. A negative voltage and current pulse will be obtained when the magnetic field decreases from 88 to 0 G. The output signal also shows high stability and repeatability in six cycles.

5.5 Applications

5.5.1 Touch and Motion Detection

The human skin based TENG is used to fabricate a pressure sensor array, which is useful in human–machine interfacing, robotics, and security systems. As shown in Figure 5.20a–c, when one of the pixels in the pressure sensor array is touched with a small pressure, the output voltage is about 1.5 V (Figure 5.20a); as the pressure become larger, the output voltage of the corresponding pixels becomes larger (Figure 5.20b,c). Figure 5.20d illustrates the optical image of the device as the 6th and 11th pixels are touched; by the corresponding voltage demonstrated in the mapping image (Figure 5.20e), the pressure applied to the sensor is estimated to be 4.9 kPa. The mapping image demonstrated in Figure 5.20f is when all pixels are touched by human hand; it is obvious that all pixels can work normally and the pressures applied to channel 1 to channel 8 are larger than the rest of the channels. The flexible pressure sensor array can realize a real-time detection of touching action, which is a desirable character for robots and prosthetic devices.

Figure 5.20g–i shows the application of an electronic skin that consists of an LED as the display unit and electret film-enhanced TENG as the pressure sensing unit. As shown in Figure 5.20g, when the pressure sensors in the matrix are touched by a finger covered by nitrile glove, the corresponding LEDs will glow (indicating the corresponding position is touched). Figure 5.20i indicates that the pressure sensor array can track the motion trail of a detection target. When the object moves along the route marked in Figure 5.20i (back side of the panel), the LEDs in the corresponding route will be lit up. The electronic skin can not only realize a self-powered and visible function but also can realize an unperceivable detection of the target object for the LEDs and the detected targets are separated by the substrate.

5.5.2 Strain Detection

As shown in Figure 5.21a, the TiO_2 nanoparticles silicone composite films along with some LEDs are fixed on a glove; when worn on a human hand, the glove can detect the gesture of fingers by detecting the light from the LED. Without strain, the light intensity is the weakest status; the strain will be larger as the fingers bend more, which will lead to an increase of light intensity. The detected absolute radiation intensity will be larger under a larger strain (Figure 5.21b); the absolute radiation varies from 0.3 to 1.8 mW/cm^2 as the strain causes the finger to become larger. The dynamic response of the strain sensor is illustrated in Figure 5.21c; according to the absolute radiation intensity–time curve, the rise time of the strain sensor to the bending of a finger is estimated to be less than 4.5 seconds. Thus, the glove can be used for precise motion control of robotics of some other mechanical devices.

The optical image of a strain sensor array by taking advantage of PTE effect of graphene–ecoflex film can be found in Figure 5.21d; the strain sensor has six channels both in the X and Y directions, which enables the detection of two-dimensional strains. At the original status, the current mapping figures in the X and Y directions are quite uniform (Figure 5.21e,f); the output currents of those 12 channels are as high as 300 nA. When the strain sensor array is under a force along the Z direction, a deformation in the graphene–ecoflex film will be induced and channels at the middle of the array possess larger strain (Figure 5.21g). The corresponding current mapping is demonstrated in Figure 5.21h,i; under large strain, the color of the mapping figure turns into blue;

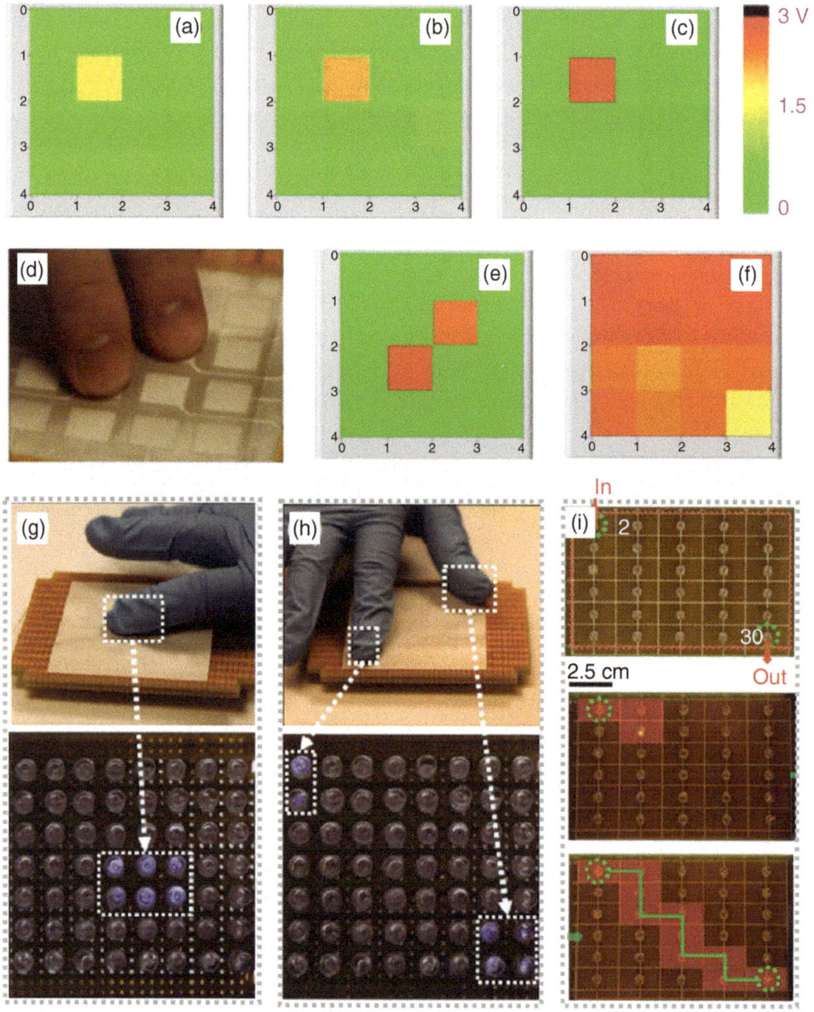

Figure 5.20 Application of some pressure sensors. (a–c) Mapping figures obtained when one of the pixels of a pressure sensor array is touched under small (a), middle (b), and large pressure (c). Source: Reproduced with permission from Yang et al. [8]. Copyright 2013, American Chemical Society. (d) Photograph of the pressure sensor array when two of the pixels are touched by fingers. Source: Reproduced with permission from Yang et al. [8]. Copyright 2013, American Chemical Society. (e) Mapping figure obtained when those two pixels demonstrated in (d) are touched. Source: Reproduced with permission from Yang et al. [8]. Copyright 2013, American Chemical Society. (f) Measured mapping figure when all the pixels are touched. Source: Reproduced with permission from Yang et al. [8]. Copyright 2013, American Chemical Society. (g,h) Measured mapping figure when one and two fingers touch the corresponding pixels of an electret film-enhanced TENG based pressure sensor. Source: Reproduced with permission from Yang et al. [9]. Copyright 2014, American Chemical Society. (i) The electret film-enhanced TENG based pressure sensor array used to track the motion route of a target object. Source: Reproduced with permission from Yang et al. [9]. Copyright 2014, American Chemical Society.

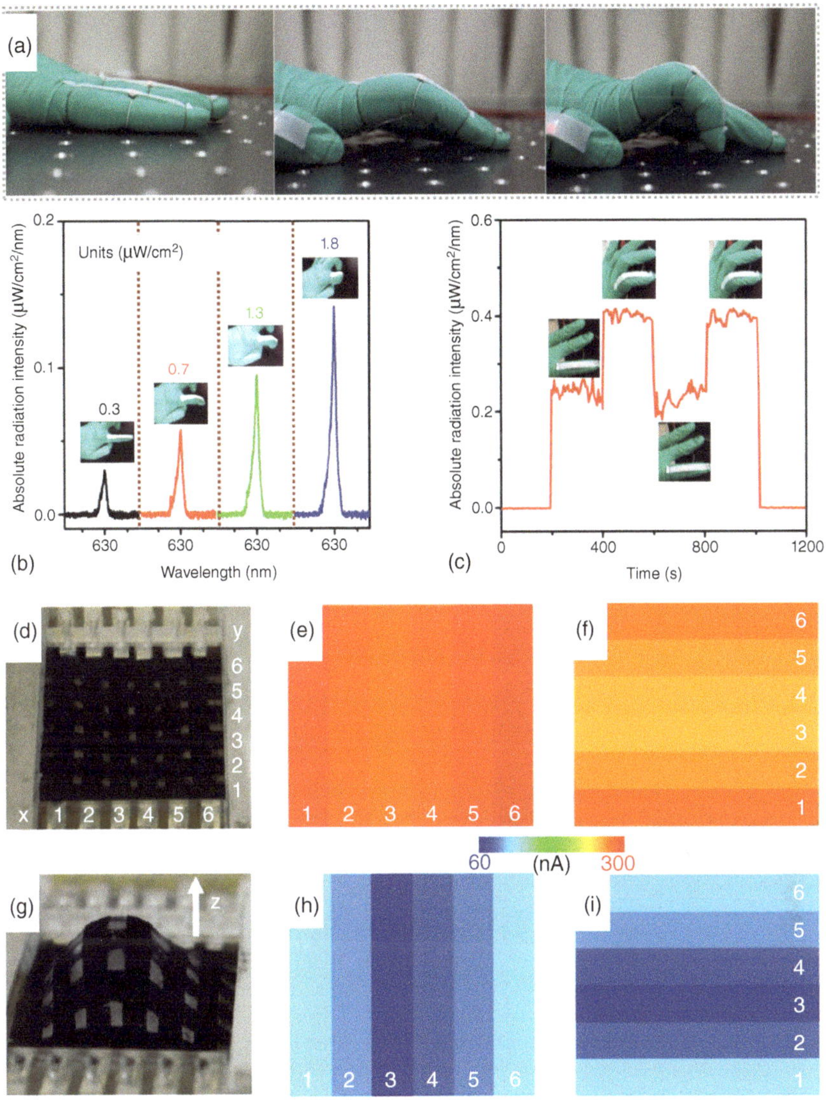

Figure 5.21 Application of some strain sensors. (a) Photograph of TiO_2 nanoparticles–silicone composite film based strain sensor when fixed on human fingers with different gestures. Source: Reproduced with permission from Zhai et al. [19]. Copyright 2017, John Wiley and Sons. (b) Absolute radiation intensities of an LED covered by a TiO_2 nanoparticles–silicone film fixed on human finger with different gestures. Source: Reproduced with permission from Zhai et al. [19]. Copyright 2017, John Wiley and Sons. (c) Dynamic response of the strain sensor to finger gesture. Source: Reproduced with permission from Zhai et al. [19]. Copyright 2017, John Wiley and Sons. (d–f) Optical image of a 3D strain sensor based on graphene–ecoflex film without deformation and the corresponding current mapping figure. Source: Reproduced with permission from Zhai et al. [6]. Copyright 2019, Elsevier. (g–i) Optical image of a 3D strain sensor based on graphene–ecoflex film with a larger deformation and corresponding current mapping figure. Source: Reproduced with permission from Zhai et al. [6]. Copyright 2019, Elsevier.

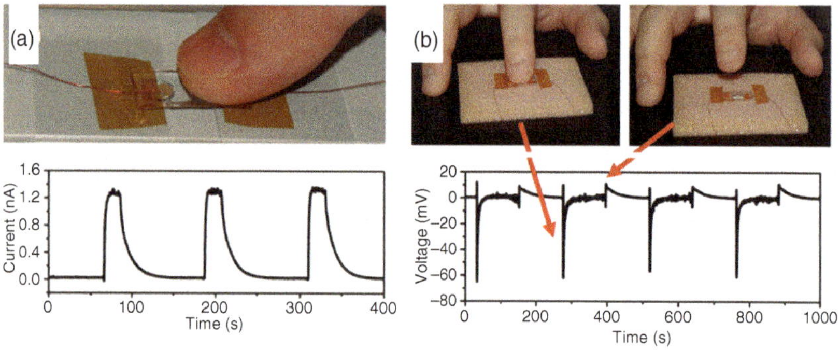

Figure 5.22 Stability of some photodetectors. (a) Optical image of an Sb-doped ZnO microwire based temperature sensor when one side of the device is touched by human finger, and the corresponding output signal. Source: Reproduced with permission from Yang et al. [16]. Copyright 2012, American Chemical Society. (b) Optical image of a PZT microwire based temperature sensor when touched by human finger, and the corresponding output signal. Source: Reproduced with permission from Yang et al. [5]. Copyright 2012, American Chemical Society.

the color of channel 3 and channel 4 are darker than that of other channels, which indicates that the strain of channel 3 and channel 4 is larger than that of other channels.

5.5.3 Temperature Detection

Figure 5.22a shows the photograph of an Sb-doped ZnO microwire based temperature sensor; the microwire is connected by two copper wires through silver paste, then, the whole device is packed with a thin layer of transparent PDMS film, which can improve the robustness of the system. The temperature sensor is used to sense the touch of a finger; when one side of the temperature sensor is touched, a sharp increase of the output current will be observed due to the thermoelectric effect of the Sb-doped ZnO microwire. When the finger periodically touches the temperature sensor, a periodical output current will be observed and the maximum output current is about 1.25 nA (Figure 5.22b).

The photograph of a temperature sensor by utilizing a PZT microwire is demonstrated in Figure 5.22b; similarly, the two ends of the PZT microwire are connected to copper wire through silver paste and then the whole device is packed with a thin layer of PDMS. The temperature of the device will increase when a finger touches the sensor; then, the output voltage will increase sharply thanks to the pyroelectric effect of the PZT (−60 mV). When the finger moves off from the temperature sensor, a reverse pulse will be observed due to the decrease of the temperature.

5.5.4 Image Sensor

Figure 5.23a illustrates the optical image and schematic light illumination of an ITO/BTO/Ag based photodetector 3 × 3 array. The output performance of the photodetector array is investigated by a multichannel testing system. The effective area of each pixel is 2.5 × 2.5 mm. As shown in Figure 5.23a, by utilizing a mask, we can realize a local light illumination (five pixels in a "+" shape of the array is under 405 nm light illumination). The voltage mapping according to Figure 5.23a is illustrated in Figure 5.23b, indicating that the photodetector array has the potential to work as an image sensor. The optical image of a polarized BFO based photodetector array and the voltage mapping obtained are shown in Figure 5.23b; the upper common electrode is ITO while the other separated electrodes of all pixels are a thin layer of Ag possessing an effective area of 3 × 3 mm; silver paste is utilized to connect the copper wires and the ITO electrode. The voltage mapping of the polarized BFO photodetector array under no light illumination and only one pixel of light illumination are shown at the bottom of the optical image demonstrated in Figure 5.23c. Figure 5.23d is the optical image of a photodetector array based on unpolarized BFO ceramic chips; the voltage mapping of the photodetector array under various shapes of light illumination are shown in Figure 5.23e,f, which demonstrates the potential of the photodetector array in real-time dynamic imaging.

An ultraviolet photodetector array with a structure of P3HT/ZnO nanowires is shown in Figure 5.24a; those independent 16 pixels are separated by utilizing a layer of Kapton to separate the top Al electrode. The photograph of the UV photodetector is shown in Figure 5.24b; these 16 pixels are neatly arranged on the 2 × 2 cm ITO coated glass substrate. Subsequently, the photodetector is covered by masks with different shapes along with the illumination of 365 nm LED (113.4 mW/cm^2). A "7" shape pixels display a red color (the open circuit voltage is about 70 mV) under a "7" shape light illumination, while the output voltage of other pixels under dark condition is zero (Figure 5.24c,). The imaging capability of the UV photodetector array is further tested by replacing the "7" shape mask with a "+" mask; a "+" shape of voltage mapping figure is observed (Figure 5.24d). The voltage mapping figures indicate that the photodetector array possesses the potential to be an image sensor.

Figure 5.24e shows the schematic diagram of a square photodetector array (36 pixels) and the schematic light illumination diagram. An ITO coated glass serves as the substrate; then a layer of SnS thin film works as the absorber material; ZnO nanowires are used to form a p–n junction with the SnS; the 6 × 6 Al electrode separated by Kapton tapes works as the upper electrode. The current mapping figure is obtained when half of the pixels are covered by a mask. The photograph of the photodetector array is shown in Figure 5.24f, which shows that all the pixels are well aligned on a 2 × 2 cm substrate. The current mapping figure under 360 nm light illumination is illustrated in Figure 5.24g, which indicates that the short-circuit of those pixels exposed to light is negative. The mapping figure under 690 nm light illumination is shown in Figure 5.24h, which hints that the short-circuit current of those pixels when illuminated by 690 nm LED is positive. The output current's polarity exhibits a light wavelength dependence, which may be useful in some special application scenarios such as filterless color discrimination and visible light communication.

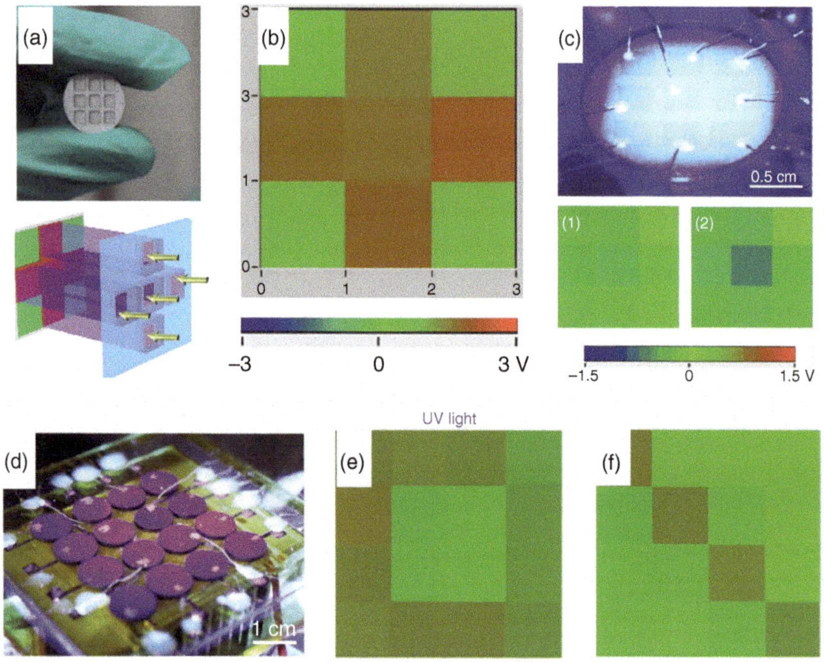

Figure 5.23 Application of some ferroelectric materials based photodetector. (a) Optical image of a BTO based photodetector array and light illumination schematic diagram. Source: Reproduced with permission from Ma et al. [11]. Copyright 2017, John Wiley and Sons. (b) Voltage mapping figure obtained according to (a). Source: Reproduced with permission from Ma et al. [11]. Copyright 2017, John Wiley and Sons. (c) Optical image of a polarized BFO based photodetector array and the voltage mapping figure obtained (1 and 2). Source: Reproduced with permission from Qi et al. [14]. Copyright 2018, John Wiley and Sons. (d–f) Optical image (d) of an unpolarized BFO based photodetector array and the voltage mapping figure obtained (e,f). Source: (e,f) Reproduced with permission from Qi et al. [13]. Copyright 2018, American Chemical Society.

5.6 Conclusion and Prospects

The development of new materials and mechanisms has promoted many new sensors, which may take the place of some traditional sensors. For instance, the newly emerging TENG can be designed as pressure sensor and magnetic field sensor owing to its high sensitivity and low cost; by utilizing the outstanding mechanical property of ecoflex and thermoelectric performance of graphene, a high performance low cost self-powered strain sensor can be fabricated; by utilizing nanotechnology, temperature can be detected by the pyroelectric effect of a PZT nanowire; image sensors can be realized by the p–n junction between N type ZnO wires and some nano thin film.

Looking forward, the upcoming Internet of Things, intelligent robots, and energy shortage have posed many challenges to high-performance self-powered sensors. New materials such as the 1D and 2D nano materials offer many opportunities to develop high-performance self-powered sensors. Besides, new

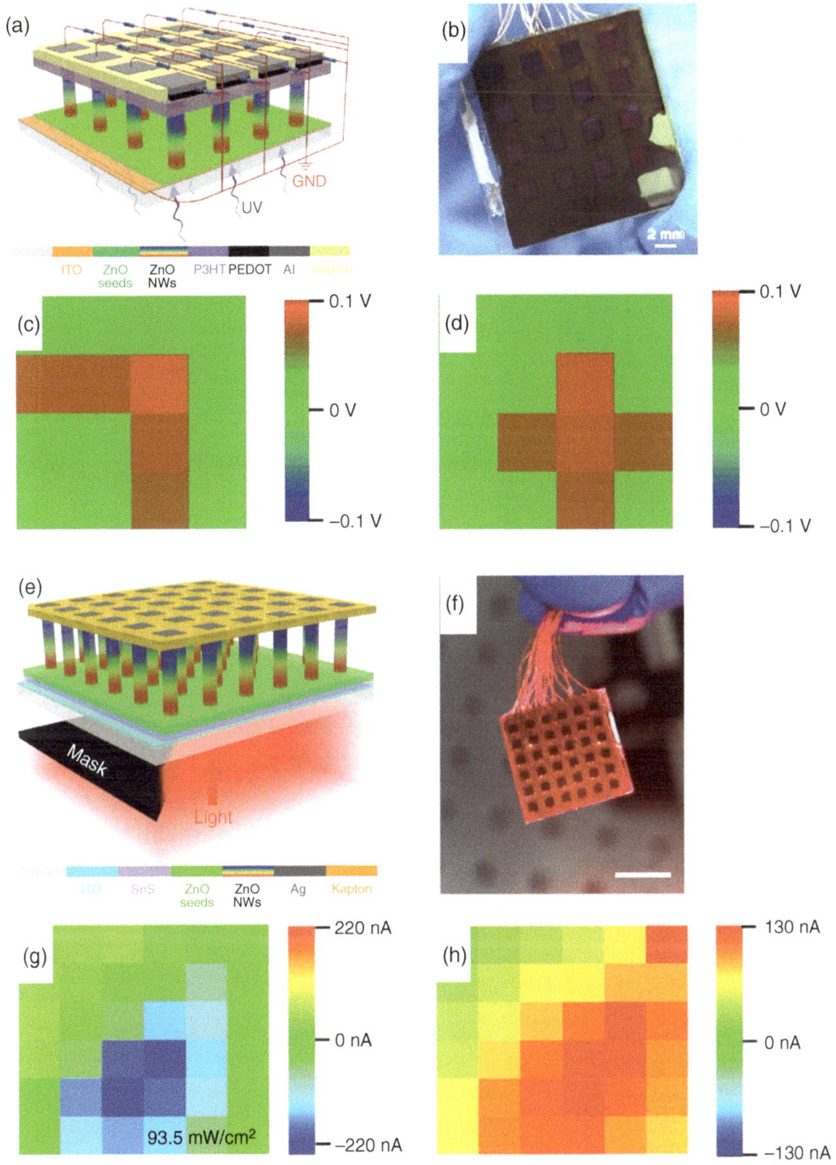

Figure 5.24 Application of some semiconductor heterojunction based photodetector. (a,b) Schematic diagram (a) and optical image (b) of a P3HT/ZnO nanowires heterojunction based photodetector array. Source: Reproduced with permission from Quyang et al. [3]. Copyright 2017, John Wiley and Sons. (c,d) Voltage mapping figure obtained of the heterojunction based photodetector array. Source: Reproduced with permission from Quyang et al. [3]. Copyright 2017, John Wiley and Sons. (e,f) Schematic diagram (e) and optical image (f) of an SnS/ZnO nanowires heterojunction based photodetector array. Source: Reproduced with permission from Quyang et al. [12]. Copyright 2018, Elsevier. (g,h) The current mapping figure obtained of the heterojunction based photodetector array. Source: Reproduced with permission from Quyang et al. [12]. Copyright 2018, Elsevier.

mechanisms such as the multi-effects coupled mechanism should be developed to boost the efficiency. Researchers should aim at developing high sensitivity, fast response speed, high stability, high energy efficiency, low cost sensors.

References

1 Jiang, Y., Wang, Y., Mishra, Y.K. et al. (2018). Stretchable CNTs–ecoflex composite as variable-transmittance skin for ultrasensitive strain sensing. *Adv. Mater. Technol.* 3: 1800248.

2 Ma, N. and Yang, Y. (2018). Boosted photocurrent in ferroelectric $BaTiO_3$ materials via two dimensional planar-structured contact configurations. *Energy* 50: 417.

3 Ouyang, B., Zhang, K., and Yang, Y. (2017). Self-powered UV photodetector array based on P3HT/ZnO nanowire array heterojunction. *Adv. Mater. Technol.* 2: 1700208.

4 Yang, Y., Lin, L., Zhang, Y. et al. (2012). Self-powered magnetic sensor based on a triboelectric nanogenerator. *ACS Nano* 6: 10378.

5 Yang, Y., Zhou, Y., Wu, J.M., and Wang, Z.L. (2012). Single micro/nanowire pyroelectric nanogenerators as self-powered temperature sensors. *ACS Nano* 6: 8456.

6 Zhang, D., Zhang, K., Wang, Y. et al. (2019). Thermoelectric effect induced electricity in stretchable graphene-polymer nanocomposites for ultrasensitive self-powered strain sensor system. *Nano Energy* 56: 25.

7 Zhao, X., Chen, B., Wei, G. et al. (2019). Polyimide/graphene nanocomposite foam-based wind-driven triboelectric nanogenerator for self-powered pressure sensor. *Adv. Mater. Technol.* 4: 1800723.

8 Yang, Y., Zhang, H., Lin, Z.-H. et al. (2013). Human skin based triboelectric nanogenerators for harvesting biomechanical energy and as self-powered active tactile sensor system. *ACS Nano* 7: 9213.

9 Yang, Y., Zhang, H., Zhong, X. et al. (2014). Electret film-enhanced triboelectric nanogenerator matrix for self-powered instantaneous tactile imaging. *ACS Appl. Mater. Interfaces* 6: 3680.

10 Zhang, D., Song, Y., Ping, L. et al. (2019). Photo-thermoelectric effect induced electricity in stretchable graphene-polymer nanocomposites for ultrasensitive strain sensing. *Nano Res.* 12: 2982.

11 Ma, N., Zhang, K., and Yang, Y. (2017). Photovoltaic–pyroelectric coupled effect induced electricity for self-powered photodetector system. *Adv. Mater.* 29: 1703694.

12 Ouyang, B., Zhang, K., and Yang, Y. (2018). Photocurrent polarity controlled by light wavelength in self-powered ZnO nanowires/SnS photodetector system. *iScience* 1: 16–23.

13 Qi, J., Ma, N., Ma, X. et al. (2018). Enhanced photocurrent in $BiFeO_3$ materials by coupling temperature and thermo–phototronic effects for self-powered ultraviolet photodetector system. *ACS Appl. Mater. Interfaces* 10: 13712.

14 Qi, J., Ma, N., and Yang, Y. (2018). Photovoltaic–pyroelectric coupled effect based nanogenerators for self-powered photodetector system. *Adv. Mater. Interfaces* 5: 1701189.

15 Song, K., Ma, N., and Yang, Y. (2017). Enhanced self-powered UV photoresponse of ferroelectric PZT materials by pyroelectric effect. *Adv. Mater. Technol.* 2: 1700221.

16 Yang, Y., Pradel, K.C., Jing, Q. et al. (2012). Thermoelectric nanogenerators based on single Sb-doped ZnO micro/nanobelts. *ACS Nano* 6: 6984.

17 Yang, Y., Lin, Z.-H., Hou, T. et al. (2012). Nanowire-composite based flexible thermoelectric nanogenerators and self-powered temperature sensors. *Nano Res.* 5: 888.

18 Song, K., Zhao, R., Wang, Z.L., and Yang, Y. (2019). Conjuncted pyro-piezoelectric effect for self-powered simultaneous temperature and pressure sensing. *Adv. Mater.* 31: 1902831.

19 Zhai, Q. and Yang, Y. (2017). Highly stretchable variable-transmittance skin for ultrasensitive and wearable strain sensing. *Adv. Mater. Technol.* 2: 1700161.

20 Zhu, L., Wang, L., Xue, F. et al. (2017). Piezo–phototronic effect enhanced flexible solar cells based on n-ZnO/p-SnS core-shell nanowire array. *Adv. Sci.* 4: 1600185.

6

Hybridizing Nanogenerators and Energy Storage Devices

6.1 Introduction

Energy is everywhere – the rapid development of modern intelligent society cannot progress without energy consumption. Meanwhile, compact and portable electronic devices are gradually popular in our daily lives. It is worth noting that most of these devices need to be powered by recharging them or using external power sources including Li-ion batteries or supercapacitors owing to limited energy storage [1–3], which also consumes much manpower, material, and financial resources. For the above reasons, designing an integrated energy collection and storage structure has gradually become an attractive research focus and hotspot. Among them, hybridizing nanogenerators and energy storage devices has developed rapidly and become one of the most outstanding and high-efficiency approach to simultaneously harvest and store energy.

At present, fast-growing nanogenerators can harvest almost all kinds of common energies in our daily lives, such as mechanical, thermal, and light energy. Wang's group invented the piezoelectric nanogenerator (PENG) in 2006 [4] and triboelectric nanogenerator (TENG) in 2012 [5], as shown in Figure 6.1. All kinds of mechanical energies occur in the external environment, including wind energy [6], droplet energy [7], wave energy [8], and even the energy of human movement [9–11]. The two kinds of nanogenerators above can effectively harvest these energies and then convert them into electric energy.

Figure 6.1a depicts the scanning electron microscopy (SEM) of ZnO nanowires (NWs), and the corresponding transmission electron microscopy (TEM) images as shown in Figure 6.1b [4]; the first ZnO NWs-based PENG can convert mechanical energy into electricity, which exhibits a superior efficiency ranging from 17% to 30%. Figure 6.1c displays the experimental diagram and procedure for electricity generation by using a conductive atomic force microscopy (AFM) tip to deform a ZnO NW with a small degree.

The first TENG was fabricated by two pieces of polymer material with different triboelectric characteristics. Figure 6.1d displays the diagram of the TENG in bending and releasing process [5]. Photographic images indicate that the TENG possesses good flexibility. Generally, the output electrical signals of the state-of-the-art TENG are greater than that of PENG, whereas TENGs always

*Corresponding author: yayang@binn.cas.cn

Hybridized and Coupled Nanogenerators: Design, Performance, and Applications,
First Edition. Ya Yang.

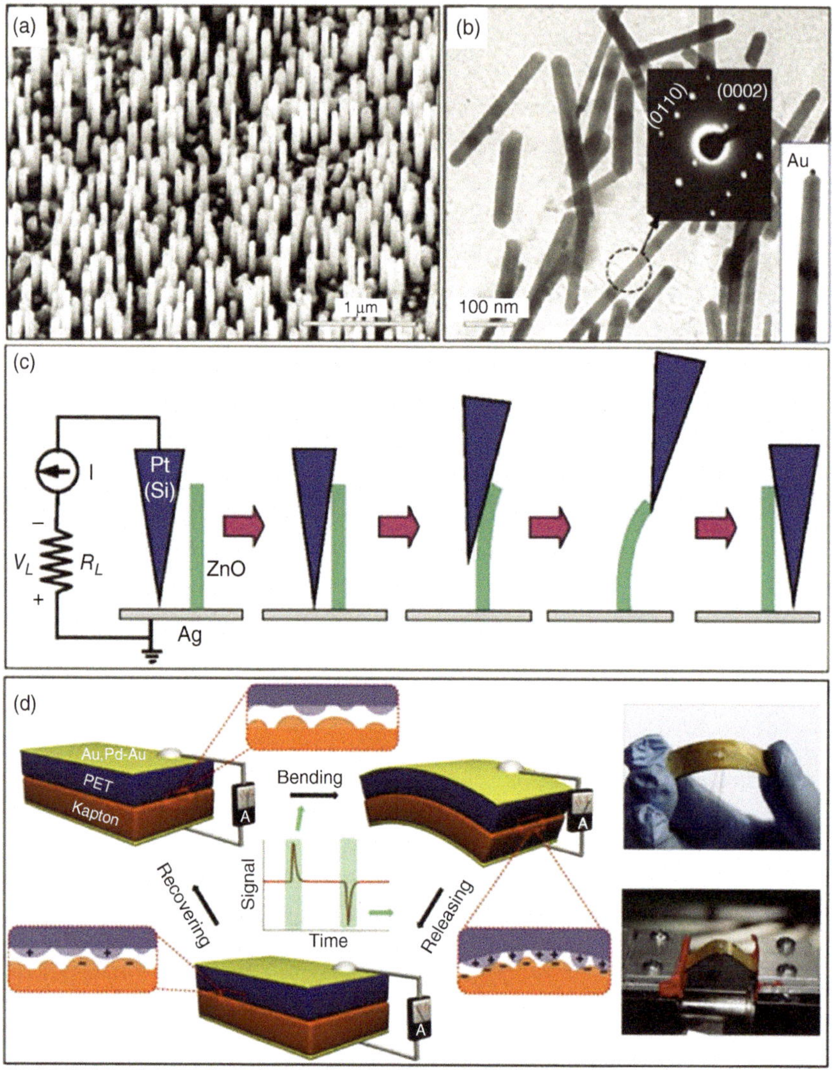

Figure 6.1 Working mechanisms of PENG and TENG. (a,b) SEM and TEM images of ZnO NWs. (c) Experimental setup and procedure for generating electricity by deforming a piezoelectric NW using a conductive AFM tip. Source: Reproduced with permission from Wang and Song [4]. Copyright 2006, The American Association for the Advancement of Science. (d) The structure diagram and photographs of the first TENG. Source: Reproduced with permission from Fan et al. [5]. Copyright 2012, Elsevier.

output high voltages and low currents [12]. For PENG, it is an efficient approach for harvesting smaller scale and low-frequency mechanical energy [13]. In addition, PENGs are usually small in size and easy to carry. Apart from being able to harvest micro-scale power sources to power small portable electric devices, with the rapid development in recent years, TENG can drive larger electronic devices by effectively harvesting some large scale mechanical energy, such as wave and wind energy. . In general, PENG and TENG have their own advantages in harvesting different forms of energy in different environments. Both PENG and TENG have been developed into a remarkable and high-efficiency approach for converting various mechanical energies into electricity.

After energy generation, energy storage systems are an equally important research hotspot. Li-ion batteries are the most popular devices for energy storage in the world [14, 15]. The first commercially available Li-ion battery with a cylindrical structure was invented by Sony Corporation in 1991 [16], as seen in Figure 6.2a. The external voltage can drive the migration of Li-ions and convert electric energy into chemical energy stored in battery owing to the electrochemical reactions occurring at the two electrodes of the battery. Figure 6.2b gives a schematic diagram of the corresponding mechanism with a brief structure [2]. In the discharge process, carbon anode and $LiCoO_2$ cathode react according to Eqs. (6.1, 6.2), respectively:

$$C + xe^- + xLi^+ = Li_xC \quad \text{(anode)} \tag{6.1}$$

$$LiCoO_2 - xe^- - xLi^+ = Li_{1-x}CoO_2 \quad \text{(cathode)} \tag{6.2}$$

After decades of rapid development, great improvement in the performance of Li-ion battery has been achieved in meeting the needs of a variety of portable electronic devices for a long time, such as notebook computers and cellular phones.

Apart from Li-ion battery, supercapacitors and electrochemical capacitors (ECs) are also regarded as effective and common energy storage devices [19]. Compared to conventional dielectric capacitors and rechargeable second batteries ECs, or supercapacitors, have many remarkable properties including long lifetime, excellent charging/discharging performance, and power density [3, 20, 21]. Figure 6.2c shows a typical diagram of a supercapacitor, indicating that energy storage occurs at the junction of the electrolyte and electrodes [16]. Supercapacitors possess symmetric or asymmetric structures, and the symmetric devices consist of two same electrodes, as seen in Figure 6.2d [17]. Similarly, the asymmetric supercapacitors consist of two different electrodes. In addition, supercapacitors can also be distinguished in other ways such as the electrolyte, storage mechanism, cell structure, and electrode material [18]. Supercapacitors consist of a pseudocapacitor and an electric double-layer capacitor in terms of storage mechanisms; two main kinds of supercapacitors are shown in Figure 6.2e,f [18]. It is noted that supercapacitors and Li-ion batteries possess two electrodes in contact with an electrolyte.

At present, energy harvesting technologies are based on nanogenerators or solar cells for collecting one or more kinds of energies among mechanical, thermal, and light energies. Energy storage technologies based on Li-ion batteries or supercapacitors have all been developed rapidly. The structure design and performance improvement of hybridizing energy collection and energy storage devices integration have attained satisfactory achievements in the recent several years. In addition, self-charging power textiles have made tremendous progress in wearable electronics for simultaneous energy harvest and storage. These hybridizing

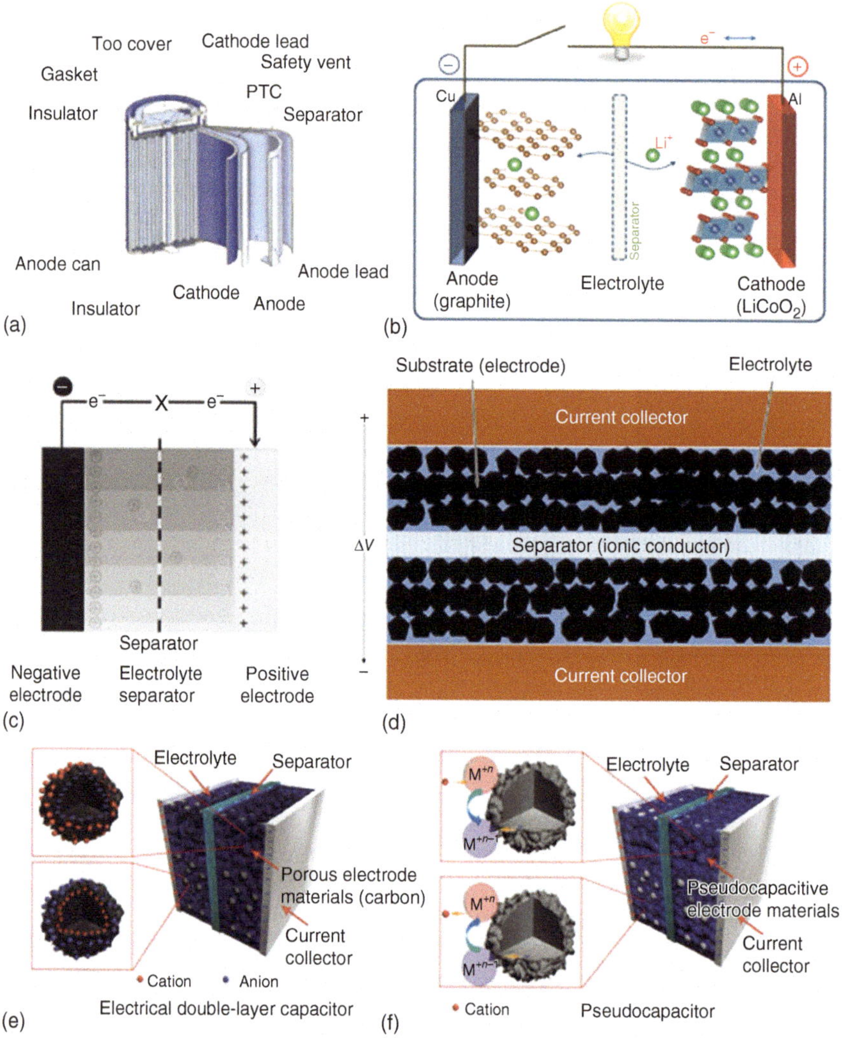

Figure 6.2 Structure diagrams of Li-ion battery and supercapacitor. (a) The first commercial Li-ion battery. Source: Reproduced with permission from Winter and Brodd [16]. Copyright 2004, American Chemical Society. (b) Schematic illustration of the first Li-ion battery ($LiCoO_2/Li^+$electrolyte/graphite). Source: Reproduced with permission from Goodenough and Park [2]. Copyright 2013, American Chemical Society. (c) Representation of an electrochemical capacitor (supercapacitor). Source: Reproduced with permission from Winter and Brodd [16]. Copyright 2004, American Chemical Society. (d) Symmetric supercapacitor. Source: Reproduced with permission from González et al. [17]. Copyright 2016, Elsevier. (e,f) A pseudocapacitor and a hybrid capacitor. Source: Reproduced with permission from Zhong et al. [18]. Copyright 2015, Royal Society of Chemistry.

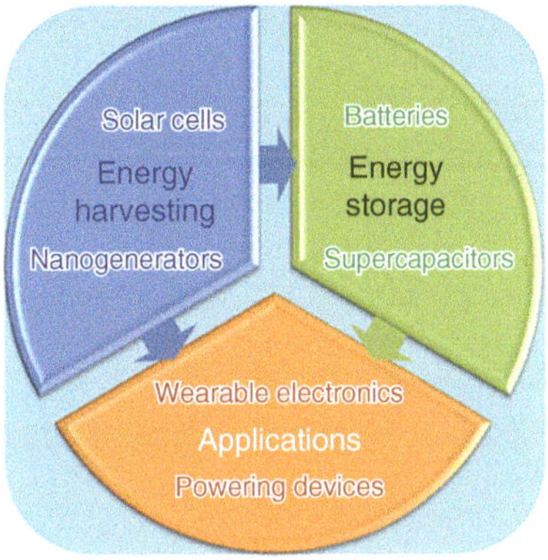

Figure 6.3 The relationship among energy harvesting, storage, and applications.

energy collection and storage devices show a huge application prospect for driving several electronic devices with small power, degrading wastewater, and generating hydrogen, etc. As shown in Figure 6.3, irrespective of the direct or indirect application through energy harvest or storage system, respectively, the ultimate purpose is to apply it to different fields. Therefore, reasonably and controllably regulating the relationship among energy harvest, storage, and applications will show huge application prospects in an intelligent society.

In this chapter, the research background is first introduced, followed by a summary of working mechanisms of four hybrid nanogenerators and energy storage devices. In particular, the progress in the development of materials, structure, performance, and applications are evaluated and summarized comprehensively. Focus will be on materials preparation and structural designs for nanogenerators and Li-ion batteries or supercapacitors. Finally, the opportunities, challenges, potential applications, and future development direction are discussed and explored.

6.2 Working Mechanisms

Hybrid nanogenerators and energy storage devices may have obvious difference in materials or structures, whereas those devices with similar type have the same working mechanisms essentially. There are four typical working mechanisms corresponding to hybrid PENGs or TENGs and Li-ion battery or supercapacitors respectively. This section focuses on summarizing the working mechanisms of four kinds of typical hybrid nanogenerators and energy storage devices consisting of hybrid PENGs (TENGs) and Li-ion batteries (supercapacitors).

6.2.1 Piezoelectric Nanogenerator-Based Energy Storage Devices

6.2.1.1 PENG-Based Li-ion Batteries

In 2012, Xue and coworkers first demonstrated a self-charging hybrid PENG and Li-ion battery for directly harvesting mechanical energy and simultaneously storing it to be chemical energy, which does not need the intermediate process of first converting mechanical energy into electric energy [22]. Owing to its excellent piezoelectricity, poly(vinylidene fluoride) (PVDF) film replaces the traditional polyethylene (PE) separator of Li-ion battery. In addition, the PVDF film is a superior Li-ion conductor, which makes PVDF to be the electrolyte base and electrodes binder of Li-ion batteries [23, 24]. The external mechanical strain will produce a piezoelectric potential that plays an important role as a charge pump and drives Li-ions flow between two electrodes.

The detailed working processes of self-charging devices are displayed in Figure 6.4 [22]. First, in a discharged state, the device is kept at the original fabricated structure. As seen from Figure 6.4a, the materials $LiCoO_2$, TiO_2 nanotube arrays, and $LiPF_6$ act as the cathode, anode, and electrolyte, respectively. When the self-charging device is subjected to severe compressive strain, an internal electric field is generated in the piezoelectric PVDF separator, which builds a positive (negative) potential at the cathode (anode), as seen in Figure 6.4b. And then, the piezoelectric potential will drive Li-ions migration across the piezo-separator film from the cathode to anode corresponding to the charging process, as observed from Figure 6.4c. The corresponding electrochemical oxidation–reduction reactions at the two electrodes are as follows:

$$\mathrm{LiCoO_2} \rightarrow \mathrm{Li_{1-x}CoO_2} + x\mathrm{Li^+} + x\mathrm{e^-} \quad \text{(cathode)} \tag{6.3}$$

$$\mathrm{TiO_2} + x\mathrm{Li^+} + x\mathrm{e^-} \rightarrow \mathrm{Li_xTiO_2} \quad \text{(anode)} \tag{6.4}$$

Under continuous mechanical deformation, Li^+ ions will migrate, driven by the generated piezoelectric potential until a stable condition is reached where the reestablished chemical equilibriums can balance the piezo-potential, as depicted in Figure 6.4d, and the self-charging process ceases under the new equilibrium above. When the compressive strain is removed, the piezoelectric potential disappears and then breaks the electrostatic equilibrium, as seen from Figure 6.4e, which will make partial Li^+ ions migrate back into the cathode and rebuild a balanced Li-ion distribution (Figure 6.4f).

In this piezoelectric-driven self-charging process above, the piezoelectric PVDF film is equivalent to the direct current (DC) power supply for traditional Li-ion battery. Though some specific mechanisms are different, both the PVDF film and the DC power supply can be considered as charge pumps. For the conventional charging method, in order to keep charge balance, the DC power supply drives electrons and Li-ions to flow in the external circuit and in the battery, respectively. Thus, the electrochemical reactions of charging process take place on the two electrodes. For a self-charging cell, the piezoelectric material only drives the migration of Li-ions rather than electrons, which also completes the charging process of the device.

6.2.1.2 PENG-Based Supercapacitors

As a kind of quickly developed hybrid nanogenerators and energy storage devices, PENG-based supercapacitors also show great application prospect similar to PENG-based batteries. Researchers have designed a self-charging supercapacitor power cell (SCSPC) that can produce electrochemical energy by directly converting mechanical energy [19]. Manganese oxide (MnO_2) exhibits

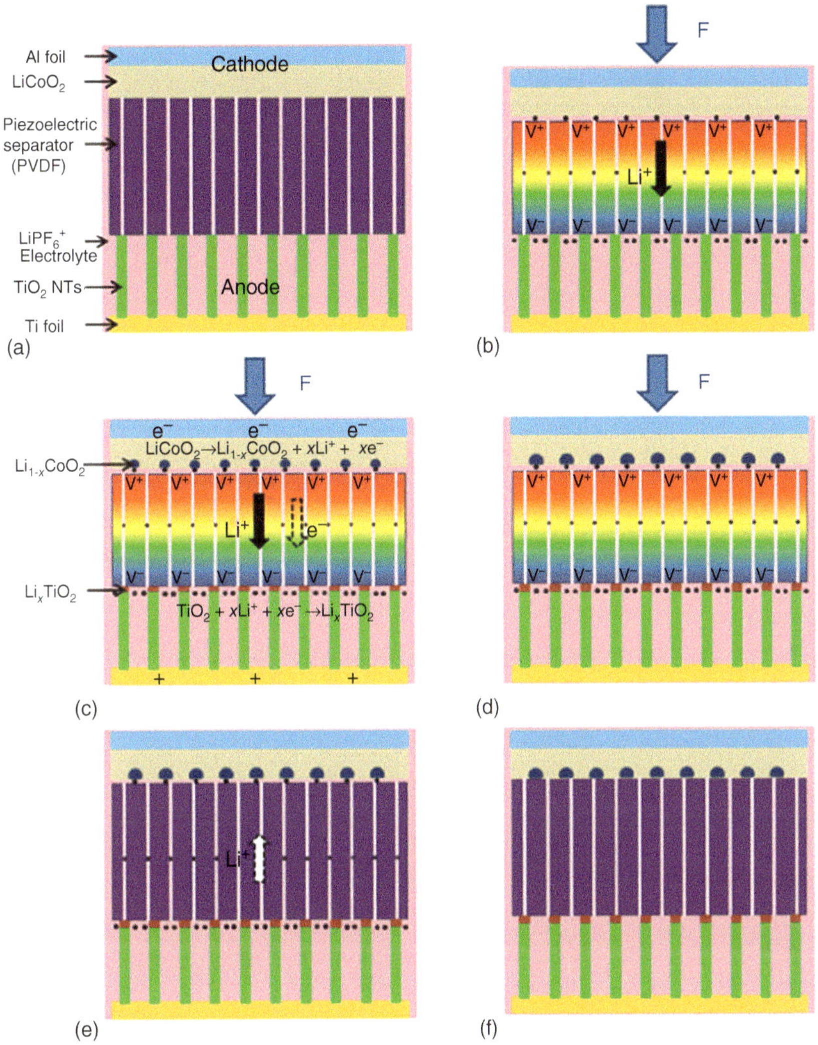

Figure 6.4 (a) Schematic illustration of the self-charging power cell in discharged state. (b) When a compressive stress is applied onto the device, the piezoelectric separator layer (e.g. PVDF film) creates a piezopotential. (c) Under the driven of the piezoelectric field, the Li ions migrate from the cathode to the anode, leading to the corresponding charging reactions at the two electrodes. (d) The status where chemical equilibrium of the two electrodes is re-established and the self-charging process ceases. (e) When the applied force is released, a portion of the Li ions will diffuse back to the cathode. (f) This electrochemical system reaches a new equilibrium and completes a cycle of self-charging. Source: Reproduced with permission from Xue et al. [22]. Copyright 2012, American Chemical Society.

high specific capacitance, excellent electrochemical activity, and low cost, which makes it an ideal electrode material [25–27]. In addition, PVA/H_3PO_4 and PVDF-ZnO composite film act as electrolyte and separator of the SCSPC. We take the SCSPC as a typical example to discuss its working mechanism in detail. Compared with PENG-based batteries, the piezoelectric potential in the separator induced by external mechanical deformation also appears in the SCSPC, which is the same in terms of this point. However, there are still great differences in some specific mechanisms.

Figure 6.5 shows the working mechanism diagram dependent on the electrochemical reaction [19]. Figure 6.5a displays the structure diagram of SCSPC. At first, the SCSPC is in a discharge state without any mechanical deformation (Figure 6.5a), and there is no electrochemical reaction. When a compressive deformation is applied on the device, piezoelectric potential will occur across the PVDF-ZnO film due to piezoelectric effect [22, 28, 29], as shown in Figure 6.5b. Under the effect of piezo-potential, cationic (H^+) will move to balance the piezo-potential in PVDF-ZnO film, which will rebuild a new chemical equilibrium by generating a redox reaction at the electrode surfaces, as shown in Figure 6.5c. The piezoelectric potential drives the H^+ ions to migrate and insert into the lattice of MnO_2 electrode, which leads to the reduction reaction and produces positive charges at the Al electrode. The oxidation reaction generates electrons owing to the desertion of H^+ ions. The electrochemical reactions may be expressed as Eq. (6.5):

$$MnO_2 + H^+ + e^- \leftrightarrow MnOOH \tag{6.5}$$

As seen from Figure 6.5d, when the compressive deformation is removed, the piezoelectric potential will disappear. The electrochemical balance of the device is correspondingly disrupted again. To compensate and rebuild a new electrochemical balance, partial ions will move back, which completes one charging cycle. Under continuing compressive stress to the SCSPC, the charge cycles occur repetitively.

6.2.2 Triboelectric Nanogenerator-Based Energy Storage Devices

Since the first TENG was invented in 2012 [5], TENGs have developed rapidly and become an increasingly popular research hotspot in academia owing to the availability of materials, relatively simple device structure, and superior output performance. Meanwhile, TENG-based energy storage devices have developed rapidly and become one of the most outstanding and high-efficiency approaches to simultaneously harvest and store energy. Therefore, we will take several representative jobs as examples to discuss working mechanisms of corresponding devices.

6.2.2.1 TENG-Based Lithium-Ion Batteries

Li-ion battery always needs to be periodically charged owing to limited capacities for continuously powering electronic devices. In 2017, Zhao and coworkers designed a triboelectrification-enabled Li-ion battery with self-charging function

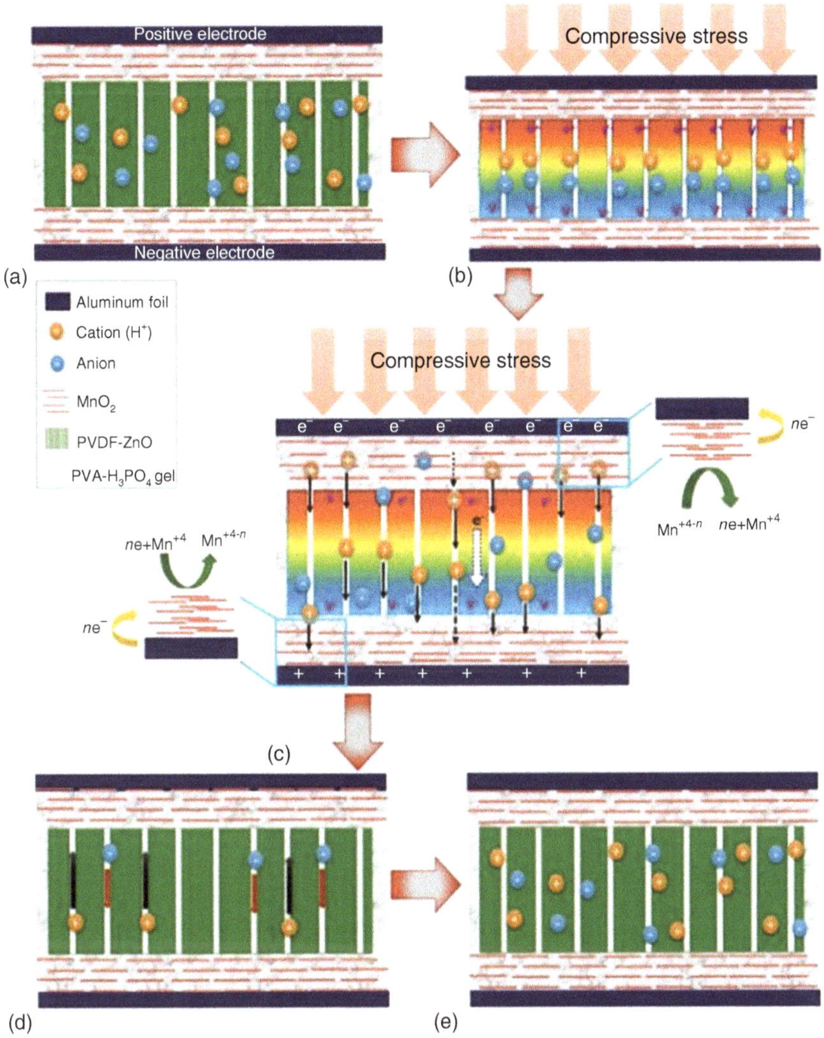

Figure 6.5 (a) Fabricated SCSPC at the discharge condition. (b) A piezoelectric field in separator film is generated due to the mechanical deformation applied on the device. (c) Under the piezoelectric field, the H^+ ions will migrate to the negative electrode, leading to the corresponding charging reactions at the two electrodes. (d) When external deformation is released, it causes a reverse reaction on both sides of the electrode, and ions are relocated into the original position. (e) Completed self-charging cycle.. Source: Reproduced with permission from Ramadoss et al. [19]. Copyright 2015, American Chemical Society.

by a TENG with a novel structure consisting of two Li-ion battery units, and successfully realized energy harvest and storage of wind-induced vibration energy [30]. The Li-ion batteries were fabricated by using $LiMn_2O_4$ and carbon as cathode and anode, respectively.

The self-charging working diagrams are shown in Figure 6.6 [30]. The adjacent copper electrodes are connected to a transformer for increasing output current

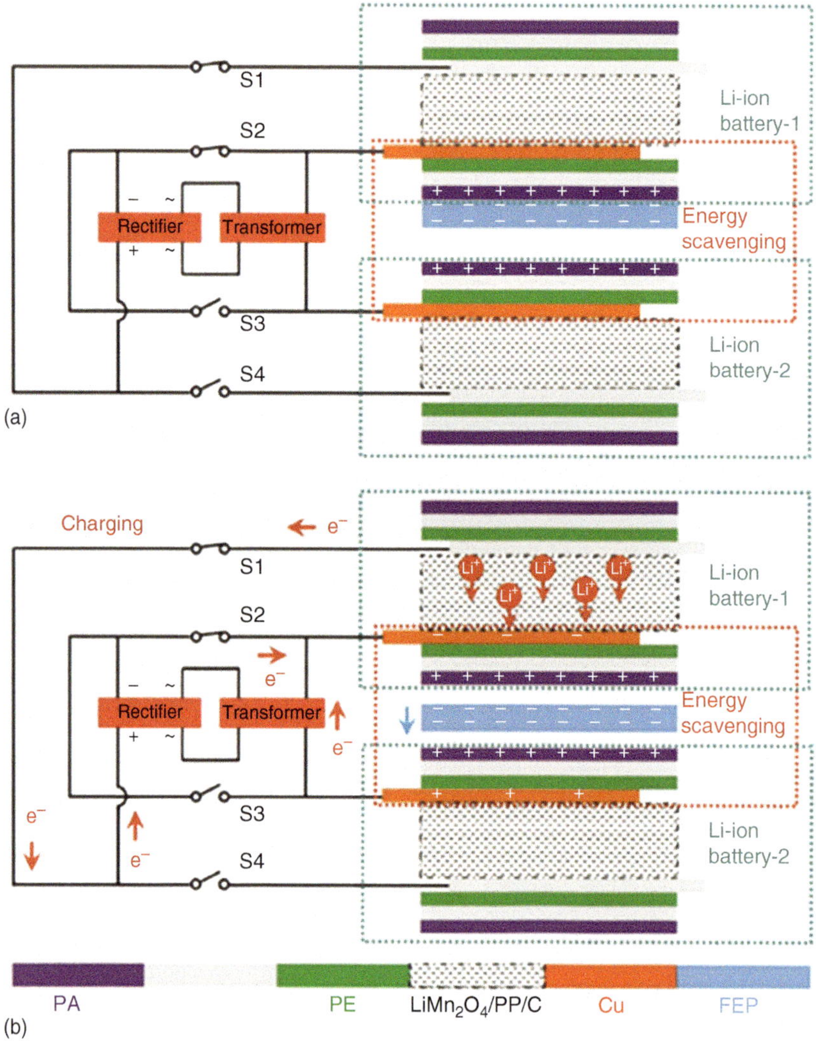

Figure 6.6 Working diagram of self-charging Li-ion batteries for energy scavenging and storage, with no output for the energy scavenging unit (a), the flow of electrons in the circuit (b). Source: Reproduced with permission from Zhao et al. [30]. Copyright 2017, John Wiley and Sons.

and a rectifier for converting the AC output signals into DC signals. Four switches were used to connect two batteries and the rectifier. As illustrated in Figure 6.6a, at the beginning, the switches S1 and S2 were closed. There is no output current signal observed in the energy harvest unit when the top Li-ion battery contacts FEP film. When the FEP film is kept away from the top battery, the generated electrons will migrate from the bottom to the upper Cu electrode, as seen from Figure 6.6b. With the help of a transformer and a rectifier, the generated voltage induces the inflow of electrons from cathode to anode for the top Li-ion battery.

The electrochemical reactions occur at two electrodes; the electrochemical reaction at the cathode can be expressed as Eq. (6.6):

$$LiMn_2O_4 \leftrightarrow Li_{1-x}Mn_2O_4 + xLi^+ + xe^- \quad (6.6)$$

The migration of Li^+ ions will lead to the corresponding electrochemical reaction expressed as Eq. (6.7) at the anode electrode:

$$6C + xLi^+ + xe^- \leftrightarrow Li_xC_6 \quad (6.7)$$

When the FEP moves upward, the generated electricity can charge the top Li-ion battery. Under the periodic mechanical vibration of FEP film between two batteries induced by wind energy, both the energy harvest and storage were simultaneously realized. Moreover, when switches S3 and S4 were turned on and S1 and S2 were turned off, the generated electricity can charge the bottom Li-ion battery. This device with two Li-ion batteries could supply more electricity to meet more power consumption. When one Li-ion battery is out of electric power, another battery can be used to drive electronic devices continuously by switching on the four switches.

In the following text, we will continue to discuss another representative TENG-based energy storage device for efficiently harvesting mechanical vibration energy [31]. The device consists of two layers as displayed in Figure 6.7a. At first, when a compressive stress is applied on the device, TiO_2 nanotubes and the FEP film make contact closely, which induces different type of charges on the FEP film and TiO_2 nanotubes (Figure 6.7b). Then, the separation between the FEP film and TiO_2 nanotubes causes the flow of electrons due to electrostatic induction, as seen in Figure 6.7c. On reaching electrostatic equilibrium, the flow of electrons disappears. Likewise, the electrostatic equilibrium will be broken when the compressive stress is applied again, which induces electrons to flow in the external circuit to rebuild an electrostatic equilibrium (Figure 6.7d).

Figure 6.7e demonstrates the charging diagram of the device with the help of a TENG-charged capacitor. Li^+ ions flow from $LiMn_2O_4$ film to TiO_2 nanotubes driven by internal electric field. The decrease of Li^+ ions induces the chemical reaction at the $LiMn_2O_4$ cathode for compensation, as shown in Eq. (6.8):

$$LiMn_2O_4 \rightarrow Li_{1-x}Mn_2O_4 + xLi^+ + xe^- \quad (6.8)$$

The corresponding chemical reaction at the TiO_2 anode is as in Eq. (6.9):

$$TiO_2 + xLi^+ + xe^- \rightarrow Li_xTiO_2 \quad (6.9)$$

Li^+ ions migrate from $LiMn_2O_4$ to TiO_2 until electrostatic equilibrium, that is, until the voltages of the battery and the power source are the same. The discharging process is displayed in Figure 6.7f. The corresponding redox reaction occurs at the two electrodes.

$$Li_xTiO_2 \rightarrow TiO_2 + xLi^+ + xe^- \quad \text{(anode)} \quad (6.10)$$

$$Li_{1-x}Mn_2O_4 + xLi^+ + xe^- \rightarrow LiMn_2O_4 \quad \text{(cathode)} \quad (6.11)$$

The chemical reactions can induce electrons flow from Ti to Al electrode and Li^+ ions migration from TiO_2 anode to $LiMn_2O_4$ cathode responding to the discharging process.

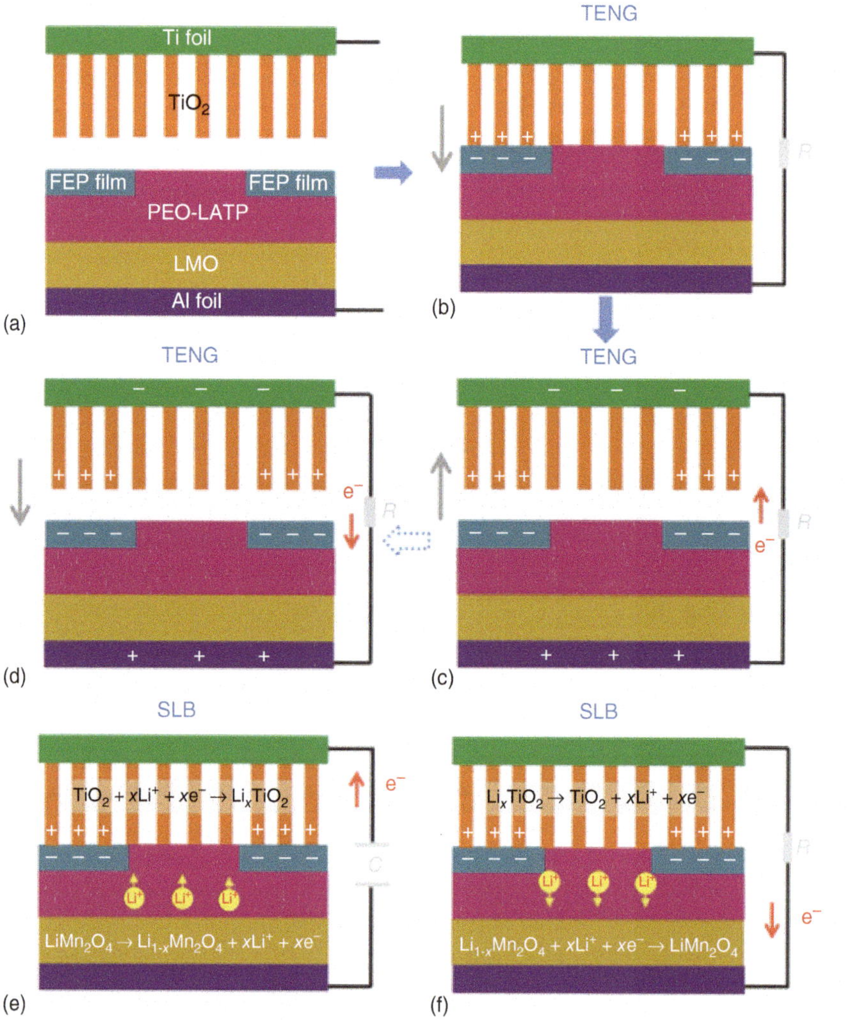

Figure 6.7 Working mechanism of the CP device as a TENG and a Li-ion battery. (a) Schematic diagram of the CP device. (b) Contact of TiO_2 nanotubes and FEP film. (c,d) Electric generation for the CP device as a TENG when the TiO_2 nanotubes are moved up and down. (e,f) The charging and discharging processes for the CP device as a Li-ion battery. Source: Reproduced with permission from Liu et al. [31]. Copyright 2017, John Wiley and Sons.

The hybrid device can act as a TENG, and also as a solid Li-ion battery. However, the high AC signals of the TENG cannot be directly used to charge the storage device. To meet the needs of practical applications, a feasible circuit was designed consisting of a rectifier, a transformer, a capacitor, and several switches. Figure 6.8 depicts the self-charging process. According to Figure 6.8a, the device can effectively collect energy as a TENG and store it in the capacitor. In addition, the self-charging device works at a 28.3 Hz frequency and can effectively charge

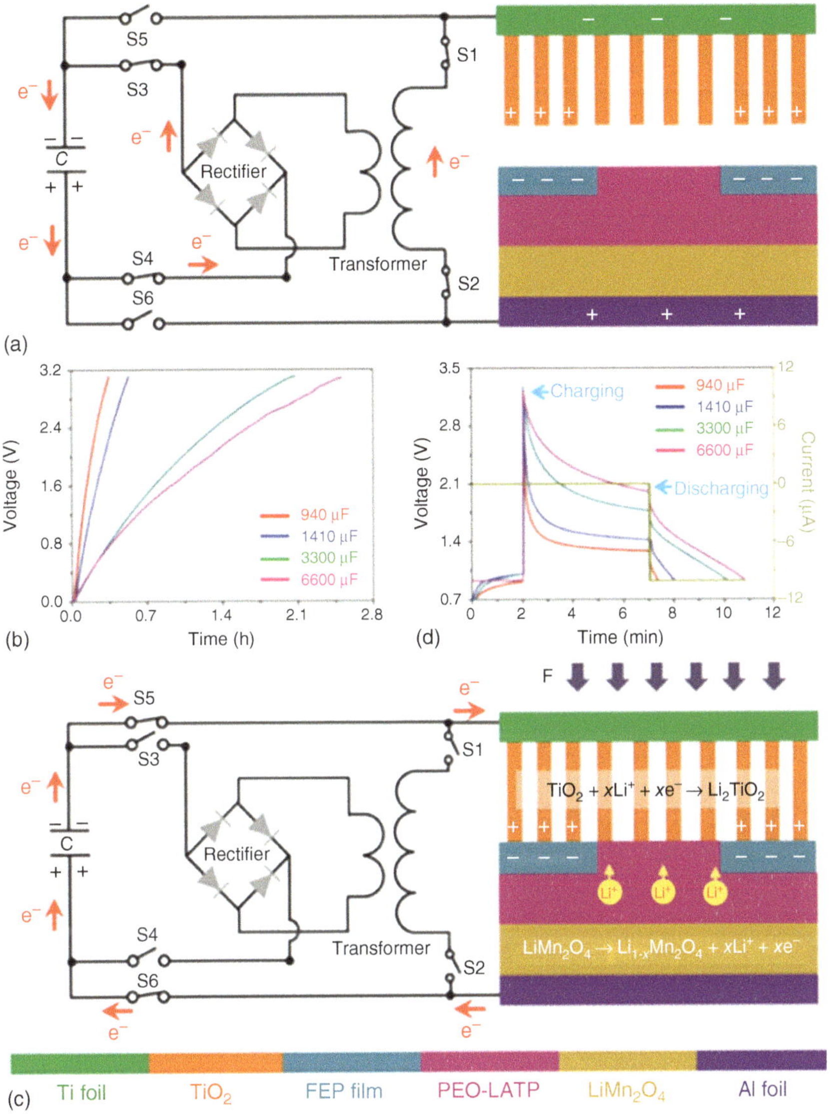

Figure 6.8 Self-charging process of the CP device by using a designed circuit. (a) The electricity transfer from the device to the capacitor. (b) Charging curves of the different capacitors by using the CP device as a TENG. (c) The electric storage process from the external capacitor to the device as a Li-ion battery. (d) Charging–discharging curves of the CP device as an SLB by using the different capacitors charged by the CP device as a TENG. Source: Reproduced with permission from Liu et al. [31]. Copyright 2017, John Wiley and Sons.

four capacitors with different capacities, as seen in Figure 6.8b; larger the capacity, more the charging time.

As displayed in Figure 6.8c, when the switches S5 and S6 are turned on and S1–S4 are turned off, the capacitor can charge the solid Li-ion battery. The charging curve of the solid Li-ion battery charged by the TENG-charged capacitor is shown in Figure 6.8d. When the charging switches are turned on, it causes a rapidly increasing voltage up to 3 V. However, the voltage decreases gradually with the increase of contact time. Under the same charging time, the larger capacity of the battery will induce a higher charging voltage by a capacitor. For example, 940 and 6600 μF capacitors were used to charge the battery, corresponding to the voltages of 1.3 and 2 V in 5 minutes, all with the same charging voltage, respectively. To characterize the charged electric energy, the discharging processes at a 0.01 mA constant current are measured. Taking the 6600 μF capacitor as an example, it was found that the device experiences the voltage variation from 2 to 0.9 V for about 240 seconds after being charged in 5 minutes, which suggests that the device can efficiently store electric energy. According to the analysis above, it can be concluded that the device can efficiently produce and store electricity by itself.

6.2.2.2 TENG-Based Supercapacitors

Like Li-ion batteries, supercapacitors also have become increasingly important devices for storing electric energy. However, supercapacitors still need regular charging owing to limited capacities for sustainably driving electronic devices. It may only reduce charging frequency by improving the energy storage capacity of supercapacitors, whereas the essential problem of periodic charging still cannot be solved. An ideal solution may be to develop self-charging supercapacitors for simultaneously harvesting external energies and storing them in the form of chemical energy in self-charging devices, rather than regular charging. However, there have been no reports about how to design self-charging supercapacitors to self-charge by generating and storing electricity simultaneously.

In 2017, Zhao et al. first reported a hybrid TENG and supercapacitors for successfully realizing the feasibility of simultaneously generating and storing electric energy by the hybrid device for solving the essential problem of periodic charging [32]. The hybrid device consists of two same flexible supercapacitors as the electrodes of the TENG and SiO_2/Ag-doped polyamide (PA) composite as the vibration membrane with an antibacterial function. Every supercapacitor consists of two electrodes coated on stainless steel foil and a PBI separator. As a TENG, the device can output a voltage/current of about 106 V/30 μA. The voltage of the supercapacitor increases up to 1 V in 33 seconds by using the wind-induced electric energy with the help of a transformer and a rectifier, where two serial supercapacitors can be charged easily for directly powering an LED.

Figure 6.9 depicts the mechanism diagram for a self-charging hybrid device. A rectifier and a transformer were used to enhance the output current and convert the AC output signals to DC signals, respectively. Four switches were used to easily control the connection or disconnection of the top or bottom supercapacitors. As shown in Figure 6.9a, at first, switches S1, S2 and S3, S4 were turned on and off, respectively. The contact between the PA membrane and the top supercapacitor

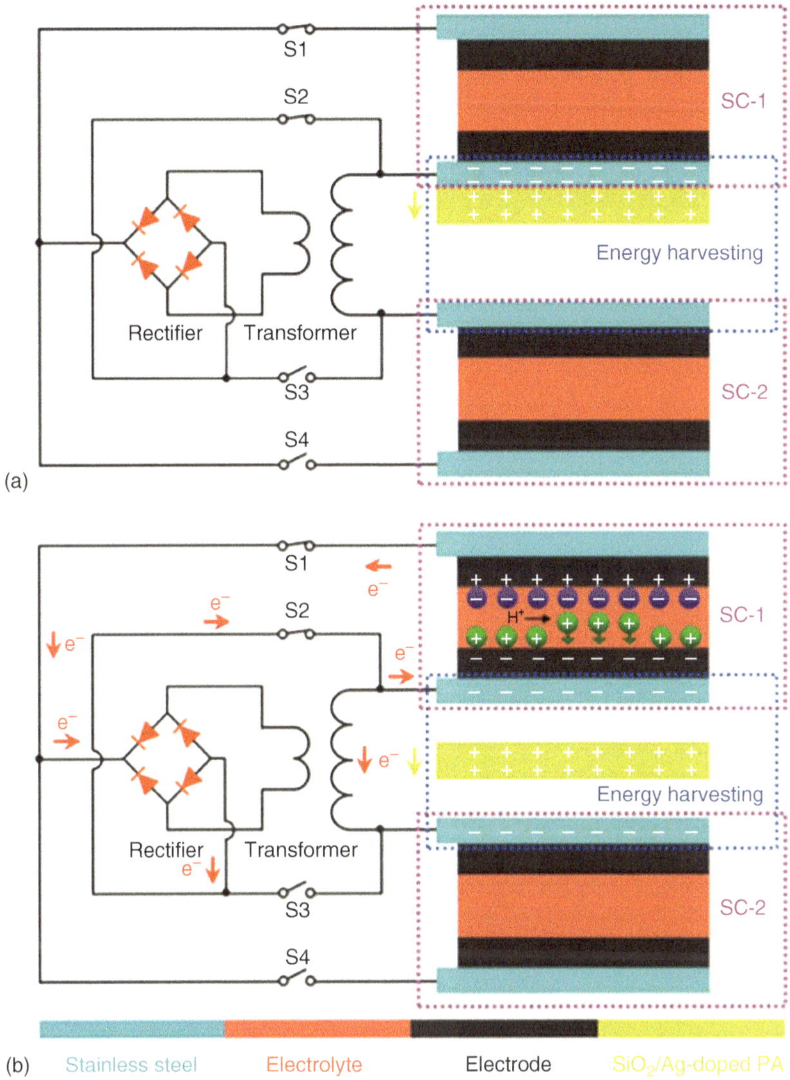

Figure 6.9 Schematic diagrams of self-charging process for the two supercapacitors without (a) and with (b) the working of the energy scavenging unit. Source: Reproduced with permission from Zhao et al. [32]. Copyright 2017, Elsevier.

will induce different charges on the top supercapacitor and PA film without the flow of electrons. And then, the separation between the PA film and the top supercapacitor can induce the migration of electrons, as shown in Figure 6.9b. The device can effectively charge the top supercapacitor by harvesting mechanical energy with the help of the transformer and the rectifier. In the two supercapacitors, a conducting polymer PBI acts as the solid electrolyte, which only H^+ can pass through. The potential difference drives H^+ to flow to the carbon electrode through PBI matrix. These absorbed H^+ ions at the interface and the electrons form an electric double layer (Figure 6.9b). In the charging process, electrons flow from the positive electrode and accumulate at the negative electrode, which can cause an increasing voltage between two electrodes until reaching an electrochemical equilibrium and realizing the self-charging functions. When switches S3 and S4 were turned on and S1 and S2 turned off, the bottom supercapacitor can be efficiently charged by harvesting mechanical energy for the same reason.

Compared with one supercapacitor, two supercapacitors can supply more electric energy to meet higher power consumption requirements. When the electrical energy of one supercapacitor is used up, another supercapacitor can be used to ensure continuous power supply by controlling the switches. At the same time, the self-charging device can harvest external energy to charge that supercapacitor without electrical energy.

6.3 Materials

Materials, energy, and information, as the three pillars of the development of modern society, have played a crucial role in technology and industry. Among them, materials are the foundation of all devices and equipment that make energy harvest or storage and information communication possible. The lack of high-performance materials in all walks of life will hinder the progress and development of our society. In a like manner, high-performance materials are vital for the improvement of hybrid nanogenerators and energy storage devices. With rapid development of materials, new nanomaterials have shown a great application prospect in energy devices owing to larger specific surface area and special nanostructures. Compared with the commercial products, new nanomaterials demonstrate a higher and reversible capacity. During the last few years, scientific researchers have carried out much research and made important and huge progress. It is noted that both the nanomaterials and traditional materials are vital to energy crisis and environmental problems. Herein, typical materials for PENGs and TENGs, Li-ion battery, and supercapacitors are discussed and determined in this section.

6.3.1 Typical Piezoelectric Materials

As the important functional materials, piezoelectric materials show excellent application prospects in energy harvest and conversion. Generally, piezoelectric materials mainly consist of organic and inorganic materials. PVDF and PZT, ZnO, $BaTiO_3$ are the most typical organic and inorganic piezoelectric materials, respectively. Additionally, polymer–piezoelectric composites and hybrid organic–inorganic piezoelectric composites are also a vital part of piezoelectric materials. Thereinafter, several representative and different types of piezoelectric materials have been discussed.

Based on Li-ion battery and PENG, Xue and coworkers fabricated a self-charging power cell (SCPC) by using an organic piezoelectric PVDF film as separator [22]. Figure 6.10a gives the SEM picture of a sandwich-structural

SCPC. As the most important component, a polarized PVDF film is in the middle of TiO_2 nanotube arrays and $LiCoO_2$ cathode as the separator with a thickness of about 120 μm. This piezoelectric PVDF film mainly consists of a β phase, which can generate strong piezoelectric property [22]. Later on, researchers continued to develop a hybrid PENG-based energy storage device by optimizing the structure of the device and piezoelectric PVDF film (Figure 6.10b) [29].

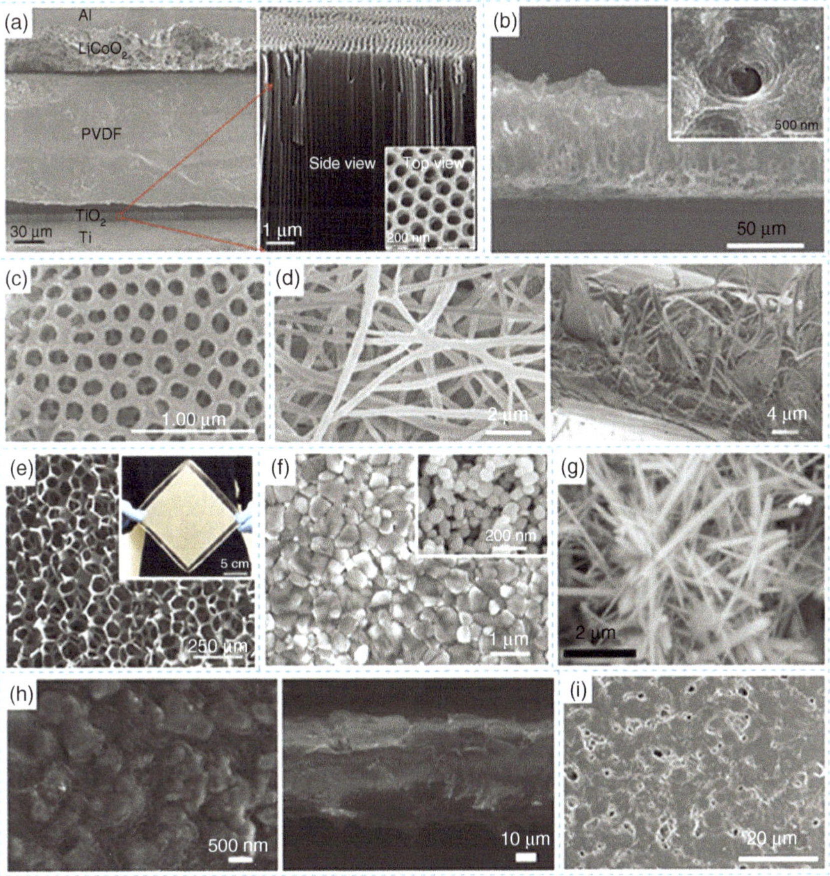

Figure 6.10 SEM images of some typical piezoelectric materials. (a) PVDF film. Source: Reproduced with permission from Xue et al. [22]. Copyright 2012, American Chemical Society. (b) PVDF/CuO composite film. Source: Reproduced with permission from Xue et al. [29]. Copyright 2013, Royal Society of Chemistry. (c) Mesoporous PVDF film. Source: Reproduced with permission from He et al. [33]. Copyright 2017, Elsevier. (d) PDVF NWs–PDMS composite film. Source: Reproduced with permission from Wang et al. [34]. Copyright 2016, John Wiley and Sons. (e) 3D interconnected PZT. Source: Reproduced with permission from Zhang et al. [35]. Copyright 2018, Royal Society of Chemistry. (f) BTO nanoparticles. Source: Reproduced with permission from Song et al. [36]. Copyright 2019, John Wiley and Sons. (g) ZnO nanorod. Source: Reproduced with permission from Ma et al. [37]. Copyright 2019, Elsevier. (h) PVDF–PZT. Source: Reproduced with permission from [38]. Copyright 2014, IOP Publishing. (i) PVDF–ZnO. Source: Reproduced with permission from Ramadoss et al. [19]. Copyright 2015, American Chemical Society.

Owing to the evaporation of solvent in the preparation process [39, 40], there are many pores produced and chaotically distributed on the surface of the PVDF film, as shown in the inset of the enlarged SEM. These pores can promote the migration of Li-ions as the ionic conduction paths. The CuO nanoarray inserted into the PVDF film will enhance the interface area of the hybrid device structure through intimate contact between the CuO nanoarray and PVDF, leading to a more effective use of the piezo-potential.

On the basis of the above research, Xue and coworkers further designed a new and all-solid-state SCPC by using a mesoporous PVDF-$LiPF_6$ film [33]; the corresponding SEM image is seen in Figure 6.10c. The preparation of solid composite piezo-electrolyte in turn experienced three steps: synthesis of PVDF film; polarization of PVDF film; deposition of $LiPF_6$. We can see that these pores are uniformly distributed with the diameter of about 200 nm similar to the size of SiO_2 microspheres as template, which leads to remarkable piezoelectric effect and also benefits the migration of Li-ion as the ionic conduction path [24].

To acquire a flexible piezoelectric material and then fabricate flexible piezoelectric devices, Wang et al. prepared a highly transparent and flexible PDVF NWs–PDMS composite film that exhibits excellent triboelectric, piezoelectric, and pyroelectric effects [34]. Figure 6.10d depicts the SEM pictures of piezoelectric materials. The unordered PVDF NWs with the diameters of 200 to 500 nm are synthesized through electro-spinning method. Piezoelectric ceramic lead zirconate titanate (PZT) has higher Curie temperature and excellent dielectric and piezoelectric properties, and the prepared PZT ceramic film has excellent mechanical strength, high piezoelectric property, and chemical stability [41]. Figure 6.10e shows the SEM image of 3D cellular-structured PZT ceramic microfoams and the photograph of large-area PDMS filled flexible PZT–PDMS composite (inset of Figure 6.10e) [35]. Unlike the composites filled with the low-dimensional ceramic filler, the 3D interconnected structure of the PZT foam promotes highly effective load transfer, which does not depend on the loading directions and relative stiffness of material constituents in the composite. These features significantly improve the piezoelectric property under various deformations such as compression, stretching, and bending.

Ferroelectrics with a complex perovskite structure, such as $BaTiO_3$ (BTO), often exhibit superior response to strain and temperature owing to the remarkable piezoelectricity and pyroelectricity, respectively. Based on the above reason, a 4×4 sensing array system was fabricated by BTO ferroelectric ceramic by sintering BTO nanoparticles for simultaneously sensing the change of strain and temperature [36]. Figure 6.10f displays the SEM image of BTO nanoparticles after being sintered with the grain size of 500–800 nm. The inset shows the morphology of unsintered BTO nanoparticles with the size of 60 nm. To design and fabricate SCPC or a temperature–pressure sensor, piezoelectric materials can also be used for dye decomposition through photocatalysis and piezocatalysis. Compared with photocatalysis as one of the most advanced and popular oxidation technologies [42, 43], piezocatalysis is a relatively new approach where piezoelectric materials produce a mass of electric charges by harvesting external vibration energy and achieve dye decomposition by redox. Ma and coworkers reported a new method to efficiently improve the decomposition ratio of acid orange-7 solution through the harvest of vibration and light energy based on photocatalysis and piezocatalysis of piezoelectric ZnO nanorods [37]. Figure 6.10g presents the SEM image of a ZnO nanorod with an average length of 4 μm.

All these piezoelectric materials mentioned above consist of one kind of organic or inorganic piezoelectric materials that exhibit excellent piezoelectric property. Additionally, scientific researchers have found that organic–inorganic piezoelectric composites may possess more significant piezoelectricity by optimizing the structure and composition of composite piezoelectric materials. A Li-ion battery was designed by using piezoelectric PVDF–PZT nanocomposite as a piezo-separator with self-charging function [38]; Figure 6.10h gives SEM images of the nanocomposite film. The nanocomposite film can provide a higher piezo-potential than pure PVDF film owing to the enhancement effect of PZT in PDVF film. Researchers fabricated a piezoelectric effect induced self-charging supercapacitor based on a PVDF–ZnO nanocomposite as the piezo-separator and MnO_2 as the positive and negative electrodes [19]. Figure 6.10i gives the SEM surface image of the piezo-separator, showing a disorderly distribution of ZnO NWs in PVDF matrix. Under continuous human palm impact, the self-charging supercapacitor can directly harvest mechanical energy for directly driving some low-power electronic devices and converting it into electrochemical energy.

6.3.2 Typical Triboelectric Materials

In the last seven years, TENGs have attracted tremendous attention on improving output performance of TENGs through reasonable structure optimization and materials selection of devices. For TENG-based energy storage devices, the selection of materials during the fabrication process should be considered to adapt the demand of TENGs as well as energy storage devices, leading to relative high performance of the devices by coordinating energy harvest unit and storage unit. Here, we have established several typical triboelectric materials for TENG-based energy storage devices.

Liu et al. fabricated a hybrid TENG and solid Li-ion battery for mechanical energy harvest and storage by using TiO_2 nanotubes as both the triboelectric layer and anode of Li-ion battery and LATP nanoparticles as the solid electrolyte [31]. Figure 6.11a shows the SEM image of TiO_2 nanotubes and LATP nanoparticles, indicating that the TiO_2 nanotubes possess the same holes size (length) of 120 nm (5.3 μm). LATP nanoparticles (average size: 150 nm) are used to fabricate the solid electrolyte film with PEO.

As we know, most TENGs always deliver a high voltage and low current [12]. To acquire high current and output power, researchers have designed high-performance TENGs based on Ag nanoparticles or NWs for scavenging wind energy, respectively. Figure 6.11b displays the SEM image of Ag nanoparticles with uniform nanoparticles size of about 50 nm (see the inset) [44]. Ag nanoparticles were deposited on a photographic paper as both positive triboelectric material and electrodes; the wind-induced TENG may deliver a high output performance owing to the excellent electrical conductivity of Ag nanoparticles. In addition, Ag nanoparticles-based TENG may possess longer

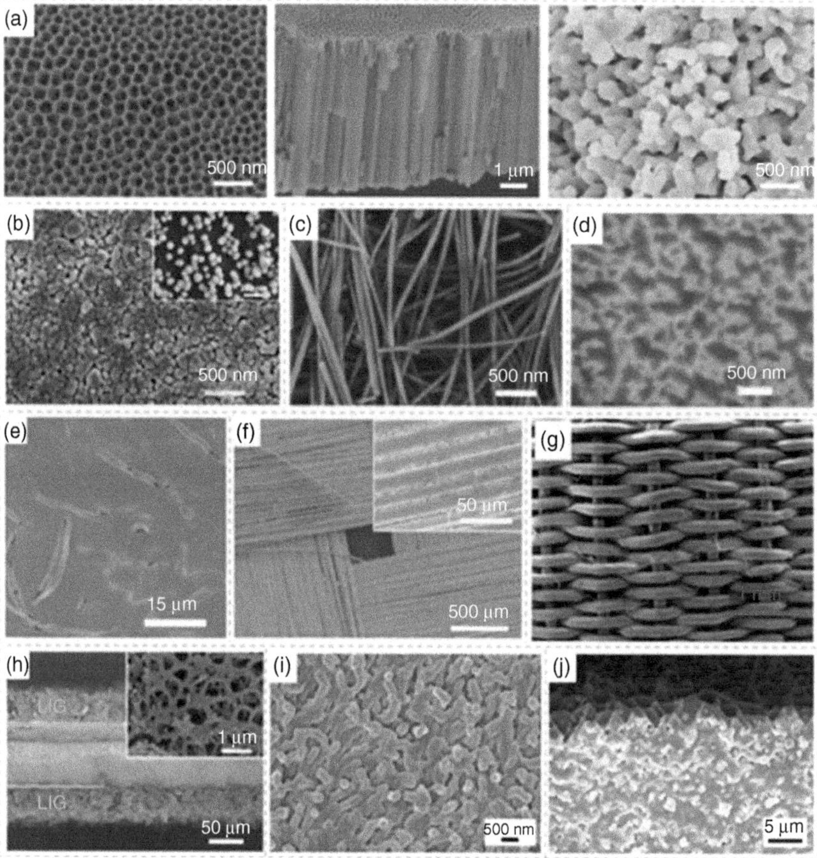

Figure 6.11 SEM images of some typical triboelectric materials. (a) TiO_2 nanotubes. Source: Reproduced with permission from Liu et al. [31]. Copyright 2017, John Wiley and Sons. (b,c) Ag nanoparticles and NWs deposited on a photographic paper. Source: Reproduced with permission from Jiang et al. [44, 45]. Copyright 2017 and 2018, American Chemical Society. (d) Etched FEP film. Source: Reproduced with permission from Yang et al. [6]. Copyright 2013, American Chemical Society. (e) PVB NWs/PDMS composite film. Source: Reproduced with permission from Quan et al. [46]. Copyright 2015, American Chemical Society. (f) Ag NWs coated conductive fabric. Source: Reproduced with permission from Zhang et al. [47]. Copyright 2016, American Chemical Society. (g) Freestanding fabric. Source: Reproduced with permission from Chen et al. [48]. Copyright 2018, Elsevier. (h) Cross-sectional image of LIG–PI–LIG. Source: Reproduced with permission from Luo et al. [49]. Copyright 2015, Springer. (i) PDMS NWs array. Source: Reproduced with permission from Yang et al. [50]. Copyright 2013, American Chemical Society. (j) $BaTiO_3$–PDMS composite film. Source: Reproduced with permission from Yang et al. [51]. Copyright 2013, Royal Society of Chemistry.

cycle life and performance stability due to strong oxidation resistance compared with Al and Cu [52]. Likewise, Ag NWs-based TENG was also fabricated with high performance, and Figure 6.11c shows that the diameter (length) of Ag NWs is 50–80 nm (10 μm) [45].

As the most common negative triboelectric material, FEP film is widely used to fabricate TENGs. To obtain high triboelectric charge density, it is effective to create the surface nanostructure of FEP by an inductively coupled plasma method. The etched FEP film surface is rough, as shown in Figure 6.11d [6]. Two Al foils are used as electrodes of TENGs. The transfer of charge occurs between the FEP film and Al foil owing to their periodic contact and separation. At initial contact, electrons will migrate from Al to FEP on the basis of triboelectric series [53], which leads to different charges. Quan and coworkers designed a hybrid nanogenerator consisting of six electromagnetic generators and a TENG, and the hybrid device can convert wrist motion energy into electricity for driving a watch [46]. The composite film consisting of polyvinyl butyral (PVB) NWs and PDMS was prepared and used as both triboelectric materials and watch belt, and the addition of PVB NWs enhances the elastic modulus of pure PDMS film from 4.4 to 6.6 MPa. Figure 6.11e displays the surface SEM picture of the composite [46].

With rapid development of wearable electronics and with an intelligent society, flexibility is an increasingly important property of the new energy devices. To obtain excellent flexibility, researchers have fabricated TENGs by using flexible composite or textile and designing other flexible structures. Figure 6.11f displays the SEM image of conductive fabric by coating Ag NWs on glass fibers and the enlarged surface view of the fabric is shown in the inset [47]. Ag NWs with high aspect ratio allow electrons to migrate rapidly owing to a decrease in the number of NW junctions [54]. As a triboelectric layer, conductive fabric/PDMS composites were used to fabricate a stretchable hybrid nanogenerator that can produce triboelectric and electromagnetic electricity through simple and periodic stretched/released movement based on both triboelectrification and electromagnetic induction. Additionally, Chen and coworkers fabricated a power textile with self-charging function for wearable electronics through traditional weaving craft [48]. Figure 6.11g shows the SEM image of a freestanding fabric TENG electrode textile (scale bar: 1 mm), indicating an average diameter of about 200 μm of micro-cable structure.

By hybridizing a micro-supercapacitor (MSC) and a TENG, a power unit with excellent self-charging function and high flexibility was designed for energy harvest and storage. The self-charging device was fabricated by using laser-induced graphene (LIG) as the electrodes of TENGs and MSC [49]. Figure 6.11h displays the SEM image of the device. Both sides of the PI substrate are covered with LIG, and the inset reveals a regular and porous microstructure, which can obviously enhance the performances of MSC array and TENG [49]. Under periodic contact–separation process of the two electrodes, the LIG–TENG produces a voltage of 168 V. After 40 000 cycles, the voltage signal of the LIG–TENG unit is almost unchanged, indicating remarkable durability and stability. The generated electric energy from LIG–TENG by harvesting ambient vibration energy can be stored in the MSC component. This self-charging supercapacitor may provide a new approach to developing a new energy device for simultaneous energy harvest, conversion, and storage.

As a common polymer with high flexibility and transparency, PDMS is often used as matrix or encapsulation material for flexible devices. By using PDMS or its composites as electrodes with a modified surface structure, researchers designed a hybrid energy cell. This hybrid device can harvest thermal energy, mechanical energy, and electrochemical energy [50, 51]. Researchers designed the first hybrid energy cell by hybridizing a TNEG and a pyroelectric nanogenerator [50]. Flexible PDMS NWs array was used to design the TENG; the SEM image of the NWs array (average diameter of 200 nm) is shown in Figure 6.11i. A $BaTiO_3$–PDMS composite film was prepared and used to fabricate a TENG as an energy harvest unit of a hybrid energy cell [51]. Figure 6.11j displays the SEM picture of the composite film above.

6.3.3 Materials for Lithium-Ion Batteries

As energy storage units of hybrid devices, the storage property of batteries determined by materials and structure may influence the performance of hybrid energy harvest and storage devices. Nowadays, nanomaterials are used to fabricate Li-ion batteries for self-charging power devices due to higher reversible capacity and larger specific surface area compared with commercial products. Here, we conclude several typical nanomaterials for Li-ion batteries, and provide a guideline for hybrid nanogenerators and Li-ion battery.

An SCPC was fabricated by using CuO nanoarrays as the anode [29]. The morphology of CuO nanoarrays is presented in Figure 6.12a, indicating uniform distribution with grass-like nanostructure. From the inset, we can see that the CuO nanoarrays have an average length of about 60 μm. In the working process of the SCPC for harvesting mechanical energy, the responding chemical reactions occur at the CuO anode, as shown in Eq. (6.12):

$$CuO + xLi^+ + xe^- \leftrightarrow Li_xCuO \tag{6.12}$$

In addition, the CuO nanoarray can enhance the interface area of the hybrid device structure through intimate contact between CuO nanoarray and PVDF under external compressive stress, leading to a more effective use of the piezo-potential.

As an important material, $LiCoO_2$ is widely and successfully used for commercial Li-ion batteries. Figure 6.12b gives the SEM images of carbon nanotubes (CNTs) (lengths: 5–15 μm, diameters: 40–60 nm) and the SCPC with sandwich structure [38]. The SCPC was fabricated by using CNTs, $LiCoO_2$, and PVDF–PZT film as the anode, cathode, and piezo-separator. Moreover, by using the $LiCoO_2$ nanoparticles as cathode, Kim et al. also reported an SCPC for simultaneous energy conversion and storage of mechanical energy [55]. Figure 6.12c shows SEM images of $LiCoO_2$ nanoparticles and artificial graphite. In the process of energy conversion and storage, under an external mechanical stress across the film, the piezo-potential is generated and drives the diffusion of Li-ions, leading to corresponding electrochemical reactions at CNTs (graphite) anode electrode expressed as Eq. (6.7) and at $LiCoO_2$ cathode as Eq. (6.13):

$$LiCoO_2 \leftrightarrow Li_{1-x}CoO_2 + xLi^+ + xe^- \tag{6.13}$$

Zhao et al. developed a new triboelectrification-enabled self-charging Li-ion battery using $LiMn_2O_4$ and carbon NWs as cathode and anode electrodes, respectively [30]. Figure 6.12d presents the SEM images of the annealed $LiMn_2O_4$ and carbon NWs; the inset is the SEM image of the microstructure

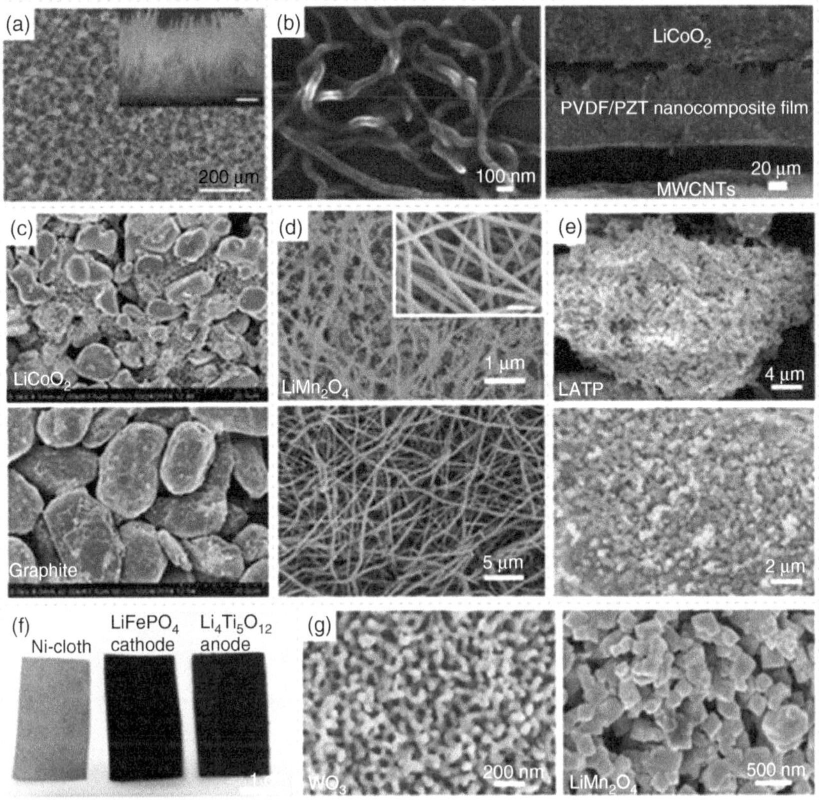

Figure 6.12 The SEM images of typical materials for Li-ion batteries. (a) CuO nanoarrays. Source: Reproduced with permission from Xue et al. [29]. Copyright 2013, Royal Society of Chemistry. (b) CNTs and $LiCoO_2$. Source: Reproduced with permission from Zhang et al. [38]. Copyright 2014, IOP Publishing. (c) $LiCoO_2$. Source: Reproduced with permission from Kim et al. [55]. Copyright 2015, Elsevier. (d) $LiMn_2O_4$ NWs. Source: Reproduced with permission from Zhao et al. [30]. Copyright 2017, John Wiley and Sons. (e) LATP and $LiMn_2O_4$ nanoparticles. Source: Reproduced with permission from Gao et al. [56]. Copyright 2017, Elsevier. (f) $LiFeO_4$ and $Li_4Ti_5O_{12}$. Source: Reproduced with permission from Pu et al. [57]. Copyright 2015, John Wiley and Sons. (g) WO_3 and $LiMn_2O_4$ nanoparticles. Source: Reproduced with permission from Jiang et al. [44]. Copyright 2017, American Chemical Society.

of $LiMn_2O_4$ precursor NWs by electro-spinning technology. It can be seen that the precursor NWs with smooth surfaces have diameters of 150–260 nm, larger than that of annealed $LiMn_2O_4$ NWs consisting of well-connected $LiMn_2O_4$ with sizes ranging from 20 to 70 nm. Researchers also designed a hybrid energy device consisting of a TENG and a Li-ion battery for harvesting wind energy [56], where $LiMn_2O_4$ and LATP nanoparticles were used as cathode and electrolyte membrane, respectively. Figure 6.12e shows the SEM images of LATP nanoparticles with several hundred nanometers and $LiMn_2O_4$ with the particle size ranging from 250 nm to 1 µm. In the charging/discharging process, electrochemical reactions at the $LiMn_2O_4$ cathode of Li-ion battery can be expressed as Eq. (6.14):

$$LiMn_2O_4 \leftrightarrow Li_{1-x}Mn_2O_4 + xLi^+ + xe^- \tag{6.14}$$

Like $LiCoO_2$, $LiFePO_4$ also has made great research progress as the cathode material of commercial batteries with high specific capacity, intrinsic thermal safety, and others [58–60]. Researchers have designed a wearable power unit by integrating a flexible Li-ion battery belt and whole-textile TENG-cloth for energy conversion and storage [57]. The flexible Li-ion battery consists of $LiFePO_4$-coated Ni-cloth as cathode and $Li_4Ti_5O_{12}$-coated Ni-cloth as anode; the corresponding photographs of the cathode and anode are shown in Figure 6.12f. The flexible device can attach on several common positions for energy harvest and storing it in Li-ion battery simultaneously. Jiang and coworkers also designed a power unit with self-charging function by using WO_3 and $LiMn_2O_4$ as anode and cathode electrode, respectively [44]. Figure 6.12g shows SEM images of electrode materials, presenting the nanoporous structure of WO_3 film and a nanoscale structure of $LiMn_2O_4$ with the nanoparticle diameter smaller than 500 nm.

6.3.4 Materials for Supercapacitors

Hybrid nanogenerators and supercapacitors for simultaneous energy harvest and storage have developed rapidly and shown a hugely positive influence in new energy field and intelligent electronics. In this section, we identify several typical nanomaterials for hybrid nanogenerators and energy storage devices based on supercapacitors.

Researchers designed a self-charging supercapacitor by using active MnO_2 NWs with excellent electrochemical performance as positive and negative electrodes [19]. Figure 6.13a displays the SEM image, indicating a smooth and uniform nanostructure. MnO_2 nanomaterial exhibits remarkable performance consisting of high specific capacitance (1400 F/g) compared with various other metal oxides, which makes it a promising material for supercapacitors [26, 27]. In addition, various MnO_2 nanomaterials were synthesized and characterized [26]. The capacitance properties of the prepared MnO_2 with different crystal types as electrodes of supercapacitors are also investigated, indicating a promising applicability in energy harvest and storage. Fe_2O_3 nanomaterials also show excellent electrochemical performance in supercapacitors [61]; the supercapacitor exhibits a large area specific capacitance up to 2775 mF/cm^2 by using porous Fe_2O_3 nanospheres anchored on activated carbon fiber as electrode material. Figure 6.13b shows the SEM images of Fe_2O_3–carbon fiber and Fe_2O_3 nanospheres with size of 20–80 nm.

The fiber-shaped supercapacitor unit has an average length of 10 cm and shows a remarkable specific capacitance (1.9 mF/cm) by weaving with two $RuO_2{\cdot}xH_2O$ coated CFs, leading to an effective and excellent property of energy storage. RuO_2 shows a high pseudocapacitance in aqueous electrolytes and excellent electrochemical property, making it an important material for ECs in recent years [68]. Based on RuO_2 as active material of electrodes of supercapacitors, Chen and coworkers designed a self-charging power textile as wearable

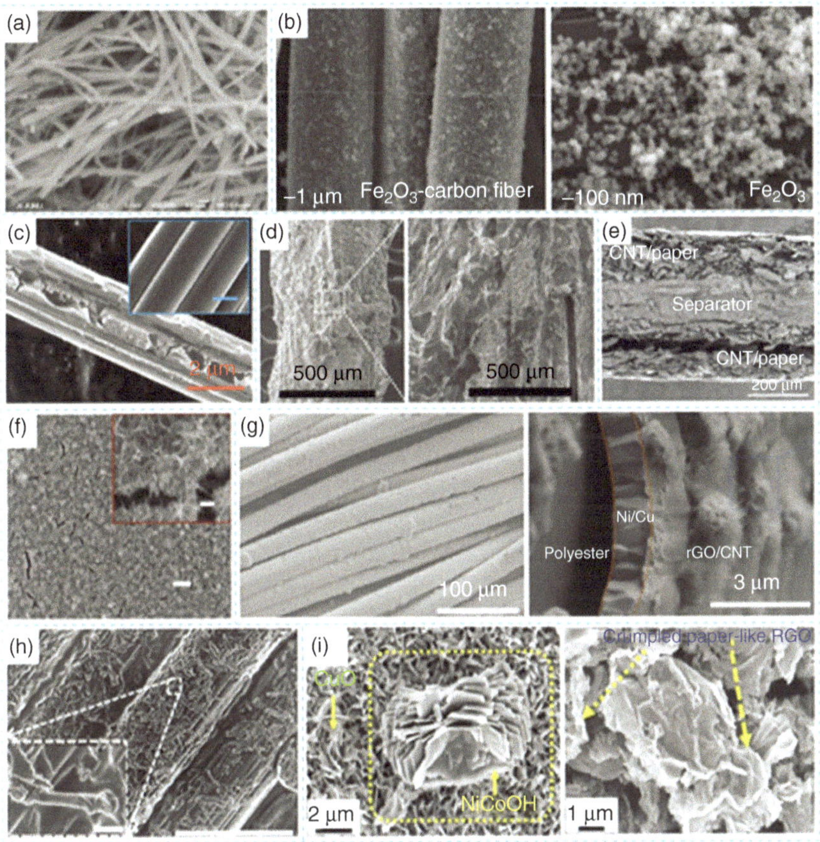

Figure 6.13 The SEM images of typical materials for supercapacitors. (a) MnO_2 NWs. Source: Reproduced with permission from Ramadoss et al. [19]. Copyright 2015, American Chemical Society. (b) Fe_2O_3–carbon fiber and Fe_2O_3 nanospheres. Source: Reproduced with permission from Li et al. [61]. Copyright 2019, Elsevier. (c) RuO_2 coated carbon fiber. Source: Reproduced with permission from Chen et al. [48]. Copyright 2018, Elsevier. (d) rGO–Ni-coated polyester yarn. Source: Reproduced with permission from Pu et al. [62]. Copyright 2015, John Wiley and Sons. (e) CNT/paper composites. Source: Reproduced with permission from Song et al. [63]. Copyright 2016, Royal Society of Chemistry. (f) AgNWs/NiO nanocomposites. Source: Reproduced with permission from Qin et al. [64]. Copyright 2018, John Wiley and Sons. (g) NiCo–BOH coated pristine fiber. Source: Reproduced with permission from Liu et al. [65]. Copyright 2019, John Wiley and Sons. (h) PEDOT: PSS/CNF-coated yarn. Source: Reproduced with permission from Dong et al. [66]. Copyright 2017, American Chemical Society. (i) CuO–NiCoOH and RGO. Source: Reproduced with permission from Maitra et al. [67]. Copyright 2017, Elsevier.

device [48]. Figure 6.13c shows the SEM pictures of RuO_2 coated CF and bare CF in the inset, indicating that the surface of carbon fiber changes from smooth to rough after being covered with RuO_2 layer.

By integrating all flexible TENG and supercapacitors, researchers fabricated a wearable power textile unit with excellent self-charging function [62]. The yarn supercapacitor was composed of polyester yarns as work electrodes using Ni as a coated layer and rGO as active material. Figure 6.13d displays the SEM image of rGO–Ni-coated polyester yarn and corresponding enlarged morphology. It can be seen that the yarn is covered with thick rGO layer, presenting a pristine morphology of Ni-coated polyester fibers.

Song and coworkers designed a sandwich-structural power unit with self-charging function based on a flexible sandwich-shaped supercapacitor and wrinkled PDMS based TENGs [63]. The CNT/paper composites are used as two electrodes of the supercapacitor and prepared through a drop-drying process of dropping CNT ink on a filter paper several times. Figure 6.13e gives the SEM image of the solid-state supercapacitor with sandwich structure, and the thickness of CNT/paper composites electrode is about 100–200 μm. As-prepared AgNWs/NiO nanocomposite was used as electrode material of the hybrid self-charging supercapacitor power package [64]. Figure 6.13f shows the surface morphology of the 3D porous structural nanocomposites, indicating high specific surface areas, which could benefit the full use of active material owing to enormous electroactive sites. In addition, conjugated AgNWs with various directions in the NiO framework can enhance the conductivity of nanocomposites.

By using rGO/CNT-coated conductive pristine and Ni–Co bimetallic oxyhydroxide (NiCo BOH) coated pristine fibers, researchers have fabricated yarn asymmetric supercapacitors (Y-ASC) as a self-charging power textile [65]. Figure 6.13g shows the SEM image of NiCo BOH-coated fibers and cross-section SEM image of rGO/CNT-coated fibers. The flexible solid-state supercapacitors are robust enough to fabricate a wearable textile. A stretchable knitting power textile with self-charging function was designed based on all-yarn structural TENGs and supercapacitors. The supercapacitor was fabricated using two carbon fiber bundles as electrodes coated with PEDOT: PSS and carbon nanofiber (CNF) [66]; the corresponding SEM image of the surface of PEDOT:PSS/CNF-coated yarn is displayed in Figure 6.13h. Maitra et al. reported a self-charging asymmetric SCPC by using NiCoOH–CuO and rGO as electrodes of this device based on a natural bio-piezoelectric separator [67]. Figure 6.13i presents the SEM image of NiCoOH–CuO with hierarchical nanostructure and crumpled paper-like rGO.

6.4 Devices Structure and Design

Hybrid nanogenerators and energy storage devices have been developed quickly for various applications in many fields. According to the different applications, the corresponding hybrid devices with different structures were fabricated. The structure and design of hybrid devices with different functions were determined and are discussed in this section.

6.4.1 Piezoelectric Nanogenerator-Based Energy Storage Devices

Figure 6.14a displays the structure diagram of the first SCPC based hybrid PENG and Li-ion battery [22]. The SCPC with sandwich structure consists of three main

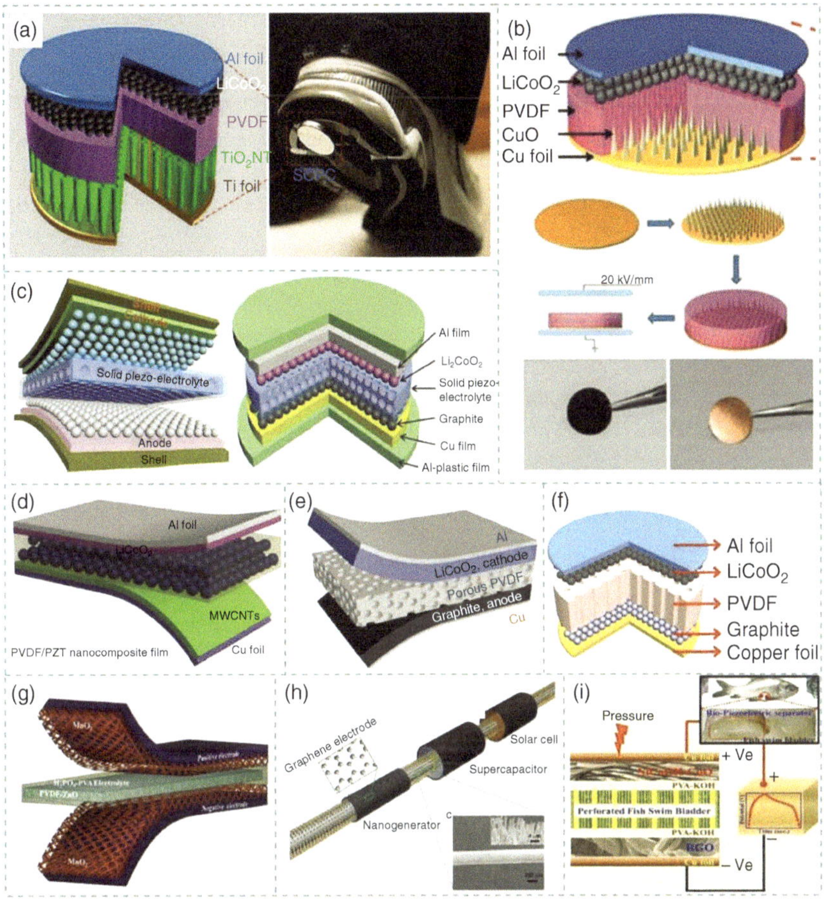

Figure 6.14 Structure diagrams of PENG-based energy storage devices. (a) The first SCPC-based hybrid PENG and Li-ion battery. Source: Reproduced with permission from Xue et al. [22]. Copyright 2012, American Chemical Society. (b) A novel SCPC based on piezoelectric CuO/PVDF film as both the separator and the anode. Source: Reproduced with permission from Xue et al. [29]. Copyright 2013, Royal Society of Chemistry. (c) Flexible SCPC. Source: Reproduced with permission from He et al. [33]. Copyright 2017, Elsevier. (d) Self-charging Li-ion battery based on piezoelectric PVDF–PZT nanocomposite film. Source: Reproduced with permission from Zhang et al. [38]. Copyright 2014, IOP Publishing. (e) Porous PVDF film-based SCPC. Source: Reproduced with permission from Kim et al. [55]. Copyright 2015, Elsevier. (f) SCPC based on mesoporous PVDF film. Source: Reproduced with permission from Xing et al. [69]. Copyright 2014, Elsevier. (g) Self-charging supercapacitor based on MnO_2 NWs as electrodes. Source: Reproduced with permission from Ramadoss et al. [19]. Copyright 2015, American Chemical Society. (h) Fiber-based hybrid multi-energy device. Source: Reproduced with permission from Bae et al. [70]. Copyright 2011, John Wiley and Sons. (i) Self-charging asymmetric supercapacitor (SCASC) power cell with a natural bio-piezoelectric separator. Source: Reproduced with permission from Maitra et al. [67]. Copyright 2017, Elsevier.

components: TiO_2 nanotube anode, organic piezoelectric PVDF separator, and $LiCoO_2$ cathode; by improving the structure of the device, the authors developed an advanced energy storage device with two main components: $LiCoO_2$-based composite as cathode and piezoelectric CuO/PVDF film as both the separator and the anode [29], as shown in Figure 6.14b. The CuO nanoarray inserted into the PVDF film will enhance the performance of the hybrid device. Later on, a new all-solid-state SCPC was further fabricated and the corresponding structure diagram is shown in Figure 6.14c [33]. This structure maybe similar to the structures in Figure 6.14a,b, and the difference is that this PVDF–$LiPF_6$ film with mesoporous structure can serve as a piezo-separator and also act as the electrolyte.

By optimizing the structure and composition of composite piezoelectric materials, researchers prepared an organic–inorganic piezoelectric composite with higher piezoelectricity. By using piezoelectric PVDF–PZT film, researchers designed the first self-charging Li-ion battery [38]. Figure 6.14d gives the structure diagram of the piezoelectric-driven self-charging device. Piezoelectric PZT can enhance the piezo-potential across the nanocomposite film compared to the pure PVDF film.

By using the $LiCoO_2$ nanoparticles as cathode, researchers reported an SCPC for simultaneous energy harvest and storage [55], as shown in Figure 6.14e. Porous structure can prevent capacity degradation and stabilize device structure, so the porous PVDF film can effectively improve total performance of SCPCs as piezo-separator. Figure 6.14f presents a schematic image of an integrated SCPC with sandwich structure [69]. ZnO NWs were used as template for synthesizing a mesoporous piezo-separator PVDF film. By harvesting external mechanical energy, the SCPC can be charged stably and efficiently. It is a great development that piezo-separators with porous structures for high-performance SCPC could bring a research upsurge about designing new materials. In addition, the study above may provide a useful strategy to solve several limited issues of the most advanced SCPCs through porous structure design.

The structure diagram of a piezoelectric-driven self-charging supercapacitor is shown in Figure 6.14g [19], and we can see that the device consists of three components: MnO_2 NWs as the two electrodes and PVDF-ZnO nanocomposite as the piezoelectric separator. Compared with conventional supercapacitors, this self-charging supercapacitor can reduce and even avoid the leakage of electrolyte and improve security from short-circuit owing to the use of H_3PO_4–PVA as solid electrolyte. ZnO NWs are uniformly distributed in PVDF matrix for enhancing the piezoelectric effect. Under continuous human palm impact, the self-charging device can directly power some low-power electronic devices by converting mechanical energy into electricity or electrochemical energy.

In 2011, researchers demonstrated the first fiber-based multi-energy device for simultaneous harvest and storage of mechanical energy and solar energy [70]. Figure 6.14h displays the structure diagram of this hybrid device by hybridizing a nanogenerator, a solar cell, and a supercapacitor all with graphene as electrodes. ZnO NWs array as piezoelectric material can obtain electric energy from mechanical energy. The produced electrical energy can be stored in the supercapacitor from a nanogenerator and solar cell by harvesting mechanical and solar energies, respectively. In addition, the study will provide an applicable methodology to develop other fiber-based electronic devices.

Inspired by fish swim bladder, researchers designed an environment friendly, all-solid-state, wearable self-charging asymmetric supercapacitoc (SCASC) power cell based on a bio-piezoelectric separator [67]. Figure 6.14i shows the structure diagram and the impact of external pressure on the power cell, and the inset presents the photograph of a fish swim bladder. The power cell consists of a positive electrode (NiCoOH–CuO), a negative electrode (rGO), and a perforated fish swim bladder soaked in PVA–KOH gel electrolyte as natural bio-piezoelectric separator. The electrode materials with novel nanostructures could exhibit superior rate capability, structural stability, and remarkably promote the interactions between electrodes and electrolyte.

6.4.2 Triboelectric Nanogenerator-Based Energy Storage Devices

To realize the harvest and conversion of natural wind energy, researchers fabricated a self-charging device based on hybrid Li-ion batteries and TENGs, as seen in Figure 6.15a [30]. This device consists of one FEP film as triboelectric material, and two Li-ion battery units act as two electrodes of TENGs and also can be used for storing electric energy. Likewise, a similar structure of self-charging device is presented in Figure 6.15b [56]; a flexible FEP film acts as the triboelectric material and stores electric energy by wrapping a Li-ion battery. Wind energy-induced periodic contact/separation between the FEP film and two electrodes will generate electric energy and store it in batteries simultaneously, which depends on triboelectrification effect and electrostatic induction. Figure 6.15c shows a hybrid device based on a TENG and a solid Li-ion battery by sharing the same electrodes for efficiently harvesting mechanical energy [31]. The top layer of this device is TiO_2 nanotubes on a Ti foil, and the bottom layer consists of circular FEP, PEO–LATP, and $LiMn_2O_4$ films. The external mechanical energy can be harvested and stored simultaneously by periodic compressive pressure.

The fabrication and design of hybrid energy devices has gradually become a research hotspot for harvesting multiple energies. Figure 6.15d shows the multilayer structure diagram of a hybrid cell [71], and this device consists of a TENG-based polymer film, a Si-based solar cell, and a thermoelectric cell based Bi_2Te_3. Figure 6.15e displays a structure diagram of a hybrid power cell consisting of a TENG and an electrochemical cell (EC) [51]. TENG was fabricated using a $BaTiO_3$–PDMS film, and EC was fabricated using the spontaneous redox reactions at Cu/Al electrodes in NaCl solution. The hybrid energy cell can drive biosensors and personal electronics by the collection of mechanical energy and chemical energy simultaneously.

Researchers reported a self-charging power unit with a sandwich structure that depended on a PDMS-based TENG and a solid-state supercapacitor based on CNT/paper [63]. Figure 6.16a presents the diagram of a sandwich-shaped self-charging device, and the CNT/paper composites are used as electrodes of the supercapacitor. Figure 6.16b shows the structure diagram of another self-charging device, which includes two components: electrochromic supercapacitor (top layer) and hybrid nanogenerators (bottom layer) [64]. AgNWs/NiO

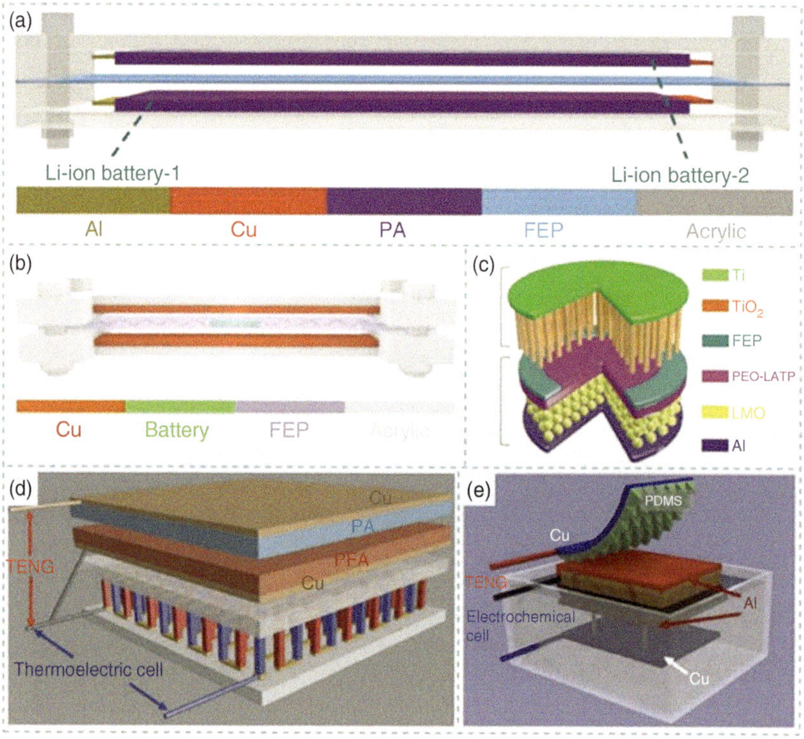

Figure 6.15 Structure diagrams of TENG-based energy storage devices. (a) Hybrid TENGs and Li-ion batteries. Source: Reproduced with permission from Zhao et al. [30]. Copyright 2017, John Wiley and Sons. (b) Self-charging device for harvesting wind energy. Source: Reproduced with permission from Gao et al. [56]. Copyright 2017, Elsevier. (c) Hybrid solid Li-ion battery and TENG. Source: Reproduced with permission from Liu et al. [31]. Copyright 2017, John Wiley and Sons. (d) Hybrid energy cell for harvesting mechanical, thermal, and solar energies. Source: Reproduced with permission from Yang et al. [71]. Copyright 2013, Royal Society of Chemistry. (e) Hybrid power cell consisting of a TENG and an electrochemical cell. Source: Reproduced with permission from Yang et al. [51]. Copyright 2013, Royal Society of Chemistry.

nanocomposites were used as electrode materials of the supercapacitor. The hybrid nanogenerator was designed by a P(VDF-TrFE)/Ag-based PENG and a contact-separation mode TENG based on PTFE/Ag structure. The two nanogenerators share the same Ag electrode and form an arched shape by vertical stack.

Figure 6.16c shows the structure diagram of a self-charging device based on coaxial SC and TENG fibers [72]. Carbon fibers are used as the electrodes of both the TENG and SC. Consisting of a one-dimensional TENG outside and a supercapacitor inside, the coaxial fiber can realize simultaneous harvesting and storage of mechanical energy. Moreover, the device can provide long-time application in daily life owing to its excellent stability. Song and coworker also designed a stretchable and wearable device with excellent self-charging function based on all-flexible TENGs, supercapacitors, and fiber-shaped dye-sensitized solar cells (FDSSCs) [73]. Figure 6.16d shows the diagram of a bracelet with self-charging function, which can effectively harvest mechanical energy and solar

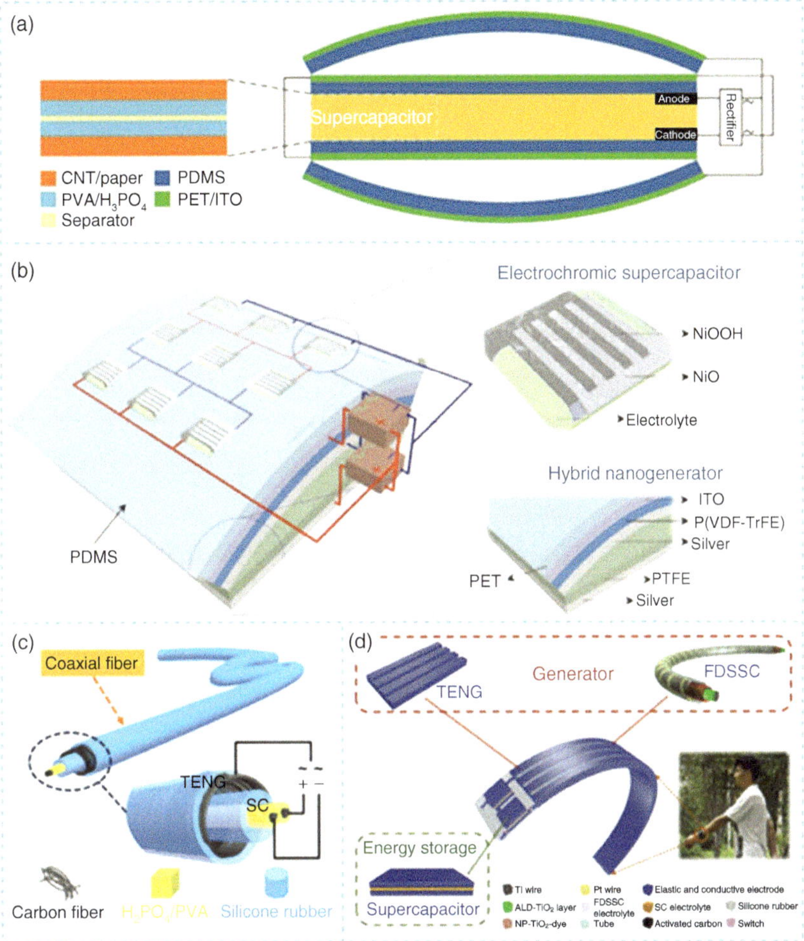

Figure 6.16 Structure diagrams of TENG-based energy storage devices. (a) A sandwich-shaped self-charging device. Source: Reproduced with permission from Song et al. [63]. Copyright 2016, Royal Society of Chemistry. (b) Self-charging power package. Source: Reproduced with permission from Qin et al. [64]. Copyright 2018, John Wiley and Sons. (c) Self-charging power fabric based coaxial TENG and SC fiber. Source: Reproduced with permission from Yang et al. [72]. Copyright 2018, American Chemical Society. (d) Self-charging power bracelet. Source: Reproduced with permission from Song et al. [73]. Copyright 2019, Elsevier.

energy through TENGs and FDSSCs, respectively. The hybrid structure greatly increases the working time of the device and extends the working environment in day and night.

6.5 Performance

The ultimate aim of reasonable material selection and structure optimization is to realize practical applications in many fields by improving the performance of hybrid nanogenerators and energy storage devices. In this section, the charging/discharging performances of hybrid energy devices are determined and discussed.

6.5.1 The Charging/Discharging Performance of Lithium-ion Batteries and Supercapacitors Based on Piezoelectric Nanogenerators

Figure 6.17a depicts the typical charging/discharging curve of an SCPC [22]. The self-charging device can be charged with an increasing voltage from 327 to 395 mV under the effect of a periodic pressure of 45 N. In the discharging process, the voltage gradually decreased to the initial voltage in 130 seconds with a discharging current (1 μA). The SCPC is separated into an individual Li-ion battery and a PENG for comparing the charging efficiency. It can be seen that under the same charging condition and same charging time, the voltage of SCPC obviously increases by ~65 mV compared with that of the Li-ion battery (~10 mV). Another SCPC can harvest mechanical energy, and simultaneously store the produced electricity in SCPC [33]. As seen in Figure 6.17b, SCPC can be charged with an increasing voltage from the original value to 473 mV easily. On discharge with a discharging current of 5 μA, the corresponding voltage is reduced to the original voltage of 25 mV in 85 seconds, corresponding to a stored capacity of 0.118 μAh.

A typical charging curve of SCPC consists of three regions: a compressive stress-induced self-charging region, a discharge region, and an equilibrium region, as displayed in Figure 6.17c [55]. In the charging process, SCPC can be charged with voltage increasing from 1.2 to 1.4 V with a compressive mechanical energy of 282 mJ for 200 seconds (frequency: 1 Hz). Then the device reaches the discharging process with a constant current, which leads to decrease of voltage corresponding to a discharge capacity of 0.4 μAh. Figure 6.17d also displays a charging/discharging process of an SCPC [69], and the voltage of SCPC increases from the original value of 160 mV to a high value of 299 mV in 250 seconds. And then, at a current of 3 μA in the discharge process, the voltage decreases from 299 mV to the original voltage in 207 seconds, which corresponds to a stored capacity of 0.173 μAh.

Figure 6.17e measures and shows the self-charging process of a piezoelectric effect-driven SCSPC [19]. Under periodic palm impact for 300 seconds, the voltage continues to increase from 35 to 145 mV. When the impact is removed, the stored energy in SCPC can sustain for 150 seconds. Soon afterwards, the voltage starts to increase when the palm impact is applied again. Figure 6.17f depicts the self-charging–discharging curves of an SCPC based on bio-piezoelectric separator [67]. Under continuous finger impact for 80 seconds with the force/frequency of 16.4 N/1.65 Hz, the device obtains an increasing voltage from 130.1 to a high value of 281.3 mV. When the finger is removed, the voltage will decrease back to the original voltage in 145.5 seconds. In addition, the influence of applied frequencies was also demonstrated. Lower frequency corresponds to a smaller voltage.

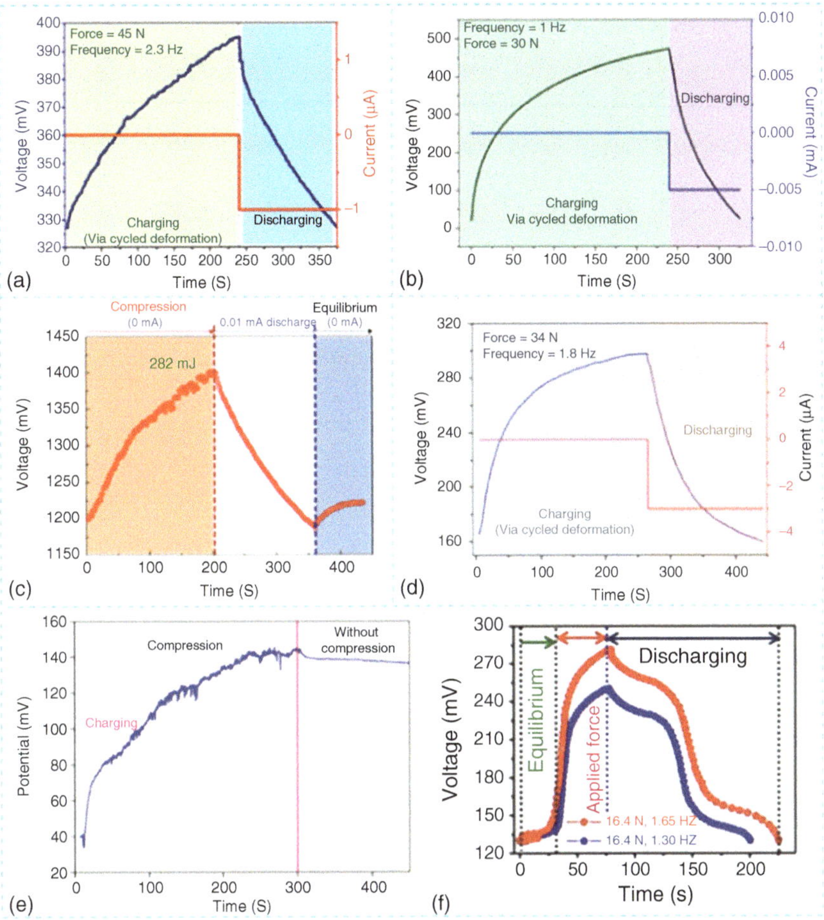

Figure 6.17 Charging–discharging curves of PENG-based energy storage devices. (a) Self-charging/discharging curve of SCPC. Source: Reproduced with permission from Xue et al. [22]. Copyright 2012, American Chemical Society. (b) Charging–discharging curves of all-solid-state SCPC. Source: Reproduced with permission from He et al. [33]. Copyright 2017, Elsevier. (c) A typical self-charging curve of self-charging power cell. Source: Reproduced with permission from Kim et al. [55]. Copyright 2015, Elsevier. (d) Hybrid energy cell for harvesting mechanical, thermal, and solar energies. Source: Reproduced with permission from Xing et al. [69]. Copyright 2014, Elsevier. (e) Self-charging process of the piezoelectric effect-driven SCSPC. Source: Reproduced with permission from Ramadoss et al. [19]. Copyright 2015, American Chemical Society. (f) Self-charging/discharging curves of flexible asymmetric supercapacitor power cell under continuous human finger impact with different force. Source: Reproduced with permission from Maitra et al. [67]. Copyright 2017, Elsevier.

6.5.2 The Charging/Discharging Performance of Triboelectric Nanogenerator-Based Lithium-Ion Batteries and Supercapacitors

By hybridizing an solid Li-ion battery (SLB) and a TENG with sharing common electrodes, researchers designed a convoluted power (CP) device, and the

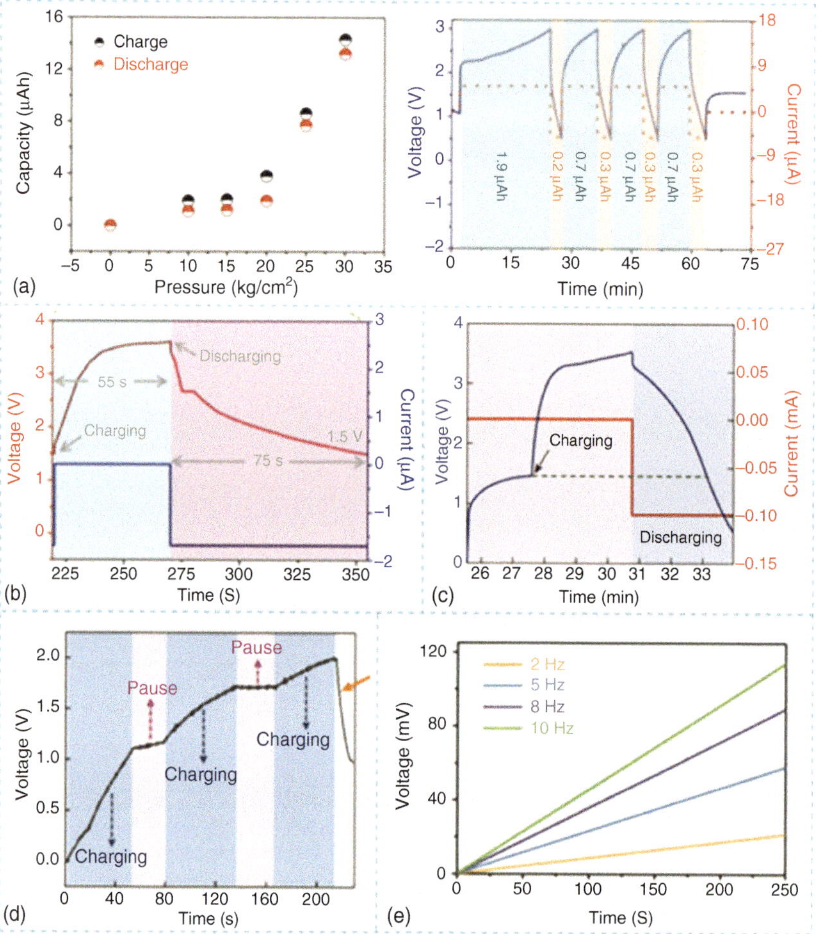

Figure 6.18 Charging–discharging curves of TENG-based energy storage devices. (a) Charging–discharging capacities of the CP device after 100 cycles under different loading pressures, and under the pressure induced by a human body. Source: Reproduced with permission from Liu et al. [31]. Copyright 2017, John Wiley and Sons. (b) Charging and discharging curves of the self-charging device at wind speed of 24.6 m/s. Source: Reproduced with permission from Gao et al. [56]. Copyright 2017, Elsevier. (c) Self-charging and discharging curves of triboelectrification-enabled self-charging Li-ion battery. Source: Reproduced with permission from Zhao et al. [30]. Copyright 2017, John Wiley and Sons. (d) The charging and working curve of two pairs of 4-series connected self-charging SCs. Source: Reproduced with permission from Yang et al. [72]. Copyright 2018, American Chemical Society. (e) Self-charging charging curves of the sandwiched SC at various frequencies. Source: Reproduced with permission from Song [63]. Copyright 2016, Royal Society of Chemistry.

charging/discharging performances were also demonstrated under different working conditions. According to the charging/discharging curves of the SLB in Figure 6.18a [31], it can be confirmed that the increase of loading pressures will bring increased charging/discharging capacities, taking the loading pressure of 30 kg/cm^{-2} as an example, which corresponds to a charging (discharging) capacity of 14.3 (13.2) μAh. If the loading pressure was replaced by an adult with a weight of 55 kg, the CP device can achieve a charging capacity of 1.9 μAh and discharging capacity of 0.2 μAh in the first cycle. After that, the charging (discharging) capacity reaches stable values of 0.7/0.3 μAh, which suggests an excellent practicability and feasibility of the CP device in wearable electronic devices.

The charging/discharging curves of a hybrid TENG and SLB are displayed in Figure 6.18b [56]. The voltage of SLB increased from 1.5 to 3.6 V in 55 seconds by harvesting the wind energy (24.6 m/s). Likewise, under the discharging process with a constant current, decreasing the voltage from 3.6 to 1.5 V corresponds to a stored electric capacity of 0.035 μAh. Figure 6.18c shows the charge–discharge process of a triboelectrification-enabled self-charging Li-ion battery for harvesting and storing wind energy [30]. After being discharged, a concentration polarization was generated owing to the retardation effect of Li-ions migration, which leads to a slowly increasing voltage. After the charging process of Li-ion battery, the voltage decreased from 3.5 to 1.5 V corresponding to a stored electric capacity of 3.9 μAh. The self-charging Li-ion battery exhibits excellent stability after 50 cycles corresponding to a high coulomb efficiency of 90%.

Figure 6.18d displays the charging curve of self-charging SCs [72]. The SCs were charged by palm impact on the coaxial fibers. When the palm impact is removed, the voltage of SCs will appear as a plateau without leakage and show excellent stability. The voltage of SCs increased again when continuing to impact the coaxial fibers. After the SCs were charged up to 2 V, an electronic watch was powered successfully by the SCs as a power cell, leading to decrease of the voltage, which suggests the feasibility of self-charging SCs as wearable electronics. Figure 6.18e shows the initial charging curves of a sandwich-structural self-charging power unit (SCPU) by parallel-connected TENGs for harvesting vibration energy with different frequencies [63]. The voltage of the SC increases up to 120 mV in the charging process under vibration with a frequency of 10 Hz, presenting excellent charging efficiency. Under a periodic compressive impact of 8 Hz for charging the SC, the voltage increased to 900 mV in 3 hours corresponding to steadily increasing stored charges.

Four-series self-charging TFSCs can be charged by a TF–TENG by finger sliding or pressing, as seen in Figure 6.19a [74]. Finger motions-induced mechanical energy can be harvested and stored in the TFSC. Under the finger motion of fast and slow sliding (0.8 and 0.5 m/s), the voltage of SCPF increased from 0 to 2.5 V in 2094 and 3518 seconds, respectively. For finger pressing, it takes about 6100 seconds for the voltage to increase from 0 to 2.5 V. In addition, the SCPF can directly power a scientific calculator sustainably, which suggests a feasible approach in intelligent electronics and self-powered sensing systems. Figure 6.19b gives the charging curve of an MSC arrays-based self-charging device [49], and by harvesting mechanical energy, the device is charged steadily from 0 to 3 V in 117 minutes. Likewise, as energy storage unit of the highly elastic self-charging power system, two-series fabricated supercapacitors were charged by a silicone-based TENG for harvesting mechanical energy at various frequencies, as presented in Figure 6.19c [73]. The energy storage unit was charged from 0 to 1.75 V within 100 seconds, and the inset shows that the higher

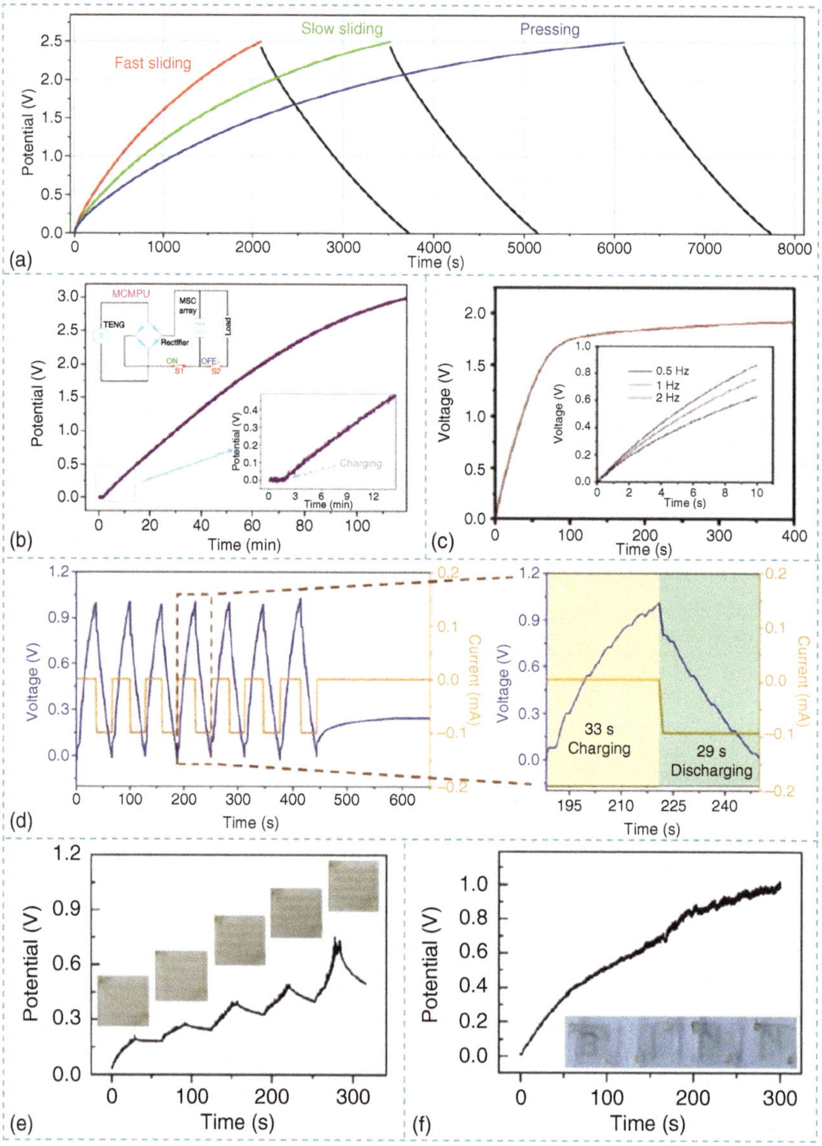

Figure 6.19 Charging–discharging curves of TENG-based energy storage devices. (a) Charging–discharging curves of TFSCs by the TF-TENG at fast and slow sliding, and pressing by finger. Source: Reproduced with permission from Luo et al. [74]. Copyright 2016, American Chemical Society. (b) Charging curve of the MSC array component in a self-charging power system. Source: Reproduced with permission from Luo et al. [49]. Copyright 2015, Springer. (c) Charging curve of hybrid TENGs and supercapacitors. Source: Reproduced with permission from Song et al. [73]. Copyright 2019, Elsevier. (d) The charging–discharging curves of the self-charging supercapacitor. Source: Reproduced with permission from Zhao et al. [32]. Copyright 2017, Elsevier. (e,f) Real-time charging process of the electrochromic μ-SC in self-charging power package. Source: Reproduced with permission from Qin [64]. Copyright 2018, John Wiley and Sons.

the frequency, the faster the charging rate. The charging–discharging curves of the self-charging device are shown in Figure 6.19d [32], and the voltage of the supercapacitor increased up to 1 V in 33 seconds self-charging process. At discharging process with a constant current, the voltage of the supercapacitor decreases from 1 to 0 V in 29 seconds.

The charging curve of a self-charging power package consisting of an electrochromic μ-SC through continuous palm impact is illustrated in Figure 6.19e [64]. The insets of Figure 6.19e record the color variation of the electrodes induced by voltage change of the μ-SC. The final charged voltage will increase and remain at 0.15 V with a continuous palm impact. When continuing the palm impact, the μ-SC was self-charged again with a corresponding color change of the device. A"BINN"-shaped μ-SC can easily be charged up to 1 V through hybrid piezo/triboelectric effect, as presented in Figure 6.19f [64]. The inset shows several small alphabets with dark color corresponding to a positively charged state of the μ-SCs, which suggests that the color change can be used as the indicator of the charging state of the smart electrochromic power package in the self-charging process.

6.6 Applications

Based on nanogenerator-based energy storage devices, researchers have attained great achievements in hybrid storage energy devices for simultaneous energy harvest and storage. At present, hybrid storage energy devices are widely used for driving electronic devices, self-powered sensing systems, and even electrochemistry reactions.

6.6.1 Powering Electronic Devices

A flexible SCPC based on PENG and Li-ion battery can power a smart watch, a sports bracelet, and LEDs by harvesting/storing body-motion energy, as displayed in Figure 6.20a [33]. The electricity can be stored in the SCPC as a power source, and can also directly drive electronic devices. Figure 6.20b shows that a CP device successfully lights up a green LED by harvesting palm impact-induced mechanical energy [31]. A hybrid nanogenerator can power an electronic watch by effectively harvesting biomechanical energy of a person's motions [46], including the vibration and shake of a wrist, as seen from Figure 6.20c.

The self-powered asymmetric SCPC can drive four LEDs with instantaneous lightning upon periodic finger impact on the device. Likewise, a green LED, small electronic watch, calculator, LCD screen, and portable speaker, as shown in Figure 6.20d [67] – all these devices can be powered by the self-charging device for harvesting external mechanical energy. Figure 6.20e shows a calculator powered by an SCPF by harvesting the mechanical energy of finger sliding [74]. In addition, the SCPF can sense a person's sliding characteristics and intelligent sliding unlock system of a mobile phone. A commercial hygrothermograph and an electronic watch are driven successfully by a flexible SCPU and power fabric, respectively [49, 72], as presented in Figure 6.20f,g, respectively.

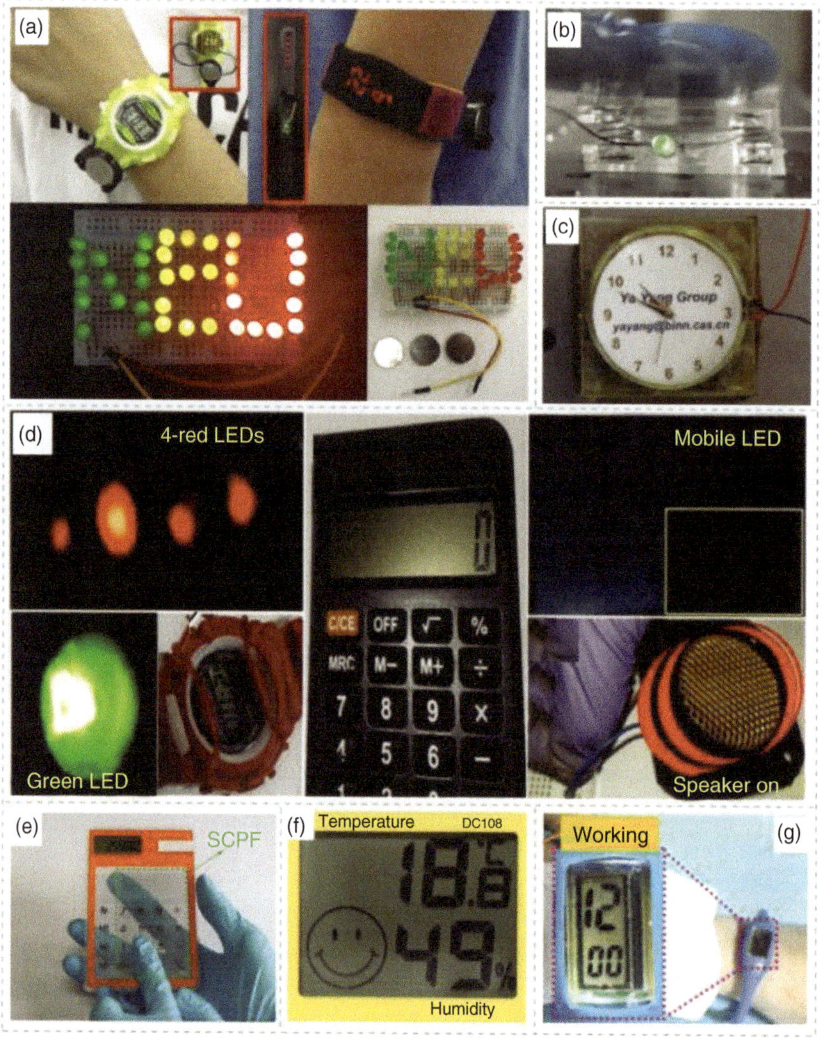

Figure 6.20 Hybrid nanogenerators and energy storage devices can drive small power electronic devices. (a) An all-solid-state SCPC can power smart watch, sports bracelet, and LEDs. Source: Reproduced with permission from He et al. [33]. Copyright 2017, Elsevier. (b) A green LED can be lit up by a CP device by harvesting palm impact-induced mechanical energy. Source: Reproduced with permission from Liu et al. [31]. Copyright 2017, John Wiley and Sons. (c) Hybrid nanogenerator for harvesting wind energy for powering an electronic watch. Source: Reproduced with permission from Quan et al. [46]. Copyright 2015, American Chemical Society. (d) Self-powered asymmetric SCPC can drive LEDs, small electronic wrist watch, commercial digital calculator, mobile LCD screen, and portable speaker. Source: Reproduced with permission from Maitra et al. [67]. Copyright 2017, Elsevier. (e) The SCPF can power a calculator by harvesting the mechanical energy of finger sliding. Source: Reproduced with permission from Luo et al. [74]. Copyright 2016, American Chemical Society. (f) A commercial hygrothermograph is driven successfully by a flexible self-charging power unit. Source: Reproduced with permission from Luo et al. [49]. Copyright 2015, Springer. (g) An electronic watch can be driven successfully by a self-charging power fabric based on micro-supercapacitors and TENGs. Source: Reproduced with permission from Yang et al. [72]. Copyright 2018, American Chemical Society.

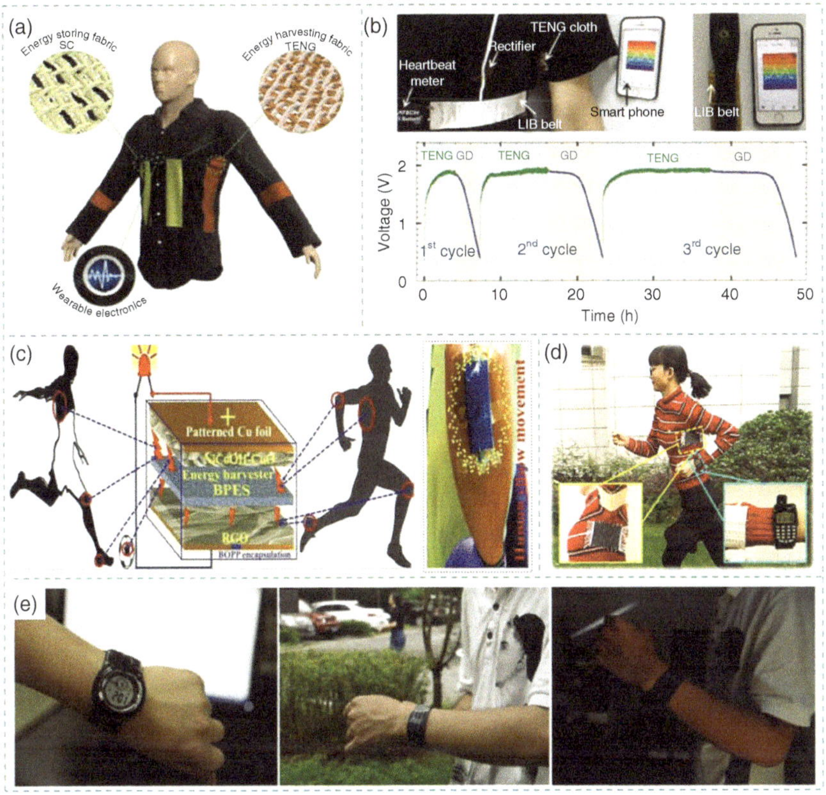

Figure 6.21 Hybrid nanogenerators and energy storage devices work as wearable electronic devices. (a) Self-charging power textile as wearable energy device for harvesting mechanical energy from human motion. Source: Reproduced with permission from Pu et al. [62]. Copyright 2015, John Wiley and Sons. (b) A wearable power unit can power a heartbeat meter for real-time heartbeat sensing. Source: Reproduced with permission from Pu et al. [57]. Copyright 2015, John Wiley and Sons. (c) A natural bio-piezoelectric induced SCASC device as wearable device at different positions of the human body. Source: Reproduced with permission from Maitra et al. [67]. Copyright 2017, Elsevier. (d) SCPT drives an electric watch by swing arm during running motion. Source: Reproduced with permission from Chen et al. [48]. Copyright 2018, Elsevier. (e) A self-charging powered system as wearable device works during the day and the night. Source: Reproduced with permission from Song et al. [73]. Copyright 2019, Elsevier.

6.6.2 Wearable/Portable Electronics

Hybrid nanogenerators and energy storage devices have played important roles in flexible electronics as wearable/portable electronic devices. Figure 6.21a shows a wearable power textile with self-charging function for simultaneous energy harvest and storage [62]; the generated energy from human motion through TENG fabrics is stored for powering electronic devices, which avoids the unstable output of intermittent energy generation. A wearable power unit was designed for harvesting the energy of human motions, which can power a heartbeat meter for real-time heartbeat sensing [57], as presented in

Figure 6.21b. The charging/discharging curve indicates that the wearable power unit shows high capability and stability, making it possible to power portable electronics as a self-powered power source.

Figure 6.21c shows the diagram of several feasible application areas and functions of a natural bio-piezoelectric driven SCASC device [67], and the photograph of the device attached to the human elbow for detecting elbow movement by measuring the electrical signals produced. With a novel SCPT as wearable electronics shown in Figure 6.21d, the electrical energy produced can continuously drive an electronic watch by arm swinging [48]. The bigger the speed of arm swing, the more the electrical energy produced. A self-charging device with high elasticity was also fabricated as wearable device for powering an electronic watch. Figure 6.21e shows the photographs of the device working during the day and the night for harvesting solar and mechanical energy [73].

6.6.3 Other Applications

Apart from harvesting external energy to store in energy storage unit or to directly drive electronic devices, the hybrid nanogenerators and energy storage devices can be used in electrochemical reactions or signal detection. Figure 6.22a displays the working diagram for self-powered electrodegradation, and the degradation percentage of methyl orange solution increases with the increase of degradation time by an electrocatalytic oxidation process [50]. The photographs of the solution in the inset show different colors before and after electrodegradation. A hybrid energy cell can directly split water for hydrogen (H_2) generation, as shown in Figure 6.22b [71]. From the photographs of H_2 collection tube, we can see that the H_2 volume gradually increases by water splitting. By fitting the experiment data, we can see that there is an exponential relationship between the volume of the H_2 produced and the splitting time, whereas the production speed decreases gradually due to the decrease of output power of the hybrid cell.

Figure 6.22c shows photographs of a smart window fabricated by a self-charging power unit [63]. Under periodic compressive stress, three serially connected SCPUs were charged and used as power supply for driving a calculator and controlling the color change of an electrochromic device as a smart window. As shown in Figure 6.22d, four TENGs are arranged along four different directions [6], which can power electronic devices, and realize self-powered detection of wind speed and direction with high sensitivity as a wind vector sensor.

6.7 Conclusions and Prospects

In conclusion, hybridizing nanogenerators and energy storage devices has made great progress for high-efficiency energy harvest and storage due to the rapid development in recent years, which will show increasingly great significance in energy crisis and environment pollution. Considering the performances of hybrid

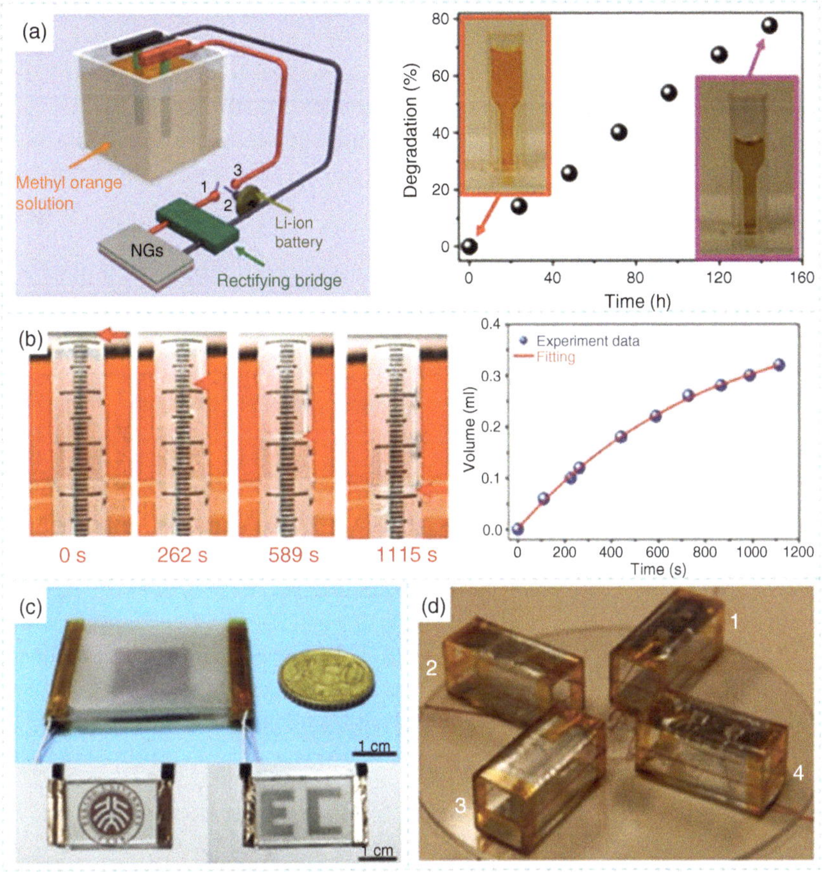

Figure 6.22 Hybrid nanogenerators and energy storage devices for electrochemical reaction and other applications. (a) A hybrid energy cell for self-powered degradation of methyl orange. Source: Reproduced with permission from Yang et al. [50]. Copyright 2013, American Chemical Society. (b) Self-powered water splitting to generate H_2. (c) Smart window fabricated by a self-charging power unit. Source: Reproduced with permission from Yang et al. [71]. Copyright 2013, Royal Society of Chemistry. (d) Self-powered wind vector sensor system for sensing wind speed and direction. Source: Reproduced with permission from Yang et al. [6]. Copyright 2013, American Chemical Society.

energy devices that depend on the combined effect between energy harvest units and energy storage units, various hybrid devices based on PENGs or TENGs and Li-ion batteries or supercapacitors have been designed and demonstrated for harvesting and storing different kinds of mechanical, thermal, and solar energies. The development progress including working mechanisms, materials selection, structure design, performance and applications, are systematically discussed and summarized. These hybrid energy devices exhibit remarkable performance in harvesting/storing energy for directly driving electronic devices and electrochemical fields, and also improve the development of wearable electronics and self-powered sensing systems.

Despite these advancements and advantages, there are still several challenges and difficulties. Researchers need to drastically improve the performance of hybrid nanogenerators and energy storage devices owing to relatively lower output power of the reported devices. Additionally, structure and performance are still controllable for matching complex application scenarios in practice. As wearable or even implantable electronic devices, the safety, compatibility, and even easiness among the prior concerns are all challenging issues and need to be considered in the fabrication process. The cost of devices will also be an important consideration factor in commercialization. In terms of the growing advancements and more attention in this research community, hybridizing nanogenerators and energy storage devices will certainly be highly efficient and viable approaches for practical applications in future.

References

1 Tarascon, J.-M. and Armand, M. (2001). Issues and challenges facing rechargeable lithium batteries. *Nature* 414: 359.
2 Goodenough, J.B. and Park, K.-S. (2013). The Li-ion rechargeable battery: a perspective. *J. Am. Chem. Soc.* 135: 1167.
3 Zhu, Y., Murali, S., Stoller, M.D. et al. (2011). Carbon-based supercapacitors produced by activation of graphene. *Science* 332: 1537.
4 Wang, Z.L. and Song, J. (2006). Piezoelectric nanogenerators based on zinc oxide nanowire arrays. *Science* 312: 242.
5 Fan, F.-R., Tian, Z.-Q., and Wang, Z.-L. (2012). Flexible triboelectric generator. *Nano Energy* 1: 328.
6 Yang, Y., Zhu, G., Zhang, H. et al. (2013). Triboelectric nanogenerator for harvesting wind energy and as self-powered wind vector sensor system. *ACS Nano* 7: 9461.
7 Liu, Y., Sun, N., Liu, J. et al. (2018). Integrating a silicon solar cell with a triboelectric nanogenerator via a mutual electrode for harvesting energy from sunlight and raindrops. *ACS Nano* 12: 2893.
8 Chen, B.D., Tang, W., He, C. et al. (2017). Water wave energy harvesting and self-powered liquid-surface fluctuation sensing based on bionic-jellyfish triboelectric nanogenerator. *Mater. Today* 21: 88.
9 Liu, R., Kuang, X., Deng, J. et al. (2018). Shape memory polymers for body motion energy harvesting and self-powered mechanosensing. *Adv. Mater.* 30: 1705195.
10 Pu, X., Guo, H., Chen, J. et al. (2017). Eye motion triggered self-powered mechnosensational communication system using triboelectric nanogenerator. *Sci. Adv.* 3: e1700694.
11 Wang, J., Li, S., Yi, F. et al. (2016). Sustainably powering wearable electronics solely by biomechanical energy. *Nat. Commun.* 7: 12744.
12 Cao, X., Jie, Y., Wang, N., and Wang, Z.L. (2016). Triboelectric nanogenerators driven self-powered electrochemical processes for energy and environmental science. *Adv. Energy Mater.* 6: 1600665.

13 Chen, J., Yang, J., Li, Z. et al. (2015). Networks of triboelectric nanogenerators for harvesting water wave energy: a potential approach toward blue energy. *ACS Nano* 9: 3324.
14 Chan, C.K., Peng, H., Liu, G. et al. (2008). High-performance lithium battery anodes using silicon nanowires. *Nat. Nanotech.* 3: 31.
15 Bruce, P.G., Scrosati, B., and Tarascon, J.M. (2008). Nanomaterials for rechargeable lithium batteries. *Angew. Chem. Int. Edit.* 47: 2930.
16 Winter, M. and Brodd, R.J. (2004). What are batteries, fuel cells, and supercapacitors? *Chem. Rev.* 104: 4245.
17 González, A., Goikolea, E., Barrena, J.A., and Mysyk, R. (2016). Review on supercapacitors: technologies and materials. *Renew. Sust. Energ. Rev.* 58: 1189.
18 Zhong, C., Deng, Y., Hu, W. et al. (2015). A review of electrolyte materials and compositions for electrochemical supercapacitors. *Chem. l Soc. Rev.* 44: 7484.
19 Ramadoss, A., Saravanakumar, B., Lee, S.W. et al. (2015). Piezoelectric-driven self-charging supercapacitor power cell. *Acs Nano* 9: 4337.
20 Liu, C., Yu, Z., Neff, D. et al. (2010). Graphene-based supercapacitor with an ultrahigh energy density. *Nano lett.* 10: 4863.
21 Chou, J.-C., Chen, Y.-L., Yang, M.-H. et al. (2013). RuO_2/MnO_2 core-shell nanorods for supercapacitors. *J. Mater. Chem. A* 1: 8753.
22 Xue, X., Wang, S., Guo, W. et al. (2012). Hybridizing energy conversion and storage in a mechanical-to-electrochemical process for self-charging power cell. *Nano lett.* 12: 5048.
23 Fransson, L., Eriksson, T., Edström, K. et al. (2001). Influence of carbon black and binder on Li-ion batteries. *J. Power Sources* 101: 1.
24 Chiang, C.-Y., Shen, Y., Reddy, M.J., and Chu, P.P. (2003). Complexation of poly(vinylidene fluoride): LiPF6 solid polymer electrolyte with enhanced ion conduction in 'wet' form. *J. Power Sources* 123: 222.
25 Yuan, C., Hou, L., Yang, L. et al. (2011). Facile interfacial synthesis of flower-like hierarchical a-MnO 2 sub-microspherical superstructures constructed by two-dimension mesoporous nanosheets and their application in electrochemical capacitors. *J. Mater. Chem.* 21: 16035.
26 Sung, D.-Y., Kim, I.Y., Kim, T.W. et al. (2011). Room temperature synthesis routes to the 2D nanoplates and 1D nanowires/nanorods of manganese oxides with highly stable pseudocapacitance behaviors. *J. Phys. Chem. C* 115: 13171.
27 Wei, W., Cui, X., Chen, W., and Ivey, D.G. (2011). Manganese oxide-based materials as electrochemical supercapacitor electrodes. *Chem. Soc. Rev.* 40: 1697.
28 Xue, X., Deng, P., He, B. et al. (2014). Flexible self-charging power cell for one-step energy conversion and storage. *Adv. Energy Mater.* 4: 1301329.
29 Xue, X., Deng, P., Yuan, S. et al. (2013). CuO/PVDF nanocomposite anode for a piezo-driven self charging lithium battery. *Energ. Environ. Sci.* 6: 2615.
30 Zhao, K., Yang, Y., Liu, X., and Wang, Z.L. (2017). Triboelectrification-enabled self-charging lithium-ion batteries. *Adv. Energy Mater.* 7: 1700103.
31 Liu, X., Zhao, K., Wang, Z.L., and Yang, Y. (2017). Unity convoluted design of solid li-ion battery and triboelectric nanogenerator for self-powered wearable electronics. *Adv. Energy Mater.* 7: 1701629.

32 Zhao, K., Qin, Q., Wang, H. et al. (2017). Antibacterial triboelectric membrane-based highly-efficient self-charging supercapacitors. *Nano Energy* 36: 30.

33 He, H., Fu, Y., Zhao, T. et al. (2017). All-solid-state flexible self-charging power cell basing on piezo-electrolyte for harvesting/storing body-motion energy and powering wearable electronics. *Nano Energy* 39: 590.

34 Wang, S., Wang, Z.L., and Yang, Y. (2016). A one-structure-based hybridized nanogenerator for scavenging mechanical and thermal energies by triboelectric–piezoelectric–pyroelectric effects. *Adv. Mater.* 28: 2881.

35 Zhang, G., Zhao, P., Zhang, X. et al. (2018). Flexible three-dimensional interconnected piezoelectric ceramic foam. *Energ. Environ. Sci.* 11: 2046.

36 Song, K., Zhao, R., Wang, Z.L., and Yang, Y. (2019). Conjuncted pyro-piezoelectric effect for self-powered simultaneous temperature and pressure sensing. *Adv. Mater.* 31: e1902831.

37 Ma, J., Ren, J., Jia, Y. et al. (2019). High efficiency bi-harvesting light/vibration energy using piezoelectric zinc oxide nanorods for dye decomposition. *Nano Energy* 62: 376.

38 Zhang, Y., Zhang, Y., Xue, X. et al. (2014). PVDF-PZT nanocomposite film based self-charging power cell. *Nanotechnology* 25: 105401.

39 Nunes-Pereira, J., Lopes, A., Costa, C. et al. (2012). Porous membranes of montmorillonite/poly(vinylidene fluoride-trifluorethylene) for Li-ion battery separators. *Electroanalysis* 24: 2147.

40 Ferreira, A., Silva, J., Sencadas, V. et al. (2010). Poly[(vinylidene fluoride)-*co*-trifluoroethylene] membranes obtained by isothermal crystallization from solution. *Macromol. Mater. Eng.* 295: 523.

41 Yang, A., Wang, C.A., Guo, R. et al. (2010). Porous PZT ceramics with high hydrostatic figure of merit and low acoustic impedance by TBA-based gel-casting process. *J. Am. Ceram. Soc.* 93: 1427.

42 Ma, J., Wu, Z., Luo, W. et al. (2018). High pyrocatalytic properties of pyroelectric $BaTiO_3$ nanofibers loaded by noble metal under room-temperature thermal cycling. *Ceram. Int.* 44: 21835.

43 Yagub, M.T., Sen, T.K., Afroze, S., and Ang, H.M. (2014). Dye and its removal from aqueous solution by adsorption: a review. *Adv. Colloid Inter. Sci.* 209: 172.

44 Jiang, Q., Chen, B., Zhang, K., and Yang, Y. (2017). Ag nanoparticle-based triboelectric nanogenerator to scavenge wind energy for a self-charging power unit. *ACS Appl. Mater. Inter.* 9: 43716.

45 Jiang, Q., Chen, B., and Yang, Y. (2018). Wind-driven triboelectric nanogenerators for scavenging biomechanical energy. *ACS Appl. Energ. Mater.* 1: 4269.

46 Quan, T., Wang, X., Wang, Z.L., and Yang, Y. (2015). Hybridized electromagnetic-triboelectric nanogenerator for a self-powered electronic watch. *ACS Nano* 9: 12301.

47 Zhang, K., Wang, Z.L., and Yang, Y. (2016). Conductive fabric-based stretchable hybridized nanogenerator for scavenging biomechanical energy. *ACS Nano* 10: 4728.

48 Chen, J., Guo, H., Pu, X. et al. (2018). Traditional weaving craft for one-piece self-charging power textile for wearable electronics. *Nano Energy* 50: 536.

49 Luo, J., Fan, F.R., Jiang, T. et al. (2015). Integration of micro-supercapacitors with triboelectric nanogenerators for a flexible self-charging power unit. *Nano Res.* 8: 3934.

50 Yang, Y., Zhang, H., Lee, S. et al. (2013). Hybrid energy cell for degradation of methyl orange by self-powered electrocatalytic oxidation. *Nano Lett.* 13: 803.

51 Yang, Y., Zhang, H., Chen, J. et al. (2013). Simultaneously harvesting mechanical and chemical energies by a hybrid cell for self-powered biosensors and personal electronics. *Energ. Environ. Sci.* 6: 1744.

52 Wang, J., Xu, Y., Sun, X. et al. (2007). Polypyrrole films electrochemically doped with dodecylbenzenesulfonate for copper protection. *J. Electrochem. Soc.* 154: C445.

53 Shashoua, V.E. (1958). Static electricity in polymers. I. Theory and measurement. *J. Polym. Sci.* 33: 65.

54 Ye, S., Rathmell, A.R., Chen, Z. et al. (2014). Metal nanowire networks: the next generation of transparent conductors. *Adv. Mater.* 26: 6670.

55 Kim, Y.-S., Xie, Y., Wen, X. et al. (2015). Highly porous piezoelectric PVDF membrane as effective lithium ion transfer channels for enhanced self-charging power cell. *Nano Energy* 14: 77.

56 Gao, T., Zhao, K., Liu, X., and Yang, Y. (2017). Implanting a solid Li-ion battery into a triboelectric nanogenerator for simultaneously scavenging and storing wind energy. *Nano Energy* 41: 210.

57 Pu, X., Li, L., Song, H. et al. (2015). A self-charging power unit by integration of a textile triboelectric nanogenerator and a flexible lithium-ion battery for wearable electronics. *Adv. Mater.* 27: 2472.

58 Sun, C., Rajasekhara, S., Goodenough, J.B., and Zhou, F. Monodisperse porous $LiFePO_4$ microspheres for a high power Li-ion battery cathode. *J. Am. Chem. Soc.* 133 (7): 2132–2135.

59 Liu, J., Conry, T.E., Song, X. et al. (2011). Nanoporous spherical $LiFePO_4$ for high performance cathodes. *Energ. Environ. Sci.* 4: 885.

60 Bai, P., Cogswell, D.A., and Bazant, M.Z. (2011). Suppression of phase separation in $LiFePO_4$ nanoparticles during battery discharge. *Nano lett.* 11: 4890.

61 Li, J., Wang, Y., Xu, W. et al. (2019). Porous Fe_2O_3 nanospheres anchored on activated carbon cloth for high-performance symmetric supercapacitors. *Nano Energy* 57: 379.

62 Pu, X., Li, L., Liu, M. et al. (2015). Wearable self-charging power textile based on flexible yarn supercapacitors and fabric nanogenerators. *Adv. Mater.* 28: 98.

63 Song, Y., Cheng, X., Chen, H. et al. (2016). Integrated self-charging power unit with flexible supercapacitor and triboelectric nanogenerator. *J. Mater. Chem. A* 4: 14298.

64 Qin, S., Zhang, Q., Yang, X. et al. (2018). Hybrid piezo/triboelectric-driven self-charging electrochromic supercapacitor power package. *Adv. Energy Mater.* 8: 1800069.

65 Liu, M., Cong, Z., Pu, X. et al. (2019). High-energy asymmetric supercapacitor yarns for self-charging power textiles. *Adv. Funct. Mater.* 29: 1806298.

66 Dong, K., Wang, Y.-C., Deng, J. et al. (2017). A highly stretchable and washable all-yarn-based self-charging knitting power textile composed of fiber triboelectric nanogenerators and supercapacitors. *ACS Nano* 11: 9490.
67 Maitra, A., Karan, S.K., Paria, S. et al. (2017). Fast charging self-powered wearable and flexible asymmetric supercapacitor power cell with fish swim bladder as an efficient natural bio-piezoelectric separator. *Nano Energy* 40: 633.
68 Rao, R.R., Kolb, M.J., Halck, N.B. et al. (2017). Towards identifying the active sites on RuO_2 (110) in catalyzing oxygen evolution. *Energ. Environ. Sci.* 10: 2626.
69 Xing, L., Nie, Y., Xue, X., and Zhang, Y. (2014). PVDF mesoporous nanostructures as the piezo-separator for a self-charging power cell. *Nano Energy* 10: 44.
70 Bae, J., Park, Y.J., Lee, M. et al. (2011). Single-fiber-based hybridization of energy converters and storage units using graphene as electrodes. *Adv. Mater.* 23: 3446.
71 Yang, Y., Zhang, H., Lin, Z.-H. et al. (2013). A hybrid energy cell for self-powered water splitting. *Energ. Environ. Sci.* 6: 2429.
72 Yang, Y., Xie, L., Wen, Z. et al. (2018). Coaxial triboelectric nanogenerator and supercapacitor fiber-based self-charging power fabric. *ACS Appl. Mater. Inter.* 10: 42356.
73 Song, W., Yin, X., Liu, D. et al. (2019). A highly elastic self-charging power system for simultaneously harvesting solar and mechanical energy. *Nano Energy* 65: 103997.
74 Luo, J., Tang, W., Fan, F.R. et al. (2016). Transparent and flexible self-charging power film and its application in a sliding unlock system in touchpad technology. *ACS Nano* 10: 8078.

7

Pyroelectric and Thermoelectric Nanogenerators

7.1 Introduction

To meet the needs of the process of industrialization in modern society, increasing fossil energy consumption leads to serious energy crisis and environment degradation. According to the statistics in 2015 [1], information about energy consumption can be obtained as shown in Figure 7.1. Of the total energy consumption by society, 86% still depends on fossil fuel consisting of oil, natural gas, and coal, as shown in Figure 7.1a; new energy sources including hydro, nuclear, and renewable sources just make up 14% of the total energy consumption. In addition, the energy demand will be significantly rising in the next two decades (Figure 7.1b). It is reported that nearly half the thermal energy is wasted and dissipated in air in industrial production, and extra energy will further be consumed for cooling industrial equipment [2]. Therefore, developing clean and sustainable green energies becomes an ideal approach to relieve energy crisis and environmental pollution [3–5]. Clean energy sources have developed rapidly and show superior power generation with high efficiency including solar cell, nuclear power, and wind-driven generators in the past few decades [6–8]. With the development of scientific technology in recent 10 years, several new energy harvest technologies, such as piezoelectric [9], triboelectric [10], pyroelectric [11], and thermoelectric (TE) [12] nanogenerators, have exhibited excellent performance in energy harvest and conversion.

All these energy harvest technologies mentioned above show excellent performance for energy harvest and possess respective advantages and disadvantages, although the power generated is negligible compared with the huge thermal energy waste. In reality, developing a novel approach to effectively reusing waste thermal energy is an important industrial and scientific field nowadays [13]. Pyroelectric nanogenerator (PyENG) and thermoelectric nanogenerator (TEG) can directly generate electricity by harvesting heat energy based on pyroelectric effect [14] and TE effect [15], respectively, which has gradually become a key research field and has acquired much attention from scientific researchers.

In our daily life, heat sources as an important source can be collected for driving some electronic devices with small power. TEG can generate electricity by harvesting thermal energy owing to temperature difference-induced charge

*Corresponding author: yayang@binn.cas.cn

Hybridized and Coupled Nanogenerators: Design, Performance, and Applications,
First Edition. Ya Yang.

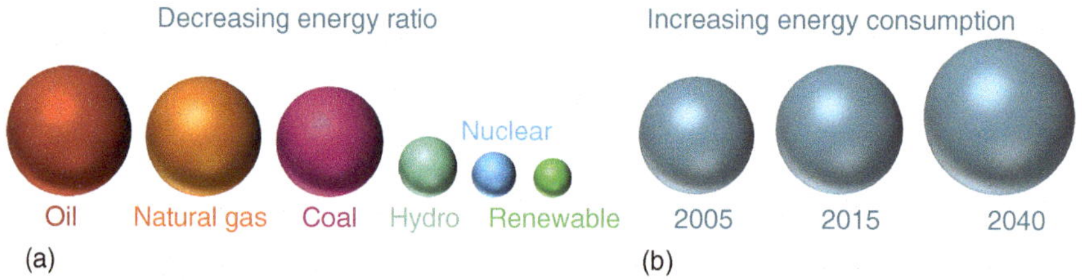

Figure 7.1 Energy consumption. (a) The share of total energy consumption in 2015. (b) Statistic and estimated global energy consumption by source of 2005, 2015, and 2040.

carrier migration across devices. After rapid development, the performances of TEGs have improved obviously, and various structures have been designed to adapt to different application conditions. However, atmospheric temperatures are often almost uniform with a time-dependent temperature fluctuation; TEG cannot work in this case and PyENG becomes a feasible energy technology for harvesting heat energy in daily life [11]. Both PyENGs and TEGs have seen huge development and played an important role in different fields and environment. With the development of intelligent electronic devices, flexibility, small volume, and light weight have gradually become important and popular characteristics. To gain these excellent characteristics, scientific researchers have carried out a lot of work to improve the performance of PyENGs and TEGs through the preparation and selection of materials and structure optimization. Apart from powering small power electronic devices [13, 16], PyENGs and TEGs also have enormous potential applications in wearable electronics [17, 18], self-powered sensing systems [19, 20], and even electrochemical reactions [21, 22].

In this chapter, the research background of thermal energy harvest is first introduced, followed by a comparison of the working mechanisms of PyENGs and TEGs. Then, progress in the development of PyENGs and TEGs, including materials, structure, performance, and application, is systematically evaluated, respectively. Focus will be on materials selection and structure designs. Finally, the future development and challenges are also discussed and evaluated.

7.2 Working Mechanisms

Based on pyroelectric and TE effects, PyENGs and TEGs can directly convert heat into electricity, respectively. With the development of PyENGs and TEGs, the structure design and material selection of these devices have greatly improved in recent years. The working mechanisms and working conditions of PyENGs and TEGs are discussed and contrasted in this section.

7.2.1 Pyroelectric Nanogenerators

PyENGs can convert heat energy into electricity owing to change of spontaneous polarization within pyroelectric materials, which results from the oscillation of

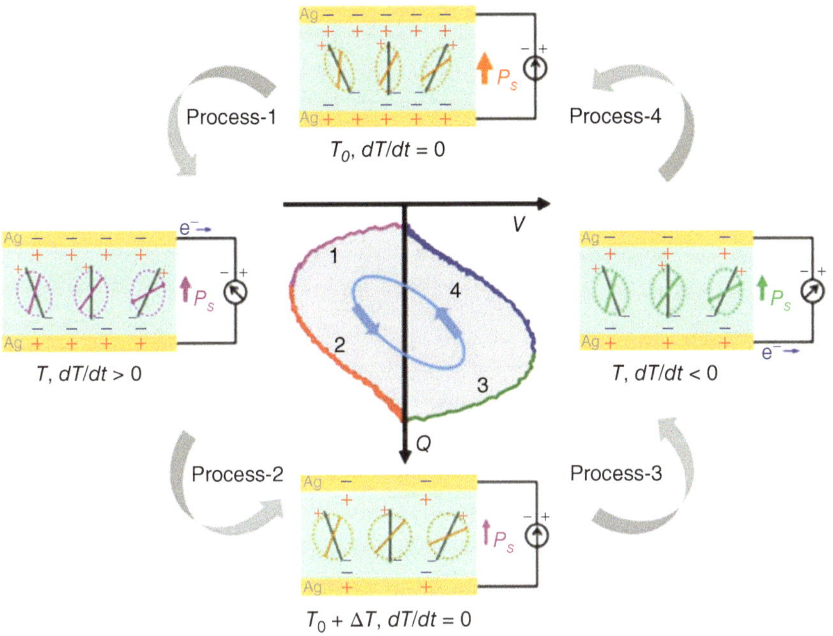

Figure 7.2 The working mechanism and corresponding charge/discharging process of a PVDF-based PyENG. Source: Reproduced with permission from Zhang et al. [14]. Copyright 2019, Elsevier.

electric dipoles induced by a time-dependent temperature fluctuation [11, 14]. Here, we take a typical polyvinylidene fluoride (PVDF) film as example to illustrate the working mechanism of PyENGs systematically, and two surfaces of the polarized PVDF film were covered with silver layer as electrodes.

Figure 7.2 shows the Q–V diagram and brief charging/discharging process of a PVDF film-based PyENG [14]. At a low temperature T_0 ($dT/dt = 0$) as an initial state, there is no flow of electrons due to oscillation equilibrium of electric dipoles in PVDF. In process-1, the oscillation degree of electric dipoles increases with the increase of temperature from T_0 to $T(dT/dt > 0)$, which will weaken spontaneous polarization, break the electrical balance, and then drive the migration of electrons. When the PVDF film reaches a high temperature value of $T_0 + \Delta T(dT/dt = 0)$ through process-2, the new equilibrium state of oscillation of electric dipoles cannot generate electricity. Likewise, the decrease of temperature from $T_0 + \Delta T$ to $T(dT/dt < 0)$ will enhance spontaneous polarization and induce reverse electron flow (process-3). When the temperature reaches T_0, the electric dipoles return to the initial state and finish the full charge/discharging process (process-4).

Figure 7.2 also presents the Q–V diagram in the charge/discharging process. The PyENG based on PVDF film is charged from process-1 to process-2 by heating the device from T_0 to $T_0 + \Delta T$, and the corresponding voltage increases and

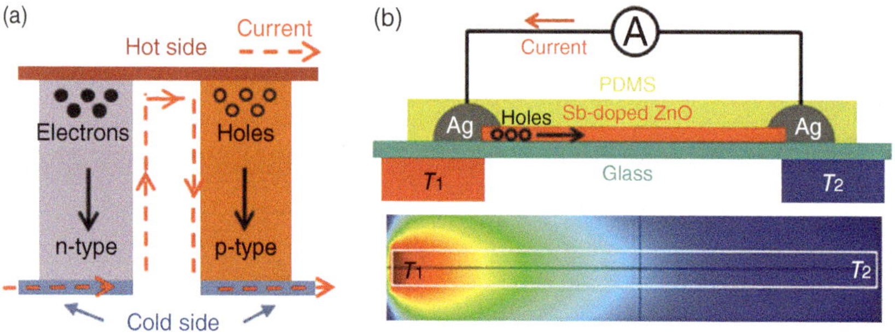

Figure 7.3 The working mechanism and temperature distribution of TEGs. (a) Carriers migration of typical π-type TE device. Source: Reproduced with permission from Zhang et al. [13]. Copyright 2019, John Wiley and Sons. (b) Thermally induced holes migration of a ZnO micro/nanobelts-based TEG and corresponding temperature distribution. Source: Reproduced with permission from Yang et al. [12]. Copyright 2012, American Chemical Society.

then decreases. Subsequently, the PyENG is discharged by cooling the temperature from $T_0 + \Delta T$ to T_0 with the output voltage being reversely increased and then decreased in process-3 and process-4, respectively.

7.2.2 Thermoelectric Nanogenerators

In 1821, T. Seebeck first discovered the TE effect through which TE materials can generate electricity by harvesting thermal energy due to the ΔT across TE materials.

Figure 7.3a shows the typical diagram of a TE device with π-type structure based on p-/n-type materials, and the corresponding flow directions of carriers are also presented [13]. When one side of the π-type TE device is heated and another one is cooled, ΔT will induce the migration of both holes (h^+) and electrons (e^-) from the hot to the cold side. The flow of carriers in p-/n-type TE materials will generate TE potentials with the same direction, and the working circuit will be formed when the external circuit is connected. The working diagram of a TEG and the corresponding temperature distribution are presented in Figure 7.3b [12]. The ΔT across the TEG induces migration of h^+ from the side with high temperature (T_1) to another side with low temperature (T_2). The working mechanism is simple, although TEGs have received increasing attention and possess a large number of relative research reports in the academic community compared with pyroelectric devices in recent years.

7.2.3 Comparison of Pyroelectric and Thermoelectric Nanogenerators

Through the above induction and analysis of corresponding working mechanisms, PyENGs and TEGs are understood systemically. Both of them can convert heat energy into electricity, whereas the work environments and structures of devices are different owing to the difference in working mechanisms.

The differences between PyENGs and TEGs are contrasted through the following aspects: First, their working mechanisms are different. PyENGs

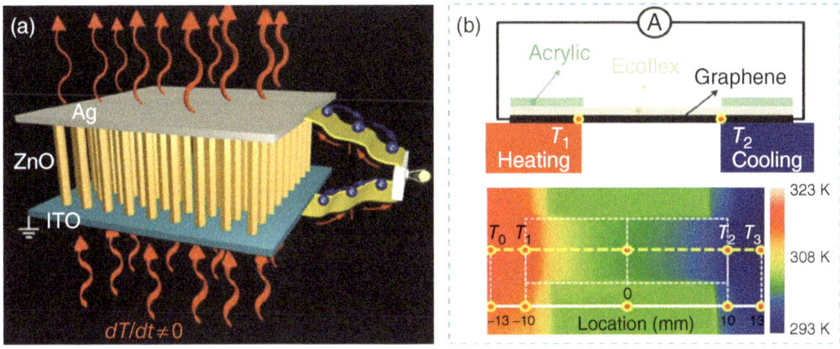

Figure 7.4 The comparison between PyENGs and TEGs. (a) Working diagram of a PyENG based on ZnO nanorods array. Source: Reproduced with permission from Yang et al. [11]. Copyright 2012, American Chemical Society. (b) Structure diagram of a graphene-based TEG and the corresponding temperature distribution. Source: Reproduced with permission from Zhang et al. [19]. Copyright 2019, John Wiley and Sons.

depend on temperature variation-induced variation of spontaneous polarization (pyroelectric effect), and TEGs are based on migration of carriers from the side with high temperature to another side with low temperature due to ΔT (TE effect). Second, the work environments of PyENGs and TEGs are different. The time-dependent temperature fluctuation of the former environment will lead to overall temperature variation of PyENGs, as shown in Figure 7.4a [11], and the temperatures at each position on the device are the same, whereas the latter induces a temperature gradient across TEGs, and the temperatures at each position on TEGs are different. The last, the structures of the two kinds of devices are different, which results from the difference in the working mechanisms and environments above. In addition, although the working mechanism is relatively simple, the structure design of TEGs is complex owing to the ΔT across TEGs, which make TEGs work in several specific conditions, as presented in Figure 7.4b [19]. For PyENGs, owing to the same temperature variation of each position on the devices in an environment with spatially uniform temperature outdoor in our daily life, it is relatively easy to fabricate PyENGs for thermal energy harvest and conversion.

7.3 Progress of Pyroelectric Nanogenerators

From now on, the development progress of PyENGs and TEGs including materials selection, structure design, performance, and applications are evaluated and discussed comprehensively and systemically.

7.3.1 Typical Pyroelectric Materials

First, typical pyroelectric materials are evaluated including inorganic and organic materials. As we know, nanomaterials possess higher specific surface area than the commercial products and show large application potential in energy harvest, conversion, and storage [13, 19, 23]. Here, several representative pyroelectric materials have been selected, discussed, and summarized.

Lead zirconate titanate (PZT) is an important pyroelectric material and shows excellent performance with a huge pyroelectric coefficient [24]; several micro-PZT with different morphologies were used to fabricate PyENGs for energy harvest and conversion. Yang et al. fabricated a PyENG as a temperature sensor based on a PZT microwire packaged by transparent polydimethylsiloxane (PDMS) [20]. Figure 7.5a presents the scanning electron microscopic (SEM) image of a microwire with a diameter (length) of 2 μm (10 μm). In addition, another kind of PyENG was designed based on a PZT film and displays excellent pyroelectric performance for powering a green LED [16]. Figure 7.5b shows the SEM image in cross section of a pyroelectric device, presenting a thickness of about 180 μm, and the enlarged SEM image shows that the film includes many crystal grains with a grain size of about 2–5 μm.

Although PZT materials possess a high pyroelectric coefficient, element lead is harmful for humans and the environment, and so it is important to explore nontoxic and easily available pyroelectric materials for energy application. Yang et al. first prepared nontoxic $KNbO_3$ nanowires-based PyENGs [25]. The nanowire with a diameter of 150 nm has a single-crystalline structure with [011] as the growth direction. Figure 7.5c presents the SEM pictures of $KNbO_3$ nanowire and its PDMS composite, and we can see that the addition of PDMS will improve the flexibility of PyENGs. Similar to $KNbO_3$ nanowires, $NaNbO_3$ nanomaterials including nanorods, nanosheets, and nanocubes also show superior pyroelectric property and can decompose dye wastewater through electrochemical redox reaction [26]. Figure 7.5d displays the SEM image of hydrothermally synthesized pyroelectric $NaNbO_3$ nanosheets. By harvesting natural heat energy, the $NaNbO_3$ nanoparticles show excellent room temperature pyrocatalytic effect.

A newly designed photodetector based on pyroelectric $BaTiO_3$ and $BiFeO_3$ film shows a coupled photovoltaic–pyroelectric effect under 405 and 450 nm light illumination, respectively [27, 28]. The $BaTiO_3$ film was fabricated by using $BaTiO_3$ nanoparticles with a diameter of about 100 nm, and the SEM image is shown in Figure 7.5e [27]. At the sintering temperature of 1200 °C, the optimized sample exhibited a higher relative density of 95.4%. Figure 7.5f presents the cross-sectional microstructure of $BiFeO_3$ film, and the relative density of the bismuth ferrite (BFO) film sintered at 850 °C is about 86% [28]. Wu et al. synthesized pyroelectric $BiFeO_3$ nanoparticles with a grain size of 60–150 nm, as displayed in Figure 7.5g [29]. Under a room temperature cold–hot alternation, $BiFeO_3$ nanoparticles show excellent pyro-catalysis owing to pyroelectric effect.

ZnO nanomaterials are important semiconductor materials and possess remarkable performance [9]. Researchers fabricated a ZnO nanowire array-based PyENG on indium tin oxide (ITO) substrate. Figure 7.5h presents the SEM image of ZnO nanowires [11], showing an average diameter (length) of 200 nm (2 μm). Researchers have fabricated a high-performance PyENG based on SbSeI nanowires for low temperature waste heat recovery; Figure 7.5i displays the morphology of nanowires with the length ranging from 20 to 50 nm [30]. A high-performance $Ba_{0.7}Sr_{0.3}TiO_3$ (BST) pyroelectric nanomaterial was prepared through a hydrothermal process, as shown in Figure 7.5j [31], and

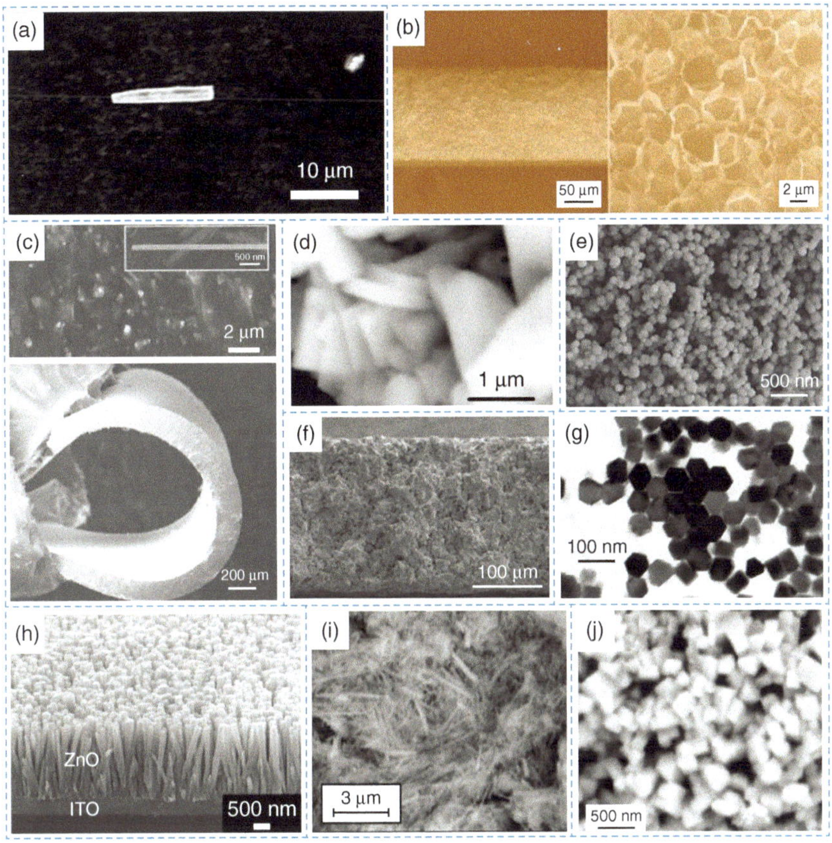

Figure 7.5 The SEM or TEM images of typical inorganic pyroelectric materials. (a) A PZT microwire. Source: Reproduced with permission from Yang et al. [20]. Copyright 2012, American Chemical Society. (b) The cross section and surface of a PZT film. Source: Reproduced with permission from Yang et al. [16]. Copyright 2012, American Chemical Society. (c) $KNbO_3$ nanowires and PDMS composite. Source: Reproduced with permission from Yang et al. [25]. Copyright 2012, John Wiley and Sons. (d) $NaNbO_3$ nanosheets. Source: Reproduced with permission from You et al. [26]. Copyright 2018, Elsevier. (e) $BaTiO_3$ nanoparticles. Source: Reproduced with permission from Ma et al. [27]. Copyright 2017, John Wiley and Sons. (f) BFO film. Source: Reproduced with permission from Qi et al. [28]. Copyright 2017, John Wiley and Sons. (g) BFO nanoparticles. Source: Reproduced with permission from Wu et al. [29]. Copyright 2016, Royal Society of Chemistry. (h) ZnO nanorods array. Source: Reproduced with permission from Yang et al. [11]. Copyright 2012, American Chemical Society. (i) ZnO nanowires. Source: Reproduced with permission from Mistewicz et al. [30]. Copyright 2019, Elsevier. (j) BST nanoparticles. Source: Reproduced with permission from Xu et al. [31]. Copyright 2018, Royal Society of Chemistry.

the cube-shaped BST nanoparticles have a uniform size of 200 nm. Based on pyro-catalysis induced by pyroelectric effect, BST provides great potential in hydrogen evolution depending on cold–hot energy in environment with near room temperature.

In addition to the typical inorganic pyroelectric materials discussed above, there are many important organic pyroelectric materials. Both of them possess excellent performance and exhibit enormous application prospects for thermal energy harvest and conversion. As one of the most popular and important materials with excellent pyroelectric performance, pure PVDF and its composites including PVDF/polymer matrix, PVDF/inorganic pyroelectric materials, and PVDF/organic pyroelectric materials have been chosen to be discussed and evaluated in this section. Several novel PyENGs based on freestanding PVDF were fabricated for effectively converting thermal energy into electricity [32–34].

A PyENG was designed using a freestanding PVDF film and graphene ink electrode and Al foil electrode, and Figure 7.6a shows the cross-sectional structure of the PyENG [32]; it can be seen that the thicknesses of PVDF and graphene ink are 52 and 9 μm, respectively. The asymmetric structure of PyENG exhibits high thermal radiation absorbance owing to the appearance of graphene ink, which will enhance the conversion efficiency of available thermal energy for generating electricity based on pyroelectric effect. Figure 7.6b displays the photograph of a flexible PyENG and enlarged SEM photograph of carbon nanotubes (CNTs) electrode [33], indicating the length and diameter of CNT to be about 10–20 μm and 50 nm, respectively. By simply attaching to the outside of a beaker, the PyENG can generate electricity by harvesting the heat energy from some chemical exothermic process. Researchers designed a flexible hybrid nanogenerator by using PVDF membranes as piezoelectric and pyroelectric material. Figure 7.6c shows the cross-sectional view of the device mainly consisting of four layers [34]. The thermoplastic polyurethanes (TPU), carbon nanofiber membrane (CNFM), and CNTs layers serve as the substrate and the two electrodes of the hybrid device, respectively.

To improve the flexibility of devices, mixing pyroelectric materials and flexible polymer matrix for forming a pyroelectric-polymer composite will be a feasible approach. Wang and coworkers designed a highly transparent hybrid nanogenerator with excellent flexibility, where a PVDF film shows excellent piezoelectric and pyroelectric properties. As triboelectric material of the triboelectric nanogenerator, the composite film consists of PVDF nanowires and PDMS [35]. Figure 7.6d displays the photograph of the PVDF film with remarkable flexibility and transparency. The inset of Figure 7.6d shows the SEM picture of PDMS composite film with randomly oriented PVDF nanowires with the diameters from 200 to 500 nm. Figure 7.6e displays the SEM image of a composite including PVDF and ZnO [36]; the pyroelectric performance of the nanocomposite will be obviously enhanced owing to the combined effect between the pyroelectric materials PVDF and ZnO. Based on $CH_3NH_3PbI_3$ incorporated electrospun PVDF nanofibers, a hybrid piezo-/PyENG was fabricated for simultaneously harvesting mechanical energy and thermal energy [37]. Figure 7.6f presents the SEM image of PVDF-$CH_3NH_3PbI_3$ nanocomposite, showing that the nanofibers have an approximate diameter of 145 nm. The pyroelectric property of PVDF can be obviously enhanced owing to the addition of $CH_3NH_3PbI_3$. Likewise, another hybrid piezoelectric-PyENG was designed, and the structure diagram and SEM images of PDMS-CNTs composite and P(VDF-TrFE) are displayed in Figure 7.6g [38]. The hybrid nanogenerator that consists of a micro-patterned P(VDF-TrFE) acts as pyroelectric and piezoelectric layers, and graphene nanosheets and PDMS-CNTs composite as the two electrodes of the nanogenerator.

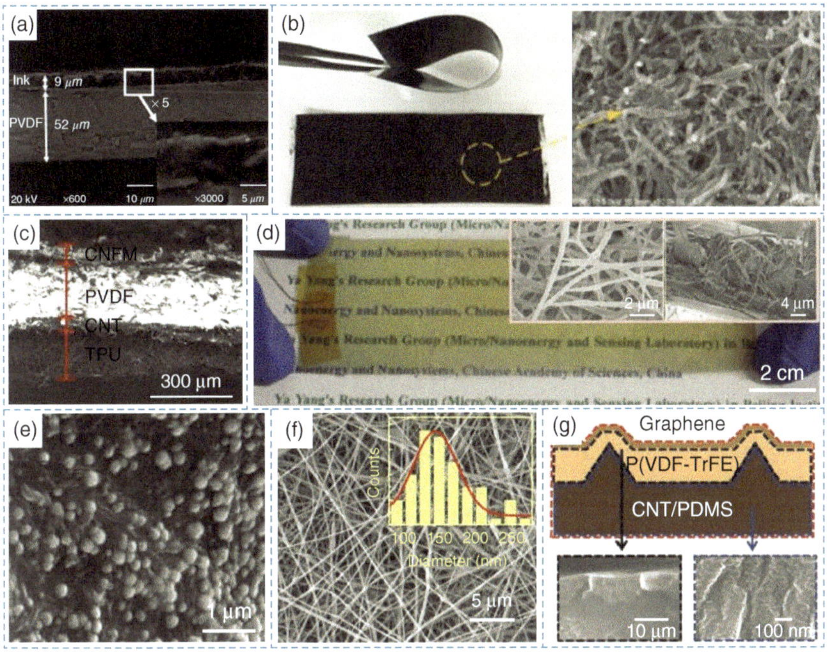

Figure 7.6 The SEM images of typical organic pyroelectric materials. (a) The cross section of a PVDF film. Source: Reproduced with permission from Zabek et al. [32]. Copyright 2017, American Chemical Society. (b) The photograph of PDVF-CNTs film and microstructure of CNTs. Source: Reproduced with permission from Zhao et al. [33]. Copyright 2017, Elsevier. (c) The cross section of a PVDF film with a sandwich structure. Source: Reproduced with permission from You et al. [34]. Copyright 2018, Royal Society of Chemistry. (d) PVDF nanowires and PDMS composite. Source: Reproduced with permission from Wang et al. [35]. Copyright 2016, John Wiley and Sons. (e) PVDF/ZnO composite. Source: Reproduced with permission from Thakur et al. [36]. Copyright 2018, Elsevier. (f) PVDF–$CH_3NH_3PbI_3$ nanocomposite. Source: Reproduced with permission from Sultana et al. [37]. Copyright 2019, American Chemical Society. (g) P(VDF-TrFE) and PDMS–CNTs composite. Source: Reproduced with permission from Lee et al. [38]. Copyright 2014, John Wiley and Sons.

7.3.2 Structure Design

Considering the differences in thermal energy sources and various application fields, the corresponding structures of PyENGs are diverse. In addition, the designs of hybrid nanogenerators for simultaneously harvesting thermal energy and other energies have been increasingly popular in academic community and practical applications. Here, the structure and design of PyENGs and hybrid devices based on PyENGs with different functions are evaluated and discussed in this section.

Figure 7.7a presents a typical structure diagram of a PyENG fabricated by placing and fixing a single PZT microwire, which is entirely packaged by transparent PDMS and exhibits excellent flexibility [20]. The thin PDMS layer not only realizes excellent flexibility but also avoids external contamination or corrosion and the influence of airflow. Based on a PVDF film with excellent performance of piezoelectricity and pyroelectricity after being polarized, a flexible and hybrid nanogenerator was fabricated for harvesting thermal and mechanical energies. Figure 7.7b displays the structure diagram [35]. The one-structure-based hybrid device with the same ITO electrodes can simultaneously deliver pyroelectric, triboelectric, and piezoelectric signals based on a heated temperature and wind-induced vibration energy. Figure 7.7c shows the diagram of a PyENG fabricated by $KNbO_3$ nanowires [25], and Ag and ITO layers serve as the two electrodes of the PyENG. This PyENG can harvest the direct thermal energy and indirect thermal energy induced by sunlight illumination owing to photothermal effect.

A PyENG was designed based on a PZT film, and Figure 7.7d gives the photograph of the flexible pyroelectric nanogenerator (PENG) with the length/width of 21/12 mm, respectively [16]. Figure 7.7e shows the diagram of a PyENG fabricated by using a pyroelectric $BaTiO_3$ film with two electrodes of ITO and Ag layers [27], which can realize the photodetection of 405 nm light illumination based on photovoltaic–pyroelectric coupled effect. Based on ZnO nanowires-based PyENG, researchers designed self-powered photoelectric devices for sensitive ultraviolet [39] and near-infrared [40] photosensing. Figure 7.7f shows the side-view diagram of the device and principle diagram of the self-powered photodetector [39]; ZnO nanowires covered with $CH_3NH_3PbI_3$ layer form the most important part of the photosensing device. Likewise, the structure diagram and photograph of the near-infrared photoelectric device are shown in Figure 7.7g [40]. This research will provide a feasible and satisfactory method to realize photon sensing by using PyENG-based photodetectors.

Figure 7.8a displays the structure diagram of a hybrid plasmonic–pyroelectric nanogenerator for detecting spectrally selective mid-wavelength infrared [41], and the device consists of a uniaxial ZnO nanowires layer, grown on a Pt template, also as bottom electrode and Au microholes array as top electrode. An innovatively designed stretchable PyENG was reported by using PDMS, Ag/AgNWs electrode with high stretchability, and micropatterned P(VDF-TrFE), as shown in Figure 7.8b [42], which successfully realized the enhancement of power generation owing to a coupling enhancement of piezoelectricity and pyroelectricity. Figure 7.8c presents the photograph and corresponding diagram (inset) of a PZT film-based hybrid piezoelectric-PyENG with excellent flexibility and Pt, $LaNiO_3$, and Ni-Cr acting as the top electrode, bottom electrode, and metal foil substrate, respectively [43]. The flexible hybrid nanogenerator can effectively scavenge vibration-induced mechanical energy and thermal energy of gradually changing temperature fluctuation. In addition, it can work normally in harsh environment including highly relative humidity and pH, etc.

Park and coworkers have designed a poly(ethylene glycol)-*block*-poly(propylene glycol)-*block*-poly(ethylene glycol)-poly(3,4-ethylenedioxythiophene) (PP-PEDOT) composite film-based photothermally driven PyENG-TEG. The PP-PEDOT composite film acts as a pyroelectric material for efficiently harvesting solar energy. Figure 7.8d displays the device structure and mechanism diagram [44]; we can see that the near infrared (NIR) irradiation will weaken spontaneous polarization and induce the flow of electron in the external circuit. Figure 7.8e shows the structure diagram of a hybrid nanogenerator based on piezoelectric–pyroelectric effect, and the corresponding photograph is in the

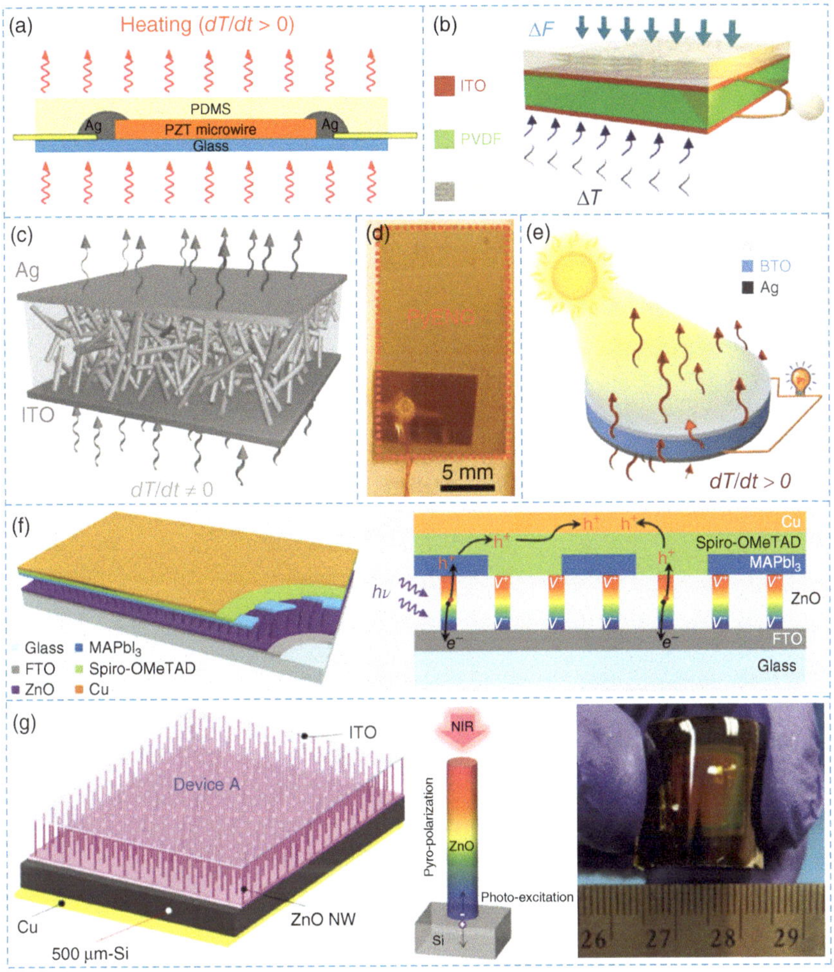

Figure 7.7 Structure diagrams of several PyENGs. (a) A PZT-based PyENG. Source: Reproduced with permission from Yang et al. [20]. Copyright 2012, American Chemical Society. (b) Hybrid pyroelectric–piezoelectric nanogenerator based on PVDF film. Source: Reproduced with permission from Wang et al. [35]. Copyright 2016, John Wiley and Sons. (c) A $KNbO_3$ nanowires-based PyENG. Source: Reproduced with permission from Yang et al. [25]. Copyright 2012, John Wiley and Sons. (d) A PyENG based on PZT film. Source: Reproduced with permission from Yang et al. [16]. Copyright 2012, American Chemical Society. (e) A PyENG based on pyroelectric $BaTiO_3$ film. Source: Reproduced with permission from Ma et al. [27]. Copyright 2017, John Wiley and Sons. (f,g) Two kinds of PyENG-based NIR photodetectors. Source: (f) Reproduced with permission from Wang et al. [39]. Copyright 2015, Springer Nature; (g) Reproduced with permission from Wang et al. [40]. Copyright 2017, American Chemical Society.

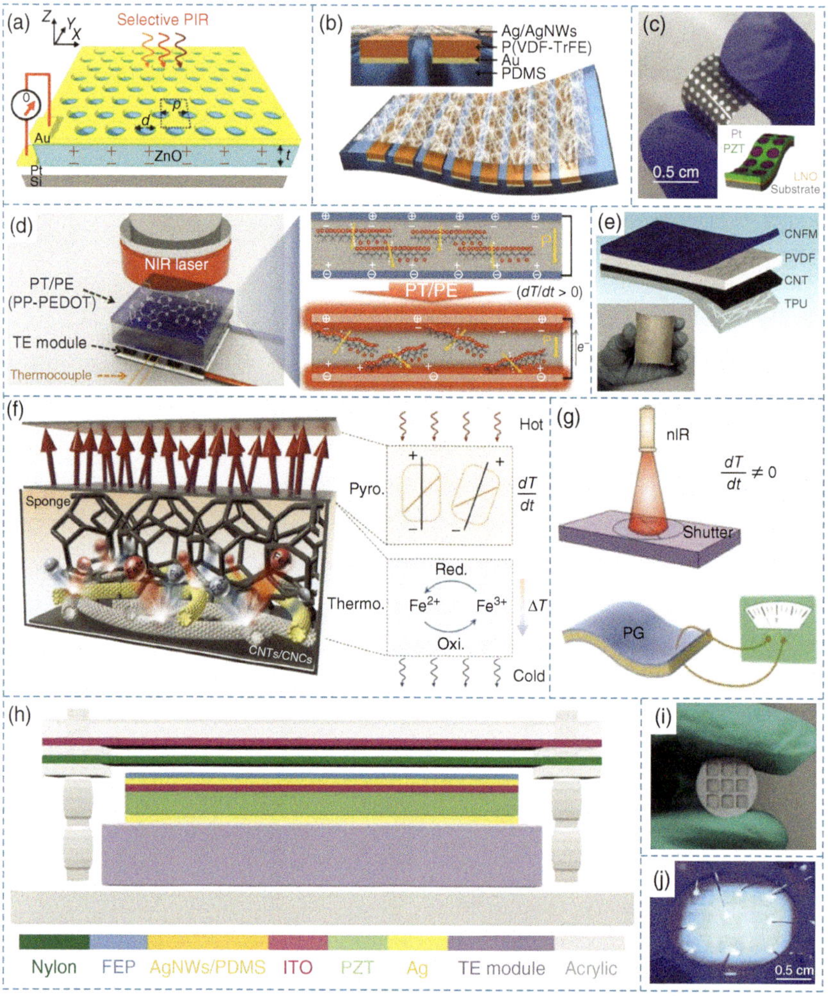

Figure 7.8 Structure diagrams of PyENGs and PyENG-based hybrid nanogenerators. (a) A hybrid plasmonic-pyroelectric nanogenerator based on ZnO nanowires layer. Source: Reproduced with permission from Dao et al. [41]. Copyright 2016, American Chemical Society. (b) A flexible PyENG based on micropatterned P(VDF-TrFE). Source: Reproduced with permission from Lee et al. [42]. Copyright 2015, John Wiley and Sons. (c) A flexible hybrid piezoelectric–pyroelectric nanogenerator based on a PZT film. Source: Reproduced with permission from Ko et al. [43]. Copyright 2016, American Chemical Society. (d) A pyroelectric–thermoelectric nanogenerator driven by photothermal energy. Source: Reproduced with permission from Park et al. [44]. Copyright 2015, American Chemical Society. (e) A self-powered hybrid piezoelectric–pyroelectric nanogenerator. Source: Reproduced with permission from You et al. [34]. Copyright 2018, Royal Society of Chemistry. (f) PVDF-based hybrid nanogenerator. Source: Reproduced with permission from Ding et al. [45]. Copyright 2018, John Wiley and Sons. (g) An infrared-driven non-contact PyENG with a sandwich structure of CNT/PVDF/CNT layer. Source: Reproduced with permission from Zhao et al. [46]. Copyright 2016, Royal Society of Chemistry. (h) Hybrid nanogenerator. Source: Reproduced with permission from Zhang et al. [47]. Copyright 2016, John Wiley and Sons. (i,j) PyENGs based on BTO and BFO pyroelectric layer with multi-channel structure. Source: (i,j) Reproduced with permission from Ma et al. [27] and Qi et al. [28]. Copyright 2017, John Wiley and Sons.

inset [34]. PVDF film acts as a piezoelectric layer and also serves as a pyroelectric layer, CNFM and CNTs act as upper and bottom electrodes, respectively, and the hybrid nanogenerator can realize high-efficiency energy harvest and conversion. In addition, researchers reported a hybrid pyroelectric and thermogalvanic nanogenerator for efficiently harvesting heat energy with low grade. The diagrams of structure and mechanism are displayed in Figure 7.8f [45]; we can see that the device consists of two ferroelectric PVDF films and a PU sponge soaked with $Fe^{2+}/Cl^{-}/Fe^{3+}$ electrolyte as a block. When light irradiates the surface of the device, the CNT/CNC (cellulose nanocrystal) composite layer will convert light into heat and generate a temperature increase across the electrolyte, which will induce the redox of Fe^{2+} and Fe^{3+} and then produce a thermogalvanic potential between the two electrodes (thermocell voltage).

Figure 7.8g displays a novel infrared-driven PyENG based on CNT/PVDF/CNT layer for harvesting the photothermal energy [46], which realizes long-distance energy transfer owing to photothermal effect and excellent penetrability of infrared light without direct contact. Zhang et al. designed a one-structure-based nanogenerator based on multifunctional PZT for realizing the coupled enhancement of multi-effects by integrating a PyENG, a TPiENG, and a photovoltaic cell into one device sharing the same electrodes; the structure diagram is seen in Figure 7.8h [47]. The hybrid device based on multi-effects can individually or simultaneously collect solar, mechanical, and thermal energies. Based on a photovoltaic–pyroelectric coupled effect, researchers have fabricated multi-channel self-powered photodetectors based on barium titanate (BTO) and BFO as pyroelectric layer for detecting 405 and 450 nm light illumination; Figure 7.8i [27] and Figure 7.8j [28] show the photographs of BTO and BFO-based photodetectors with similar structures, respectively.

7.3.3 Performance of Pyroelectric Nanogenerators

After rapid development in recent years, the output performances of PyENGs have achieved obvious improvement for successfully realizing various functions in different fields. However, owing to the relatively lower output power of the reported devices, the performance of PyENGs needs to be drastically improved to expand the broad application fields. In this section, the output performances of various PyENGs are evaluated and analyzed.

A PyENG fabricated by PZT microwire for detecting temperature change was reported as a temperature sensor with self-powered function. The performance of the device was researched by attaching on a heater with a controllable temperature [20]. The periodic change of the heater's temperature from 296 to 333 K will generate a cyclic temperature change rate in the PyENG, as shown in Figure 7.9a, which will output cyclic pulse signals, and the sharp voltage (current) pulses are about 60 mV (0.6 nA). The output signals were researched and it was found that output electrical signals linearly increase with increasing temperature change rate. As a temperature sensor, the device can detect the temperatures of the heat source with a minimum detection limit of 0.4 K.

As displayed in Figure 7.9b [20], increase of temperature of the heat source will cause a linear increase of temperature change rate, which leads to linear increase

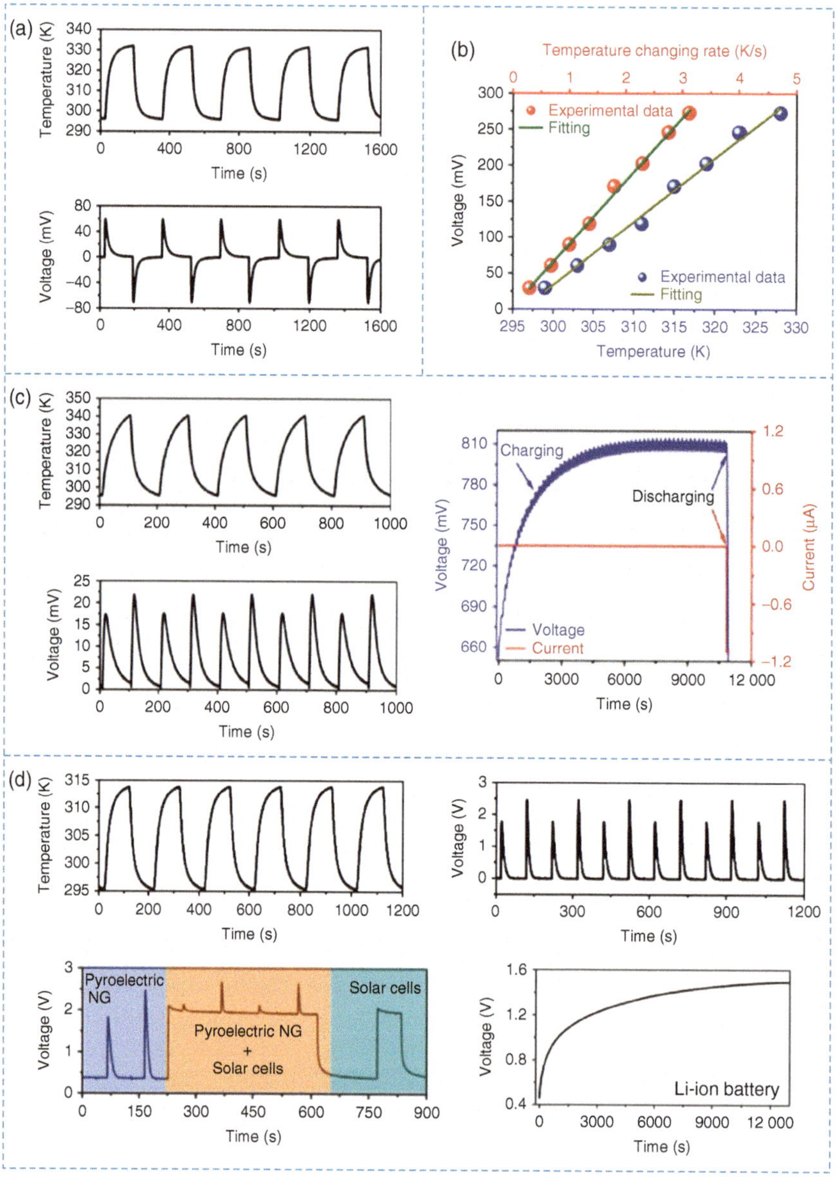

Figure 7.9 Pyroelectric performances of PyENGs. (a, b) Time-dependent temperature curve and corresponding output voltage signal (a), and the dependence of the output voltage on the temperature changing rate and the temperature of the heat source (b). Source: (a,b) Reproduced with permission from Yang et al. [20]. Copyright 2012, American Chemical Society. (c) The cyclic temperature variation and corresponding output voltage of a flexible PyENG, and the charging and discharging curves of a Li-ion battery by using the PyENG. Source: Reproduced with permission from Yang et al. [16]. Copyright 2012, American Chemical Society. (d) The pyroelectric output performance of a single PyENG and hybrid nanogenerator, and the charging curve of a Li-ion battery by using the PyENG. Source: Reproduced with permission from Yang et al. [48]. Copyright 2013, American Chemical Society.

of the output voltage of the device. Owing to the excellent pyroelectric property of PZT nanomaterials, the authors further fabricated a flexible PyENG based on PZT nanoparticles for harvesting thermal energy. As shown in Figure 7.9c [16], under a cyclic temperature of the PyENG ranging from 295 to 340 K, a cyclically positive output voltage was obtained through a rectifier. Under periodic temperature change of 45 K, the PyENG generates a corresponding pulse voltage/current of 22 V/430 nA. The electric energy generated from temperature fluctuation can be stored in a battery. Figure 7.9c presents the charging/discharging curve, and we can see that the voltage increases from 650 to 810 mV in the charging process by using the PyENG as a power source. In the charging process, many peaks occur and correspond to the pulse voltage produced under a working frequency of 0.005 Hz.

Under a periodic temperature variation process, as shown in Figure 7.9d, PVDF film-based PyENG can harvest thermal energy and output cyclic positive electrical signal after rectification [48]. The pulse signals of 2.5 V/24 nA and 3.2 V/31 nA correspond to the process of increased (from 295 to 314 K) and decreased (from 314 to 295 K) temperature, respectively. The authors further fabricated a hybrid nanogenerator based on the PyENG and several solar cells and realized simultaneous thermal energy and solar energy harvest. Under 100 mW/cm^2 light illumination, solar cells can generate a voltage (current) of 2 V (6 μA). Figure 7.9d displays the output voltage curve, and the pyroelectric output signals were rectified. The thermal and solar energies can be harvested. In addition, the generated electricity can charge a battery with the charged voltage increased up to 1.5 V easily.

By using ZnO nanowire arrays, researchers fabricated a PyENG for harvesting and converting heat energy [11]. The output electrical signals of PyENG gradually increase with increasing temperature change rates, as seen in Figure 7.10a. The temperature changes from room temperature to high values of 309, 314, and 320 K will generate corresponding output voltage peaks of 10, 15, and 18 mV, respectively. The stability of the PyENG was also measured, as displayed in Figure 7.10a; the voltage signal is stable without obvious decline for 200 cycles within six hours.

Based on p-Si/ZnO NW heterostructure, a light-triggered PyENG was designed for NIR photosensing. As displayed in Figure 7.10b [40], under NIR illumination with the light power density of 4.8 mW/cm^2, the induced positive sharp pyroelectric current peak at the moment appeared as high as 1.05×10^{-3} A with a fast response of 15 μs. When the NIR was turned off , the negative pyroelectric current signal also appeared with the response time of 21 μs. This research provides a novel approach to realize high-performance NIR sensing.

As wearable device, a small PyENG can be used to detect different breathing conditions by harvesting thermal energy from human breath. By harvesting the airflow from human breath in the environment with an ambient temperature of about 5 °C, the PyENG produces an output voltage (current) of 42 V (2.5 μA) [49]. In Figure 7.10c, the positive signal of PyENG was generated after rectification under temperature fluctuation induced by harvesting the airflow from human breath. The generated electric energy can charge a 10.0 μF capacitor, and the voltage of the capacitor can increase up to 1 V within 18 seconds.

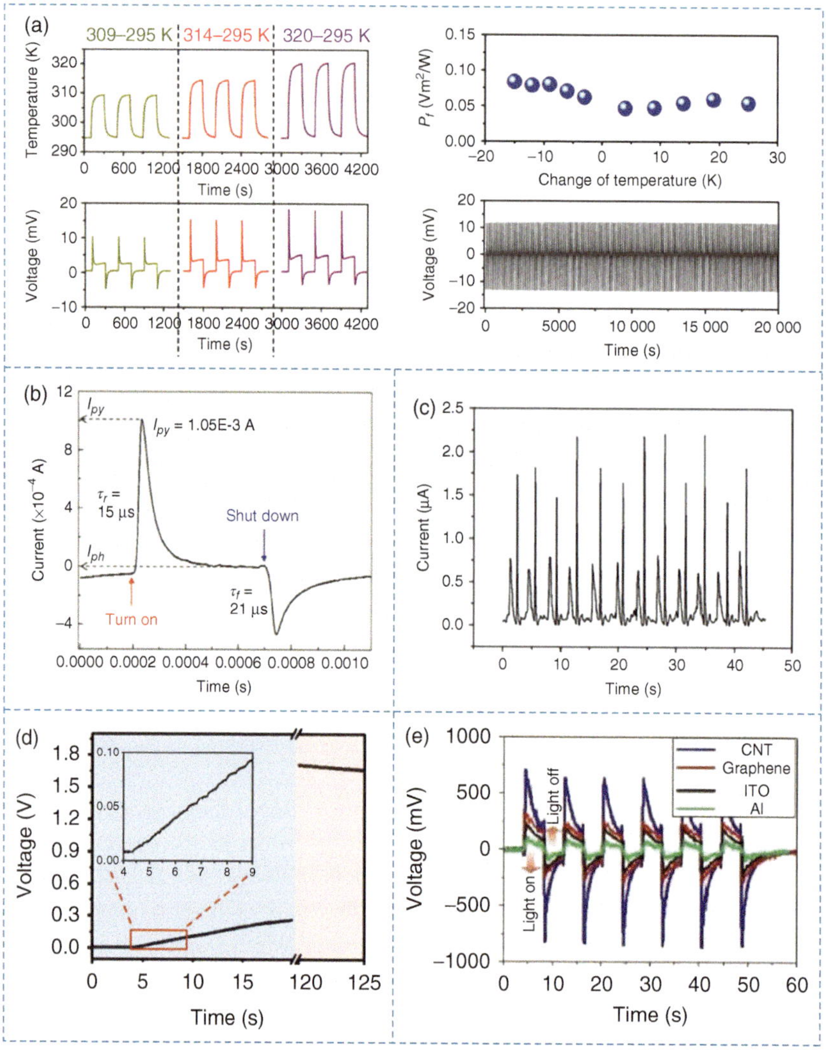

Figure 7.10 Pyroelectric output performances of PyENGs. (a) The output voltage and cycling stability of a PyENG based on ZnO nanorods. Source: Reproduced with permission from Yang et al. [11]. Copyright 2012, American Chemical Society. (b) One typical cycle of short-circuit current and time curve under periodical NIR illumination. Source: Reproduced with permission from Wang et al. [40]. Copyright 2017, American Chemical Society. (c) The positive output current after rectification under temperature fluctuation induced by human respiration. Source: Reproduced with permission from Xue et al. [49]. Copyright 2017, American Chemical Society. (d) The charging curve of a capacitor using a stretchable PyENG. Source: Reproduced with permission from Lee et al. [42]. Copyright 2015, John Wiley and Sons. (e) The output voltages of an NIR-induced PyENG with CNT, graphene, ITO, and Al as top electrode respectively. Source: Reproduced with permission from Zhao et al. [46]. Copyright 2016, Royal Society of Chemistry.

On changing the temperature from room temperature to 317 K, the PyENG can output a voltage (current density) of 2.5 V (570 nA/cm^2). The produced electricity can charge a capacitor, as shown in Figure 7.10d [42]; it can be seen that the voltage is increased up to 1.7 V in 120 seconds. Figure 7.10e displays the output voltages of an infrared-driven PyENG with different materials as electrode with a temperature variation frequency of 0.125 Hz [46]. Under the same light illumination, the pyroelectric property of the PyENG with CNT as electrode is the best with the voltage of 800 mV owing to the highest temperature difference of CNT. The performance of the PyENG with Al/PVDF/Al sandwich is the worst with a voltage of 100 mV.

7.3.4 Applications for Sensing and Electrochemistry

As an important energy harvest technology, PyENGs have developed rapidly and exhibited a large potential in many fields. At present, many kinds of PyENGs and PyENGs-based hybrid nanogenerators are widely used for driving electronic devices, self-powered sensing system for detecting different signals, and improving electrochemistry reactions, by harvesting thermal energy from the external environment.

Figure 7.11a shows photographs of a PyENG as a temperature sensor [20]; when the sensor is first touched by a finger and then the finger removed, a large negative and small positive output voltage peaks can be observed, respectively. The temperature sensor shows excellent performance with a fast response less than 0.9 seconds and minimum detection limit of 0.4 K. Under a small temperature variation rate, the output pyroelectric energy of a PyENG can directly drive an LCD. An enlarged output voltage peak is shown in Figure 7.11b [16]; it can be seen that the generated electricity can drive an LCD in region-2 for more than 60 seconds, where the voltage is larger than the working voltage of the LCD. In addition, the PyENG can charge a Li-ion battery as an electric source. Figure 7.11c [42] and Figure 7.11d [50] present the photographs of some LEDs powered by a stretchable PyENG under different strain conditions and a water vapor-driven PyENG, respectively.

As a wearable device, a small PyENG can be integrated in a respirator, as shown in Figure 7.11e [49], by harvesting the thermal energy of human breath; an LCD and several blue LEDs can be powered. The temperature of human health is stable relatively, so the PyENG can detect the temperature of ambient environment through the difference of output pyroelectric signals. A flexible hybrid energy generator attached on a respirator was used to detect physiological signals. The flexible device can deliver different output voltage signals corresponding to different breathing conditions by harvesting thermal energy, as shown in Figure 7.11f [51], and the temperatures of air exhaled from the human body and the ambient environment are about 37 and 15 °C, respectively. Based on the variation of voltage signals generated by pyroelectric effect, the weak, normal, and deep breath conditions can be monitored.

As a temperature sensor, the temperature of hot water can be detected by PyENG by harvesting thermal energy of water vapor, as presented in Figure 7.11g [52]. It can precisely detect the temperatures of five fingers owing to different

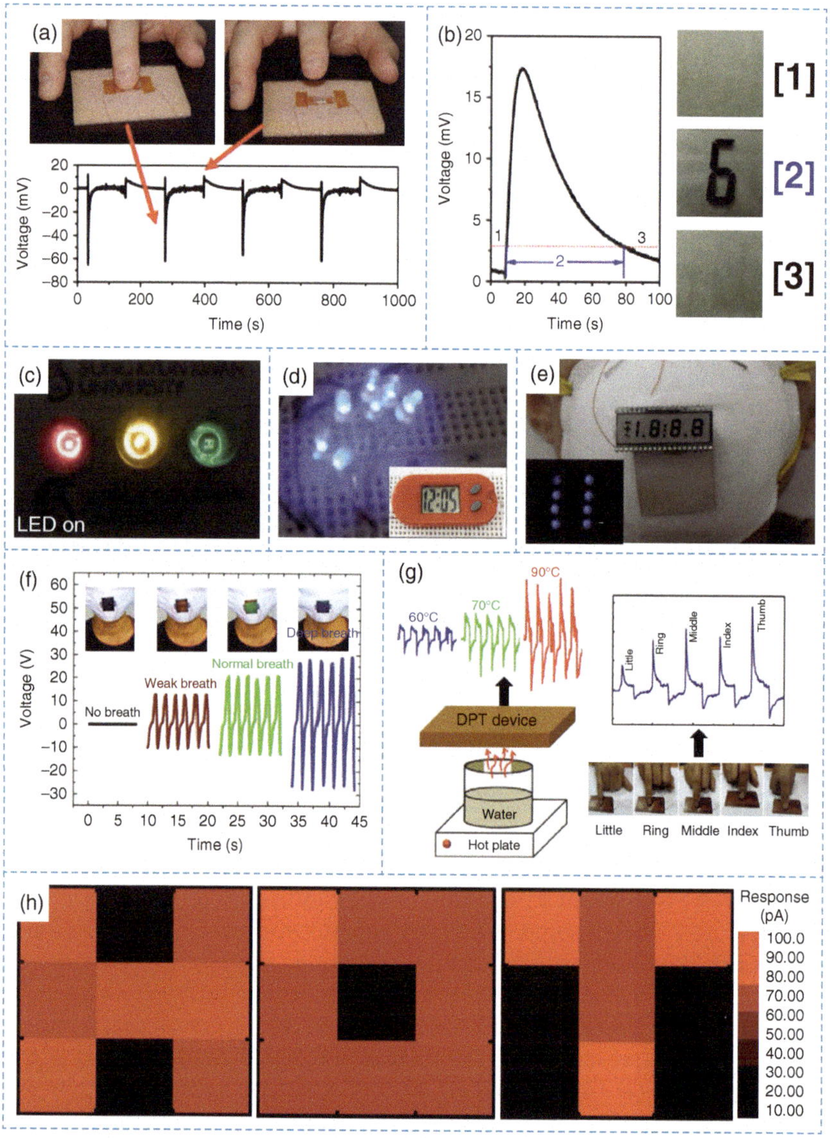

Figure 7.11 A variety of applications of PyENGs. (a) A PZT micro/nanowires-based PyENG can monitor the temperature of human skin. Source: Reproduced with permission from Yang et al. [20]. Copyright 2012, American Chemical Society. (b) The generated electricity can drive an LCD. Source: Reproduced with permission from Yang et al. [16]. Copyright 2012, American Chemical Society. (c) Several LEDs can be powered by a stretchable PyENG for harvesting thermal energy. Source: Reproduced with permission from Lee et al. [42]. Copyright 2015, John Wiley and Sons. (d) A self-sustaining PyENG can drive LEDs by harvesting the thermal energy of water vapor. Source: Reproduced with permission from Gao et al. [50]. Copyright 2016, Elsevier. (e) A PyENG as wearable device for driving an LCD or LEDs under temperature fluctuation from human breathing. Source: Reproduced with permission from Xue et al. [49]. Copyright 2017, American Chemical Society. (f) Different breath conditions can be monitored by a PyENG. Source: Reproduced with permission from Sun et al. [51]. Copyright 2018, Elsevier. (g) A PyENG for sensing temperatures of hot water and human body. Source: Reproduced with permission from Sultana et al. [52]. Copyright 2018, Elsevier. (h) Response of pixels in the GPIR detector array corresponding to current distribution mapping. Source: Reproduced with permission from Kulkarni et al. [53]. Copyright 2014, John Wiley and Sons.

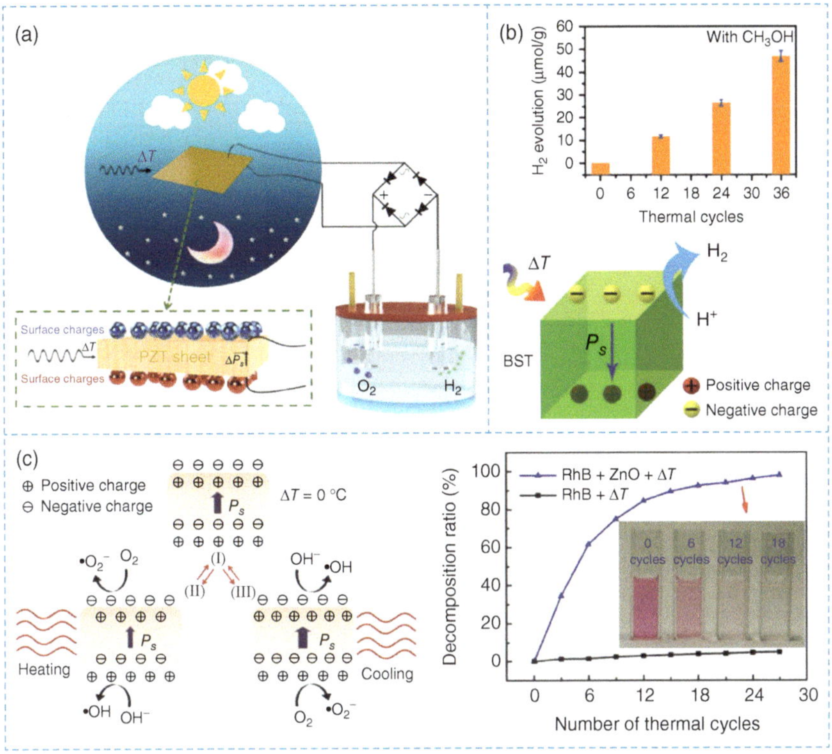

Figure 7.12 PyENGs for electrochemical reactions. (a) A PZT sheets-based PyENG for water splitting. Source: Reproduced with permission from Zhang et al. [21]. Copyright 2019, Elsevier. (b) The pyro-catalysis of BST nanoparticles for water splitting. Source: Reproduced with permission from Xu et al. [31]. Copyright 2018, Royal Society of Chemistry. (c) Pyroelectric effect of ZnO nanorods for RhB decomposition. Source: Reproduced with permission from Qian et al. [54]. Copyright 2017, Elsevier.

active touch areas. Figure 7.11h [53] shows the different pixels response of pyroelectric effect in a graphene-pyroelectric infrared (GPIR) detector array, and the incoming radiation on the surface of the GPIR detector forms a letter "HOT," which demonstrates a pixelated array of the infrared detector and provides a feasible method for the imaging of thermal scenes.

In addition to power generation and self-powered sensing, PyENGs and pyroelectric effect have also exhibited remarkable application potential in electrochemical reactions. Based on a pyroelectric PZT sheet, a PyENG was fabricated as an external source under the process of hot–cold thermal cycles for producing hydrogen by harvesting the wasted thermal energy in nature, as shown in Figure 7.12a [21]. It can be easily understood that different electrical charges are first generated on the surface of the PyENG owing to the effect of thermal fluctuation. Then the generated electrical energy after being rectified acts as the DC power source, which induces migration of electrons and reduces the H^+ to H_2 on the cathode surface, and the production of O_2 occurs at the anode.

Based on the pyroelectric effect of BST nanoparticles as pyro-catalyst, Huang and coworkers demonstrated a high-efficiency method of direct hydrogen generation by harvesting the thermal energy under cyclic room temperature fluctuations of cold–hot alternation from 298 to 323 K. As presented in Figure 7.12b [31], the produced hydrogen is about 46.89 mmol/g with the addition of methanol after 36 thermal cycles corresponding to an average hydrogen generation rate of 1.30 mmol/g per cycle. According to the working mechanism diagram of pyro-catalytic hydrogen generation, it can be seen that temperature change of pyroelectric BST nanoparticles will generate positive and negative charges that react with water molecules H_2O to produce O_2 and H_2.

Besides hydrogenxs generation, pyroelectric materials also show excellent improvement effect in sewage decomposition. ZnO nanorods with high pyroelectric effect are used to enhance the degradation rate of Rhodamine B (RhB) at room temperature owing to the coupling effect of pyroelectric effect and electrochemical oxidation. As shown in Figure 7.12c [54], under 22–62 °C heating–cooling cycles, the variation of spontaneous polarization in ZnO nanoparticles will induce the generation of the active species hydroxyl ($^{\cdot}OH$) and superoxide radicals ($^{\cdot}O_2^-$). Then RhB molecule reacts with strongly oxidative $^{\cdot}O_2^-$ and $^{\cdot}OH$ to produce CO_2 and H_2O. It can be seen that after 27 cycles, the addition of ZnO nanorods with high pyro-catalytic effect results in a high decomposition rate of ~98.15% larger than that of the condition without ZnO nanorods for RhB solution.

7.4 Progress of Thermoelectric Nanogenerators

The detailed description above gives a picture of the development progress of PyENGs. Subsequently, the development progress of TEGs is discussed, including materials, structure, performance, and applications.

7.4.1 Typical Thermoelectric Materials

Nanomaterials exhibit excellent performance compared with commercial products owing to higher specific surface area and possess large application potential in energy harvest and conversion of TEGs. We discuss and summarize several typical TE materials in this section.

As a kind of important functional nanomaterials, ZnO nanomaterials with various morphologies not only possess excellent piezoelectric and semiconducting properties, but also exhibit remarkable TE performance [9, 12, 55]. A single Sb-doped ZnO micro/nanobelt-based TEG was fabricated for self-powered temperature sensing through high-efficiency harvest and conversion of thermal energy [12]; Figure 7.13a displays the SEM photograph and energy dispersive spectrometer (EDS) analysis of the Sb-doped single ZnO nanobelt. A ZnO nanorods-based thermoelectric-enhanced solar cell was designed as a photovoltaic and TE system based on thermo-phototronic effect [55]; Figure 7.13b presents the SEM image of ZnO, showing the length of 700 nm. As hexagonal nonmetallic crystals, tellurium (Te) nanomaterials possess remarkable TE performance depending on high power factor and low thermal conductivity

[63, 64]. Te nanowires were used for fabricating a flexible TEG that shows excellent TE property; Figure 7.13c shows the transmission electron microscopic (TEM) image of Te nanowires [56]. In addition to nanowires, Te nanomaterials with other morphologies also possess excellent TE performance, including Te nanoparticles, nanorods, and nanotubes, as shown in Figure 7.13d [57].

As important materials on earth, carbon-based materials have made great progress and show a hugely positive influence in the development of modern society. Figure 7.13e–h shows the TEM image of C_{60} nanoparticles [58], SEM images of porous carbon film [59], graphene–ecoflex composite [19], and CNTs [60], respectively; several high-performance TEGs were fabricated based on these carbon materials respectively for thermal harvest and conversion. Du et al. fabricated a flexible TEG by using polyester fabric coated with PEDOT:PSS (poly(styrenesulfonate)), as shown in Figure 7.13i [61], which can be easily deformed into any desired shapes. As wearable device, it can harvest the heat of human body owing to the ΔT between the human body and the ambient environment. Through automatic integration of silicon nanowire arrays, an all-silicon TEG was reported for power generation by harvesting thermal energy [62]; the SEM image of silicon nanowires is presented in Figure 7.13j.

Except for the typical TE materials of metallic oxide, nonmetallic substances, and other materials above, metallic compounds based on Te, sulfur (S), and selenium (Se) respectively are an important class of TE materials. N-type Bi_2Te_3 nanoplates were prepared, and the lateral size and thickness of these nanoplates are several hundred nanometers and 20 nm respectively, as shown in Figure 7.14a [65]. The TE property of Bi_2Te_3 can be obviously enhanced through engineering nanostructure. The performance of TE materials has been improved by doping some active materials, as shown in Figure 7.14b [66]; the TE performance of multi-walled carbon nanotube (MWCNT) doped Bi_2Te_3 nanocomposite was greatly improved and better than that of pure Bi_2Te_3 nanoflowers. A series of Te-based metallic compounds including Ag_2Te [67], SnTe [68], and Sb_2Te_3 [69] nanomaterials were prepared with excellent TE performance, as seen in Figure 7.14c–e. Based on an efficient, simple, and inexpensive preparation process, these Te-based nanomaterials with high TE performance have a large potential application in power generation through thermal energy harvest and conversion.

Figure 7.14f shows the SEM image of two-dimensional semimetallic TiS_2 nanosheets with excellent TE performance [70], which was obtained by a chemical exfoliation process and a deposition process of Al^{3+} bridged TiS_2 nanosheets. The flexible TiS_2 TE films exhibit huge potential applications in wearable electronics. Similarly, WS_2 and $NbSe_2$ nanosheets as TE materials were respectively prepared by a chemical exfoliation process, as seen in Figure 7.14g [71]; the lateral size and thickness of the nanosheets are 300–500 and 1–3 nm, respectively. SnSe TE materials have developed rapidly and showed excellent TE performance in recent years [72–74]. Figure 7.14h,i show the SEM image of SnSe nanoplates and the photograph of SnSe crystal cut along the three different axes respectively. It is demonstrated that the SnSe nanoplates and crystal exhibit high TE power factor at 550 K [72] and possess a superhigh ZT of 2.6 ± 0.3 at 923 K [73].

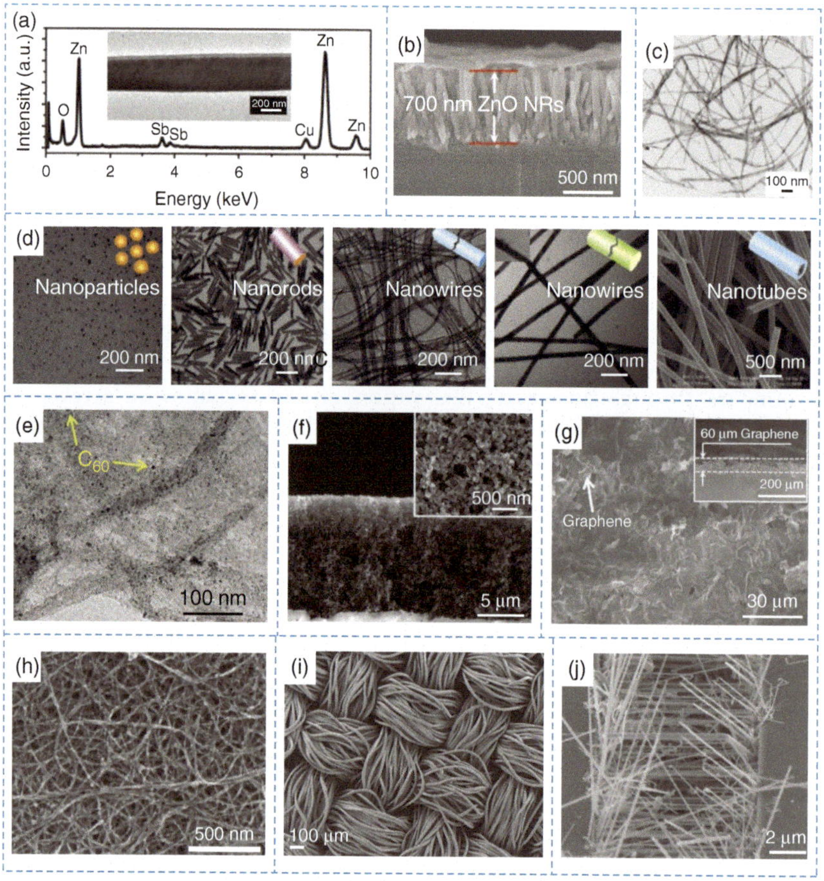

Figure 7.13 The SEM or TEM images of typical thermoelectric materials. (a) A single ZnO nanowire. Source: Reproduced with permission from Yang et al. [12]. Copyright 2012, American Chemical Society. (b) The cross section of a ZnO nanorods array. Source: Reproduced with permission from Zhang et al. [55]. Copyright 2017, John Wiley and Sons. (c) Te nanowires. Source: Reproduced with permission from Yang et al. [56]. Copyright 2012, Springer Nature. (d) Te nanomaterials with different morphologies. Source: Reproduced with permission from Liu et al. [57]. Copyright 2016, John Wiley and Sons. (e) TEM image of C_{60} nanomaterial. Source: Reproduced with permission from Wang et al. [58]. Copyright 2018, Royal Society of Chemistry. (f) Porous carbon film. Source: Reproduced with permission from Liu et al. [59]. Copyright 2018, John Wiley and Sons. (g) Few-layer graphene–ecoflex nanocomposite film. Source: Reproduced with permission from Zhang et al. [19]. Copyright 2019, John Wiley and Sons. (h) CNTs. Source: Reproduced with permission from Zhou et al. [60]. Copyright 2017, Springer Nature. (i) PEDOT:PSS coated polyester fabrics. Source: Reproduced with permission from Du et al. [61]. Copyright 2015, Springer Nature. (j) Silicon nanowires. Source: Reproduced with permission from Fonseca et al. [62]. Copyright 2016, IOP Publishing.

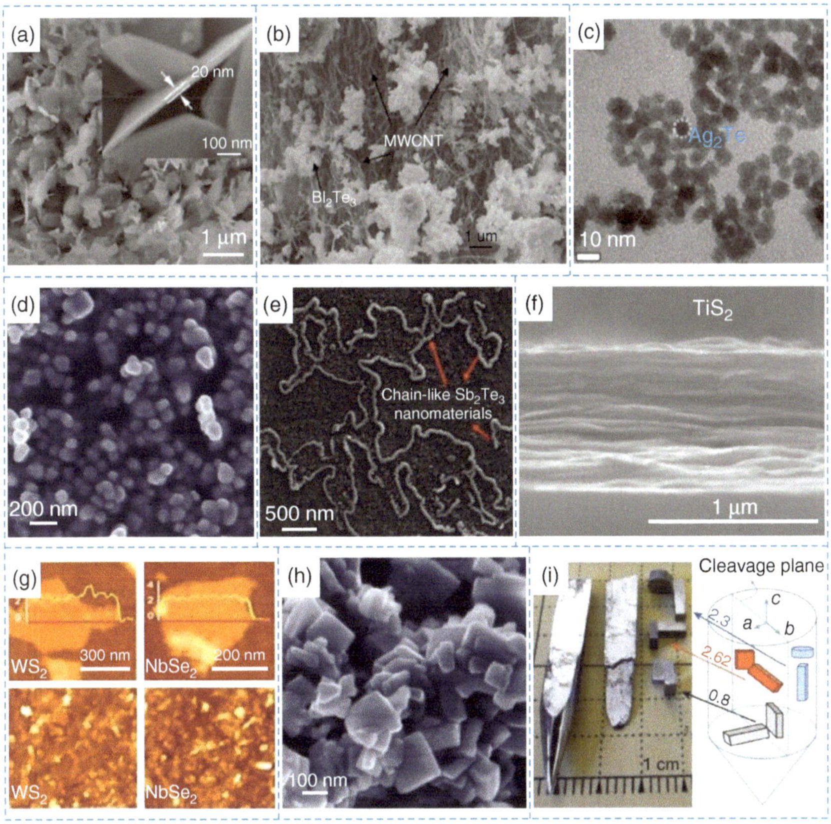

Figure 7.14 The micromorphologies of typical thermoelectric materials. (a) Typical Bi_2Te_3 nanoplates. Source: Reproduced with permission from Yang et al. [65]. Copyright 2015, American Chemical Society. (b) Bi_2Te_3-MWCNTs nanocomposite. Source: Reproduced with permission from Kumar et al. [66]. Copyright 2017, John Wiley and Sons. (c) TEM images of Ag_2Te nanoparticles. Source: Reproduced with permission from Yang et al. [67]. Copyright 2017, John Wiley and Sons. (d) SnTe nanoparticles. Source: Reproduced with permission from Han et al. [68]. Copyright 2017, MDPI Publishing. (e) Sb_2Te_3 nanowires. Source: Reproduced with permission from Nithiyanantham et al. [69]. Copyright 2015, Royal Society of Chemistry. (f) TiS_2-nanosheet restacked film. Source: Reproduced with permission from Zhou et al. [70]. Copyright 2017, American Chemical Society. (g) Atomic force microscopic (AFM) images of chemically exfoliated WS_2 and $NbSe_2$ nanosheets. Source: Reproduced with permission from Oh et al. [71]. Copyright 2016, Royal Society of Chemistry. (h) SnSe nanoplates. Source: Reproduced with permission from Han et al. [72]. Copyright 2016, John Wiley and Sons. (i) SnSe crystal. Source: Reproduced with permission from Zhao et al. [73]. Copyright 2014, Springer Nature.

7.4.2 Structure Design

In recent years, various kinds of TEGs with increasingly improved TE performance have been fabricated for thermal harvest and conversion. With the booming development of flexible electronics, more and more researchers have attained great achievements in the flexibility of TEGs for meeting the demand of intelligent devices. Here, the structure and design of TEGs are evaluated and discussed in this section.

A lampshade-type TEG was designed for harvesting the thermal energy from light illumination; as displayed in Figure 7.15a, the TEG consists of 12 couples of p-/n-type TE modules [75]. The cold side is suspended in air. Another side of the TEG is attached on the surface of the shield of a light bulb as the hot side. When the light is on for several minutes, the TEG can generate TE voltage owing to the ΔT across the device. Figure 7.15b displays the structure diagram of a flexible TEG including three main parts: Te/PEDOT:PSS composites as TE layer, MoS_2/PU composite film as photothermal layer, and polymer polyethylene terephthalate (PET) as matrix [76]. The SEM images show the structures of MoS_2 nanoclusters and Te nanowires, respectively. Based on Ag_2Te nanowires and PEDOT:PSS composites as TE materials, as shown in Figure 7.15c [77], a flexible TEG was fabricated and can harvest the thermal energy by attaching on the surface of a cup filled with hot water.

Figure 7.15d [60] and Figure 7.15e [58] show the conventional π-type TE modules consisting of different p-/n-type TE materials on flexible polymer matrix, which results in a markedly high TE performance. Figure 7.15f displays a fabric-based TEGs array by connecting a five-strip TEG in series with PEDOT:PSS coated polyester fabric as TE material [61], which will promote the development of novel clothing for power generation and enhance the feasibility of self-powered wearable electronics. Through an inkjet-printing process, a TEG was printed onto a flexible PET substrate including some graphene legs connected by silver, and the photographs in Figure 7.15g indicate that the TEG has excellent flexibility [78]. Through a wet-spinning process as shown in Figure 7.15h, the CNTs/PEDOT:PSS composite-based fibers were divided into p-type TE fibers and n-type TE fibers through special chemical treatment [79]. A flexible TEG with high TE performance was designed based on 12 pairs of TE modules, and the inset displays the TEG picture in Figure 7.15h. To gain high flexibility and stretchability of TEG based on freestanding PEDOT:PSS conducting polymer Figure 7.15i [80], the kirigami-structured TEGs are designed and can obtain a repeatable strain of 200% by a laser cutting process.

Figure 7.16a displays the photograph of a transparent and flexible TEG based on polycrystalline CuI as TE material deposited on flexible PET substrate by sputtering [81]. The TEG shows excellent flexibility and exhibits great application potential in flexible transparent electronics. Based on several pairs of n-/p-type carbon nanotube yarn, as shown in Figure 7.16b [82], the fabricated TEG with excellent flexibility and ultralow weight exhibits high power density. Paper is an essential material and has huge application potential in flexible electronics. Through simple micromachining and microfabrication techniques, researchers reported a paper-based TEG with the deposited Bi_2Te_3-Sb_2Te_3 as TE pairs, as shown in Figure 7.16c [83]; the flexible TEG shows remarkable mechanical stability and adaptability.

Integrating several p-/n-type TE pairs in a flexible matrix will be an efficient approach for high TE output power and excellent flexibility of TEGs. Based on the idea above, several flexible TEGs with different functions were designed and researched. The TEG can exhibit high output power and flexibility with a flexible

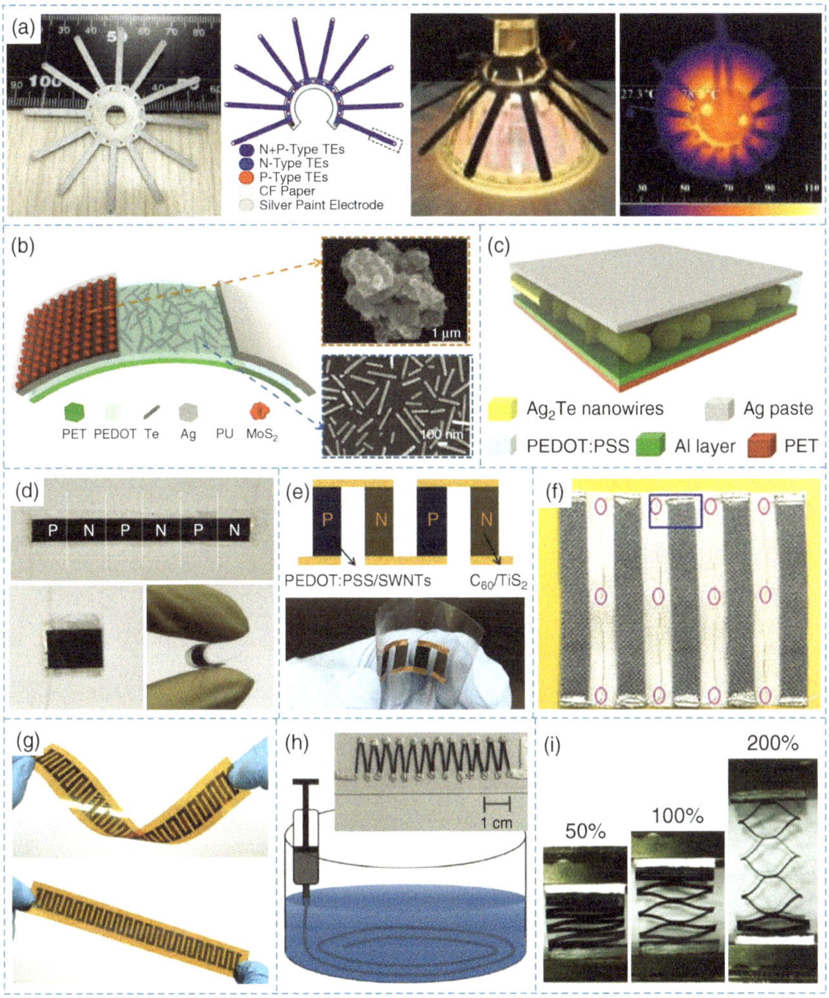

Figure 7.15 Several typical TEGs with different structures. (a) A lampshade-type TEG based on several PN TE pairs. Source: Reproduced with permission from Jin et al. [75]. Copyright 2017, American Chemical Society. (b) Flexible TEG based on Te/PEDOT:PSS composite. Source: Reproduced with permission from He et al. [76]. Copyright 2018, John Wiley and Sons. (c) A sandwich-structural TEG based on Ag_2Te. Source: Reproduced with permission from Jao et al. [77]. Copyright 2017, Electrochemical Society. (d,e) Flexible TEG based on p-/n-type CNTs pairs (d), PEDOT:PSS/SWNTs and C_{60}/TiS_2 (e). Source: (d) Reproduced with permission from Zhou et al. [60]. Copyright 2017, Springer Nature; (e) Reproduced with permission from Wang et al. [58]. Copyright 2018, Royal Society of Chemistry. (f) Wearable TEGs based on several PEDOT:PSS strips. Source: Reproduced with permission from Du et al. [61]. Copyright 2015, Springer Nature. (g) Few layer graphene-based flexible TEG. Source: Reproduced with permission from Juntunen et al. [78]. Copyright 2018, John Wiley and Sons. (h) Flexible TEG based on CNTs/PEDOT:PSS composite fibers through a wet-spinning process. Source: Reproduced with permission from Kim et al. [79]. Copyright 2018, Elsevier. (i) Stretchable TEG through a kirigami-like technology. Source: Reproduced with permission from Guan et al. [80]. Copyright 2018, American Chemical Society.

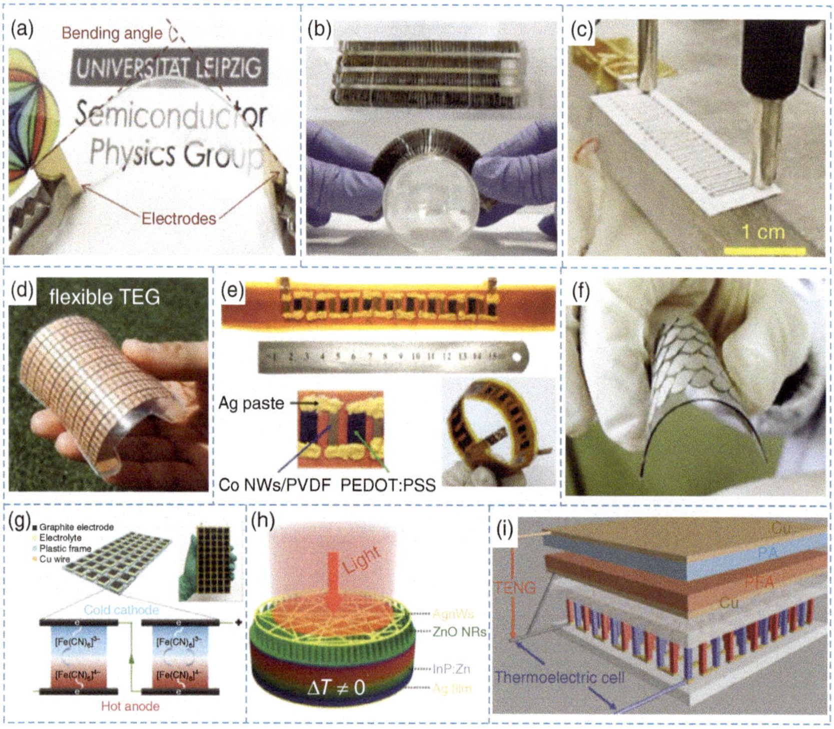

Figure 7.16 The photographs and structure diagrams of several TEGs. (a) A flexible and transparent TEG based on polycrystalline CuI as TE material. Source: Reproduced with permission from Yang et al. [81]. Copyright 2017, Springer Nature . (b) Flexible TEG based on 60 pairs of n- and p-doped CNTY. Source: Reproduced with permission from Choi et al. [82]. Copyright 2017, American Chemical Society. (c) A paper-based TEG based on Bi_2Te_3-Sb_2Te_3 as TE pairs. Source: Reproduced with permission from Rojas et al. [83]. Copyright 2016, Elsevier. (d,e) Flexible TEG based on p-/n-type Bi_2Te_3 (d), Co NWs/PVDF and PEDOT:PSS as TE pairs (e). Source: Reproduced with permission from Kim et al. [84]. Copyright 2016, American Chemical Society. Source: Reproduced with permission from Chen et al. [85]. Copyright 2018, John Wiley and Sons. (f) Photograph of a flexible TEG module. Source: Reproduced with permission from Jung et al. [86]. Copyright 2016, Elsevier. (g) Structure diagram and photograph of an aqueous TEGs module. Source: Reproduced with permission from Duan et al. [87]. Copyright 2018, Springer Nature . (h) Structure diagram of a thermo-phototronic effect enhanced solar cell. Source: Reproduced with permission from Zhang et al. [55]. Copyright 2017, John Wiley and Sons. (i) Hybrid nanogenerator based on a TEG. Source: Reproduced with permission from Yang et al. [22]. Copyright 2013, Royal Society of Chemistry.

polymer as a substrate, as seen in Figure 7.16d [84]. Based on Co NWs/PVDF and PEDOT:PSS as TE pairs, a flexible TEG was fabricated as displayed in Figure 7.16e [85]; the TE performance can be greatly improved by the use of a magnetic field, which can lead to Co nanowires aligned with the same direction in the PVDF matrix. Based on p-/n-type Bi/Te TE modules, a PDMS-based TEG with excellent flexibility was designed, as shown in Figure 7.16f [86], and the electrical conductivity of the device was improved by the addition of CNTs. After over 1000 bending cycles, the TEG module still exhibits outstanding TE stability.

It is challenging to deliver high output power by harvesting low-grade thermal energy; researchers designed an aqueous TEG with a large power density by adding guanidinium and urea in aqueous electrolytes containing $[Fe(CN)_6]^{4-}/[Fe(CN)_6]^{3-}$ [87]; the TEG exhibits a high Seebeck coefficient, and the structure diagram is shown in Figure 7.16g. Based on InP/ZnO heterostructure with ZnO nanorod as an optoelectronic layer and Zn-doped InP as the TE layer, a hybrid solar cell was designed as displayed in Figure 7.16h [55]. The thermo-phototronic coupled effect of InP/ZnO heterostructure can enhance the transfer of interfacial charges and then improve the performance of the solar cell. Figure 7.16i shows the structure diagram of a hybrid energy device including a triboelectric nanogenerator, solar cell, and TEG [22], which can directly convert the thermal, mechanical, and solar energies into electricity, simultaneously or individually.

7.4.3 Thermoelectric Performance

To date, the performances and structures of TEGs and TEG-based hybrid nanogenerators have made tremendous improvement and progress in different application fields. In this section, the output performances of various kinds of TEGs and TEG-based hybrid nanogenerators with different application fields are evaluated and discussed.

Figure 7.17a gives the curve between generated voltage and steady temperature difference of a π-type TEG fabricated by p-/n-type continuous CNT network as TE pairs [60]. With an increasing ΔT of the device, the TE voltage increases linearly, which corresponds to a high Seebeck coefficient of 410 μV/K. According to Figure 7.17a, the device can generate a TE voltage (current) of 11.3 mV (0.9 mA) under a ΔT of 27.5 K, which corresponds to a maximum power output of 2.51 mW by calculation. A single ZnO microbelt-based TEG can generate output voltage (current) of 10 mV (190 nA) under a ΔT of 30 K. According to Figure 7.17b, the TEG exhibits a high power factor and Seebeck coefficient of 3.2×10^{-4} Wm/K^2 and −350 μV/K, respectively [12]. Based on Te-nanowire/P3HT composite as TE material, the TEG with high flexibility delivers the voltage/current of 14 mV/254 nA under a ΔT of 55 K corresponding to the Seebeck coefficient of 285 μV/K (Figure 7.17c) [56]. In addition, large output voltage of 38 mV was generated by two serial-connected TEGs.

When a flexible TEG was illuminated by an IR laser for about 50 seconds, the TE voltage increased up to 1.2 mV as a stable value, as seen in Figure 7.17d [76]. In addition, under solar illumination, the TEG will deliver a TE voltage of 1.48 mV. Figure 7.17e gives the relationship between TE voltage and applied strain of a graphene-based TEG under different temperature differences [19]; we can see that the voltage is almost unchanged with an increasing strain, which indicates that voltage is only dependent on ΔT across the device. Figure 7.17f presents the current–voltage curves of several TEG-based hybrid nanogenerators in series connection under different conditions [22]. The output voltage (current) is up to 3.5 V (30 mA) under light illumination. On this basis, when the applied heat energy is close to the TEG, the total voltage (current) increases up to about 5.2 V (34 mA).

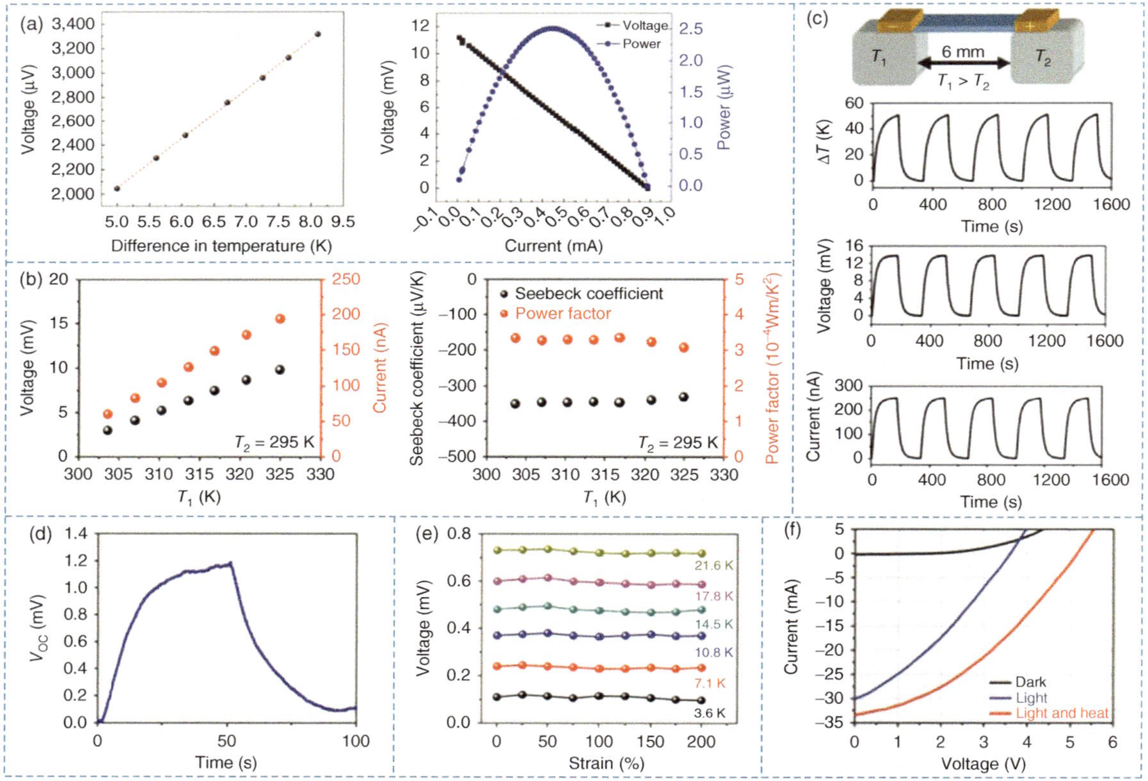

Figure 7.17 TE performances of several TEGs. (a) Output signals of a flexible TEG based on TE CNTs pairs. Source: Reproduced with permission from Zhou et al. [60]. Copyright 2017, Springer Nature . (b) Output voltage and current of a fabricated device under different temperatures, and the corresponding Seebeck coefficients and power factors. Source: Reproduced with permission from Yang et al. [12]. Copyright 2012, American Chemical Society. (c) Structure diagram of Te nanowires-based TEG and output electric signals. Source: Reproduced with permission from Yang et al. [56]. Copyright 2012, Springer. (d) The output voltage of a flexible TEG under light illumination. Source: Reproduced with permission from He et al. [76]. Copyright 2018, John Wiley and Sons. (e) The variation of output voltages with increasing strain. Source: Reproduced with permission from Zhang et al. [19]. Copyright 2018, John Wiley and Sons. (f) Current–voltage characteristics of a hybrid energy cell under different conditions. Source: Reproduced with permission from Yang et al. [22]. Copyright 2013, Royal Society of Chemistry.

Researchers designed an evaporation-driven new TEG owing to water flow in a porous carbon film, and the vapor pressure difference can directly produce electricity by harvesting heat energy. According to Figure 7.18a [59], the increasing voltage gradually changes with the increasing ΔT. When a ΔT of 12.8 °C is applied, the voltage increases up to 0.8 mV for over three hours without attenuation. When the cylinder of the TEG was pumped with a 5.8 kPa vacuum pressure, the Seebeck coefficient greatly increases from 78 to 210 mV/K. A fabric-based TEGs array has higher TE performance and excellent flexibility, as seen in Figure 7.18b [61], and the increase in the number of TEG units leads to the linear increase of output voltage. The fabricated TEG including 60 couples of n-/p-type carbon nanotube yarn (CNTY) can generate a large TE voltage density of 1.2 V/g corresponding to the Seebeck coefficient of 30 mV/K, as seen in Figure 7.18c [82]. A Ag_2Te-based TEG and corresponding voltage curve are shown in Figure 7.18d [77], indicating a linearly increasing output voltage with a high Seebeck coefficient of 100 μV/K. An aqueous TEG module can deliver a high output signal by harvesting thermal energy in low grade, and the curves of voltage, current, and ΔT are displayed in Figure 7.18e [87]. Under the ΔT of 18 K, the TEG modules can generate a high voltage/current of 3.4 V/1.2 mA and directly drive an LED array. Based on organic TE materials, researchers designed a high-performance TEG with excellent stability, as seen in Figure 7.18f [88], and the flexible TEG exhibits an accurate temperature resolution of <0.1 K.

7.4.4 Applications in Various Fields

With the development of TEGs in recent years, the performances of various kinds of TEGs have been increasingly improved and enhanced for huge potential applications in many fields, including power generation, self-powered sensors, and wearable, even implantable, electronics.

Two flexible TEGs with a similar structure were fabricated based on Te nanowires and ZnO micro/nanobelts, respectively, as shown in Figure 7.19a,b [12, 56]. When a finger touches one side of the TEG as self-powered temperature sensor, it can monitor small temperature differences with a fast response. Owing to excellent flexibility, a thermo-phototronic effect enhanced solar cell can act as a wearable electronic device. As seen in Figure 7.19c [55], under light illumination, it can power a temperature sensor as a power source by harvesting solar energy. Figure 7.19d shows the photographs of a graphene-based TEG and multi-channel TEGs system and corresponding applications [19]. As a temperature sensor, the TEG can detect the finger-bend of different degrees and the motion of throat swallowing, which are based on strain-induced current variation. In addition, the multi-channel TEGs system as a stretchable 3D strain sensor system can monitor the strain with different directions according to the corresponding current mappings. Figure 7.19e displays a flexible TEG attaching on a stone and a transparent glass of a window in ambient environment [76]; it can efficiently convert solar energy into electricity based on photothermal and TE effects. Based on Ag_2Te nanowires and PEDOT:PSS composite TE materials, a sandwich structure flexible TEG can detect the temperature of hot water, as seen in Figure 7.19f [77]. Figure 7.19g shows a dual-function flexible TEGs array

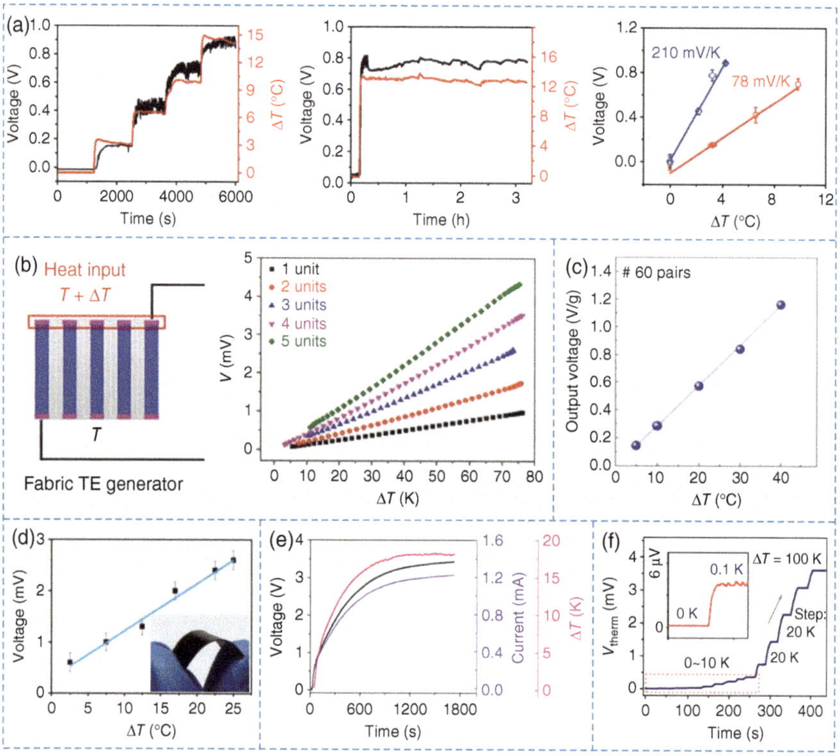

Figure 7.18 Output thermoelectric signals of TEGs. (a) Stability of output electrical signal and voltage variation under different pressure conditions. Source: Reproduced with permission from Liu et al. [59]. Copyright 2018, John Wiley and Sons. (b) Output voltages under different temperatures. Source: Reproduced with permission from Du et al. [61]. Copyright 2015, Springer Nature. (c) Voltage–ΔT curve of TEG. Source: Reproduced with permission from Choi et al. [82]. Copyright 2017, American Chemical Society. (d) Generated voltage and the corresponding ΔT. Source: Reproduced with permission from Jao et al. [77]. Copyright 2015, Electrochemical Society. (e) Increasing voltage and current with the increase of ΔT. Source: Reproduced with permission from Duan et al. [87]. Copyright 2018, Springer Nature. (f) Output voltage variation with increasing temperature difference from 0 to100 K. Source: Reproduced with permission from Zhang et al. [88]. Copyright 2015, Springer Nature.

as a temperature and pressure sensor based on high-performance organic TE materials with porous microstructure; when a hand presses the device, it can simultaneously sense the temperature and pressure [88].

Wearable electronic devices have received great attention and become popular both in academic community and vast user population. Here, several TEGs as wearable electronic device by harvesting the thermal energy of human body are evaluated and discussed, as presented in Figure 7.20a–e [59, 82, 87, 89, 90]. Figure 7.20a shows a TEG consisting of 240 PN TE pairs attached to the human arm and the arm of a hand [82]. There will be a ΔT occurring between the human body and the external environment, and generate a TE voltage of 54 mV. An aqueous TEGs module was designed for thermal energy harvest from human body heat, as shown in Figure 7.20b [87]. The device is covered by a hand for several

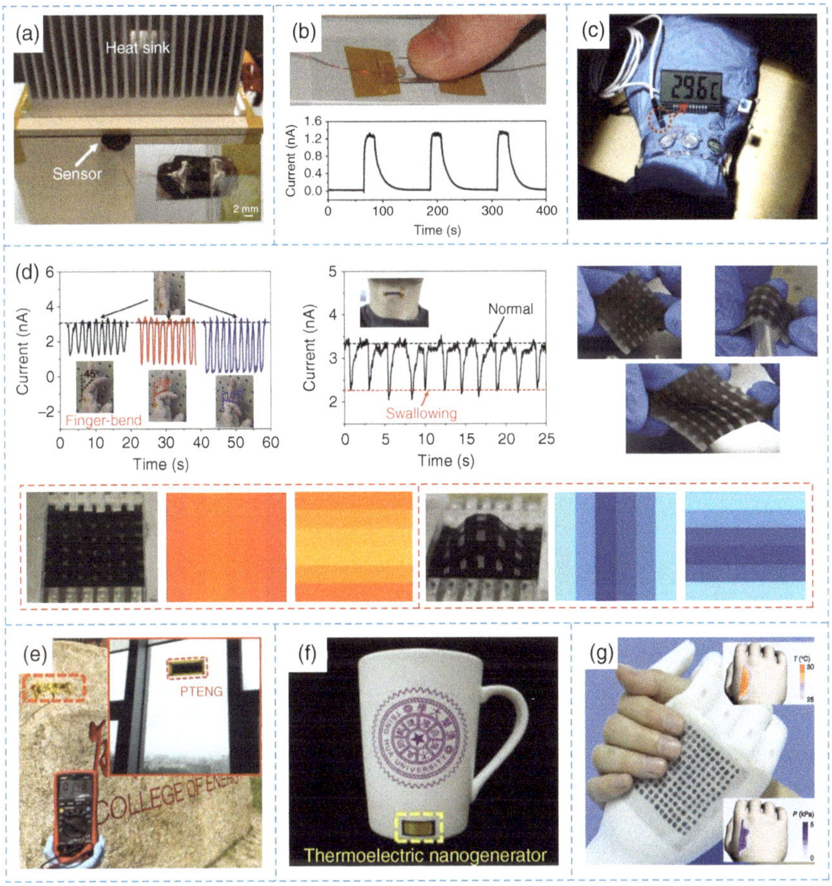

Figure 7.19 Various applications of TEGs. (a) Photograph of a flexible TEG as a temperature sensor. Source: Reproduced with permission from Yang et al. [56]. Copyright 2012, Springer. (b) A ZnO micro/nanobelts-based TEG for temperature sensing of human body. Source: Reproduced with permission from Yang et al. [12]. Copyright 2012, American Chemical Society. (c) Photograph of a hybrid solar cell for powering a digital temperature sensor as a power source. Source: Reproduced with permission from Zhang et al. [55]. Copyright 2017, John Wiley and Sons. (d) Stretchable TEG based on few-layer graphene–ecoflex composite for temperature and strain sensing. Source: Reproduced with permission from Zhang et al. [19]. Copyright 2019, John Wiley and Sons. (e) A photothermal driven TEG for efficiently harvesting solar energy. Source: Reproduced with permission from He et al. [76]. Copyright 2018, John Wiley and Sons. (f) Photograph of a flexible TEG with sandwich structure attached to the external surface of a cup for monitoring water temperature. Source: Reproduced with permission from Jao et al. [77]. Copyright 2015, Electrochemical Society. (g) A dual-function TEGs module for simultaneously detecting pressure and temperature. Source: Reproduced with permission from Zhang et al. [88]. Copyright 2015, Springer Nature.

minutes, which generates a high TE voltage of 0.3 V. For the same consideration with similar structures, three other TEGs as wearable electronic devices were measured. By harvesting human body heat, these TEGs could generate a high voltage, as seen in Figure 7.20c [89] and Figure 7.20d [59]. The generated electricity of a TEG from human body heat can directly power some small power electronic device including temperature sensors and cardiotachometers, as shown in Figure 7.20e [90].

A TEG-based hybrid cell can generate electricity and store it in a battery by harvesting from mechanical and thermal energies simultaneously. In addition, the hybrid device can be directly used to produce hydrogen (H_2) by water splitting. As presented in Figure 7.20f [22], the volume of the generated H_2 increased linearly with the increase of time with a rate of 4×10^4 ml/s. Figure 7.20g shows the structure diagram and the corresponding photograph of a hybrid device consisting of a 3D organic sponge and a TE module [91]. Under sunlight illumination, the sponge as a high-efficiency solar absorber can promote photothermal water evaporation owing to the 3D structure of the sponge. A high ΔT occurs between the upper surface and the bottom surface of the TE module owing to photothermal effect of the sponge, and simultaneously induces electricity production due to TE effect. This study will provide a novel strategy for acquiring electricity supply and fresh water in remote areas by harvesting inexhaustible solar energy.

Implantable electronics have become one of the hottest research fields and has made great progress, which promotes the development of health monitoring and medical treatment. Experimental results were obtained by implanting a small TEG into a piece of pork for simulating human body thermal state, as shown in Figure 7.20h [92]. A small TEG module was put in the pork between the tissue and fat layer; the temperature of the pork was kept at 37 °C corresponding to the temperature of the human body. When reaching thermal equilibrium, the temperature difference (0.5 K) occurs across the TEG corresponding to the TE voltage of 3.3 mV, which provides the possibility of TEG as an implantable device, and has an important guiding role in the medical application of implantable devices.

7.5 Conclusions and Prospects

With the aggravation of energy crisis and environment pollution, PyENG and TEGs have received huge attention from the academic community and played increasingly important roles in thermal energy harvest and conversion. Additionally, great progress and important accomplishments have been achieved in thermal energy harvesting and conversion from ambient environment, which promotes the reuse of waste heat with different forms through PyENG and TEGs. In this chapter, the research progress in PyENG and TEGs has been evaluated and discussed in detail including materials, structures, performance, and applications. The two kinds of nanogenerators for thermal harvest and conversion can act as self-powered sensing systems and directly drive electronic devices as wearable electronic devices.

The structure designs and performances of PyENG and TEGs have been developed rapidly. However, most of the reported research is just limited to

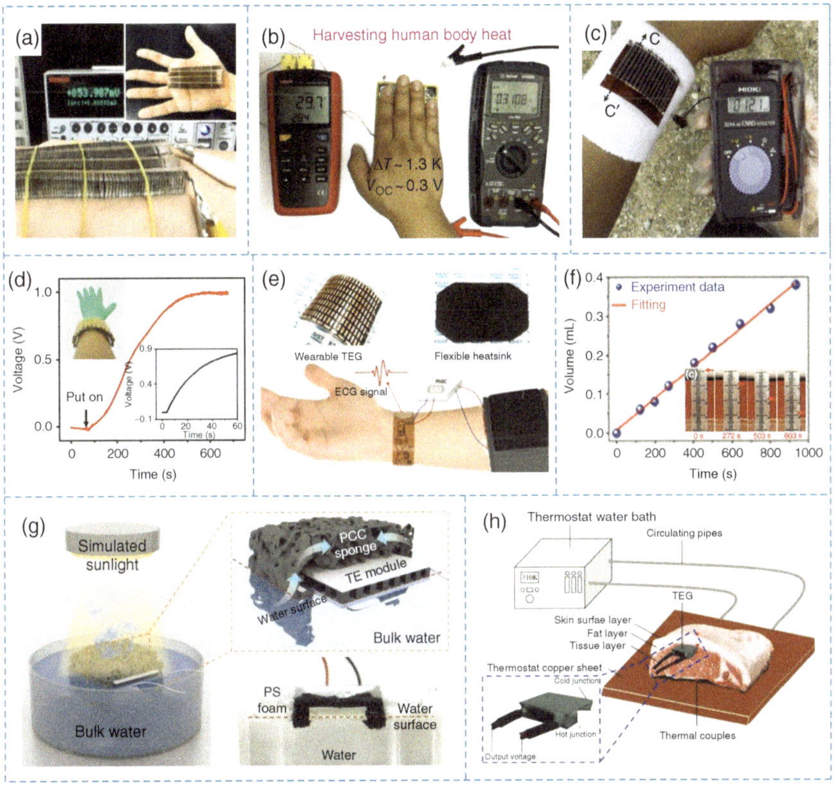

Figure 7.20 A variety of applications of TEGs. (a) Photograph of a flexible TEG as a wearable device for harvesting thermal energy from human body heat. Source: Reproduced with permission from Choi et al. [82]. Copyright 2017, American Chemical Society. (b) An aqueous TEGs module for harvesting human body heat. Source: Reproduced with permission from Duan et al. [87]. Copyright 2018, Springer Nature. (c–e) Three TEGs for harvesting human body heat as wearable device respectively. Source: (c) Reproduced with permission from We et al. [89]. Copyright 2014, Elsevier. (d) Reproduced with permission from Liu et al. [59]. Copyright 2018, Wiley. (e) Reproduced with permission from Kim et al. [90]. Copyright 2018, American Chemical Society. (f) A hybrid cell for water splitting. Source: Reproduced with permission from Yang et al. [22]. Copyright 2013, Royal Society of Chemistry. (g) A hybrid device consisting of a 3D organic sponge and a TEG module for simultaneous power generation and water evaporation. Source: Reproduced with permission from Zhu et al. [91]. Copyright 2019, John Wiley and Sons. (h) A TEG module as an implantable device for harvesting the thermal energy of a pork with a constant temperature of about 37 °C corresponding to the temperature of human body. Source: Reproduced with permission from Yang et al. [92]. Copyright 2007, IOP Publishing.

the laboratory, far from practical application; so increasingly work needs to be done to significantly improve the overall performance of PyENG and TEGs. In addition, the cost of devices needs to be kept within acceptable limits of broad users in future commercialization. Last but not least, as wearable or even implantable electronics, the security of devices is the most important factor. Likewise, compatibility and ease of use of devices all need be considered in the fabrication process. Hopefully, in the near future the growing progress will certainly make PyENG and TEGs to be feasible and highly efficient approaches for thermal energy harvest and conversion in practical applications.

References

1 Doman L. E. and Arora V. (2016). International Energy Outlook. *U.S. Energy Information Administration.*
2 Zheng, X.F., Liu, C.X., Yan, Y.Y., and Wang, Q. (2014). A review of thermoelectrics research – recent developments and potentials for sustainable and renewable energy applications. *Renewable Sustainable Energy Rev.* 32: 486.
3 Han, G., Chen, Z.G., Drennan, J., and Zou, J. (2014). Indium Selenides: structural characteristics, synthesis and their thermoelectric performances. *Small* 10: 2747.
4 Tan, G., Zhao, L.D., and Kanatzidis, M.G. (2016). Rationally designing high-performance bulk thermoelectric materials. *Chem. Rev.* 116: 12123.
5 Zhu, T., Liu, Y., Fu, C. et al. (2017). Compromise and synergy in high-efficiency thermoelectric materials. *Adv. Mater.* 29: 1605884.
6 Burschka, J., Pellet, N., Moon, S.J. et al. (2013). Sequential deposition as a route to high-performance perovskite-sensitized solar cells. *Nature* 499: 316.
7 Blaabjerg, F., Liserre, M., and Ma, K. (2011). Power electronics converters for wind turbine systems. *IEEE T. Ind. Appl.* 48: 708.
8 Marcus, G.H. (2000). Considering the next generation of nuclear power plants. *Prog. Nucl. Energy* 37: 5.
9 Wang, Z.L. and Song, J. (2006). Piezoelectric nanogenerators based on zinc oxide nanowire arrays. *Science* 312: 242.
10 Fan, F.-R., Tian, Z.-Q., and Lin Wang, Z. (2012). Flexible triboelectric generator. *Nano Energy* 1: 328.
11 Yang, Y., Guo, W., Pradel, K.C. et al. (2012). Pyroelectric nanogenerators for harvesting thermoelectric energy. *Nano Lett.* 12: 2833.
12 Yang, Y., Pradel, K.C., Jing, Q. et al. (2012). Thermoelectric nanogenerators based on single Sb-doped ZnO micro/Nanobelts. *ACS Nano* 6: 6984.
13 Zhang, D., Wang, Y., and Yang, Y. (2019). Design, performance, and application of thermoelectric Nanogenerators. *Small* 1805241.
14 Zhang, K., Wang, Y., Wang, Z.L., and Yang, Y. (2019). Standard and figure-of-merit for quantifying the performance of pyroelectric nanogenerators. *Nano Energy* 55: 534.
15 Yang, L., Chen, Z.-G., Dargusch, M.S., and Zou, J. (2017). High performance thermoelectric materials: progress and their applications. *Adv. Energy Mater.* 8: 1701797.
16 Yang, Y., Wang, S., Zhang, Y., and Wang, Z.L. (2012). Pyroelectric nanogenerators for driving wireless sensors. *Nano Lett.* 12: 6408.
17 Zeng, W., Shu, L., Li, Q. et al. (2014). Fiber-based wearable electronics: a review of materials, fabrication, devices, and applications. *Adv. Mater.* 26: 5310.

18 Suarez, F., Nozariasbmarz, A., Vashaee, D., and Öztürk, M.C. (2016). Designing thermoelectric generators for self-powered wearable electronics. *Energy Environ. Sci.* 9: 2099.
19 Zhang, D., Zhang, K., Wang, Y. et al. (2019). Thermoelectric effect induced electricity in stretchable graphene-polymer nanocomposites for ultrasensitive self-powered strain sensor system. *Nano Energy* 56: 25.
20 Yang, Y., Zhou, Y., Wu, J.M., and Wang, Z.L. (2012). Single micro/nanowire pyroelectric nanogenerators as self-powered temperature sensors. *ACS Nano* 6: 8456.
21 Zhang, Y., Kumar, S., Marken, F. et al. (2019). Pyro-electrolytic water splitting for hydrogen generation. *Nano Energy* 58: 183.
22 Yang, Y., Zhang, H., Lin, Z.-H. et al. (2013). A hybrid energy cell for self-powered water splitting. *Energy Environ. Sci.* 6: 2429.
23 Zhao, K., Wang, Y., Han, L. et al. (2009). Nanogenerator-based self-charging energy storage devices. *Nano-Micro Lett.* 11: 19.
24 Lang, S.B. (2005). Pyroelectricity: from ancient curiosity to modern imaging tool. *Phys. Today* 58: 31.
25 Yang, Y., Jung, J.H., Yun, B.K. et al. (2012). Flexible pyroelectric nanogenerators using a composite structure of lead-free $KNbO_3$ nanowires. *Adv. Mater.* 24: 5357.
26 You, H., Wu, Z., Wang, L. et al. (2018). Highly efficient pyrocatalysis of pyroelectric $NaNbO_3$ shape-controllable nanoparticles for room-temperature dye decomposition. *Chemosphere* 199: 531.
27 Ma, N., Zhang, K., and Yang, Y. (2017). Photovoltaic-pyroelectric coupled effect induced electricity for self-powered photodetector system. *Adv. Mater.* 29: 1703694.
28 Qi, J., Ma, N., and Yang, Y. (2017). Photovoltaic-pyroelectric coupled effect based nanogenerators for self-powered photodetector system. *Adv. Mater. Inter.* 5: 1701189.
29 Wu, J., Mao, W., Wu, Z. et al. (2016). Strong pyro-catalysis of pyroelectric $BiFeO_3$ nanoparticles under a room-temperature cold–hot alternation. *Nanoscale* 8: 7343.
30 Mistewicz, K., Jesionek, M., Nowak, M., and Kozioł, M. (2019). SbSeI pyroelectric nanogenerator for a low temperature waste heat recovery. *Nano Energy* 64: 103906.
31 Xu, X., Xiao, L., Jia, Y. et al. (2018). Pyro-catalytic hydrogen evolution by $Ba_{0.7}Sr_{0.3}TiO_3$ nanoparticles: harvesting cold–hot alternation energy near room-temperature. *Energy Environ. Sci.* 11: 2198.
32 Zabek, D., Seunarine, K., Spacie, C., and Bowen, C. (2017). Graphene ink laminate structures on poly (vinylidene difluoride) (PVDF) for pyroelectric thermal energy harvesting and waste heat recovery. *ACS Appl. Mater. Interfaces* 9: 9161.
33 Zhao, T., Jiang, W., Niu, D. et al. (2017). Flexible pyroelectric device for scavenging thermal energy from chemical process and as self-powered temperature monitor. *Appl. Energy* 195: 754.

34 You, M.-H., Wang, X.-X., Yan, X. et al. (2018). A self-powered flexible hybrid piezoelectric-pyroelectric nanogenerator based on non-woven nanofiber membranes. *J. Mater. Chem. A* 6: 3500.

35 Wang, S., Wang, Z.L., and Yang, Y. (2016). A one-structure-based hybridized Nanogenerator for scavenging mechanical and thermal energies by triboelectric-piezoelectric-pyroelectric effects. *Adv. Mater.* 28: 2881.

36 Thakur, P., Kool, A., Hoque, N.A. et al. (2018). Superior performances of in situ synthesized ZnO/PVDF thin film based self-poled piezoelectric nanogenerator and self-charged photo-power bank with high durability. *Nano Energy* 44: 456.

37 Sultana, A., Ghosh, S.K., Alam, M.M. et al. (2019). Methylammonium lead iodide incorporated poly(vinylidene fluoride) nanofibers for flexible piezoelectric-pyroelectric nanogenerator. *ACS Appl. Mater. Interfaces* 11: 27279.

38 Lee, J.H., Lee, K.Y., Gupta, M.K. et al. (2014). Highly stretchable piezoelectric-pyroelectric hybrid nanogenerator. *Adv. Mater.* 26: 765.

39 Wang, Z., Yu, R., Pan, C. et al. (2015). Light-induced pyroelectric effect as an effective approach for ultrafast ultraviolet nanosensing. *Nat. Commun.* 6: 8401.

40 Wang, X., Dai, Y., Liu, R. et al. (2017). Light-triggered pyroelectric nanogenerator based on a pn-junction for self-powered near-infrared photosensing. *ACS Nano* 11: 8339.

41 Dao, T.D., Ishii, S., Yokoyama, T. et al. (2016). Hole array perfect absorbers for spectrally selective midwavelength infrared pyroelectric detectors. *ACS Photonics* 3: 1271.

42 Lee, J.H., Ryu, H., Kim, T.Y. et al. (2015). Thermally induced strain-coupled highly stretchable and sensitive pyroelectric nanogenerators. *Adv. Energy Mater.* 5: 1500704.

43 Ko, Y.J., Kim, D.Y., Won, S.S. et al. (2016). Flexible Pb $(Zr_{0.52}Ti_{0.48})O_3$ films for a hybrid piezoelectric-pyroelectric nanogenerator under harsh environments. *ACS Appl. Mater. Interfaces* 8: 6504.

44 Park, T., Na, J., Kim, B. et al. (2015). Photothermally activated pyroelectric polymer films for harvesting of solar heat with a hybrid energy cell structure. *ACS Nano* 9: 11830.

45 Ding, T., Zhu, L., Wang, X.Q. et al. (2018). Hybrid photothermal pyroelectric and thermogalvanic generator for multisituation low grade heat harvesting. *Adv. Energy Mater.* 8: 1802397.

46 Zhao, T., Jiang, W., Liu, H. et al. (2016). An infrared-driven flexible pyroelectric generator for non-contact energy harvester. *Nanoscale* 8: 8111.

47 Zhang, K., Wang, S., and Yang, Y. (2017). A one-structure-based piezo-tribo-pyro-photoelectric effects coupled nanogenerator for simultaneously scavenging mechanical, thermal, and solar energies. *Adv. Energy Mater.* 7: 1601852.

48 Yang, Y., Zhang, H., Zhu, G. et al. (2013). Flexible hybrid energy cell for simultaneously harvesting thermal, mechanical, and solar energies. *ACS Nano* 7: 785.

49 Xue, H., Yang, Q., Wang, D. et al. (2017). A wearable pyroelectric nanogenerator and self-powered breathing sensor. *Nano Energy* 38: 147–154.

50 Gao, F., Li, W., Wang, X. et al. (2016). A self-sustaining pyroelectric nanogenerator driven by water vapor. *Nano Energy* 22: 19.

51 Sun, J.-G., Yang, T.-N., Wang, C.-Y., and Chen, L.-J. (2018). A flexible transparent one-structure tribo-piezo-pyroelectric hybrid energy generator based on bio-inspired silver nanowires network for biomechanical energy harvesting and physiological monitoring. *Nano Energy* 48: 383.

52 Sultana, A., Alam, M.M., Middya, T.R., and Mandal, D. (2018). A pyroelectric generator as a self-powered temperature sensor for sustainable thermal energy harvesting from waste heat and human body heat. *Appl. Energy* 221: 299.

53 Kulkarni, E.S., Heussler, S.P., Stier, A.V. et al. (2015). Exploiting the IR transparency of graphene for fast pyroelectric infrared detection. *Adv. Opt. Mater.* 3: 34.

54 Qian, W., Wu, Z., Jia, Y. et al. (2017). Thermo-electrochemical coupling for room temperature thermocatalysis in pyroelectric ZnO nanorods. *Electrochem. Commun.* 81: 124.

55 Zhang, K. and Yang, Y. (2017). Thermo-phototronic effect enhanced InP/ZnO nanorod heterojunction solar cells for self-powered wearable electronics. *Adv. Funct. Mater.* 27: 1703331.

56 Yang, Y., Lin, Z.-H., Hou, T. et al. (2012). Nanowire-composite based flexible thermoelectric nanogenerators and self-powered temperature sensors. *Nano Res.* 5: 888.

57 Liu, J.W., Xu, J., Hu, W. et al. (2016). Systematic synthesis of tellurium nanostructures and their optical properties: from nanoparticles to Nanorods, nanowires, and nanotubes. *ChemNanoMat* 2: 167.

58 Wang, L., Zhang, Z., Geng, L. et al. (2018). Solution-printable fullerene/TiS_2 organic/inorganic hybrids for high-performance flexible n-type thermoelectrics. *Energy Environ. Sci.* 11: 1307.

59 Liu, K., Ding, T., Li, J. et al. (2018). Thermal-electric nanogenerator based on the electrokinetic effect in porous carbon film. *Adv. Energy Mater.* 8: 1702481.

60 Zhou, W., Fan, Q., Zhang, Q. et al. (2017). High-performance and compact-designed flexible thermoelectric modules enabled by a reticulate carbon nanotube architecture. *Nat. Commun.* 8: 14886.

61 Du, Y., Cai, K., Chen, S. et al. (2015). Thermoelectric fabrics: toward power generating clothing. *Sci. Rep.* 5: 6411.

62 Fonseca, L., Santos, J.-D., Roncaglia, A. et al. (2016). Smart integration of silicon nanowire arrays in all-silicon thermoelectric micro-nanogenerators. *Semicond. Sci. Technol.* 31: 084001.

63 Lin, S., Li, W., Chen, Z. et al. (2016). Tellurium as a high-performance elemental thermoelectric. *Nat. Commun.* 7: 10287.

64 Wang, Y., Qiu, G., Wang, R. et al. (2018). Field-effect transistors made from solution-grown two-dimensional tellurene. *Nat. Electron.* 1: 228.

65 Yang, L., Chen, Z.G., Hong, M. et al. (2015). Enhanced thermoelectric performance of nanostructured Bi_2Te_3 through significant phonon scattering. *ACS Appl. Mater. Interfaces* 7: 23694.

66 Kumar, S., Chaudhary, D., Kumar Dhawan, P. et al. (2017). Bi_2Te_3-MWCNT nanocomposite: an efficient thermoelectric material. *Ceram. Int.* 43: 14976.
67 Yang, S., Cho, K., Yun, J. et al. (2017). Thermoelectric characteristics of γ-Ag_2Te nanoparticle thin films on flexible substrates. *Thin Solid Films* 641: 65.
68 Han, G., Ruizhi, Z., Srinivas, P. et al. (2017). Large-scale surfactant-free synthesis of p-type SnTe nanoparticles for thermoelectric applications. *Materials* 10: 233.
69 Nithiyanantham, U., Ede, S.R., Ozaydin, M.F. et al. (2015). Low temperature, shape-selective formation of Sb_2Te_3 nanomaterials and their thermoelectric applications. *RSC Adv.* 5: 89621.
70 Zhou, Y., Wan, J., Li, Q. et al. (2017). Chemical welding on semimetallic TiS_2 nanosheets for high performance flexible n-type thermoelectric films. *ACS Appl. Mater. Interfaces* 9: 42430.
71 Oh, J.Y., Lee, J.H., Han, S.W. et al. (2016). Chemically exfoliated transition metal dichalcogenide nanosheet-based wearable thermoelectric generators. *Energy Environ. Sci.* 9: 1696.
72 Han, G., Popuri, S.R., Greer, H.F. et al. (2016). Facile surfactant-free synthesis of p-type SnSe nanoplates with exceptional thermoelectric power factors. *Angew. Chem. Int. Ed.* 47: 6433.
73 Zhao, L.D., Lo, S.-H., Zhang, Y. et al. (2014). Ultralow thermal conductivity and high thermoelectric figure of merit in SnSe crystals. *Nature* 508: 373.
74 Zhao, L.D., Tan, G., Hao, S. et al. (2016). Ultrahigh power factor and thermoelectric performance in hole-doped single-crystal SnSe. *Science* 351: 141.
75 Jin, Q., Shi, W., Zhao, Y. et al. (2018). Cellulose fiber-based hierarchical porous bismuth telluride for high-performance flexible and tailorable thermoelectrics. *ACS Appl. Mater. Interfaces* 10: 1743.
76 He, M., Lin, Y.-J., Chiu, C.-M. et al. (2018). A flexible photo-thermoelectric nanogenerator based on MoS_2/PU photothermal layer for infrared light harvesting. *Nano Energy* 49: 588.
77 Jao, Y.T., Li, Y.-C., Xie, Y., and Lin, Z.-H. (2017). A self-powered temperature sensor based on silver telluride nanowires. *ECS J. Solid State Sc.* 6: N3055.
78 Juntunen, T., Jussila, H., Ruoho, M. et al. (2018). Inkjet printed large-area flexible few-layer graphene thermoelectrics. *Adv. Funct. Mater.* 28: 1800480.
79 Kim, J.Y., Lee, W., Kang, Y.H. et al. (2018). Wet-spinning and post-treatment of CNT/PEDOT:PSS composites for use in organic fiber-based thermoelectric generators. *Carbon* 133: 293.
80 Guan, Y.-S., Li, H., Ren, F., and Ren, S. (2018). Kirigami-inspired conducting polymer thermoelectrics from electrostatic recognition driven assembly. *ACS Nano* 12: 7967.
81 Yang, C., Souchay, D., Kneiss, M. et al. (2017). Transparent flexible thermoelectric material based on non-toxic earth-abundant p-type copper iodide thin film. *Nat. Commun.* 8: 16076.
82 Choi, J., Jung, Y., Yang, S.J. et al. (2017). Flexible and robust thermoelectric generators based on all-carbon nanotube yarn without metal electrodes. *ACS Nano* 11: 7608.

83 Rojas, J.P., Conchouso, D., Arevalo, A. et al. (2016). Paper-based origami flexible and foldable thermoelectric Nanogenerator. *Nano Energy* 31: 296.

84 Kim, S.J., Lee, H.E., Choi, H. et al. (2016). High-performance flexible thermoelectric power generator using laser multiscanning lift-off process. *ACS Nano* 10: 10851.

85 Chen, Y., He, M., Tang, J. et al. (2018). Flexible thermoelectric generators with ultrahigh output power enabled by magnetic field-aligned metallic nanowires. *Adv. Electron. Mater.* 4: 1800200.

86 Jung, K.K., Jung, Y., Choi, A.J. et al. (2016). Flexible thermoelectric generator with polydimethyl siloxane in thermoelectric material and substrate. *Curr. Appl Phys.* 16: 1442.

87 Duan, J., Feng, G., Yu, B. et al. (2018). Aqueous thermogalvanic cells with a high Seebeck coefficient for low-grade heat harvest. *Nat. Commun.* 9: 5146.

88 Zhang, F., Zang, Y., Huang, D. et al. (2015). Flexible and self-powered temperature-pressure dual-parameter sensors using microstructure-frame-supported organic thermoelectric materials. *Nat. Commun.* 6: 8356.

89 We, J.H., Kim, S.J., and Cho, B.J. (2014). Hybrid composite of screen-printed inorganic thermoelectric film and organic conducting polymer for flexible thermoelectric power generator. *Energy* 73: 506.

90 Kim, C.S., Yang, H.M., Lee, J. et al. (2018). Self-powered wearable electrocardiography using a wearable thermoelectric power generator. *ACS Energy Lett.* 3: 501.

91 Zhu, L., Ding, T., Gao, M. et al. (2019). Shape conformal and thermal Insulative organic solar absorber sponge for photothermal water evaporation and thermoelectric power generation. *Adv. Energy Mater.* 9: 1900250.

92 Yang, Y., Wei, X.J., and Liu, J. (2007). Suitability of a thermoelectric power generator for implantable medical electronic devices. *J. Phys. D: Appl. Phys.* 40: 5790.

8

Photovoltaic–Pyroelectric Coupled Effect Nanogenerators

8.1 Introduction

To address issues of energy crisis and environmental degradation, numerous efforts have been devoted to developing clean and renewable energy resources [1–3]. Solar energy has attracted tremendous attention because it is abundant, sustainable, and reliable [4–7]. To effectively scavenge solar energy, a variety of devices have been developed [8–12]. Among these devices, photovoltaic effect based nanogenerator is an outstanding approach to directly convert solar energy into electricity [13–17], showing great potential to provide electricity for people's daily lives and industrial production [18–20]. In addition, photovoltaic nanogenerators can also act as self-powered photodetectors to be utilized in the area of flame warning, air quality monitoring, and chemical/biological analysis [21–25]. Depending on the photoactive materials used, photovoltaic nanogenerators can be divided into two types, semiconductor-based photovoltaic nanogenerators and ferroelectrics-based photovoltaic nanogenerators [26–32]. To realize photogenerated carrier separation and transport, a p–n junction is usually needed to generate a built-in electric field in semiconductor-based photovoltaic nanogenerators [33–35]. A variety of n-type pyroelectric semiconductors, including wurtzite-structured zinc oxide (ZnO) and cadmium sulfide (CdS), are often utilized to construct the p–n junction because of their excellent chemical and thermal stability [36, 37]. Due to lack of central symmetry in pyroelectric materials, pyroelectric potential can be generated by changing the temperature across the pyroelectric materials [38]. Coincidentally, light can not only induce photogenerated carriers in a p–n junction device because of photovoltaic effect but also heat the device because of thermal effect resulting from light. As a result, pyroelectric potential can be generated across these n-type pyroelectric materials, modifying the built-in electric field and consequently modulating the behavior of charge carriers. Such a phenomenon is called photovoltaic–pyroelectric coupled effect [37, 39]. For ferroelectric photovoltaic nanogenerator, p–n junctions are usually not necessary [40, 41]. Nevertheless, photovoltaic–pyroelectric coupled effect can be observed in ferroelectric photovoltaic nanogenerator because of the pyroelectricity of ferroelectrics

*Corresponding author: yayang@binn.cas.cn

Hybridized and Coupled Nanogenerators: Design, Performance, and Applications,
First Edition. Ya Yang.

[42]. Photovoltaic–pyroelectric coupled effect can be employed to control the behavior of photogenerated carriers, operating through the coupling among pyroelectricity, semiconductor/ferroelectrics, and photoexcitation processes [43–45], which has been demonstrated to be an effectual method to improve photoexcited carrier separation/transport and suppress the recombination between free electrons and holes. [46–49]. Photovoltaic–pyroelectric coupled effect nanogenerators have become one of the research hotspots.

In this chapter, the fundamental theories of photovoltaic–pyroelectric coupled effect nanogenerators will be first reviewed. Afterwards, excellent photovoltaic–pyroelectric materials, recent advancements, and representative prototypes of photovoltaic–pyroelectric coupled effect nanogenerators will be introduced. Application areas of photovoltaic–pyroelectric coupled effect nanogenerators will be presented as well.

8.2 Basic Principle

8.2.1 Pyroelectric Effect

Pyroelectric phenomenon is induced by spontaneous polarization P_s and has been observed in many materials, including lead zirconate titanate (PZT), barium titanate (BTO), bismuth ferrite (BFO), ZnO, and gallium nitride (GaN) [50, 51]. Taking pyroelectric effect in polarized PZT material as an example, there is a certain amount of polarized charges on the surface of PZT when the temperature of PZT is constant. To shield the polarized charges, positive and negative charges are drawn to the charged surfaces of PZT, as illustrated in Figure 8.1a [52]. When PZT is heated, spontaneous polarization decreases, and hence, pyroelectric potential is generated due to the excess captured charges (pyroelectric charges), as shown in Figure 8.1b. Similarly, when PZT is cooled, the mean spontaneous polarization increases. As a result, more charges are attracted to the surfaces of PZT, resulting in a reversed pyroelectric potential, as displayed in Figure 8.1c. The pyroelectric current I_p is defined according to the formula $I_p = p \cdot A \cdot dT/dt$, where dT/dt stands for the rate of temperature change, A represents the effective area on the material, and p is the pyroelectric coefficient. [44].

The pyroelectric coefficient p is a vital parameter to assess the electrical conversion ability of the pyroelectric materials and can be defined according to the equation $p^{\sigma,E} = (dP_S/dT)_{\sigma,E}$, where σ and E are related to the constant stress and electric field, respectively [42]. Depending on the production mechanism, pyroelectric effect has three forms, which are primary pyroelectric, secondary pyroelectric, and tertiary pyroelectric effects. Without external electric fields, primary pyroelectric effect occurs in the situation of pinched pyroelectric materials under constant strain as well as a homogenous heating/cooling spread. As is known, thermal expansion/contraction induced by temperature change leads to strain in pyroelectric materials. Therefore, secondary pyroelectric effect is generated because strain changes the electric displacement by piezoelectric effect [53]. Tertiary pyroelectric effect is relevant to the piezoelectric effect resulting from uneven heating/cooling [54].

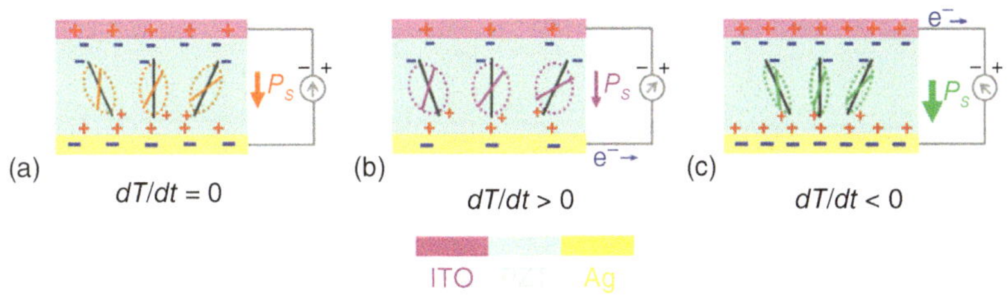

Figure 8.1 Physical mechanism of pyroelectric effect. Working principle of pyroelectric effect in PZT under constant temperature (a), heating (b), and cooling (c) conditions. Source: Reproduced with permission from Zhang et al. [52]. Copyright 2017, John Wiley and Sons.

8.2.2 Photovoltaic Effect

Photovoltaic effect can be classified as traditional photovoltaic effect and ferroelectric photovoltaic effect. For traditional photovoltaic effect, upon irradiation, semiconductors absorb photons (energy is higher than the optical bandgap of the semiconductors). Consequently, electrons can be transferred from valence to conduction band, leading to electron–hole pairs (excitons). Then, these excitons are divided into free electrons and holes if there is an asymmetric electrostatic potential (built-in electric field). Subsequently, these free electrons and holes move toward different ends of the asymmetric electrostatic potential. Consequently, photovoltaic signals are produced. Such asymmetric electrostatic potential is usually generated by constructing Schottky contact as well as p–n junction (including homojunctions and heterojunctions). Taking p–n junction as a demonstration (Figure 8.2), when p-SnS makes contact with n-ZnO, holes and electrons diffuse into ZnO and SnS, respectively. Consequently, a space charge depletion region is formed. An electric field that is directed from ZnO to SnS is built up in the depletion region. When light illuminates at the vicinity of the SnS/ZnO interface, photoexcited excitons are divided by this built-in electric field. Then, the electrons and holes move to ZnO and SnS, respectively. When the diffusion and drift motion of electrons as well as holes reach equilibrium, a photogenerated voltage traversing the SnS–ZnO junction is generated. While turning on this SnS–ZnO junction, a photocurrent is generated in the external circuit [55]. Theoretically, the open-circuit voltages are determined by the bandgap of semiconductors for p–n junction photoelectric devices.

Photovoltaic electric signals along the polarization direction are generated when ferroelectric materials are illuminated. Such a phenomenon is called the ferroelectric photovoltaic effect, which has been known for more than 50 years. To date, several possible theories are ascribed for the origin of the ferroelectric photovoltaic effect, including bulk photovoltaic as well as depolarized field effects, domain wall theory, and interfacial effect controlled by asymmetric electrostatic potential [56, 57]. The photovoltage induced by ferroelectric photovoltaic effect increases with increasing polarization strength and electrode distance. Consequently, the photovoltage caused by ferroelectric photovoltaic effect can be quite larger than the optical bandgap of ferroelectrics, which is called anomalous photovoltaic effect [58–60]. So far, ferroelectric photovoltaic

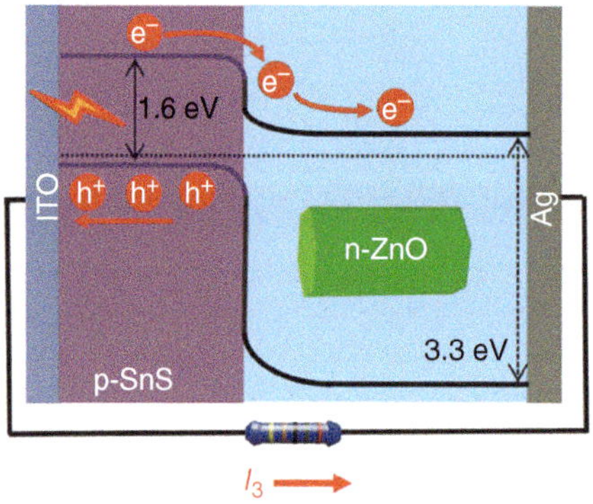

Figure 8.2 Physical mechanism of photovoltaic effect in p–n junction (p-SnS/n-ZnO) nanogenerator. Source: Reproduced with permission from Ouyang et al. [55]. Copyright 2018, Elsevier.

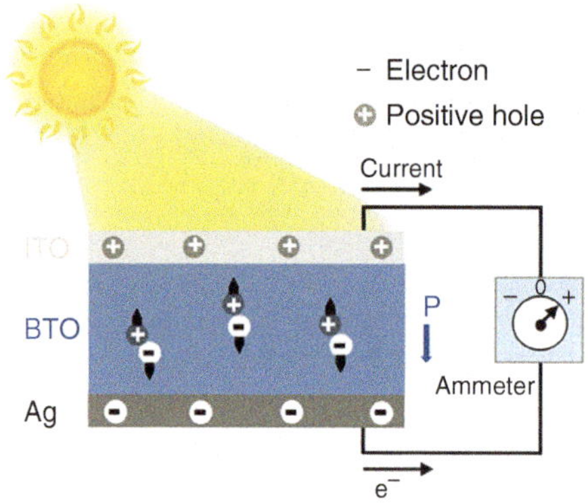

Figure 8.3 Physical mechanism of ferroelectric photovoltaic effect in BTO-based nanogenerator. Source: Reproduced with permission from Ma et al. [56]. Copyright 2017, John Wiley and Sons.

effect in a variety of ferroelectrics has been reported, including BTO, PZT, BFO, and bismuth vanadate (BVO). Taking ferroelectric BTO material as an example (Figure 8.3), when light illuminates the BTO material, electron–hole pairs are generated. These photoexcited electron–hole pairs are divided by an asymmetric electrostatic potential near BTO/Ag interface as well as a depolarized field in BTO. Subsequently, free negative and positive carriers move toward the indium tin oxide (ITO) as well as Ag sides, respectively, leading to a photocurrent in the external circuit [56].

8.2.3 Photovoltaic–Pyroelectric Coupled Effect

Photovoltaic–pyroelectric coupled effect is a coupling effect based on pyroelectric polarization, semiconductor/ferroelectrics characteristics, together with photoexcitation processes, which can modulate carrier separation, transport, and recombination. Taking p-SnS/n-ZnO heterojunction as a demonstration [55], the working principle of photovoltaic–pyroelectric coupled effect is displayed in Figure 8.4. Upon light irradiation, photoexcited negative and positive carriers move toward ITO as well as the Ag electrode due to the traditional photovoltaic effect, respectively. Consequently, photoexcited current (I_2) is generated in the external circuit. Meanwhile, a fast transient temperature increase is generated within in the ZnO nanowires due to light-thermal effect, leading to a

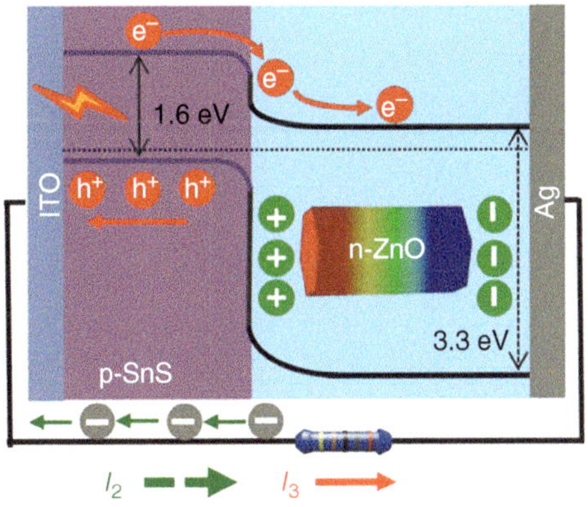

Figure 8.4 Energy-band diagram of p-SnS/n-ZnO heterojunction under illumination, showing the photovoltaic–pyroelectric coupled effect. Source: Reproduced with permission from Ouyang et al. [55]. Copyright 2018, Elsevier.

pyroelectric potential traversing the ZnO nanowires. Meanwhile, positive as well as negative polarized charges are present in the adjacent p–n junction interface and Ag electrode, respectively. To shield the polarized charges, free charges on the ITO and Ag electrodes rearrange as the temperature increases; consequently, pyroelectric current (I_3) flowing from ITO to Ag film is produced. Since light-induced pyroelectric current is in the same direction as the photocurrent, the output current is enhanced. In addition, the light-induced pyroelectric effect could efficiently and rapidly tune the built-in electric field, thus modulating the behavior of the charge carriers. Taking p-Si/n-ZnO configuration as an example [61], when the device is irradiated, the ZnO nanowires are heated, resulting in negative polarized charges distributing at the SnS/ZnO interface. Consequently, valance as well as conduction bands of ZnO nanowires near the SnS/ZnO interface get higher, significantly enhancing the photogenerated carrier transport across the heterojunction. Therefore, the output current is promoted at the moment of illumination. If irradiation is retained, the temperature of ZnO nanowires is constant. Consequently, the pyroelectric potential vanishes rapidly, retaining a stable photovoltaic current. After turning off the light, the transient temperature reduces rapidly within the ZnO nanowires; consequently, positive pyroelectric charges appear adjacent to the SnS/ZnO interface. As a result, the valence as well as conduction bands of ZnO nanowires decline, resulting in a "carrier trap" effect, which can impede the carrier transport across the p–n junction area and significantly reduce the current under dark conditions [62–64]. Afterwards, the temperature of the ZnO nanowires recovers to room temperature and stays constant, and hence, both the pyroelectric current and photocurrent disappear.

As for the ferroelectric photovoltaic effect, depending on device structures (vertical- and planar-structured nanogenerators), two types of working mechanism are reported [56, 65]. A typical nanogenerator with vertical-structured electrode configuration is presented in Figure 8.5a, which consists of a BTO film as the active material, a transparent top ITO film, and a bottom Ag film. Figure 8.5b depicts the corresponding energy-band structure illustration of the nanogenerator upon illumination. After being polarized, a polarization directed from ITO to the Ag film can be built up, leading to negative polarized charges

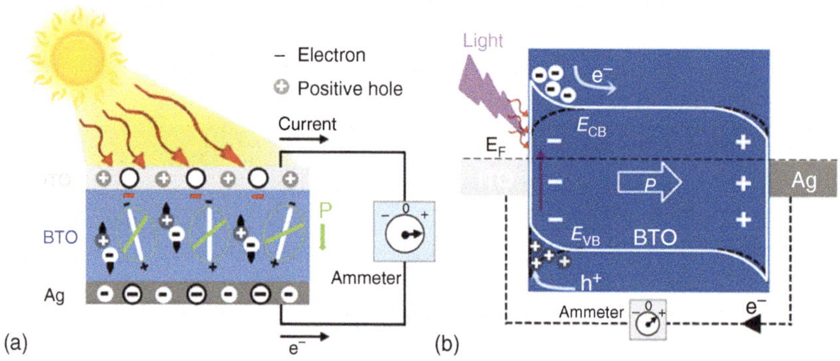

Figure 8.5 (a) Schematic cross-sectional diagram of vertical-structured ITO/BTO/Ag nanogenerator. Source: Reproduced with permission from Ma et al. [56]. Copyright 2017, John Wiley and Sons. (b) Energy-band diagram of vertical-structured ITO/BTO/Ag nanogenerator. Source: Reproduced with permission from Ma and Yang [65]. Copyright 2018, Elsevier.

and positive polarized charges distributed adjacent to the ITO as well as the Ag films, respectively. To shield the polarized charges, free charges are captured on the surfaces of the device. Upon light illumination, the BTO layer is heated by the light-induced thermal effect, resulting in acute oscillation of electric dipoles. Hence, spontaneous polarization within the BTO layer is reduced. As a consequence, the captured electrons are released and move from Ag to the ITO electrode, leading to a pyroelectric current flowing to the Ag film. Additionally, under light illumination, photons with high enough energy can be absorbed by BTO, leading to electrons being excited to the conduction band of BTO to form electron–hole pairs, which are divided via the domain wall, asymmetric electrostatic potential, and depolarization field, resulting in a photocurrent flowing from ITO to the Ag electrode. As a result, the output electricity of the nanogenerator is increased because of the photovoltaic–pyroelectric coupled effect. The temperature of BTO reaches a stable value as the light illumination is retained. Consequently, the pyroelectric current disappears rapidly, leaving behind a steady output photocurrent. When the light is turned off, the photocurrent vanishes and the temperature of BTO begins to decline, resulting in a reversed pyroelectric current. After the temperature of BTO recovers to room temperature, only dark current can be detected.

In the vertical-structured nanogenerator, part of the incident light is absorbed by the top electrode and wasted. As a result, the quantity of photogenerated carriers absorbed by the active layer is reduced, and the direct interaction between light and ferroelectrics is weakened. Therefore, the short-circuit photocurrent will be reduced in a vertical-structured nanogenerator [40]. To solve these problems, planar-structured nanogenerators have been presented. A representative nanogenerator with planar-structured configuration consists of a ferroelectric material layer (BTO) and interdigitated electrodes (ITO) on its surface, as displayed in Figure 8.6a. Although the device has a symmetric structure, the barrier heights at the ITO/BTO interfaces are different. As illustrated in Figure 8.6b, after being polarized, a depolarization field directed from the right to left is produced

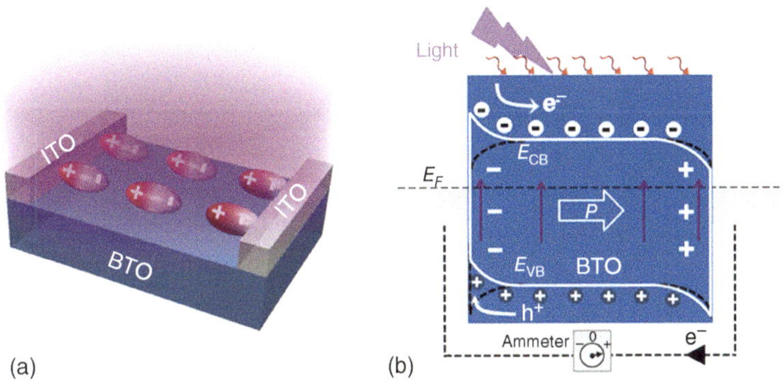

Figure 8.6 Schematic (a) and energy-band (b) diagrams of planar-structured ITO/BTO/ITO nanogenerator under illumination. Source: Reproduced with permission from Ma and Yang [65]. Copyright 2018, Elsevier.

across the BTO layer. Consequently, the valence and conduction bands of BTO on the left are bent up by the negative polarized charges, while those on the right are bent down by the positive polarized charges. Upon illumination, the temperature of the planar-structured nanogenerator increases rapidly, resulting in a pyroelectric current moving to the right ITO film. Meanwhile, some of the photons pump electrons from the valence to the conduction band of the BTO. As a result, excitons are produced. Subsequently, the excitons are divided and transported via asymmetric electrostatic potential as well as depolarization field, resulting in a photocurrent in the same direction as the pyroelectric current. Consequently, output electricity can be promoted through the coupling effect of the photovoltaic and pyroelectric effects. Notably, in the planar-structured nanogenerator, the ferroelectric BTO material between two adjacent ITO electrodes can be directly exposed to light. As a consequence, light can directly heat the BTO layer, so that the BTO material could obtain a higher rate of temperature change. Hence, the pyroelectric current of a planar-structured nanogenerator is larger than that of a vertical-structured nanogenerator.

8.2.4 Temperature Dependence of Photovoltaic–Pyroelectric Coupled Effect

Output electric signals of photovoltaic–pyroelectric coupled effect nanogenerators, especially photocurrent signals, can be greatly affected by ambient temperature [66, 67]. Taking a planar-structured ITO/BTO/ITO nanogenerator as an example, the energy-band configuration illustration of the device under low temperature, room temperature, and high temperature conditions are shown in Figure 8.7 [66]. At low temperature (below 200 K), shallow trap levels exist in the bandgap of BTO because of impurity centers. Consequently, a certain amount of free electrons can be trapped by the shallow trap levels because of freeze-out effect [68]. However, the trapped electrons are easy to be excited to the conduction band by absorbing photons with high enough energy. As a consequence, when the light is turned on, electrons in the valence band as well as electrons

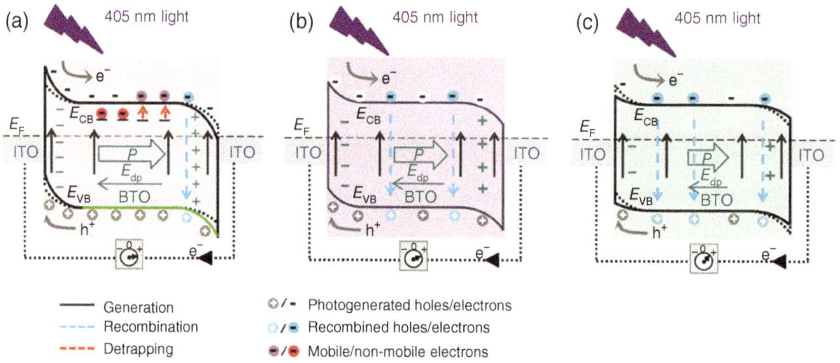

Figure 8.7 Energy-band diagram of planar-structured ITO/BTO/ITO nanogenerator at low temperature (a), room temperature (b), and high temperature (c), showing the temperature dependence of photovoltaic–pyroelectric coupled effect. Source: Reproduced with permission from Ma and Yang [66]. Copyright 2019, Elsevier.

captured by the shallow trap level can be activated to the conduction band. Therefore, the quantity of photogenerated excitons is much more at low temperature than at room temperature, resulting in the photovoltaic current being greatly improved. Moreover, low temperature effectively suppresses random oscillation of electric dipoles to enhance the mean spontaneous polarization and depolarized field of BTO. Correspondingly, more negative polarized charges and positive polarized charges distribute at the left and right ends of the BTO material, respectively. Consequently, barrier height at the left and right ITO/BTO interfaces are increased and reduced, respectively. Both the increase of barrier height and the enhancement of depolarized field can facilitate photogenerated excitons separation together with transport, and prevent separated carriers recombination, resulting in the promotion of the photocurrent. As the ambient temperature recovers to room temperature, effects of shallow trap levels have been weakened, leading to reduced photogenerated excitons. Additionally, the increased temperature also reduces the left asymmetric electrostatic potential as well as the depolarized field. Consequently, the output current decreases (Figure 8.7b). As the temperature further increases, the left barrier height gets lower and the depolarized field become much weaker; hence, the photogenerated excitons are divided effectively and more free carriers are recombined. Thus, the photocurrent is drastically reduced.

8.3 Materials

8.3.1 ZnO Nanowires

Semiconductor ZnO usually has a hexagonal wurtzite structure at ambient conditions, where planes consist of tetrahedrally coordinated Zn^{2+} as well as O^{2-} stack alternatively along the c-axis [69]. Wurtzite structured ZnO has the space group *C6mc*. Spontaneous polarization along the *c*-axis forms because of the noncentral symmetry configuration of ZnO, making it possible for ZnO to

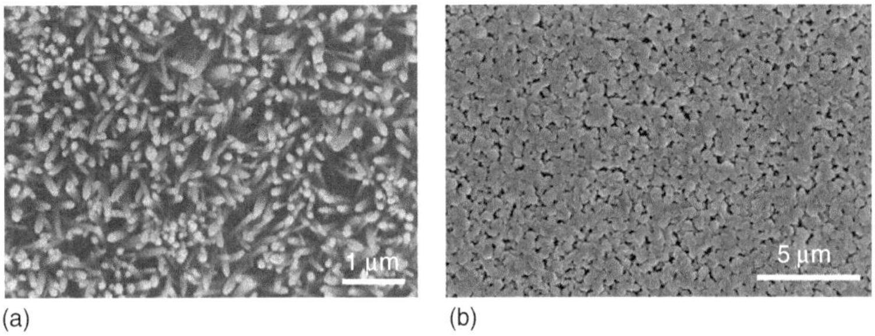

Figure 8.8 SEM images of ZnO nanowires fabricated by hydrothermal method. (a) SEM image of ZnO nanowires prepared by hydrothermal method with $ZnCl_2$ and HMTA as main starting materials. Source: Reproduced with permission from Ouyang et al. [70]. Copyright 2017, John Wiley and Sons. (b) Top-view SEM image of ZnO nanowires fabricated by hydrothermal method with $Zn(NO_3)_2$ and HMTA as main starting materials. Source: Reproduced with permission from Ouyang et al. [55]. Copyright 2018, Elsevier.

scavenge thermal fluctuation through pyroelectric effect. Meanwhile, as a direct bandgap material, ZnO has a ~3.37 eV energy bandgap, exhibiting excellent photovoltaic response to ultraviolet light. In addition, the response range of ZnO in optoelectronic field is enlarged by doping and surface engineering induced modification of the defects and surface states within ZnO. In recent years, photovoltaic–pyroelectric coupled effect in ZnO is widely investigated to improve the performance of ZnO based optoelectric devices. To obtain high quality ZnO nanowire arrays, a majority of synthesis methods have been developed, including vapor–solid–solid as well as wet chemical approach. The wet chemical approach, such as hydrothermal method and electrochemical deposition, holds the characteristics of simplicity, effectiveness, and low cost. Hydrothermal method is an effective and simple approach for aligned ZnO nanowires preparation on conductive substrates. The morphology as well as size of ZnO nanowires can be modulated by adjusting the fabrication temperature, preparation time, and reaction precursor type/concentration. A typical procedure of preparing ZnO nanowires on ITO conductive substrate was introduced by Ouyang et al. [70] by using $ZnCl_2$, hexamethylenetetramine (HMTA), and NH_4OH as the main starting materials. The processes are described in detail in the following. ITO conductive substrate was covered by a ZnO seed film. After dissolving $ZnCl_2$ (0.3365 g) and HMTA (0.3207 g) in deionized water, a certain amount of NH_4OH was poured into the solution to achieve a homogenous reaction precursor. After that, seeded ITO glass was immersed downward into the solution and reacted at 95 °C. Subsequently, this specimen was washed by using deionized water and then dried. Figure 8.8a presents the scanning electron microscopic (SEM) image of the as-prepared ZnO nanowires [70], showing that these as-synthesized ZnO nanowires (80–200 nm in diameter) were highly crystallized, well aligned, and structurally uniform. In addition, by using reaction solutions that contained $ZnNO_3{\cdot}6H_2O$ and HMTA, dense ZnO nanowires could be obtained (Figure 8.8b) [55].

8.3.2 BTO Materials

BTO is a representative ABO_3 perovskite-type ferroelectric oxide that has cubic, tetragonal, orthorhombic, and rhombohedral crystal structures. Usually, tetragonal phase and cubic phase are stable at room temperature. In cubic BTO, Ti^{4+} is octahedrally coordinated with six O^{2-} and no dipole moments are generated because of noncentral symmetry configuration. Consequently, the cubic BTO is paraelectric. A permanent electric dipole exists in tetragonal BTO because of the relative displacement of Ti^{4+} along the c-axis from its centrosymmetric position; consequently, ferroelectricity is generated. As ferroelectric material, tetragonal BTO has the ability of converting thermal energy to electricity due to its pyroelectricity. Moreover, as a wide bandgap material (3.3 eV), tetragonal BTO can transform light energy into electricity because of the photovoltaic effect. BTO ceramic disks made from BTO powders are often used as the core layer of photovoltaic–pyroelectric coupled effect nanogenerators due to its simple fabrication methods. High-purity BTO powders can be synthesized by solid-state reaction approach, sol–gel method, and electrospinning method. A traditional fabrication process of BTO ceramic disk has been introduced by Ma et al., which can be described as follows. A certain amount of polyvinyl alcohol solution (PVA, 2 wt%) was evenly mixed with 1.2 g BTO nanopowders with an average diameter of about 100 nm (Figure 8.9a) [56, 65]. And then, the mixture was pressed into a green disk with a diameter of about 20 mm. Subsequently, the green disks were sintered at 1200 °C to become compact ceramic disks. Notably, the cubic BTO powders were converted to tetragonal BTO ceramic disks through the sintering process, as confirmed in the X-ray diffraction (XRD) curves (Figure 8.9b). Additionally, the obtained BTO disks needed to be polarized because of their random spontaneous polarization [50].

8.3.3 BFO Materials

Besides BTO material, BFO is another common ABO_3 ferroelectric oxide. As a narrow bandgap ferroelectric material (2.67 eV) with large remnant polarization, BFO materials can effectively harvest visible light energy and thermal energy. BFO films and BFO disks are often utilized to construct photovoltaic–pyroelectric coupled effect nanogenerators. Compared to BFO films, BFO disks are much easier to be prepared and employed. BFO disks are usually made from BFO powders. A typical procedure of preparing BFO powders by hydrothermal method was introduced by Qi et al. [49], which can be described in the following. First, $Fe(NO_3)_3{\cdot}9H_2O$ and $Bi(NO_3)_3{\cdot}5H_2O$ (mole ratio = 1 : 1.05) were put in KOH aqueous solution and mixed evenly. Subsequently, the precursor was reacted at 180 °C in a Teflon container, and the products obtained were cleaned by dilute HNO_3 as well as deionized water to remove the impurities. The as-prepared BFO powders consisted of layered agglomerates, as shown in Figure 8.10a [49]. Generally, the as-synthesized BFO powders need to be annealed to improve their crystallinity. By evaluating the XRD patterns of annealed BFO powders, the optimal sintering temperature is determined. As illustrated in Figure 8.10b [49], single rhombohedral phase

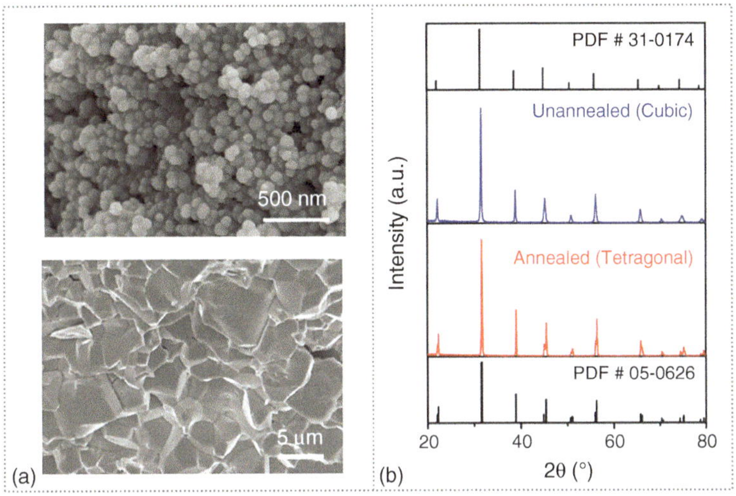

Figure 8.9 Morphology and crystal structure of BTO materials. (a) SEM images of BTO nanopowders and ceramic disk. Source: Reproduced with permission from Ma and Yang [65]. Copyright 2018, Elsevier. (b) XRD patterns of BTO materials before and after being annealed at 1200 °C. Source: Reproduced with permission from Ji et al. [50]. Copyright 2018, John Wiley and Sons.

BFO powders could be obtained at a sintering temperature of 750–850 °C. The impurity phase of $Bi_2Fe_4O_9$ appeared as the sintering temperature was increased to 875 °C. The conventional fabrication method of BFO ceramic disks was similar to that of the BTO disks. However, the sintering temperature of the BFO disks is usually at 850 °C, which is much lower than that of the BTO samples. BFO ceramic disks made by the traditional sintering approach often have relatively dense microstructure, as illustrated in Figure 8.10c.

8.3.4 Other Materials

Besides ZnO, BTO, and BFO, photovoltaic–pyroelectric coupled effect has been observed in some other materials as well. Dai et al. [71] investigated the photovoltaic–pyroelectric coupled effect derived from wurtzite structured CdS. The n-type CdS is a direct bandgap semiconductor and is widely utilized in optoelectronic devices because of its extraordinary thermal and chemical stability, narrow optical bandgap (2.42 eV), and low work function (4.4 eV). CdS is a pyroelectric material because of its noncentrosymmetric crystalline structure. When the temperature of CdS is changed, pyroelectric potential is generated along its polar direction, which can be utilized to control the behavior of photogenerated carriers. Consequently, upon illumination, the CdS-based nanogenerator can deliver electric signals related to photovoltaic–pyroelectric coupled effect. Kumar et al. [72] observed photovoltaic–pyroelectric coupled effect in vertical tin sulfide (SnS) materials. The p-type SnS material is a direct bandgap two-dimensional (2-D) material. The optical bandgap of SnS films can be modulated in the range of 1.0–1.5 eV, resulting in the excellent photoresponse

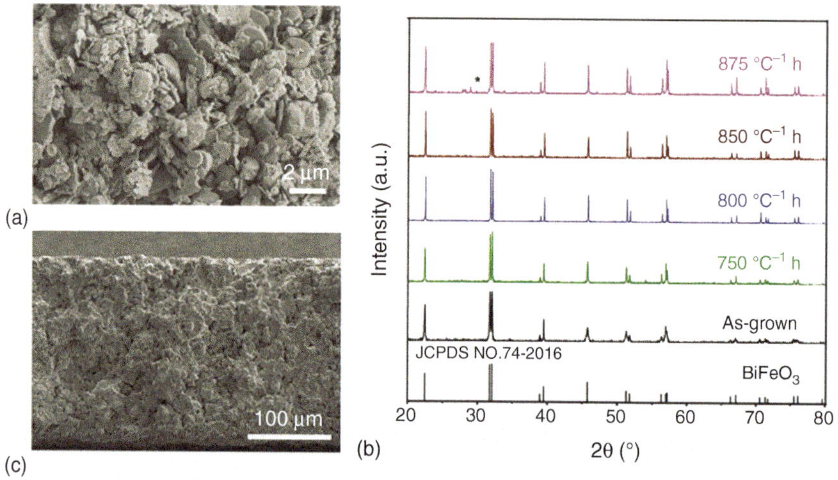

Figure 8.10 Morphology and crystal structure of BFO materials. (a) SEM image of BFO nanopowders. (b) XRD patterns of BFO powders annealed at different temperatures. (c) Cross-sectional SEM image of BFO ceramic disk. Source: Reproduced with permission from Qi et al. [49]. Copyright 2018, John Wiley and Sons.

of SnS films to visible light and near-infrared light. SnS film has high carrier mobility (1–3.8 × 10^4 cm^2 V/s) together with a high optical absorption coefficient (>10^4 cm^{-1}). Moreover, it is reported that pyroelectric potential can be generated in a noncentrosymmetric chalcogenide layered material. All the characteristics of SnS mentioned above indicate that SnS is an outstanding semiconductor for preparing photovoltaic–pyroelectric coupled effect nanogenerators. In addition, photovoltaic–pyroelectric coupled effect in ferroelectric PZT materials has also been observed.

8.4 Device Design

8.4.1 Vertical Structure

A variety of vertical-structured photovoltaic–pyroelectric coupled effect nanogenerators have been intensively investigated. Figure 8.11a presents a photovoltaic–pyroelectric coupled effect nanogenerator with P3HT/ZnO heterojunction configuration [70]. The nanogenerator was composed of a P3HT/ZnO nanowires composite layer, an ITO top electrode, an Al bottom electrode, and a poly(3,4-ethylenedioxythiophene)–poly(styrene sulfonate) (PEDOT:PSS) electron transport layer. Figure 8.11b illustrates the transverse section SEM image of the device, showing P3HT material covered on the surfaces of upper ZnO nanowires. This device was around 4 μm in thickness. Top-view SEM image of the device exhibits that Al electrode is uniformly covered on the PEDOT:PSS film, providing good path for photogenerated carriers collection, as shown in Figure 8.11c In addition, a variety of photovoltaic–pyroelectric coupled effect nanogenerators were constructed by utilizing ZnO nanowires,

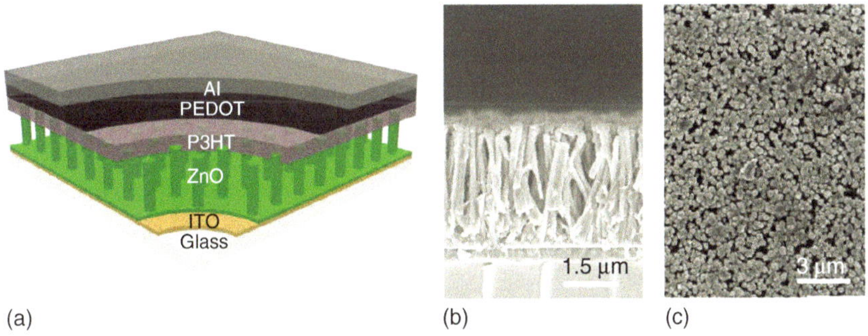

Figure 8.11 Configuration of vertical-structured photovoltaic–pyroelectric coupled effect nanogenerator based on semiconductor. (a) Schematic diagram of P3HT/ZnO heterojunction nanogenerator. (b,c) The corresponding cross-sectional (b), and top-view (c) SEM images of the device. Source: Reproduced with permission from Ouyang et al. [70]. Copyright 2017, John Wiley and Sons.

including perovskite/ZnO-heterostructured nanogenerator, PEDOT:PSS/ZnO heterojunction nanogenerator, and 2D ZnO/graphene heterojunction nanogenerator [73–77]. Dai et al. [71] designed a p-Si/n-CdS flexible nanogenerator, where numerous micropyramids were fabricated on the flexible p-Si wafer. CdS NW with length and diameter of about 0.6 μm and 50 nm were perpendicularly covered on the Si micropyramids. ITO and Al film were utilized as the top and bottom electrodes, respectively. In addition, Kumar et al. [72] fabricated an ITO/SnS/Si/Al nanogenerator, where the 2-D SnS layers were deposited by magnetron sputtering technology. The thickness of the SnS layer was about 24 nm.

A representative vertical-structured photovoltaic–pyroelectric coupled effect nanogenerator based on ferroelectrics is illustrated in Figure 8.12a [56], where a BTO ceramic disk served as the main functional material and transparent ITO (Ag) film worked as the electrodes. The thickness, piezoelectric coefficient, and relative density of the BTO film were evaluated to be about 350 μm, 375 pC/N, and 95.4%, respectively. Figure 8.12b exhibits a photovoltaic–pyroelectric coupled effect nanogenerator with vertical-structured ITO/BFO/Ag configuration [49]. The relative density of the BFO film was determined to be ∼86% by using the Archimedes method and the thickness of the BFO film was about 220 μm. The photograph of the nanogenerator is shown in Figure 8.12c, indicating the diameter of the device to be about 18.6 mm.

8.4.2 Planar Structure

A typical photovoltaic–pyroelectric coupled effect nanogenerator with planar structure is shown in Figure 8.13a [65], indicating that the device consisted of a BTO layer and a couple pairs of interdigitated ITO electrodes. Photographic image of the device indicates that the electrode had a width as well as spacing of 1 and 0.6 mm, respectively. The illuminated size of the device was evaluated to be about 2.01 cm^2. The optical profile image shows that the engraving depth was about 8 μm. The planar-structured ITO/BTO/ITO nanogenerator was polarized via a direct electric field at 2.4 kV. Zhao et al. [44] designed a radially polarized

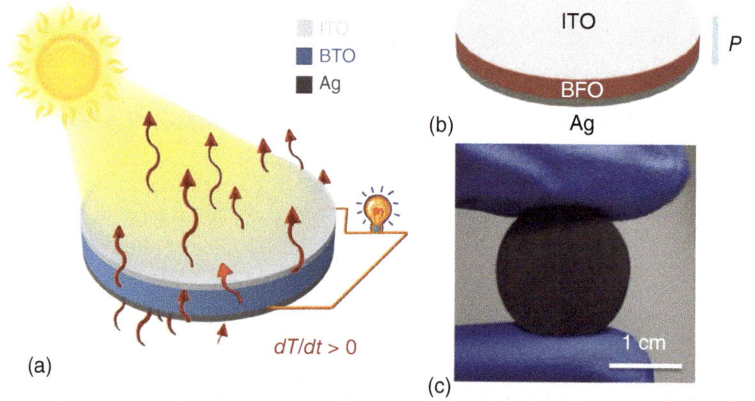

Figure 8.12 Schematic diagram of vertical-structured photovoltaic–pyroelectric coupled effect nanogenerators based on ferroelectrics. (a) Schematic diagram of ITO/BTO/Ag nanogenerator under illumination. Source: Reproduced with permission from Ma et al. [56]. Copyright 2017, John Wiley and Sons. (b,c) Schematic diagram (b) and photograph (c) of ITO/BFO/Ag nanogenerator. Source: Reproduced with permission from Qi et al. [49]. Copyright 2018, John Wiley and Sons.

planar-structured ITO/BTO/ITO nanogenerator with 10 pairs of ITO electrodes. Figure 8.13b illustrates the main fabrication procedure and a photograph of the device. As seen, a circular BTO ceramic disk covered with ITO electrode was first prepared. Second, a layer engraving process was conducted to obtain the radial ITO electrodes with a spacing of 0.6 mm. The thickness diameters of the BTO layers were about 1 and 25 mm, respectively. The nanogenerator was polarized by using a direct electric field of 2.1 kV/mm.

8.5 Performance

8.5.1 Output Electric Signals

Typical output electric signals of P3HT/ZnO heterojunction nanogenerator are illustrated in Figure 8.14a [70]. When illuminated by a 365 nm light, the output electric signals exhibited obvious two-stage behavior, which contained a peak as well as a steady plateau. The peak current/voltage (I_1, V_1) and plateau current/voltage (I_2, V_2) resulted from photovoltaic–pyroelectric coupled effect and photovoltaic effect, respectively. Under illumination (113.4 mW/cm^2), I_1, V_1, and I_2, V_2 were 108 nA, 51 mV and 95 nA, 34 mV, respectively. Obviously, photovoltaic–pyroelectric coupled effect greatly improved the produced electricity of the nanogenerator. Figure 8.14b illustrates the output current of a vertical-structured ITO/BTO/Ag nanogenerator upon 365–690 nm irradiation [56]. As is seen, current–time curves could be divided into two types. The first type can be observed under illumination with the light wavelength of 405 and 450 nm, where an obvious two-stage curve could be observed. In the beginning,

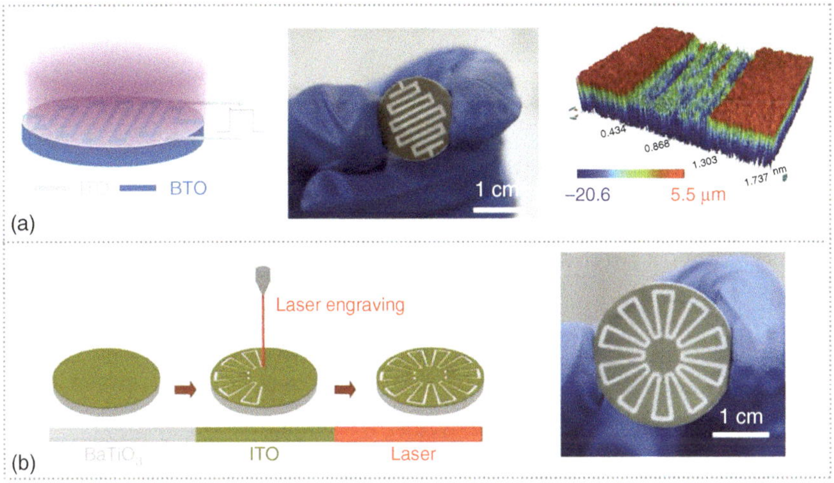

Figure 8.13 Photovoltaic–pyroelectric coupled effect nanogenerators with planar structure. (a) Schematic diagram, photograph, and optical surface profiler image of ITO/BTO/ITO nanogenerator. Source: Reproduced with permission from Ma and Yang [65]. Copyright 2018, Elsevier. (b) Schematic diagram and photograph of a radially polarized ITO/BTO/ITO nanogenerator. Source: Reproduced with permission from Zhao et al. [44]. Copyright 2018, Elsevier.

an obvious peak resulting from photovoltaic–pyroelectric effect was observed. Subsequently, the current became stabilized due to the photovoltaic effect. Considering that the photon energies of 405 and 450 nm light were only 3.06 and 2.75 eV, which is lower than the optical bandgap of the BTO material (3.3 eV), the observed photocurrent was ascribed to the oxygen vacancies in BTO. The second type of current–time curve could be observed under illumination with the light wavelength of more than 450 nm, where only a sharp current peak but no current platform could be observed due to the absence of photovoltaic current. Qi et al. [49] investigated electricity generation of a photovoltaic–pyroelectric coupled effect nanogenerator with vertical-structured ITO/BFO/Ag configuration upon 450 nm light irradiation. Figure 8.14c depicts the current–time curves obtained. Obviously, the output current showed typical two-stage characteristics, and the peak current and plateau current value are 8.8 and 2.0 nA, respectively.

8.5.2 Resistance

Output power P is dependent on resistance R of the photovoltaic–pyroelectric coupled effect nanogenerator, which can be calculated by the formula $P = I^2R$. Ouyang et al. [70] systematically investigated the resistance R of a vertical-structured p-P3HT/n-ZnO nanogenerator by changing the external loading resistances in the testing system, as illuminated in Figure 8.15a, where the output current I_1, resistance R_1, and output power P_1 are related to photovoltaic–pyroelectric coupled effect, and photocurrent I_2, resistance R_2, and output power P_2 are relevant to photovoltaic effect. As seen, I_1 and I_2 reduced with increasing loading resistance. The maximum P_1/P_2 of 0.78/0.42 nW was

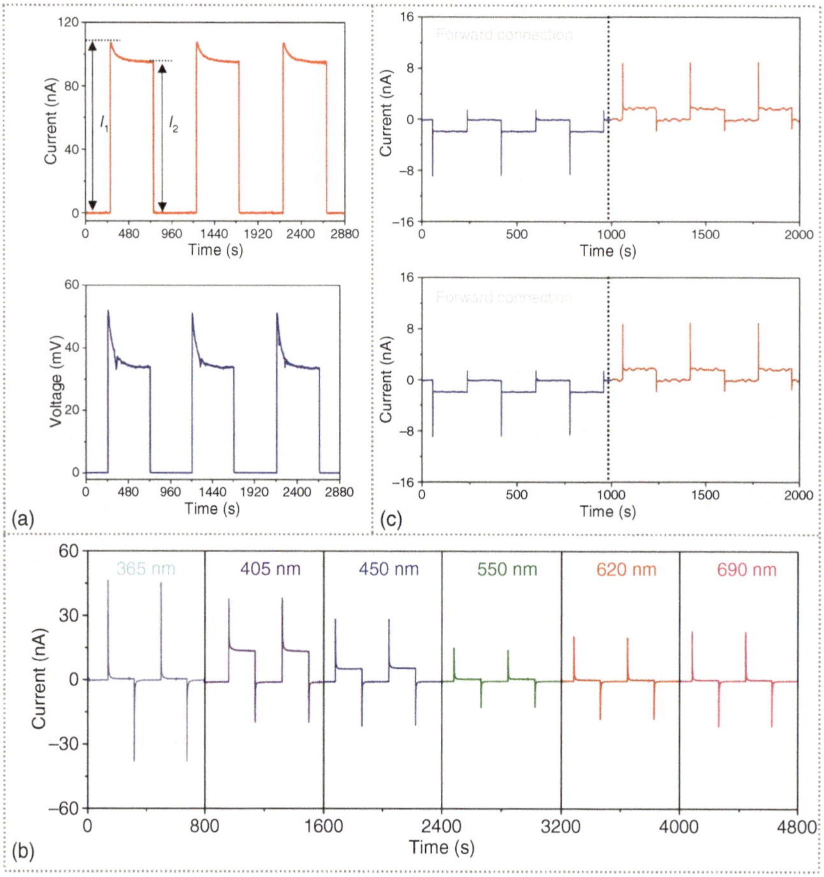

Figure 8.14 Output electric signals of vertical-structured photovoltaic–pyroelectric coupled effect nanogenerators. (a) Output current and output voltage of p-P3HT/n-ZnO nanogenerator. Source: Reproduced with permission from Ouyang et al. [70]. Copyright 2017, John Wiley and Sons. (b) Output current of ITO/BTO/Ag nanogenerator. Source: Reproduced with permission from Ma et al. [56]. Copyright 2017, John Wiley and Sons. (c) Output current and voltage of ITO/BFO/Ag nanogenerator. Source: Reproduced with permission from Qi et al. [49]. Copyright 2018, John Wiley and Sons.

obtained at the resistance R_1/R_2 of 280/200 kΩ. Figure 8.15b illustrates the output power produced by a vertical-structured ITO/BTO/Ag nanogenerator [56]. Obviously, photovoltaic–pyroelectric current and pyroelectric current together with photocurrent decreased drastically with increasing loading resistance. The largest photovoltaic–pyroelectric power was 3.2 nW, which was obtained at 40 MΩ. Notably, the largest photovoltaic–pyroelectric power was higher than the largest pyroelectric and photovoltaic power. Figure 8.15c exhibits the output power generated by a vertical-structured nanogenerator with ITO/BFO/Ag configuration [49]. As can be seen, photovoltaic–pyroelectric current and pyroelectric current, together with the photocurrent, decreased rapidly with

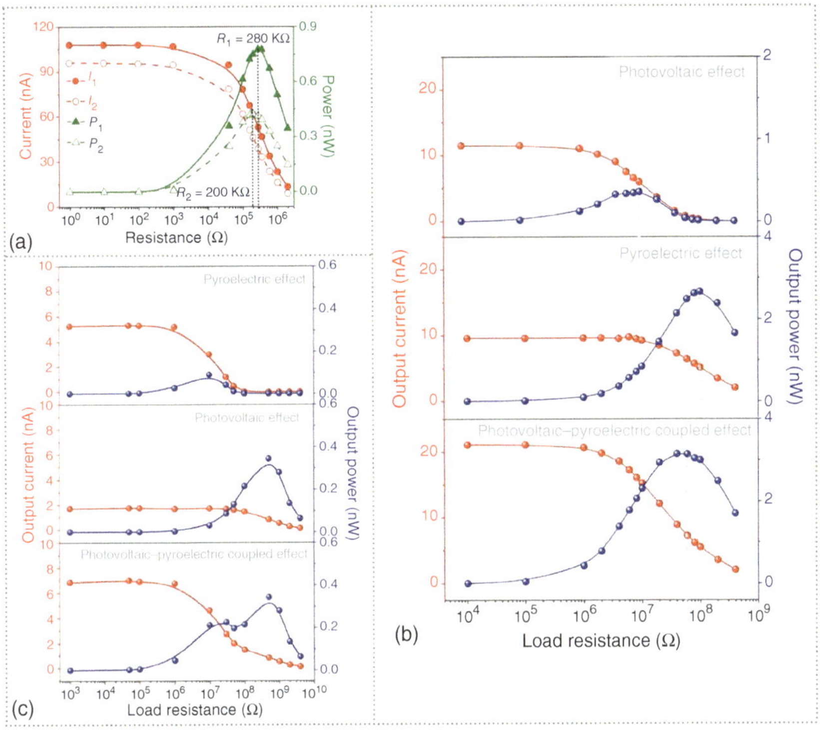

Figure 8.15 Output performance of photovoltaic–pyroelectric coupled effect nanogenerators with vertical structure. (a) Dependence of output current and power of p-P3HT/n-ZnO nanogenerator on loading resistance. Source: Reproduced with permission from Ouyang et al. [70]. Copyright 2017, John Wiley and Sons. (b) Dependence of output current and power of ITO/BTO/Ag nanogenerator on loading resistance. Source: Reproduced with permission from Ma et al. [56]. Copyright 2017, John Wiley and Sons. (c) Dependence of output current and power of ITO/BFO/Ag nanogenerator on loading resistance. Source: Reproduced with permission from Qi et al. [49]. Copyright 2018, John Wiley and Sons.

increasing resistance due to Ohmic loss, while the output power increased first and then declined. The maximum output power related to photovoltaic effect and pyroelectric effect were 0.34 and 0.09 nW. The maximum output power relevant to photovoltaic–pyroelectric coupled effect could be kept above 0.2 nW under loading resistance ranging from 10 MΩ to 1 GΩ.

8.5.3 Comparison of Vertical- and Planar-Structured Nanogenerators

The output current of vertical-structured and planar-structured photovoltaic–pyroelectric coupled effect nanogenerators have been compared by Ma and Yang [65]. Figure 8.16a,b illustrates the obtained current–time patterns of the two types of nanogenerators. In a vertical-structured ITO/BTO/Ag device, the time-dependent current curve consists of a quite sharp peak value relevant to photovoltaic–pyroelectric coupled effect and a small plateau value related to

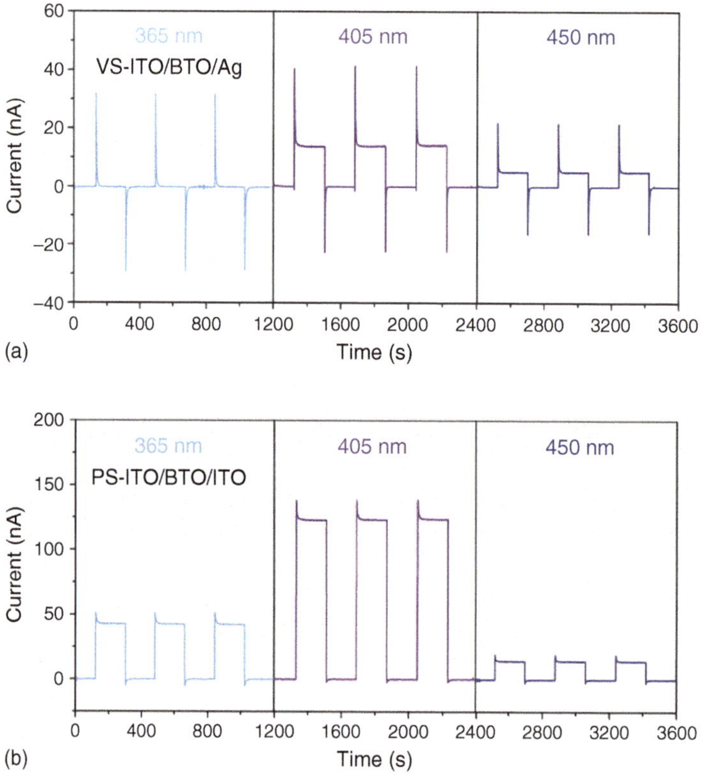

Figure 8.16 Output current of vertical-structured ITO/BTO/Ag nanogenerator (a) and planar-structured ITO/BTO/ITO nanogenerator (b) under different light wavelengths. Source: Reproduced with permission from Ma and Yang [65]. Copyright 2018, Elsevier.

photovoltaic effect. By defining the difference between peak and plateau currents as the pyroelectric current, it was found that the pyroelectric current was higher than the photocurrent, indicating that pyroelectric effect is more important than photovoltaic effect upon irradiation for the vertical-structured ITO/BTO Ag device. The phenomenon was more evident when the nanogenerator was irradiated by 365 nm light, and this behavior was ascribed to the rapid heat transfer of the thin BTO layer in the vertical-structured ITO/BTO/Ag device. Compared to the vertical-structured ITO/BTO/Ag device, the time-dependent current curve of planar-structured ITO/BTO/ITO device was quite different. As illustrated in Figure 8.16b, although the current–time pattern also showed a two-stage behavior, the peak current seemed much smaller due to the significant enhancement of the photovoltaic current platform. Such phenomenon became more evident when the nanogenerator was irradiated by 405 nm light. For a planar-structured ITO/BTO/ITO device, a considerable part of the BTO layer was not covered with ITO, so that it could be directly illuminated by light. The greatly improved current was supposed to be due to the direct reaction between the light and the BTO materials.

8.5.4 Temperature Dependence

It is reported that temperature has a vital impact on the output current in photovoltaic–pyroelectric coupled effect based ITO/BTO/ITO nanogenerators with planar structure. Ma and Yang [66] investigated the temperature-dependent (80–360 K) output current under 405 nm light illumination (131.0 mW/cm^2) and revealed the current generation mechanism. The obtained results suggested that current–time curves recorded at different temperatures could be classified into three types. The first type of current–time curve showed a two-stage behavior, which could be observed below 200 K. In the first stage, a small platform followed by an abrupt increase could be observed. In the second stage, a slow increase followed by a large plateau current could be observed. The first stage of the current–time curve was caused by the fast activation of electrons between valence and conduction bands. The second stage was caused by excitation of electrons between the shallow trap levels and the conduction band. The second type of current–time curve could be observed when the temperature ranged from 220 to 280 K, and this type of curve showed a transition process of current with increasing temperature. If the temperature increased from 220 to 240 K, only one nearly plateau current could be detected, as displayed in Figure 8.17a. As the temperature increased to 260–300 K, the current first reached a peak value upon irradiation and then gradually decreased to a steady value, as shown in Figure 8.17b. This type of current–time curve results from the coupling effect of pyroelectric effect and photovoltage effect, which is relevant to band–band as well as shallow trap levels activation. The third type of time-dependent current curve was observed when the temperature was above 300 K. A typical current–time curve measured at 360 K is presented in Figure 8.17c, showing that the current rapidly increased to a maximum value upon illumination, and subsequently decreased to a steady value. The temperature-dependent photocurrent is summarized in Figure 8.17d, showing that the photocurrent gradually increased to 393.1 nA as the temperature increased from 80 to 220 K. However, the photocurrent began to decrease as the temperature further increased. The corresponding enhancement ratio of photocurrent was in the range of 121.9–179.6% as the temperature ranged from 80 to 240 K and reduced to 66.7% at 360 K. To explain the observed phenomena, temperature-dependent remnant polarization P_r under dark and illumination conditions was investigated. Figure 8.17e shows the representative P–E hysteresis loops at 260, 300, and 360 K, respectively. Figure 8.17f illustrates the temperature dependence of remnant polarization P_r. It is found that the remnant polarization P_r increased as the temperature increased and attained its maximum value at 160 K, and then decreased as the temperature further increased. Notably, the peak photocurrent and remnant polarization P_r were obtained at different temperatures, indicating that the shallow trap levels dominate the current generation at low temperature. When the temperature was above 260 K, both remnant polarization P_r and photocurrent reduced as the temperature increased, suggesting that the photocurrent is dominated by polarization at such temperatures.

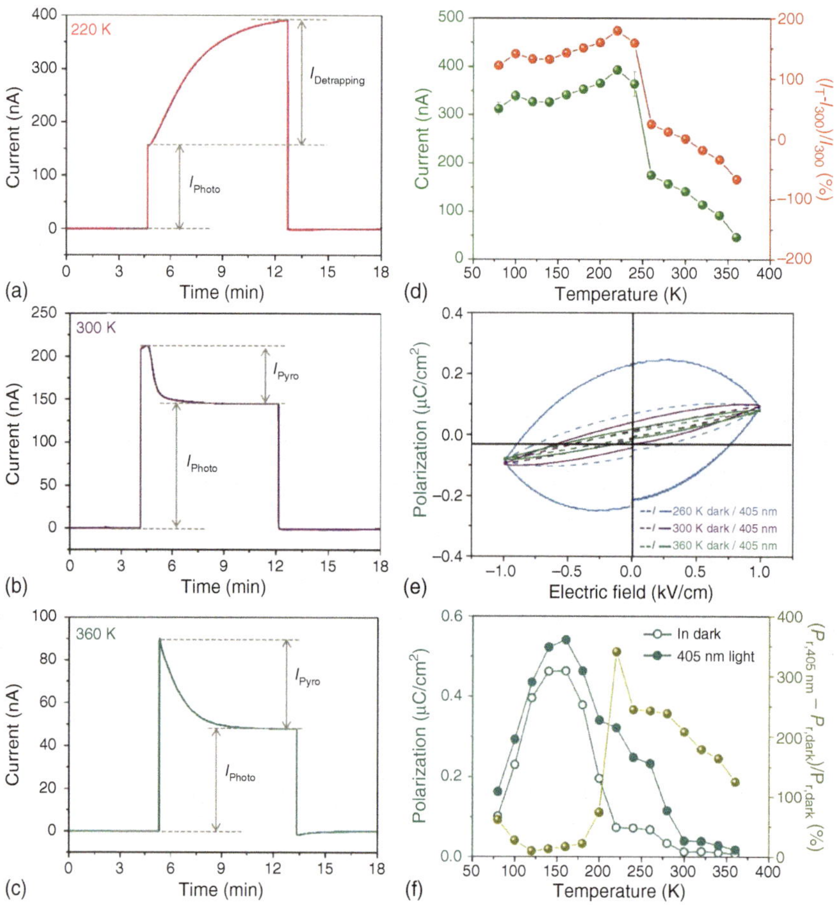

Figure 8.17 Temperature-dependent performance of photovoltaic–pyroelectric coupled effect ITO/BTO/ITO nanogenerator. (a–c) Output current of the device under low temperature (a), room temperature (b), and high temperature (c). (d) Temperature-dependent output current and the corresponding enhancement ratio with the temperature ranging from 80 to 360 K. (e) *P–E* hysteresis loops under illumination with different light intensities. (f) Temperature-dependent remnant polarization P_r under dark and illumination condition, and the corresponding enhancement ratio induced by illumination. Source: Reproduced with permission from Ma and Yang [66]. Copyright 2019, Elsevier.

8.6 Applications

Photovoltaic–pyroelectric coupled effect can modify the generation, separation, and transport of carriers together with recombination through light-induced pyroelectric effect [78], showing great potential to improve the performance of optoelectric devices, especially self-powered photodetectors. Several key parameters are usually utilized to evaluate the detection ability of photodetectors, which are responsivity *R*, photoconductive gain *G*, specific detectivity *D**, together with response speed [79].

Responsivity R reflects the capability of a photodetector to transfer incident light signals into electric signals, which can be evaluated according to the formula $R = I_{ph}/P/S$. I_{ph} stands for the obtained current of the photodetector upon illumination, S stands for the active irradiation size, and P is the light power density. Photoconductive gain G is assessed according to the formula $G = (I_{ph}/e)/(PS/h\upsilon)$, where e is the elemental charge and $h\upsilon$ stands for the energy carried by each incident photon. Specific detectivity D^* suggests the smallest detectable power of illumination and can be defined as $D^* = R/(2eI_d/S)$, where I_d is the dark current. The response speed of photodetectors can be evaluated through response time together with recovery time. Generally, the response (recovery) time is assessed by the time needed for output signals to change from 10% to 90% (90–10%) of their largest values. High response speed indicates short response and recovery time.

8.6.1 Common Photodetection

The detection ability of a p-P3HT/n-ZnO junction photodetector for monitoring 365 nm light (0.84–113.4 mW/cm^2) has been studied by Ouyang et al. [70]. Figure 8.18a illustrates the temperature as well as output electric signals of the photodetector. When the light intensity was beyond 16.6 mW/cm^2, the temperature increased significantly because of the light-caused thermal energy. Consequently, two-stage electric signals were generated due to photovoltaic–pyroelectric coupled effect. Figure 8.18b,c illustrates the output electric signals and responsivity induced by photovoltaic–pyroelectric coupled effect as well as photovoltaic effect. Obviously, I_1/V_1 (resulting from photovoltaic–pyroelectric coupled effect) together with I_2/V_2 (caused by photovoltaic effect) increased when light intensity enhanced, indicating the feasibility of the device for ultraviolet photodetection. Both the current responsivity and voltage responsivity decreased with increasing light intensity and then became stable. Notably, photovoltaic–pyroelectric coupled effect induced responsivity ($R_{I,1}$, $R_{V,1}$) was higher than photovoltaic effect induced responsivity ($R_{I,2}$, $R_{V,2}$).

Ma et al. [56] studied the performance of a vertical-structured ITO/BTO/Ag self-powered photodetector for detecting 405 nm illumination, as illustrated in Figure 8.19a. While light intensity exceeded 3.5 mW/cm^2, the thermal effect resulting from illumination got obviously enhanced, resulting in an obvious temperature increase in the photodetector. As a result, a two-stage output current was observed. In the first stage, the temperature of the photodetector increased rapidly because of irradiation, resulting in a peak current (I_1) due to the coupling effect of the photovoltaic effect together with pyroelectric effect. During the second stage, the temperature of the device reached up to a stable value; meanwhile, the current resulting from pyroelectric effect disappeared, leading to a plateau current (I_2) because of photovoltaic effect. Both I_1 and I_2 increased as the light intensity increased. According to the obtained current, R and G together with D^* were calculated and summarized in Figure 8.19b,c, where G_1 and R_1, as well as D^*_1 are related to photovoltaic–pyroelectric coupled effect, and G_2 and R_2, as well as D^*_2 are relevant to photovoltaic effect. As seen, G_1, R_1, and D^*_1 increased rapidly at the beginning and remained stable as the

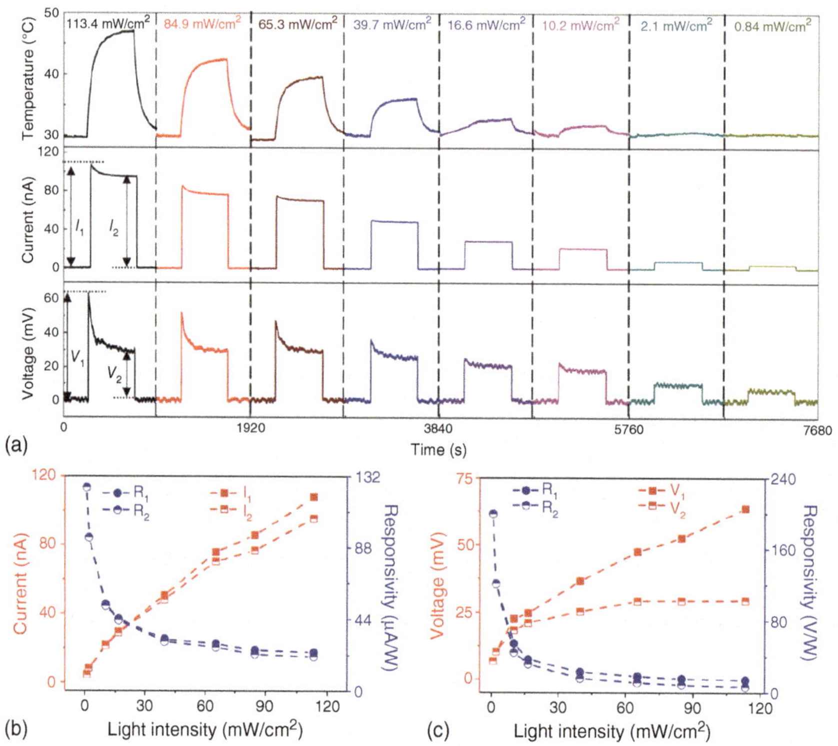

Figure 8.18 Output electric signals of P3HT/ZnO self-powered photodetector. (a) Time-dependent temperature, current, and voltage of the device under 365 nm illumination with different light intensities. (b,c) Dependence of output current and current responsivity (b), output voltage and voltage responsivity (c) on light intensity. Source: Reproduced with permission from Ouyang et al. [70]. Copyright 2017, John Wiley and Sons.

light intensity increased. However, G_2, R_2, and D^*_2 reduced as the light intensity increased. Notably, the values of G_1, R_1, and D^*_1 were 2.6 times higher than G_2, R_2, and D^*_2, indicating that the performance of the photodetector was significantly improved by photovoltaic–pyroelectric coupled effect. The performance of a vertical-structured ITO/BFO/Ag photodetector for monitoring 450 nm irradiation is illustrated in Figure 8.19d [49]. The current–time curves show a peak current (I_1) followed by a platform current (I_1). Upon illumination, the device was heated by the light-induced thermal effect, resulting in I_1 due to the photovoltaic–pyroelectric coupled effect. Subsequently, the temperature of the device became stable, leading to I_2 because of photovoltaic effect. The light intensity dependence of I_1 and I_2 is plotted in Figure 8.19e. As can be seen, the value of I_1 is higher than that of I_2. In addition, both I_1 and I_2 monotonically increase with increasing light intensity, indicating that the device had good photoresponse to the 450 nm light. The sensitivity of the photodetector was further assessed by computing responsivity (R_1, R_2), photoconductive gain (G_1, G_2), as well as specific detectivity (D^*_1, D^*_2), as shown in Figure 8.19e,f, where G_1, R_1, and D^*_1 are derived from the peak current value I_1, and G_2, R_2, and D^*_2 are calculated according to the plateau current value I_2. It is obvious that the responsivity (R_1, R_2), photoconductive gain (G_1, G_2), and specific detectivity (D^*_1, D^*_2) decreased as the light intensity increased. Additionally, compared to G_2, R_2, and D^*_2, G_1, R_1, and D^*_1 were much higher, and the enhancement could reach up to more than 978% (0.86 mW/cm^2), suggesting that the photovoltaic–pyroelectric coupled effect could significantly promote the performance of the photodetector for monitoring weak 450 nm light.

Additionally, the performance of a planar-structured ITO/BTO/ITO photodetector for monitoring 405 nm light at different temperatures has been systematically investigated. The obtained output currents of the device at 240, 300, and 360 K are illustrated in Figure 8.20a. As seen, the output current signals showed typical two-stage behavior as the temperature increased to 300 K. The corresponding light intensity dependency of the output current and responsivity are summarized in Figure 8.20b. At 300 K, the output current showed a linear increasing behavior with the light intensity increasing. As the temperature was increased to 360 K, the rate of increase of the output current reduced. Remarkably, when decreasing the temperature to 240 K, the output current exhibited a nonlinear increase and the rate of increase was much faster than that obtained at higher temperatures (300 and 360 K). It is reported that more electrons captured by shallow trap levels could be excited and contribute to the photocurrent generation at a lower temperature, which is ascribed to the faster increasing rate of the output current at 240 K. In addition, the responsivity of the device increased linearly at the temperature of 240 K; however, it reduced with light intensity at 300 and 360 K. Moreover, Ma et al. systematically studied the temperature-dependent responsivity when the light intensity was fixed at 131.0 mW/cm^2, as shown in Figure 8.20c. It was found that low temperature could effectively improve the responsivity of the device and a maximum responsivity of 1.46×10^{-6} A/W was obtained at 220 K. Moreover, the photoconductive gain and specific detectivity could be improved to 4.46×10^{-6} and 1.59×10^7 jones at 220 and 240 K, respectively. Furthermore, the performance of a radially polarized ITO/BTO/ITO photodetector with planar structure for detecting 365 nm irradiation has been reported. The response (recovery) time obtained under 62.2 mW/cm^2 was 0.5 (0.2) seconds, which could satisfy the requirement for fast photodetection (response and recovery time <1 seconds). Additionally, the responsivity and specific detectivity under 0.6 mW/cm^2 were 2.3×10^{-7} A/W and 11.9 jones, respectively. By heating the device, the responsivity and specific detectivity were promoted to 71.7×10^{-7} A/W and 378.8 jones because of the coupling effect of ferroelectric property, pyroelectricity, and photoelectricity.

8.6.2 Image Photodetection

Utilizing a P3HT/ZnO junction, Ouyang et al. [70] constructed a 4×4 array self-powered photodetector with the size of 2×2 cm to monitor the 365 nm light. Figure 8.21a illustrates the schematic diagram of the photodetector array, which was connected to a multichannel testing system. A photograph of the device is shown in Figure 8.21b. Each pixel of the device consisted of an ITO film serving as the top electrode, an active layer of ZnO nanowires infiltrated

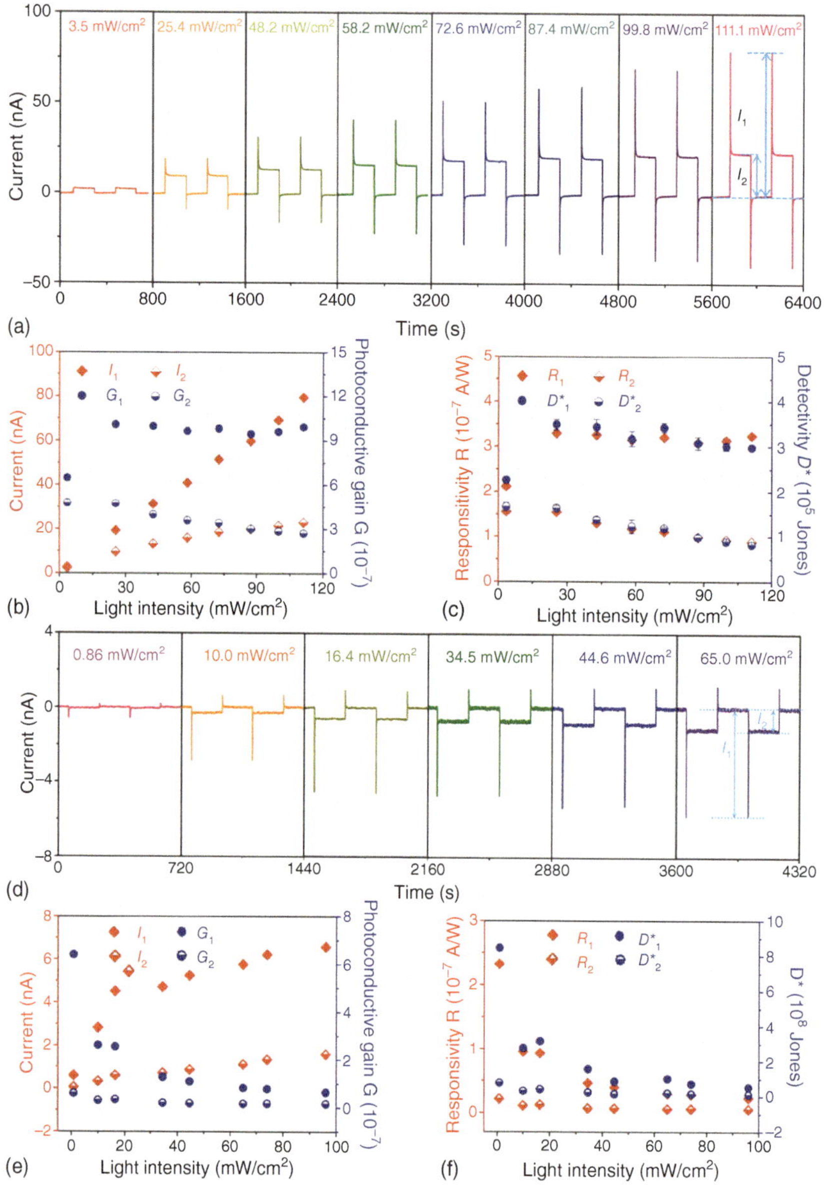

Figure 8.19 Performance of ferroelectrics-based self-powered photodetectors. (a) Output current of ITO/BTO/Ag photodetector under 405 nm illumination with different light intensities. (b,c) Dependence of output current and photoconductive gain (b), responsivity and specific detectivity (c) of the ITO/BTO/Ag device on light intensity. Source: Reproduced with permission from Ma et al. [56] Copyright 2017, Wiley. (d) Output current of a ITO/BFO/Ag photodetector under 450 nm illumination with different light intensities. (e,f) Dependence of output current and photoconductive gain (e), responsivity and specific detectivity (f) of the ITO/BFO/Ag photodetector on light intensity. Source: Reproduced with permission from Qi et al. [49] Copyright 2018, Wiley.

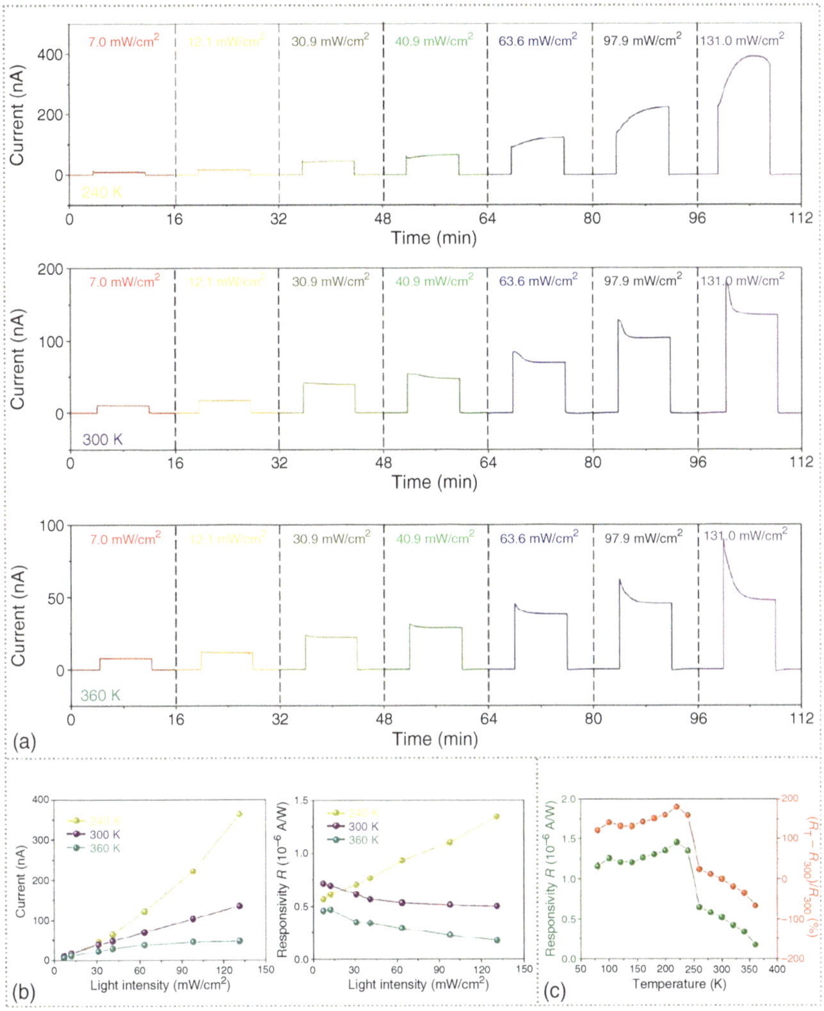

Figure 8.20 Performance of planar-structured ITO/BTO/ITO photodetector for monitoring 405 nm light. (a) Output current of the device at different temperatures. (b) Light-intensity dependence of output current and responsivity. (c) Temperature dependence of responsivity upon illumination with the light intensity of 131.0 mW/cm^2. Source: Reproduced with permission from Ma and Yang [66]. Copyright 2019, Elsevier.

with P3HT polymer, a PEDOT:PSS layer, as well as an Al bottom electrode. Each Ag electrode was connected to a multifunctional testing system through a resistance. The real-time output voltage from 16 independent pixels were measured to obtain an image figure about light irradiation. The performance of the multichannel photodetector for monitoring 365 nm light was evaluated under 113.4 mW/cm^2 irradiation and the imaging figures were recorded. Figure 8.21c displays the original state of the mapping figure, showing that no voltage signals were detected under dark condition. When the light illuminated pixel 11, the color of pixel 11 turned to red because of the produced photovoltaic–pyroelectric

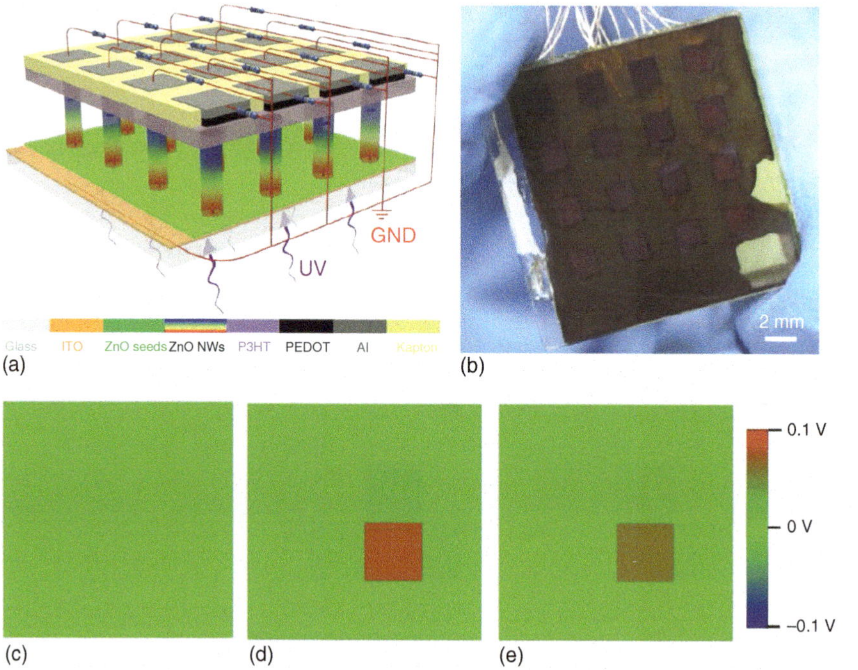

Figure 8.21 Structure and mapping image of P3HT/ZnO heterojunction photodetector with 4 × 4 array. (a,b) Schematic diagram (a) and photograph (b) of the device. (c) Mapping figure of the device under dark condition. (d,e) Image figures of the device when pixel 11 was illuminated by 365 nm light (113.4 mW/cm^2) with different illumination time. Source: Reproduced with permission from Ouyang et al. [70]. Copyright 2017, John Wiley and Sons.

voltage (Figure 8.21d). Notably, the color relevant to other pixels was still green, suggesting a good resolution of the image photodetector. With the illumination increasing, the color of pixel 11 turned to brownish red (Figure 8.21e), indicating a reduction of the output voltage. This phenomenon resulted from the two-stage characteristic of pyroelelctric–photovoltaic effect.

Based on photovoltaic–pyroelectric coupled effect, a self-powered image photodetector consisting of 3 × 3 photodetection elements for detecting 405 nm light has been constructed [56]. Configuration of the image photodetector was illustrated in Figure 8.22a, showing that the device consisted of a circular BTO layer, a top ITO film, and a bottom Ag film array. The inset figure (Figure 8.22a) illustrates an optical image of the device, exhibiting the size of each square-shaped Ag electrode to be 2.5 × 2.5 mm^2. The performance of the image photodetector was characterized by employing a multichannel testing system. The ITO film was in contact with the negative electrode of the testing instrument. Each Ag electrode was in contact with the positive electrode of the testing instrument. By using a multichannel testing system, the electric signals of the nine pixels could be recorded simultaneously. Figure 8.22b shows the time-dependent light image figure and the corresponding temperature of the photodetector when only element 3 was illuminated. As seen, photovoltage of the four stages could be observed within a 12-second illumination process. Without irradiation, no photovoltage signals could be observed, leading to the original green color in the image figure (stage

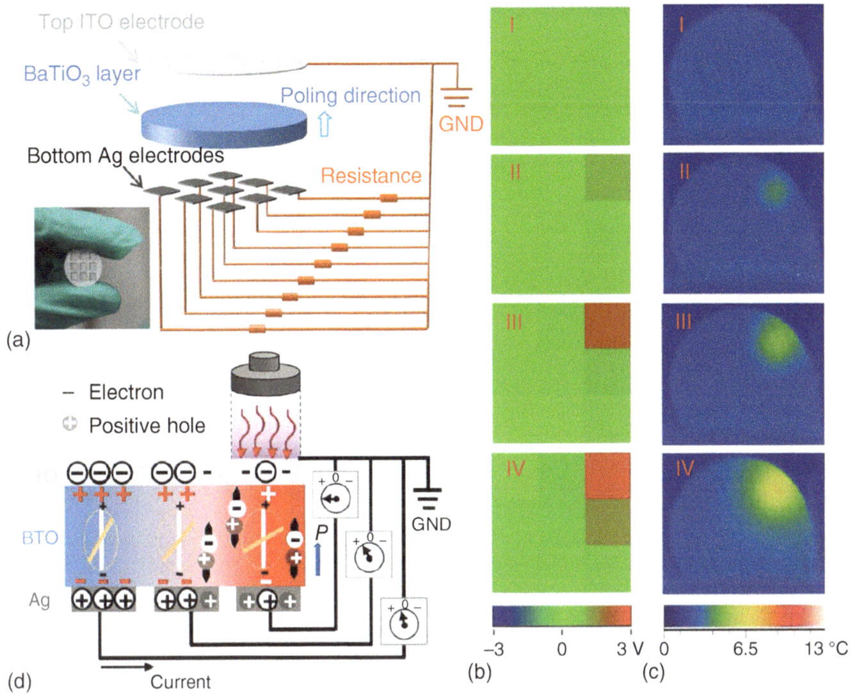

Figure 8.22 Structure and the corresponding mapping image of ITO/BTO/Ag photodetector array. (a) Schematic diagram of the image photodetector; the inset shows a photographic image of the device. (b) Time-dependent image figures of voltage when the applying illumination time is about 0 second (I), 3 seconds (II), 9 seconds (III) and 11.5 seconds (IV). (c) The corresponding temperature image related to (b). (d) Schematic diagram of the voltage generation process when only one element was illuminated. Source: Reproduced with permission from Ma et al. [56]. Copyright 2017, John Wiley and Sons.

I). When element 3 was illuminated, the color on element 3 began to change (stage II) and turned to orange (stage III). Subsequently, the color on element 3 changed to red with the illumination time increasing (stage IV); meanwhile, the color on elements 2 and 6 started to change, due to the spread of thermal energy from element 3, as shown in Figure 8.22c. Nevertheless, the mapping figure could clearly distinguish the irradiated element from the surrounding elements. The corresponding process of charge generation and transfer when only element 3 was illuminated was analyzed, as illustrated in Figure 8.22d. In dark condition, the electric dipoles of the BTO material oscillated randomly and the polarized charges were shielded by free charges. Upon irradiation, the temperature of the illuminated element increased rapidly. Consequently, the electric dipoles started to dramatically oscillate and the average spontaneous polarization was decreased, resulting in a pyroelectric voltage. Meanwhile, the light-induced excitons were divided and the free carriers obtained could be transported to generate a photovoltage. Correspondingly, a rapid output voltage in this element can be observed. With the illumination time getting longer, the thermal energy diffused to the neighbor elements, resulting in a pyroelectric voltage from these elements.

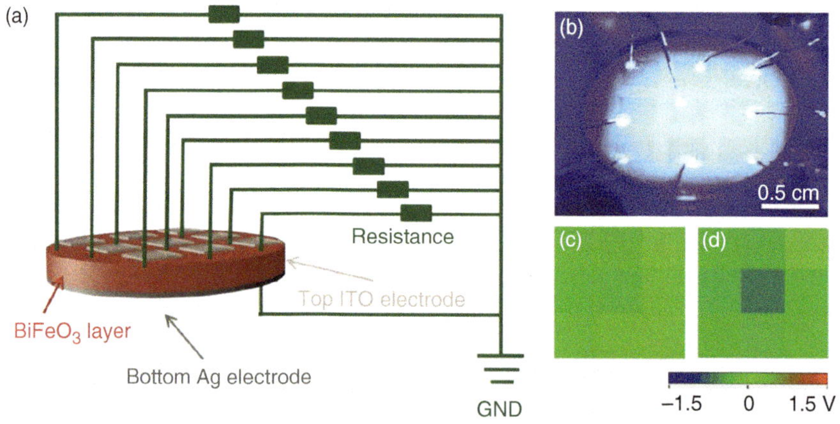

Figure 8.23 Structure and the corresponding mapping image of ITO/BFO/Ag image photodetector. (a) Schematic diagram of the device. (b) Photograph of the device illuminated by 450 nm light with the light intensity of 55.2 mW/cm^2. (c,d) Mapping images obtained from the photodetector array without illumination (c) and only pixel 5 was illuminated (d). Source: Reproduced with permission from Qi et al. [49]. Copyright 2018, John Wiley and Sons.

Furthermore, a self-powered multichannel photodetector was successfully constructed by employing BFO materials, which was able to monitor 450 nm light [49]. Figure 8.23a illustrates the configuration of the prepared photodetector array. The device consisted of a 3 × 3 transparent ITO top electrode array, a BFO layer with a thickness of about 220 μm, and a Ag bottom electrode. The size of each ITO electrode was 3 × 3 mm. Every ITO electrode was connected to the ground through a resistance. The voltage signal of every element was recorded by a multichannel testing system. By recording the voltage signals of all the pixels simultaneously, image figure indicating the light intensity and the illuminated area could be observed. Figure 8.23b illustrates a photographic image of the constructed multichannel photodetector under 55.2 mW/cm^2 light irradiation, demonstrating that all pixels generated a similar voltage. When the light was turned off, the image figure showed uniform green color (Figure 8.23c), suggesting good uniformity of the photodetector array. When only pixel 5 was irradiated, a voltage of 0.39 V from pixel 5 was detected; correspondingly, the color of pixel 5 turned blue, as shown in Figure 8.23d. Notably, the color of the adjacent pixels also changed due to the pyroelectric voltage caused by the diffusion of the light-induced thermal energy. Nevertheless, one can easily distinguish the illustrated pixel from the other pixels.

8.7 Conclusions and Prospects

In conclusion, advanced progress in the field of photovoltaic–pyroelectric coupled effect nanogenerators has been introduced. By coupling pyroelectric polarization and semiconductor/ferroelectrics characteristics together with photoexcitation processes, photovoltaic–pyroelectric coupled effect can be employed as an outstanding way to effectively improve the generation, separation, and transport of carriers, and suppress their recombination, consequently improving the performance of optoelectric devices. By utilizing the photovoltaic–pyroelectric coupled effect, the response speed, responsivity, and specific detectivity of photodetectors have been greatly improved. Despite these advancements, there are still some challenges. Further investigations in terms of materials modification, structure optimization, performance improvement, and application extension are still needed to satisfy the requirements of daily life and industry production. In addition, the working principle and characteristics of photovoltaic–pyroelectric coupled effect nanogenerators need to be further investigated in a more complicated environment for better practical applications. Nevertheless, considering that the performance of photodetector can be efficiently improved, it is hoped that photovoltaic–pyroelectric coupled effect nanogenerators would be widely utilized in the future.

References

1 Yu, Y.H., Zhang, Z., Yin, X. et al. (2017). Enhanced photoelectrochemical efficiency and stability using a conformal TiO_2 film on a black silicon photoanode. *Nat. Energy* 2: 17045.
2 Tyagi, V.V., Rahim, N.A.A., Rahim, N.A., and Selvaraj, J.A.L. (2013). Progress in solar PV technology: research and achievement. *Renewable Sustainable Energy Rev.* 20: 443–461.
3 Chen, B., Yang, Y., and Wang, Z.L. (2017). Scavenging wind energy by triboelectric nanogenerators. *Adv. Energy Mater.* 8: 1702649.
4 Polman, A., Knight, M., Garnett, E.C. et al. (2016). Photovoltaic materials: present efficiencies and future challenges. *Science* 352: aad4424.
5 Heo, J.H., Im, S.H., Noh, J.H. et al. (2013). Efficient inorganic-organic hybrid heterojunction solar cells containing perovskite compound and polymeric hole conductors. *Nat. Photonics* 7: 486–491.
6 Xu, C., Pan, C., Liu, Y., and Wang, Z.L. (2012). Hybrid cells for simultaneously harvesting multi-type energies for self-powered micro/nanosystems. *Nano Energy* 1: 259–272.
7 Kraemer, D., Poudel, B., Feng, H.P. et al. (2011). High performance flat-panel solar thermoelectric generators with high thermal concentration. *Nat. Mater.* 10: 532–538.
8 Lee, S.M., Park, K.H., Jung, S. et al. (2018). Stepwise heating in Stille polycondensation toward no batch-to-batch variations in polymer solar cell performance. *Nat. Commun.* 9: 1867.
9 Oener, S.Z., Cavalli, A., Sun, H.Y. et al. (2018). Charge carrier-selective contacts for nanowire solar cells. *Nat. Commun.* 9: 3248.
10 Lu, X.W., Jiang, P., and Bao, X.H. (2019). Phonon-enhanced photothermoelectric effect in $SrTiO_3$ ultra-broadband photodetector. *Nat. Commun.* 10: 138.
11 He, M.H., Lin, Y.J., Chiu, C.M. et al. (2018). A flexible photo-thermoelectric nanogenerator based on MoS_2/PU photothermal layer for infrared light harvesting. *Nano Energy* 49: 588–595.

12 Jin, L.F., Zhang, Y.T., Yu, Y. et al. (2018). Self-powered colloidal Wurtzite-structure quantum dots photodetectors based on photoinduced-pyroelectric effect. *Adv. Opt. Mater.* 6: 1800639.
13 Yusoff, A.R.b.M., Kim, D., Kim, H.P. et al. (2015). A high efficiency solution processed polymer inverted triple-junction solar cell exhibiting a power conversion efficiency of 11.83%. *Energy Environ. Sci.* 8: 303.
14 Yoshikawa, K., Kawasaki, H., Yoshida, W. et al. (2017). Silicon heterojunction solar cell with interdigitated back contacts for a photoconversion efficiency over 26%. *Nat. Energy* 2: 17032.
15 Li, X., Bi, D.Q., Yi, C.Y. et al. (2016). A vacuum flash-assisted solution process for high-efficiency large-area perovskite solar cells. *Science* 353: 58–62.
16 Scharber, M.C., Mühlbacher, D., Koppe, M. et al. (2006). Design rules for donors in bulk heterojunction solar cells towards 10% energy-conversion efficiency. *Adv. Mater.* 18: 789–794.
17 Liu, M., Johnston, M.B., and Snaith, H.J. (2013). Efficient planar heterojunction perovskite solar cells by vapour deposition. *Nature* 501: 395–398.
18 Chakrabartty, J., Harnagea, C., Celikin, M. et al. (2018). Improved photovoltaic performance from inorganic perovskite oxide thin films with mixed crystal phase. *Nat. Photonics* 12: 271–276.
19 Kang, G., Bae, K., Nam, M. et al. (2015). Broadband and ultrahigh optical haze thin films with self-aggregated alumina nanowire bundles for photovoltaic applications. *Energy Environ. Sci.* 8: 2650–2656.
20 Liu, J., Chen, S.S., Qian, D.P. et al. (2016). Fast charge separation in a non-fullerene organic solar cell with a small driving force. *Nat. Energy* 1: 16089.
21 Yu, R.M., Pan, C.F., Chen, J. et al. (2013). Enhanced performance of a ZnO nanowire-based self-powered glucose sensor by piezotronic effect. *Adv. Funct. Mater.* 23: 5868–5874.
22 Zhang, Y., Zhao, X.Y., Chen, J.X. et al. (2019). Self-polarized $BaTiO_3$ for greatly enhanced performance of ZnO UV photodetector by regulating the distribution of electron concentration. *Adv. Funct. Mater.* 30: 1907650.
23 Koppens, F.H.L., Mueller, T., Avouris, P. et al. (2014). Photodetectors based on graphene, other two-dimensional materials and hybrid systems. *Nat. Nanotechnol.* 9: 780.
24 Liu, C.H., Chang, Y.C., Norris, T.B., and Zhong, Z. (2014). Graphene photodetectors with ultra-broadband and high responsivity at room temperature. *Nat. Nanotechnol.* 9: 273–278.
25 Lin, Q., Armin, A., Burn, P.L., and Meredith, P. (2015). Filterless narrowband visible photodetectors. *Nat. Photonics* 9: 687–694.
26 Nie, W.Y., Tsai, H., Asadpour, R. et al. (2015). High-efficiency solution-processed perovskite solar cells with millimeter-scale grains. *Science* 347: 522–525.
27 Ji, Y., Gao, T.T., Wang, Z.L., and Yang, Y. (2019). Configuration design of $BiFeO_3$ photovoltaic devices for self-powered electronic watch. *Nano Energy* 64: 103909.
28 Bullock, J., Hettick, M., Geissbühler, J. et al. (2016). Efficient silicon solar cells with dopant-free asymmetric heterocontacts. *Nat. Energy* 1: 15031.

29 Bullock, J., Wan, Y.M., Hettick, M. et al. (2019). Dopant-free partial rear contacts enabling 23% silicon solar cells. *Adv. Energy Mater.* 9: 1803367.

30 Tune, D.D., Mallik, N., Fornasier, H., and Flavel, B.S. (2019). Breakthrough carbon nanotube-silicon heterojunction solar cells. *Adv. Energy Mater.* 10: 1903261.

31 Nechache, R., Harnagea, C., Li, S. et al. (2015). Bandgap tuning of multiferroic oxide solar cells. *Nat. Photonics* 9: 61–67.

32 Liang, L.Y., Kang, X.L., Sang, Y.H., and Liu, H. (2016). One-dimensional ferroelectric nanostructures: synthesis, properties, and applications. *Adv. Sci.* 3: 1500358.

33 Guo, Z.X., Chen, Y., Zhang, H. et al. (2018). Independent band modulation in 2D van der Waals heterostructures via a novel device architecture. *Adv. Sci.* 5: 1800237.

34 Pan, L., Wang, S.B., Xie, J.W. et al. (2016). Constructing TiO_2 p–n homojunction for photoelectrochemical and photocatalytic hydrogen generation. *Nano Energy* 28: 296–303.

35 Duttan, M. and Fukata, N. (2015). Low-temperature UV ozone-treated high efficiency radial p–n junction solar cells: N-Si NW arrays embedded in a p-Si matrix. *Nano Energy* 11: 219–225.

36 Yang, Y., Guo, W.X., Pradel, K.C. et al. (2012). Pyroelectric nanogenerators for harvesting thermoelectric energy. *Nano Lett.* 12: 2833–2838.

37 Zhu, L.P., Lin, P., Chen, B.D. et al. (2018). Piezo-phototronic and pyro-phototronic effects to enhance Cu(In, Ga)Se_2 thin film solar cells. *Nano Res.* 11: 3877–3885.

38 Katiyar, A.K., Mukherjee, S., Zeeshan, M. et al. (2015). Enhancement of efficiency of a solar cell fabricated on black Si made by inductively coupled plasma-reactive ion etching process: a case study of a n-CdS/p-Si heterojunction cell. *ACS Appl. Mater. Interfaces* 7: 23445.

39 Xu, Z.X., Zhang, Y.L., and Wang, Z.N. (2019). ZnO based photodetector: from photon detector to pyro-phototronic effect enhanced detector. *J. Phys. D: Appl. Phys.* 52: 223001.

40 Paillard, C., Bai, X., Infante, I.C. et al. (2016). Photovoltaic with ferroelectrics: current status and beyond. *Adv. Mater.* 28: 5153.

41 Yuan, Y., Xiao, Z., Yang, B., and Huang, J. (2014). Arising applications of ferroelectric materials in photovoltaic devices. *J. Mater. Chem. A* 2: 6027.

42 Bowen, C.R., Taylor, J., LeBoulbar, E. et al. (2014). Pyroelectric materials and devices for energy harvesting applications. *Energy Environ. Sci.* 7: 3836–3856.

43 Abbas, S., Kumar, M., Kim, D.W., and Kim, J. (2019). Translucent photodetector with blended nanowires-metal oxide transparent selective electrode utilizing photovoltaic and pyro-phototronic coupling effect. *Small* 15: 1804346.

44 Zhao, K., Ouyang, B.S., and Yang, Y. (2018). Enhancing photocurrent of radially polarized ferroelectric $BaTiO_3$ materials by ferro-pyro-phototronic effect. *iScience* 3: 208–216.

45 Yin, B., Zhang, H.Q., Qiu, Y. et al. (2017). The light-induced pyro-phototronic effect improving a ZnO/NiO/Si heterojunction photodetector for selectively detecting ultraviolet or visible illumination. *Nanoscale* 9: 17199–17206.

46 Kumar, M., Patel, M., Lee, G.N., and Kim, J. (2018). Light-induced all transparent pyroelectric photodetector. *ACS Appl. Nano Mater.* 1: 319–324.
47 Rana, A.K., Kumar, M., Ban, D.K. et al. (2019). Enhancement in performance of transparent p-NiO/n-ZnO heterojunction ultrafast self-powered photodetector via pyro-phototronic effect. *Adv. Electron. Mater.* 5: 1900438.
48 You, D.T., Xu, C.X., Zhang, W. et al. (2019). Photovoltaic–pyroelectric effect coupled broadband photodetector in selfpowered ZnO/ZnTe core/shell nanorod arrays. *Nano Energy* 62: 310–318.
49 Qi, J., Ma, N., and Yang, Y. (2018). Photovoltaic–pyroelectric coupled effect nanogenerators for self-powered photodetector system. *Adv. Mater. Interfaces* 5: 1701189.
50 Ji, Y., Zhang, K.W., and Yang, Y. (2018). A one-structure-based multieffects coupled nanogenerator for simultaneously scavenging thermal, solar, and mechanical energies. *Adv. Sci.* 5: 1700622.
51 Zhang, J. and Wang, C.Y. (2016). Size-dependent pyroelectric properties of gallium nitride nanowires. *J. Appl. Phys.* 19: 145102.
52 Zhang, K.W., Wang, S.H., and Yang, Y. (2017). A one-structure-based piezo-tribo-pyro-photoelectric effects coupled nanogenerator for simultaneously scavenging mechanical, thermal, and solar energies. *Adv. Energy Mater.* 7: 1601852.
53 Lubomirsky, I. and Stafsudd, O. (2012). Practical guide for pyroelectric measurements. *Rev. Sci. Instrum.* 83: 051101.
54 Lang, S.B. (2005). Pyroelectricity: from ancient curiosity to modern imaging tool. *Phys. Today* 58: 31–36.
55 Ouyang, B.S., Zhang, K.W., and Yang, Y. (2018). Photocurrent polarity controlled by light wavelength in self-powered ZnO nanowires/SnS photodetector system. *iScience* 1: 16–23.
56 Ma, N., Zhang, K.W., and Yang, Y. (2017). Photovoltaic–pyroelectric coupled effect induced electricity for self-powered photodetector system. *Adv. Mater.* 29: 1703694.
57 Ji, Y., Zhang, K.W., Wang, Z.L., and Yang, Y. (2019). Piezo–pyro–photoelectric effects induced coupling enhancement of charge quantity in $BaTiO_3$ materials for simultaneously scavenging light and vibration energies. *Energy Environ. Sci.* 12: 1231–1240.
58 Alexe, M. and Hesse, D. (2011). Tip-enhanced photovoltaic effects in bismuth ferrite. *Nat. Commun.* 2: 256.
59 Yang, M.M., Bhatnagar, A., and Alexe, M. (2015). Electronic origin and tailoring of photovoltaic effect in $BiFeO_3$ single crystals. *Adv. Electron. Mater.* 1: 1500139.
60 Spanier, J.E., Fridkin, V.M., Rappe, A.M. et al. (2016). Power conversion efficiency exceeding the Shockley–Queisser limit in a ferroelectric insulator. *Nat. Photonics* 10: 611–616.
61 Wang, Z.N., Yu, R.M., Wang, X.F. et al. (2016). Ultrafast response p-Si/n-ZnO heterojunction ultraviolet detector based on pyro-phototronic effect. *Adv. Mater.* 28: 6880–6886.

62 Kumar, M., Patel, M., Nguyen, T.T. et al. (2018). High-performing ultra-fast transparent photodetector governed by the pyro–phototronic effect. *Nanoscale* 10: 6928.

63 Ban, D.K., Patel, M., Nguyen, T.T., and Kim, J. (2019). All-transparent oxide photovoltaics: AZO embedded ZnO/NiO/AgNW band selective high-speed electric power window. *Adv. Electron. Mater.* 5: 1900348.

64 Wang, Y., Zhu, L.P., Feng, Y.J. et al. (2019). Comprehensive pyro-phototronic effect enhanced ultraviolet detector with ZnO/Ag Schottky junction. *Adv. Funct. Mater.* 29: 1807111.

65 Ma, N. and Yang, Y. (2018). Boosted photocurrent in ferroelectric $BaTiO_3$ materials via two dimensional planar-structured contact configurations. *Nano Energy* 50: 417–424.

66 Ma, N. and Yang, Y. (2019). Boosted photocurrent via cooling ferroelectric $BaTiO_3$ materials for selfpowered 405 nm light detection. *Nano Energy* 60: 95–102.

67 Ji, Y., Wang, Y.H., and Yang, Y. (2019). Photovoltaic–pyroelectric–piezoelectric coupled effect induced electricity for self-powered coupled sensing. *Adv. Electron. Mater.* 5: 1900195.

68 Wang, X.F., Yu, R.M., Peng, W.B. et al. (2015). Temperature dependence of the piezotronic and piezophototronic effects in a-axis GaN nanobelts. *Adv. Mater.* 27: 8067–8074.

69 Dulub, O., Diebold, U., and Kresse, G. (2003). Novel stabilization mechanism on polar surfaces: ZnO (0001)-Zn. *Phys. Rev. Lett.* 90: 016102.

70 Ouyang, B.S., Zhang, K.W., and Yang, Y. (2017). Self-powered UV photodetector array based on P3HT/ZnO nanowire array heterojunction. *Adv. Mater. Technol.* 2: 1700208.

71 Dai, Y.J., Wang, X.F., Peng, W.B. et al. (2018). Self-powered Si/CdS flexible photodetector with broadband response from 325 to 1550 nm based on pyro–phototronic effect: an approach for photosensing below bandgap energy. *Adv. Mater.* 30: 1705893.

72 Kumar, M., Patel, M., Kim, J., and Lim, D. (2017). Enhanced broadband photoresponse of a self-powered photodetector based on vertically grown SnS layers via the pyro–phototronic effect. *Nanoscale* 9: 19201–19208.

73 Peng, W.B., Yu, R.M., Wang, X.F. et al. (2016). Temperature dependence of pyro–phototronic effect on self-powered ZnO/perovskite heterostructured photodetectors. *Nano Res.* 9: 3695–3704.

74 Wang, Z.N., Yu, R.M., Pan, C.F. et al. (2015). Light-induced pyroelectric effect as an effective approach for ultrafast ultraviolet nanosensing. *Nat. Commun.* 6: 8401.

75 Peng, W.B., Wang, X.F., Yu, R.M. et al. (2017). Enhanced performance of a self-powered organic/inorganic photodetector by pyro–phototronic and piezo–phototronic effects. *Adv. Mater.* 29: 1606698.

76 Sahatiya, P., Shinde, A., and Badhulika, S. (2018). Pyro–phototronic nanogenerator based on flexible 2D ZnO/graphene heterojunction and its application in self-powered near infrared photodetector and active analog frequency modulation. *Nanotechnology* 29: 325205.

77 Boruah, B.D., Majji, S.N., Nandi, S., and Misra, A. (2018). Doping controlled pyro–phototronic effect in self-powered zinc oxide photodetector for enhancement of photoresponse. *Nanoscale* 10: 3451–3459.
78 Zhang, K.W., Wang, Z.L., and Yang, Y. (2016). Enhanced P3HT/ZnO nanowire array solar cells by pyro–phototronic effect. *ACS Nano* 10: 10331–10338.
79 Teng, F., Hu, K., Ouyang, W.X., and Fang, X.S. (2018). Photoelectric detectors based on inorganic p-type semiconductor materials. *Adv. Mater.* 30: 1706262.

9

Multi-effects Coupled Nanogenerators

9.1 Introduction

The construction of nanogenerators to effectively scavenge green and renewable resources for powering electrics is an increasingly crucial issue due to the horrible energy crisis in modern society [1–5]. As promising alternatives for conventional energies (fossil fuels), thermal, light, and mechanical energies have been found abundantly in the surrounding environment [6–9]. Correspondingly, a variety of nanogenerators that can individually scavenge thermal, light, biochemical, and mechanical energies have been successfully developed [10–16]. As is known, spontaneous polarization in certain materials with noncentral symmetry structure can be modulated by changing the temperature of the materials, leading to a pyropotential and a pyroelectric current, which is called the pyroelectric effect. Based on the pyroelectric effect, the pyroelectric nanogenerator (PyENG/PNG) has been fabricated and utilized to harvest thermal fluctuations to generate electricity [17–19]. Another effective approach to scavenge thermal energy is the thermoelectric nanogenerator (ThENG), which can employ temperature gradients to produce electric power depending on the thermoelectric effect [20, 21]. Strain-induced variation of spontaneous polarization in noncentral-symmetry-structured materials can generate piezopotential and piezoelectric current, which is named as the piezoelectric effect. The piezoelectric nanogenerator (PiENG/PENG) has the ability to realize transformation of mechanical energy to electrical power by utilizing the piezoelectric effect [21–23]. A triboelectric nanogenerator (TENG) is constructed by coupling triboelectric effect and electrostatic induction in one device, which is able to effectively achieve the conversion of mechanical energy to electrical power [24–27]. A photovoltaic device (PVC) plays a significant role in converting light energy into electricity by using the photovoltaic effect [28–30]. In the past few years, the nanogenerators mentioned above have been intensively investigated; however, the electric power generation abilities of these nanogenerators are limited because of their selectivity for energy. Fortunately, some materials possess multiple effects, making it possible for the materials to simultaneously scavenge various energies [31–34]. Taking ferroelectric barium titanate (BTO) materials

*Corresponding author: yayang@binn.cas.cn

Hybridized and Coupled Nanogenerators: Design, Performance, and Applications,
First Edition. Ya Yang.

as an example, piezoelectricity, pyroelectricity, and photoelectricity coexist in the materials [35, 36]. Based on multifunctional materials, multi-effects nanogenerators, including tribo-piezoelectric effect nanogenerator (TPiENG), pyro-piezoelectric effect nanogenerator, thermo-photoelectric effect nanogenerator, piezo-photoelectric effect nanogenerator, pyro-piezo-photoelectric effect nanogenerator, and tribo-piezo-pyro-photoelectric effect nanogenerator have been constructed and have demonstrated an efficient approach for synchronously scavenging multiple kinds of energies from the ambient environment [37–43]. Virtual factors to promote the electric power of multi-effects coupled nanogenerator are the selection of suitable multifunctional materials and construction of novel device configurations.

In this chapter, the basic working principle of multi-effects coupled nanogenerators will be first reviewed. Then, suitable multifunctional materials, recent advancements, and representative prototypes of multi-effects coupled nanogenerators will be introduced. Additionally, application areas of multi-effects coupled nanogenerators will be presented as well.

9.2 Materials

The multi-effects coupled nanogenerator is an excellent approach to simultaneously scavenge different types of energies. To construct highly integrated multi-effects coupled nanogenerators, various multifunctional materials have been intensively investigated, including multifunctional semiconductors, inorganic ferroelectrics, and polymeric ferroelectrics.

9.2.1 Semiconductors

Semiconductors play a vital role in the field of multi-effects coupled nanogenerators. Plenty of semiconductors, including ZnO, SnS, and SnSe, have the ability to harvest several kinds of energies. ZnO material usually serves as the absorption layer for ultraviolet light in optoelectric devices on account of its wide bandgap as well as excellent chemical stability. Furthermore, spontaneous polarization exists in ZnO material owing to its noncentral symmetry structure. Consequently, when thermal and mechanical energies are applied on the ZnO materials, pyropotential and piezopotential can be produced because of pyroelectric and piezoelectric effects, respectively. Moreover, thermoelectric effect is also observed in ZnO [39]. Much research has been focused on preparing high-quality bulk ZnO (such as ZnO ceramics) and ZnO thin films for multi-effects coupled nanogenerators. As illustrated in Figure 9.1a, Gao et al. [43] prepared porous ZnO ceramics for thermo-photoelectric effect nanogenerators. The ZnO ceramics were fabricated by dry-pressing a mixture of ZnO tetrapods and polystyrene microspheres followed by an annealing procedure at 1400 °C. The polystyrene microspheres that were utilized as pore-forming agent could be eliminated during the sintering procedure; consequently, pores with the size of 8–32 μm were uniformly distributed in the ZnO ceramics.

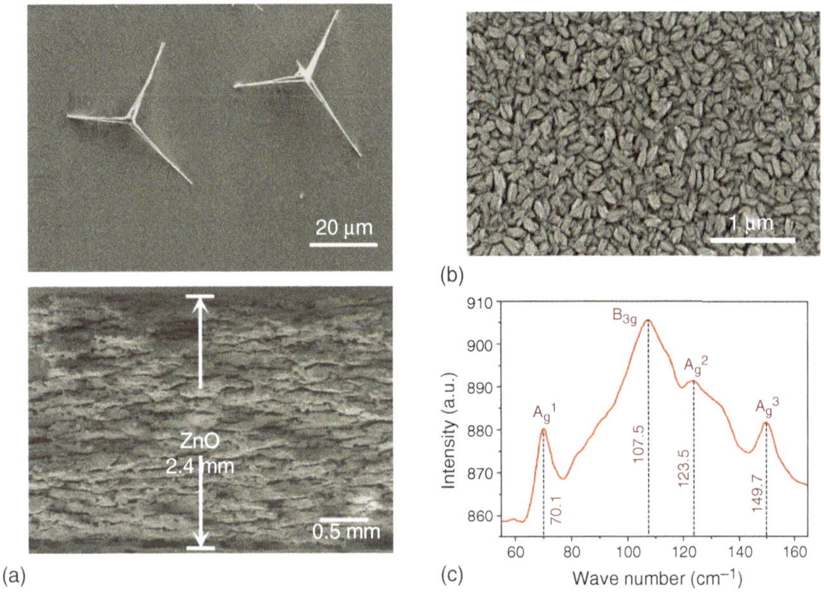

Figure 9.1 Semiconductors for multi-effects coupled nanogenerator. (a) SEM images of ZnO tetrapods and ceramics. Source: Reproduced with permission from Gao et al. [43]. Copyright 2019, John Wiley and Sons. (b) Shuttle-like SnS nanomaterial deposited on ZnO nanowires. Source: Reproduced with permission from Ouyang et al. [44]. Copyright 2018, Elsevier. (c) Raman spectra of bulk SnSe:Br material. Source: Reproduced with permission from Ouyang et al. [45]. Copyright 2019, Elsevier.

To improve thermoelectricity, Al and Sb are doped into the ZnO materials [46, 47]. Hydrothermal method is an outstanding approach to fabricate ZnO nanowire films. Han et al. [48] successfully prepared uniform ZnO nanowires on an ITO substrate for piezo-photoelectric effect nanogenerators by employing $ZnNO_3{\cdot}6H_2O$ and HMTA as raw materials via the hydrothermal approach. Besides ZnO nanowires, GaN and wurtzite-structured CdS are suitable for preparing piezo-photoelectric effect nanogenerators as well [49, 50]. In addition, SnS materials have exhibited great potential for constructing multi-effects coupled nanogenerator because of its thermoelectricity, pyroelectricity, and photoelectricity. As direct bandgap materials, the optical bandgap of SnS film can be adjusted in the range of 1.0–1.5 eV. Ouyang et al. [44] prepared a SnS film for a thermo-photoelectric effect nanogenerator through magnetron sputtering method, and a highly crystalline dense SnS film was obtained, as displayed in Figure 9.1b. Kumar et al. [51] prepared a compact orthorhombic p-SnS film on Si film via the magnetron sputtering approach and the synthesized SnS displayed stacked-layer structure. In recent years, SnSe:Br material has attracted researchers' attention because of its excellent thermoelectricity. Ouyang et al. [45] synthesized [100]-oriented bulk SnSe:Br for the thermo-photoelectric coupled nanogenerator. The prepared SnSe:Br was single crystal, as confirmed by Raman spectra in Figure 9.1c.

9.2.2 Inorganic Ferroelectrics

Ferroelectrics have drawn immense attention because of the coexistence of pyroelectricity, piezoelectricity, and photoelectricity. Inorganic ferroelectrics, especially lead zirconate titanate (PZT), BTO, and bismuth ferrite (BFO), have attracted much interest because of their large piezoelectric coefficient, anomalous photovoltaic effect, and excellent pyroelectricity. Among these ferroelectrics, PZT materials have higher Curie temperature and lower sintering temperature, and can be easily polarized, drawing considerable attention. Ko et al. [52] fabricated a flexible PZT thin film with Zr/Ti of 52/48 on a Ni–Cr substrate for a pyro-piezoelectric effect nanogenerator by spin-coating method. The synthesized perovskite-structured single phase PZT film has a thickness of about 2 μm. The piezoelectric coefficient of the PZT layer at ambient condition was 140 pC/N, which showed a declining trend as the temperature increased. The measured pyroelectric coefficient of the PZT film at room temperature was 50 nC/(cm^2 K), which exhibited a rising trend as the temperature increased. As eco-friendly ferroelectrics, multifunctional BTO materials are more suitable for people's daily life. Ferroelectric BTO material exhibits a tetragonal phase at room temperature. Generally, as a wide bandgap material (3.3 eV), BTO material can absorb light with the wavelength of $\leq$376 nm. However, the wavelength of the absorbed light can be extended to 405 nm because of the oxygen vacancy in BTO. BTO ceramics have been intensively studied for preparing various multi-effects coupled nanogenerators because of their simple preparation method and easily controllable shapes. Ji et al. [53] prepared rectangular BTO ceramics for the piezo-pyro-photoelectric effect nanogenerator. The BTO ceramics were fabricated according to a classic sintering process by employing cubic BTO powders as raw materials. The fabricated BTO ceramics exhibited a tetragonal phase and had a compacted microstructure, as illustrated in Figure 9.2a. Song et al. [55] prepared a cylindrical BTO ceramic disk for the pyro-piezoelectric effect nanogenerator. Through the dry-pressing and sintering process, tetragonal BTO ceramics consisting of closely arranged grains were formed by using BTO nanopowders as the starting material. The prepared BTO ceramics with the thickness of 850 μm had a large piezoelectric coefficient (350 pC/N). To improve the light-harvesting ability of certain multi-effects coupled nanogenerators, BFO materials have been widely investigated. As narrow bandgap ferroelectrics with high remnant polarization, the BFO material can efficiently scavenge both light energy and thermal energy. Qi et al. [54] prepared rhombohedral phase BTO ceramic disks with relatively dense microstructure for a thermo-photoelectric effect nanogenerator, where BFO powders synthesized by hydrothermal method were employed as the starting materials (Figure 9.2b). To improve the flexibility and scavenging ability of multi-effects coupled nanogenerators for mechanical energy, inorganic ferroelectrics/PDMS composite configuration was developed. Shi et al. [56] prepared a BTO/PDMS film for a tribo-piezoelectric effect nanogenerator. The BTO/PDMS sample was prepared through solution-casting approach with a mixture of BTO nanopowders and PDMS as the precursor.

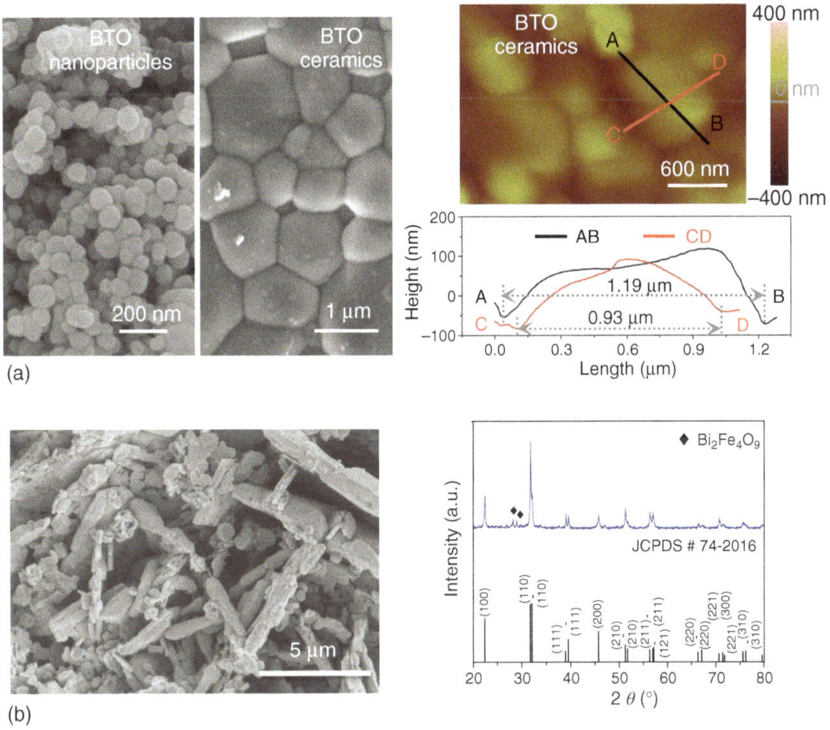

Figure 9.2 Inorganic ferroelectrics for multi-effects coupled nanogenerators. (a) SEM images and AFM topography image of BTO powders and ceramics. Source: Reproduced with permission from Ji et al. [53]. Copyright 2019, John Wiley and Sons. (b) SEM image and XRD patterns of BFO powders. Source: Reproduced with permission from Qi et al. [54]. Copyright 2018, American Chemical Society.

9.2.3 Polymeric Ferroelectrics

With rapid advances in portable electronic devices, nanogenerators with high flexibility have been intensively investigated. Highly flexible multi-effects coupled nanogenerators can be constructed by utilizing polymeric ferroelectrics, including poly(vinylidene fluoride) (PVDF) as well as its copolymer poly(vinylidenefluoride-*co*-trifluoroethylene) (P(VDF-TrFE)). PVDF together with P(VDF-TrFE) has excellent biocompatibility as well as high chemical stability. According to chain conformation, four types of crystalline phases PVDF can be achieved, which are α-phase, β-phase, σ-phase, and δ-phase PVDF. Among these four kinds of PVDF, only β-phase PVDF and σ-phase PVDF exhibit ferroelectricity, and β-phase PVDF has more excellent piezoelectricity and pyroelectricity [57]. Various approaches have been reported to fabricate β-phase PVDF, such as polarizing PVDF by a high electric field or doping metal particles into PVDF. Guo et al. [34] prepared β-phase PVDF nanofibers for the tribo-piezoelectric coupled nanogenerator by electrospinning method. The precursor solution was fabricated by evenly mixing PVDF, *N*,*N*-dimethylformamide (DMF), as well as acetone together. The average

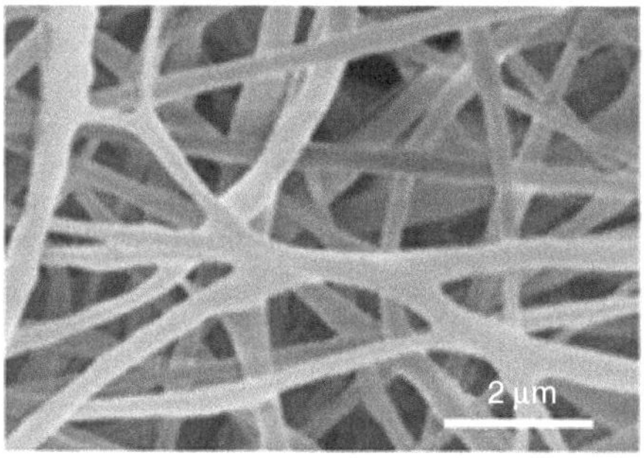

Figure 9.3 SEM image of PVDF nanofibers. Source: Reproduced with permission from Wang et al. [59]. Copyright 2016, John Wiley and Sons.

diameter of the obtained β-phase PVDF nanofibers was about 160 nm. Sultana et al. [58] synthesized $CH_3NH_3PbI_3$ incorporated β-phase PVDF material for the pyro-piezoelectric coupled nanogenerator. The $CH_3NH_3PbI_3$/PVDF composite fiber was prepared by the electrospinning approach. The precursor solution contained PVDF particles, $CH_3NH_3PbI_3$ particles, and DMF-acetone solvent. The prepared $CH_3NH_3PbI_3$/PVDF fiber had an average diameter of about 145 nm. By introducing $CH_3NH_3PbI_3$ in the precursor solution, more β-phase could form in PVDF, improving both the piezoelectricity and pyroelectricity of the PVDF film. Moreover, $CH_3NH_3PbI_3$ could effectively absorb the thermal energy induced by near-infrared light, enhancing the ability of the PVDF film to harvest thermal energy. Wang et al. [59] prepared PVDF nanofibers for the tribo-piezo-pyroelectric effect nanogenerator by the electrospinning approach. Figure 9.3 illustrates the microstructure of the obtained PVDF nanofibers, suggesting that these nanofibers were around 200–500 nm in diameter. Lee et al. [60] synthesized β-phase P(VDF-TrFE) films for pyro-piezoelectric effect nanogenerators by using spin-coating technique with precursor solution containing 20 wt% P(VDF-TrFE) and 80 wt% DMF. The P(VDF-TrFE) layers were first coated on $Ni/SiO_2/Si$ disks, and subsequently peeled off to form a flexible and uniform film with a thickness of about 7 μm.

9.3 Device Design and Working Principle

9.3.1 Electrode/Bulk Semiconductor/Electrode Structure

Multi-effects coupled nanogenerators prepared by employing bulk semiconductor materials have drawn much interest on account of their simple fabrication process and stable performances. Gao et al. [43] designed a thermo-photoelectric effect nanogenerator by using porous ZnO as the multifunctional material. Figure 9.4a presents the schematic illustration of the device under illumination, showing that the device was composed of an ITO top layer, a porous ZnO ceramic disk with the thickness of 2.4 mm, and a Ag bottom layer. The photograph indicates that the active area of this nanogenerator was ~0.35 cm^2. To obtain electrodes that were in good contact with the polyporous ZnO disk, ITO together with the Ag electrode was deposited by magnetron sputtering method. Electric signals of this nanogenerator were produced on

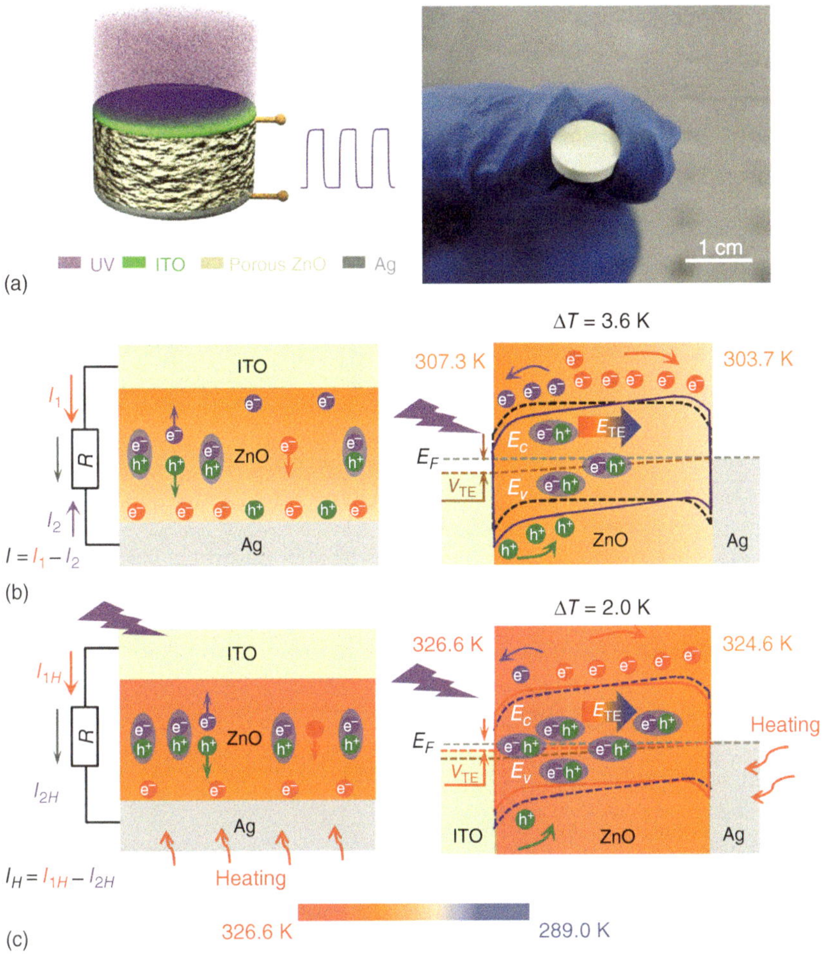

Figure 9.4 Multi-effects coupled nanogenerator based on ITO/porous ZnO/Ag structure. (a) Schematic diagram and photograph of the device. (b,c) Working mechanism of the device under light (b) and light + heating (c) conditions. Source: Reproduced with permission from Gao et al. [43]. Copyright 2019, John Wiley and Sons.

account of the synergistic effect of thermoelectric and photoelectric effects. Figure 9.4b,c illustrated the relevant working principle. A traversed schematic illustration and the corresponding energy band structure of this device upon irradiation (365 nm) at room temperature are presented in Figure 9.4b. On one hand, a stable temperature gradient ΔT formed across the device because of light-induced thermal effect, resulting in thermopotential V_{TE} across the ZnO ceramic. Such thermopotential V_{TE} could lift the conduction band together with the valence band of ZnO adjacent to the Ag electrode and lower that of ZnO near the ITO electrode. Correspondingly, the thermoelectric current I_1 flowing from ITO to the Ag electrode was observed under short-circuit condition. On the other hand, photoexcited electron–hole pairs (excitons) were able to be divided via V_{TE} to generate free electrons and holes. Subsequently, electrons moved to the ITO film; meanwhile, holes were transported to the Ag film under the force of V_{TE}, resulting in photocurrent I_2 moving to the ITO film. Therefore, the total current I of this nanogenerator was evaluated by $I = I_1 - I_2$. Notably, I and I_1 had the same polarity owing to the dominant role of the thermoelectric effect. When heating process was applied on the Ag side, the temperature gradient ΔT reduced, and correspondingly the thermopotential declined, as illustrated in Figure 9.4c. Consequently, the energy band tilt resulting from irradiation weakened, and less photoexcited excitons were divided; meanwhile, free carriers easily recombined. Hence, the thermoelectric current slightly decreased to I_{1H}, while the photoelectric current drastically reduced to I_{2H}. The decrease in the thermoelectric current was much lower than the reduction of the photocurrent, resulting in the total current increasing from I to I_H. Such enhancement of output current was ascribed to the coupling effect of thermoelectric and photoelectric effects.

Figure 9.5a presents a thermo-photoelectric effect nanogenerator, which was prepared by employing bulk n-SnSe:Br as the active layer, indicating that the coupled nanogenerator was composed of ITO top electrode, SnSe:Br single crystal as the active material, and Ag bottom electrode [45]. Asymmetric built-in electric fields formed at the ITO/SnSe:Br and SnSe:Br/Ag interfaces. Figure 9.5b-d displays the cross-sectional schematic illustration as well as working mechanism of this coupled nanogenerator. When the device was illuminated, electrons could transfer from the valence to the conduction band of this Br-doped SnSe, forming excitons. Then, the excitons were separated by the asymmetric electrostatic potential adjacent to the ITO electrode to produce free electrons and holes. Subsequently, the generated free electrons drifted to the Ag film; meanwhile, free holes moved to the ITO film. As a consequence, photocurrent I_{1L} occurred and flowed toward the Ag film (Figure 9.5b). Upon cooling, a temperature gradient was formed across SnSe:Br, resulting in electrons diffusing from the hotter ITO side toward the colder Ag film. Consequently, positive ions were left adjacent to the ITO film. As a result, a thermopotential was established and an energy band tilt occurred across the SnSe:Br, as illustrated in Figure 9.5c. Correspondingly, the thermoelectric current I_{2C} that flowed to the Ag film was detected. As shown in Figure 9.5d, when illumination and cooling processes were applied on the device simultaneously, both photocurrent I_{1L} and thermoelectric current I_{2C} were generated. Notably, the photocurrent I_{1L} and the thermoelectric current I_{2C} had the same direction. Consequently, the current produced by this nanogenerator increased on account of the integration of photocurrent I_{1L} and thermoelectric current I_{2C}. In addition, the energy band tilt could effectively improve the electron–hole separation and free carrier transport; meanwhile, reduced recombination occurred between the electrons and holes. Hence, the photocurrent I_{1L} under illumination–cooling condition was significantly increased. As a consequence, an interesting phenomenon in which the output current under illumination–cooling was much higher than the sum of the photocurrent I_{1L} and the thermoelectric current I_{2C} was observed.

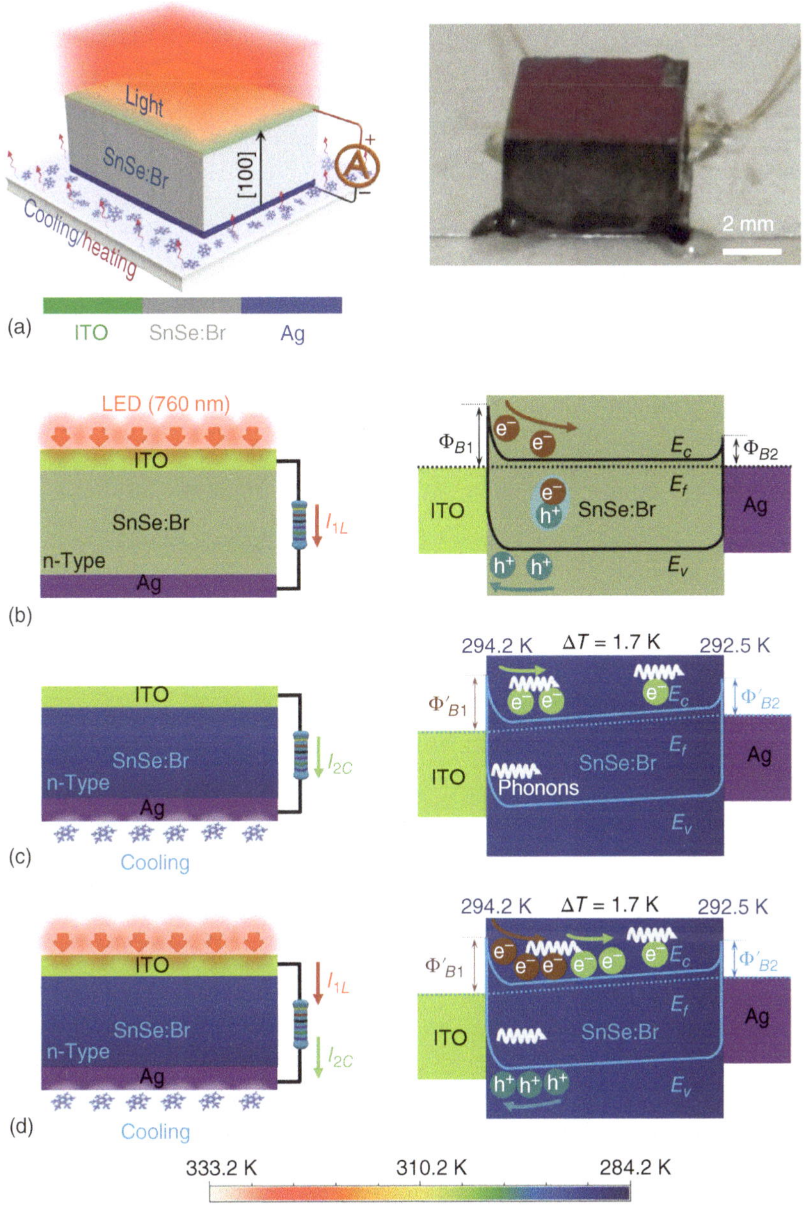

Figure 9.5 Multi-effects coupled nanogenerator based on ITO/bulk SnSe:Br/Ag structure. (a) Schematic diagram and photograph of the device. (b–d) Working mechanism of the device under illumination (b), cooling (c) and illumination + cooling (d) conditions. Source: Reproduced with permission from Ouyang et al. [45]. Copyright 2019, Elsevier.

9.3.2 Electrode/Heterojunction/Electrode Structure

An abundance of light energy exists in our surrounding environment. A nanogenerator with p–n heterojunction structure is an efficacious way to collect light energy. To simultaneously scavenge light energy and some other kinds of energies, a variety of multi-effects coupled nanogenerators based on the p–n heterojunction have been constructed. Zhang and Yang [61] designed a thermo-photoelectric effect based nanogenerator by using an InP/ZnO heterojunction, as shown in Figure 9.6a. The device was composed of the following parts: Ag nanowire layer as one electrode, Zn doped p-InP wafer with (100) orientation as the main active layer, ZnO nanowires that were covered on the InP wafer, and a compact Ag film as another electrode. The photoactive area of this nanogenerator was approximately 0.5 cm^2. Figure 9.6b illustrates transverse section SEM images of this coupled nanogenerator, suggesting that the InP wafer was approximately 350 μm in thickness and the ZnO nanowires were in tight contact with the InP wafer. These ZnO nanowires were approximately 60 nm in diameter and 700 nm in length. Staggered Ag nanowires were uniformly

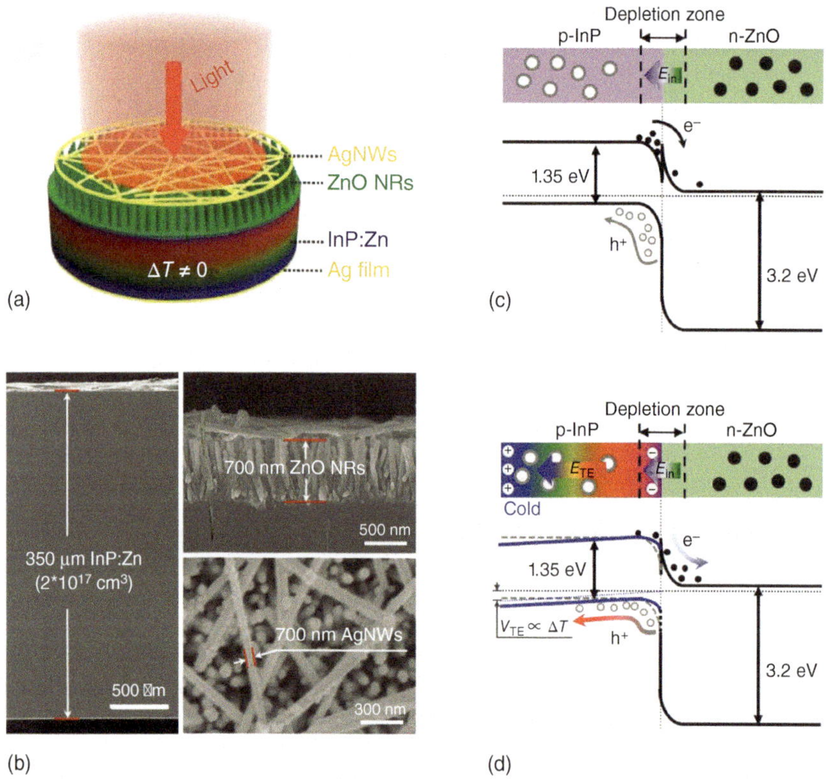

Figure 9.6 Thermo-photoelectric effect nanogenerator based on InP/ZnO heterojunction. (a,b) Schematic diagram (a) and SEM images (b) of the device. (c,d) Working mechanism of the device at room temperature (c) and low temperature. Source: Reproduced with permission from Zhang and Yang [61]. Copyright 2017, John Wiley and Sons.

deposited on this ZnO nanowire array. Figure 9.6c presents an energy-level illustration of this InP/ZnO junction under illumination. Upon illumination, excitons were produced and were divided by the built-in electric field in the space charge region to generate free carriers. Subsequently, electrons and holes drifted to the bottom and top electrodes, respectively, resulting in a photocurrent moving toward the Ag nanowire layer. However, owing to conduction band offset and bending, some electrons were trapped in the InP near the InP/ZnO interface, which could not contribute to the photocurrent. When cooling this nanogenerator, a temperature gradient across the InP wafer formed, which significantly accelerated the holes in InP diffusing toward the bottom electrode and a thermopotential occurred. As a result, the photocurrent was improved. In addition, the negative thermoelectric charges in InP near the InP/ZnO interface lifted up the energy bands of this InP wafer (Figure 9.6d). As a result, those trapped electrons were released to ZnO and the photocurrent was increased. The coupling effect of thermoelectric and photovoltaic effects could not only improve carrier transport but also suppress carrier recombination.

A piezo-photoelectric effect nanogenerator based on perovskite/ZnO heterojunction was prepared by Hu et al. [62]. The coupled device mainly consisted of a ZnO nanowire, a $CH_3NH_3PbI_3$ thin film, a thin layer of spiro-MeOTAD, and Ag electrodes. One end of this ZnO nanowire was covered with the $CH_3NH_3PbI_3$ layer. The spiro-MeOTAD layer was coated on the $CH_3NH_3PbI_3$ thin film. Silver paste coated on this spiro-MeOTAD layer as well as the other end of this ZnO nanowire worked as Ag electrodes. Without strain, photogenerated excitons were produced in the $CH_3NH_3PbI_3$ thin film upon solar illumination simulator and separated under the force of asymmetric electrostatic potential in the space charge zone. Subsequently, the separated electrons as well as holes moved toward ZnO and the spiro-MeOTAD thin film, respectively. However, it was quite difficult for the free carriers to get across the high barrier at the $CH_3NH_3PbI_3$/ZnO interface. Consequently, low output electric signals were produced. When compressive strain was employed on the coupled device, positive piezoelectric charges were produced in ZnO near the $CH_3NH_3PbI_3$/ZnO interface because of piezoelectric effect, declining the energy bands of this ZnO nanowire, consequently decreasing the barrier height at the $CH_3NH_3PbI_3$/ZnO interface. As a result, free carriers could easily get across this barrier to the Ag electrodes. Finally, the output electricity of this coupled device was promoted on account of piezo-photoelectric effect.

9.3.3 Electrode/Ferroelectrics/Electrode Structure

Figure 9.7a displays a thermo-photoelectric effect nanogenerator with ITO/BFO/Ag structure [54]. The coupled nanogenerator consisted of an ITO top electrode, a BFO ceramic disk as the thermoelectric and photoelectric material, and an Ag film as the bottom electrode. The effective area of the device was about 0.7 cm^2. The SEM images shown in Figure 9.7b,c indicate that the thickness of the BTO disk was about 1.15 mm. Song et al. [55] designed a pyro-piezoelectric effect nanogenerator by utilizing ferroelectric BTO materials. The device consisted of three parts: a circular BTO ceramic disk as the pyroelectric and piezoelectric

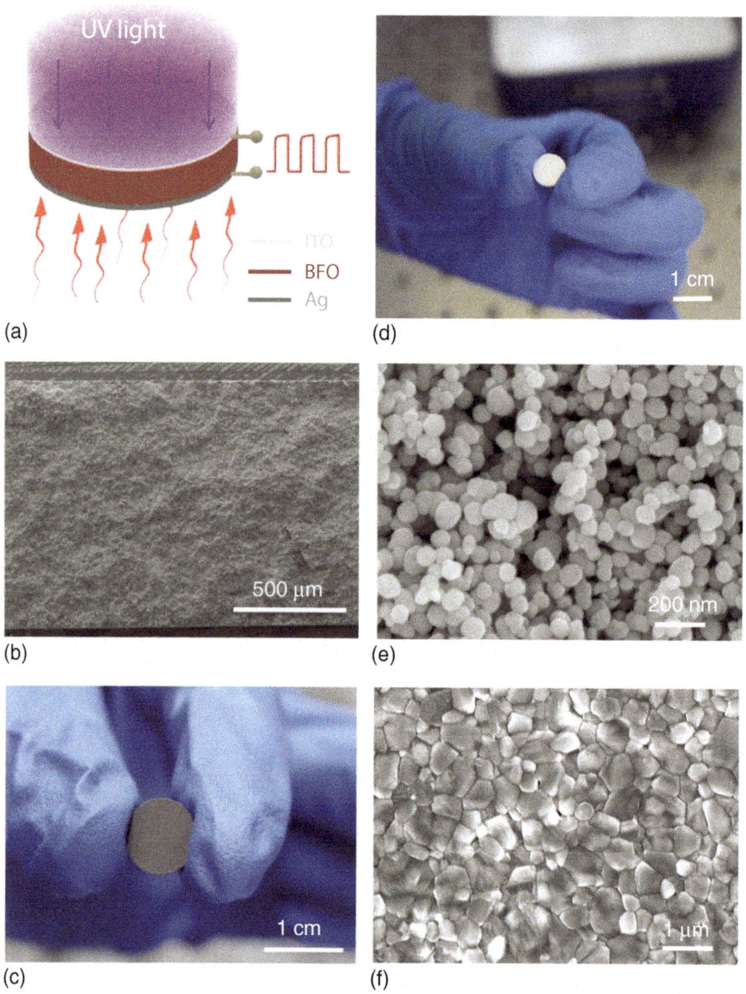

Figure 9.7 Multi-effects coupled nanogenerator based on electrode/ferroelectrics/electrode structure. (a–c) Schematic diagram (a) and SEM images (b,c) of ITO/BFO ceramics/Ag thermo-photoelectric effect nanogenerator. Source: Reproduced with permission from Qi et al. [54]. Copyright 2018, American Chemical Society. (d) Photograph of Ag/BTO ceramics/Ag pyro-piezoelectric effect nanogenerator. (e,f) SEM images of BTO powders and ceramics for the Ag/BTO ceramics/Ag device. Source: Reproduced with permission from Song et al. [55]. Copyright 2019, John Wiley and Sons.

material and Ag electrodes on both sides of the BTO disk. Figure 9.7d presents a photographic image of this device, showing that this BTO disk was approximately 10 mm in diameter. The BTO disk fabricated with BTO powders (diameter of 60 nm) had a dense microstructure (Figure 9.7e). The working principle of this coupled nanogenerator was based on the coupling effect between pyroelectric and piezoelectric effects of the ferroelectric BTO ceramic disk. In the absence of temperature and pressure variation, electric dipoles oscillated randomly around their respective alignment axes and achieved equilibrium; meanwhile, the mean spontaneous polarization in BTO was invariant. Consequently, electric signals produced by this device were zero. When pressing the device, the BTO ceramic disk was compressed, the strength of average spontaneous polarization was decreased, and a piezoelectric current moving from the top to the bottom Ag electrode was generated. When this pressure was released, a reverse piezoelectric current could be observed. If heating process was employed on the device, the random oscillations of electric dipoles in BTO were enhanced; consequently, the strength of the total spontaneous polarization decreased. Consequently, a pyroelectric current moving from the top to the bottom Ag electrode was produced. When this heating process was terminated, opposite pyroelectric signals were generated. When the heating and pressing processes were applied on the device simultaneously, the pyroelectric and piezoelectric signals were easily integrated because of the same signal polarity produced in the heating and pressing condition, resulting in significant increase of the output electric signals. Hence, employing such unique configuration and combining the pyroelectric and piezoelectric properties of BTO, the performance of the device for scavenging thermal and mechanical energies could be significantly improved. Furthermore, Lee et al. [63] designed a flexible pyro-piezoelectric effect nanogenerator that relied on a β-phase P(VDF-TrFE) film. The coupled nanogenerator mainly consisted of the following parts: a ferroelectric P(VDF-TrFE) with a thickness of about 7 μm as the multifunctional active material, a multilayer graphene top electrode, and a PDMA-CNT composite bottom electrode.

To further extend the energy scavenging ability of multi-effects coupled nanogenerators, Ji et al. [41] designed a piezo-pyro-photoelectric effect nanogenerator by using a BTO ceramic disk. As shown in Figure 9.8a, the nanogenerator was mainly composed of three layers, which were the multifunctional BTO ceramic disk, ITO, and Ag films. This BTO ceramics had the size of 38 mm × 8.3 mm × 0.3 mm. The Ag film was coated with a flexible fluorinated ethylene propylene (FEP) layer. A 0.3 g mass was fastened on the free end of the device. Photographs of this coupled nanogenerator under dark condition as well as upon irradiation indicate that the photoactive area of the device was about 2.4 cm^2. The working mechanism of the coupled nanogenerator relied on conjunction of the piezoelectricity, pyroelectricity, and photoelectricity of BTO materials. When heating/cooling process was applied on the device, the strength of spontaneous polarization of BTO changed, resulting in a pyropotential and pyroelectric current. When the vibration process acted on the device, the BTO deformed, leading to piezoelectric outputs. Upon illumination, electron–hole pairs were generated and separated to form free carriers. Then, the free electrons (holes) transported toward the ITO (Ag) electrode, leading to photoexcited outputs. Because of the light-induced thermal effect, the temperature of the device could be increased, which could reduce the strength of spontaneous polarization; consequently, the photovoltaic outputs of the device decreased. Fortunately, air flow caused by vibration could effectively decline the temperature of the device under illumination, leading to promotion of photovoltaic outputs. Therefore, the transferred charges of the device were enhanced. In addition, the negative part of the piezoelectric signals could be reduced under illumination, further increasing the transferred charges of the device.

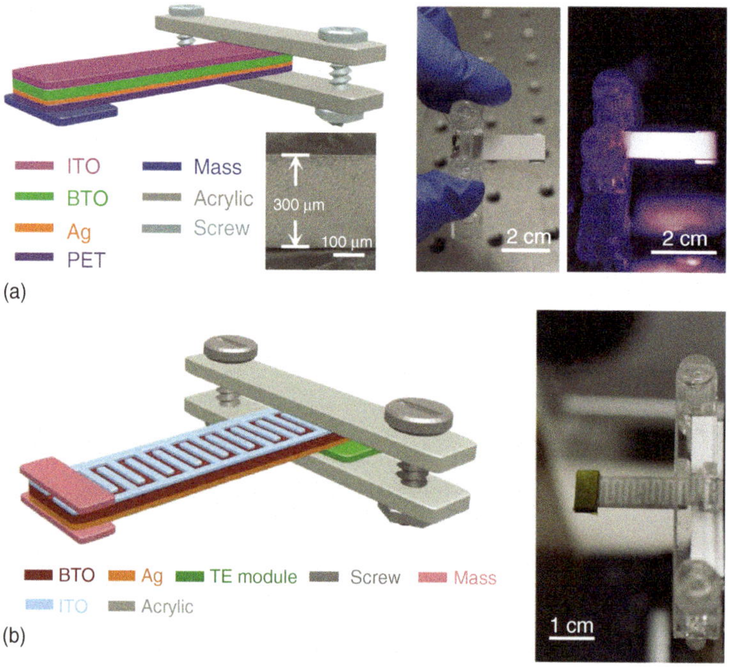

Figure 9.8 Multi-effects coupled nanogenerator based on ITO/BTO/Ag cantilever structure. (a) Schematic diagram, cross-sectional SEM, and photographic images of ITO/BTO/Ag nanogenerator. Source: Reproduced with permission from Ji et al. [41]. Copyright 2019, Royal Society of Chemistry. (b) Schematic diagram and photograph of ITO/BTO/Ag device with interdigitated ITO electrode. Source: Reproduced with permission from Ji et al. [53]. Copyright 2019, John Wiley and Sons.

In addition, a piezo-pyro-photoelectric effect nanogenerator with interdigitated electrode was fabricated [53]. Such a device could work in piezoelectric d_{33} mode and generate larger piezoelectric signals. Figure 9.8b illustrates the configuration of this coupled nanogenerator, indicating that this device mainly consisted of a rectangular BTO ceramic tablet, interdigitated ITO films as the output electrodes, and a Ag film at the bottom. One end of this coupled nanogenerator was fastened, while the other end was free. Masses were affixed onto the free end to improve the mechanical energy harvesting ability of this nanogenerator. The width and gap distance of the ITO electrodes were ~600 μm. This coupled nanogenerator was polarized with a DC electric field employed on interdigitated ITO electrodes. Since the bottom Ag film could effectively shield the DC electric field, the device could be sufficiently polarized. A photograph of the device indicates that the irradiation area of the device was around 2.04 cm^2.

9.3.4 Other Structures

Although harvesting several types of energy can be realized in multi-effects coupled nanogenerators based on a multifuctional material, it is necessary to

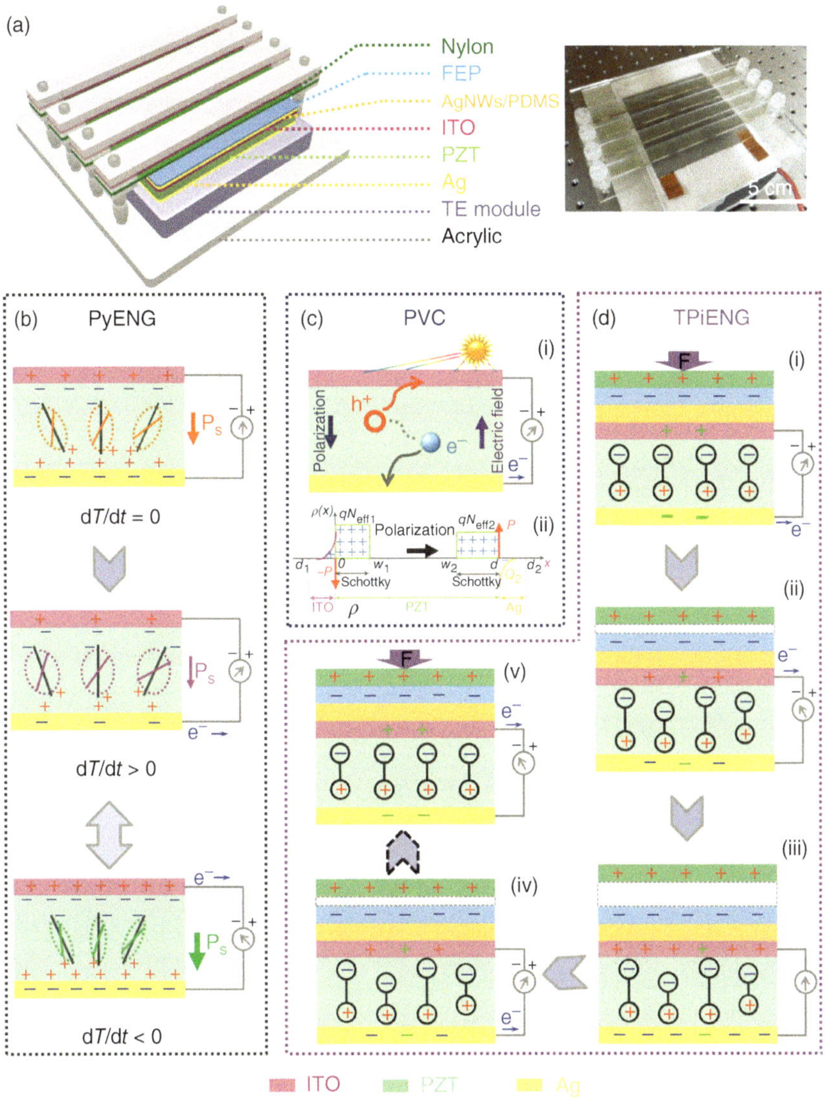

Figure 9.9 Multi-effects coupled nanogenerator based on PZT/friction material structure. (a) Schematic diagram and photograph of the device. (b) Working principle of the device. Source: Reproduced with permission from Zhang et al. [42]. Copyright 2017, John Wiley and Sons.

combine different kinds of materials to further enhance the ability of the coupled nanogenerator for scavenging multiple types of energy.

A piezo-tribo-pyro-photoelectric effect nanogenerator based on PZT/friction material structure was successfully constructed [42]. As demonstrated in Figure 9.9a, the device mainly consisted of three components: PZT ceramics, FEP and polyamide (nylon) films serving as friction materials, an electrode combined ITO film with Ag NWs/PDMS and dense Ag electrode. This PZT ceramic disk worked as the pyroelectric, photoelectric, piezoelectric material. The FEP film was pasted on the top electrode. The nylon layer was 1.3 mm above this FEP layer. The PZT worked as a PyENG and a PVC. The friction materials combined with the PZT acted as a TPiENG. Figure 9.9b–d illustrates the working mechanism of this device. Electric power production could be derived from three different physical mechanisms, including the tribo-piezoelectric effect, pyroelectric effect, and photoelectric effect. Figure 9.9b illustrates the power production process from PyENG. Without thermal fluctuation, electric dipoles oscillated randomly to a certain degree, and spontaneous polarization of PZT was constant. As a consequence, the output electric signals were zero. Upon heating, the electric dipoles oscillated violently, reducing the mean spontaneous polarization. Therefore, the flow of pyroelectric current to the ITO electrode occurred. On cooling this PZT, the strength of the spontaneous polarization greatly increased, leading to a negative pyroelectric current. As ferroelectric material, PZT has the ability to convert solar energy to electricity owing to bulk photovoltaic effect, interfacial effect, depolarization electric field, and domain wall theory. Figure 9.9c depicts the power generation mechanism of the PVC. Upon illumination, photons with proper energy could be absorbed by the PZT and excited electrons to the conduction band of this PZT, resulting in excitons. Subsequently, these excitons were divided under the force of Schottky junction and depolarization electric field to generate free electrons and holes. The free carriers were driven toward the Ag and ITO electrodes. Consequently, a photoelectric current was generated. Figure 9.9d presents the charge production process from TPiENG relevant to tribo-piezoelectric effect. At the initial state, a part of the nylon film was contacted with the FEP layer, and positive and negative charges were distributed on the nylon and FEP surface, respectively, which is shown in Figure 9.9d(i). Consequently, the output current was zero because of electrostatic equilibrium. On applying wind on the device, the nylon film applied a compressive stress periodically on the FEP. When the PZT is compressed, positive piezoelectric potential occurred across this PZT, resulting in the piezoelectric current moving toward the dense Ag film. When the nylon film was separated from the FEP film, positive induced charges were produced on the ITO film, while the electrons moved toward the compact Ag film. Consequently, a triboelectric current moving toward the ITO film was created. Meanwhile, the compressive strain on the PZT was releasing as the nylon moved apart from the FEP, leading to a decrease in polarization density. Consequently, the electrons moved from Ag toward the ITO electrode. When the nylon film was completely separated from the FEP film, the compressive stress vanished, as well as the piezoelectric current, leaving a negative current due to the triboelectric effect. On the other hand, when the nylon was in contact with the FEP, a reverse current occurred due to the triboelectric effect, which is presented in Figure 9.9d(iv). When the nylon was in complete contact with FEP, the mean spontaneous polarization increased, resulting in an inverse piezoelectric current. The working principle of this PyENG, PVC, and TPiENG is completely different from each other, making this device suitable for collecting thermal, light, as well as mechanical energies. A piezo-tribo-pyro-photoelectric

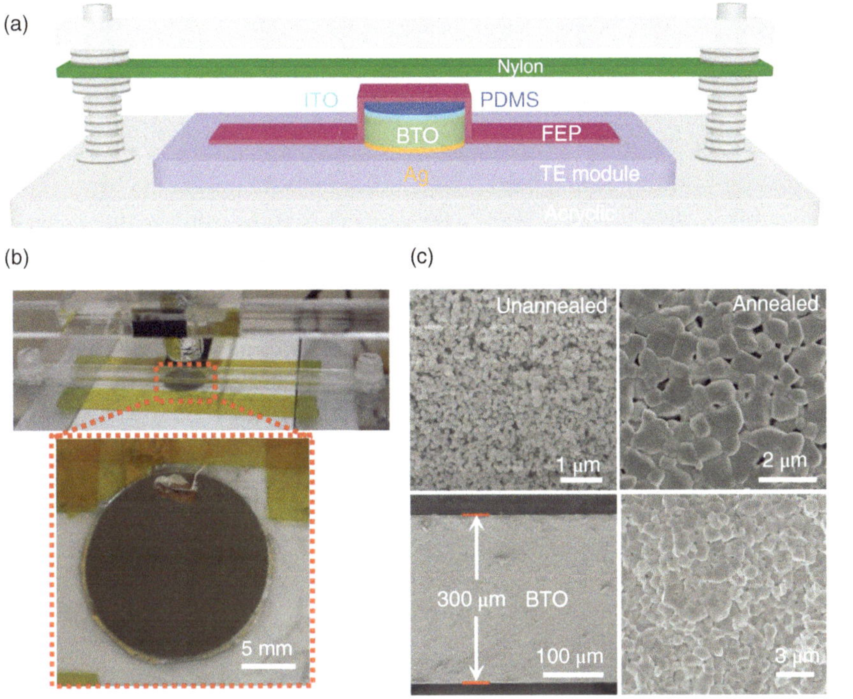

Figure 9.10 Multi-effects coupled nanogenerator based on BTO/friction material structure. (a,b) Schematic diagram (a) and photograph (b) of the device. (c) SEM images of BTO powders and ceramics. Source: Reproduced with permission from Ji et al. [64]. Copyright 2018, John Wiley and Sons.

effect nanogenerator based on BTO/friction material configuration was prepared as well [64]. Figure 9.10a,b illustrates a cross-sectional schematic diagram and photographic image of this nanogenerator. As is seen, this device was mainly composed of three parts: a circular BTO ceramic disk working as the piezoelectric, photoelectric, and pyroelectric material, an ITO film, and a Ag layer. The thickness and piezoelectric coefficient of the compact BTO ceramic disk were about 300 μm and 180 pC/N, respectively. The FEP film was attached on the ITO electrode. The nylon film served as a vibration film not only to induce triboelectrification with the PEP layer but also to offer stress on the ferroelectric BTO disk. The effective area of the device is about 2.22 cm^2. By coupling ferroelectric BTO material, electrode materials, and friction materials, a PyENG, a TPiENG, and a PVC were perfectly constructed in one structure sharing the same pair of electrodes. Moreover, a triboelectric-piezoelectric-pyroelectric effect nanogenerator combining PVDF with PVDF-PDMS was reported [59]. The coupled device was composed of a layer of polarized PVDF as active material, ITO electrodes on both sides of this PVDF, and PVDF-PDMS friction material. The whole device was not only transparent but also flexible. A TPiENG and a PyENG were successfully integrated in the coupled device, sharing the same ITO electrodes.

9.4 Performance

9.4.1 Output Characteristics for Harvesting Thermal and Light Energies

Output performance of a thermo-photoelectric effect nanogenerator with ITO/porous ZnO/Ag configuration was presented in Figure 9.11a [43]. Upon illumination, output electric signals were generated owing to the coupling effect of thermoelectric and photoelectric effects. As light intensity increased, I_1 linearly enhanced with a rate of 11.25 μA/(W cm^2), reaching up to a maximum of 4.88 μA. When individually heating/cooling the bottom of this nanogenerator, a temperature gradient ΔT was generated within the device; consequently, a two-stage time-dependent curve consisting of a peak and a plateau was observed due to the thermoelectric effect. Both the peak and plateau current showed a linear increasing behavior with the temperature gradient ΔT increasing. The increasing rates related to the peak and plateau current were 1.42 and 0.23 μA/K, respectively. Figure 9.11a presents the current signals from this device under various working conditions, where plateau output currents under illumination, heating/cooling, and heating/cooling + illumination conditions were marked as I_1, I_2, and I_3, respectively. During measurement, the light intensity was fixed at 0.408 W/cm^2 and temperature gradient ΔT was modulated in the range of −2.4 to 1.5 K. Under heating/cooling + illumination conditions, a positive I_3 could be observed. To clearly exhibit the temperature-dependent output current, I_1, I_2, $I_{3,}$ and $I_1 + I_2$ under heating/cooling + illumination conditions were summarized. As can be seen, as the temperature gradient ΔT decreased, I_3 increased while $I_1 + I_2$ decreased. When the temperature gradient ΔT changed in the range of −0.2 to −1.7 K, I_3 was much higher than $I_1 + I_2$ on account of thermo-photoelectric effect. The enhancement ratio relevant to the output current could be increased to 20.6% by decreasing the temperature gradient ΔT to 1.7 K. Furthermore, Ouyang et al. [45] investigated the output measurements of a thermo-photoelectric effect nanogenerator with ITO/SnSe:Br/Ag structure, as illustrated in Figure 9.11b. Upon 760 nm light illumination (12.63 mW/cm^2), the photocurrent I_1 and photovoltage V_1 of the coupled nanogenerator were 10.9 μA and 1.95 mV, respectively. When the cooling process with the temperature gradient of 1.5 K was applied on the coupled nanogenerator, a stable thermocurrent I_2 and V_2 of 2.3 μA and 0.72 mV could be observed, respectively. By simultaneously applying illumination and cooling processes on the coupled nanogenerator, the output electric signals were increased to 15.1 μA and 3.64 mV owing to the thermo-photoelectric effect.

Output characteristics of a thermo-photoelectric effect nanogenerator with InP/ZnO heterojunction structure under weak (7–50 Lux), moderate (~426 Lux), and strong (more than 20 000 Lux) light illumination were investigated by Zhang and Yang [61]. To realize output enhancement by thermo-photoelectric effect in the coupled device, cooling process with temperature gradient ΔT of 3.5 °C was applied on the bottom of this nanogenerator during illumination. The current obtained together with voltage is presented in Figure 9.12, in which the photoelectric current (V) was denoted by I_{photo} (V_{photo}), and the

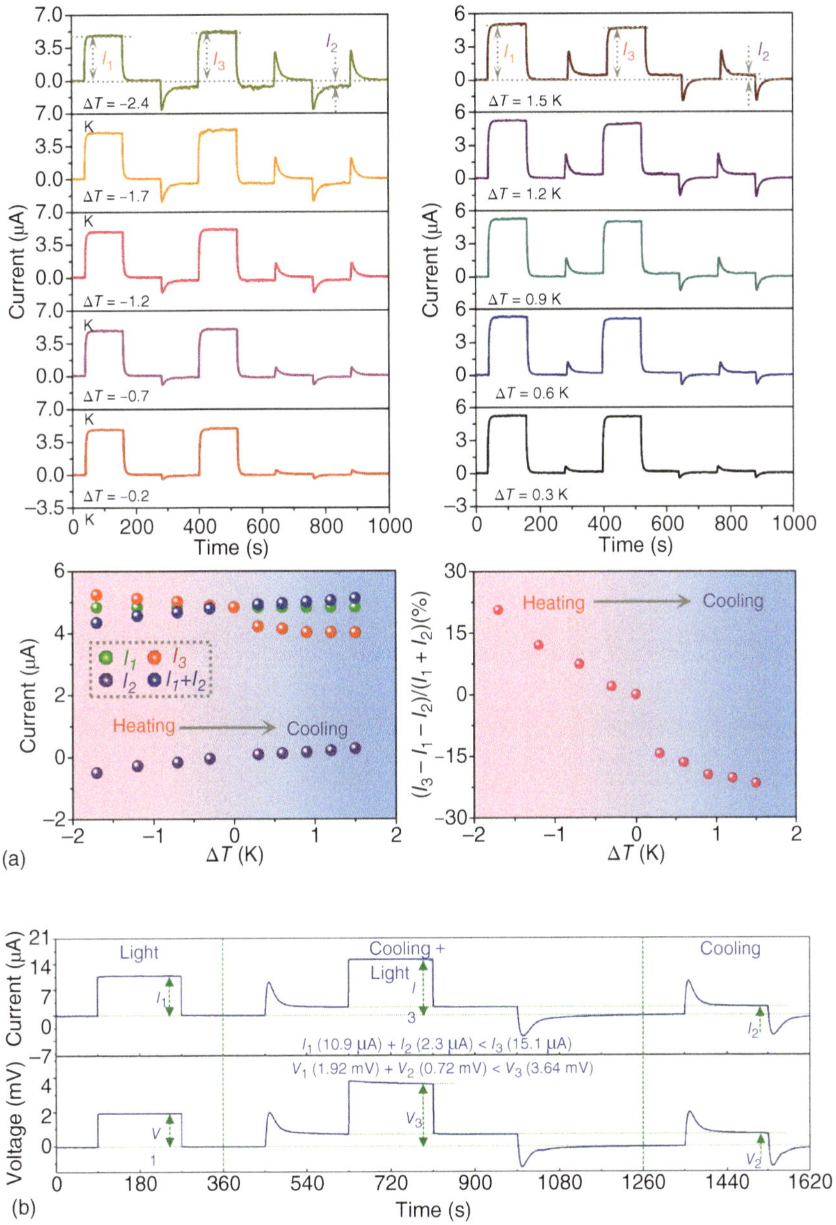

Figure 9.11 Output electric signals of ITO/bulk semiconductor/Ag nanogenerator induced by thermal and light energies. (a) Output current of ITO/porous ZnO/Ag nanogenerator. Source: Reproduced with permission from Gao et al. [43]. Copyright 2019, John Wiley and Sons. (b) Output current and output voltage of ITO/SnSe: Br/Ag nanogenerator. Source: Reproduced with permission from Ouyang et al. [45]. Copyright 2019, Elsevier.

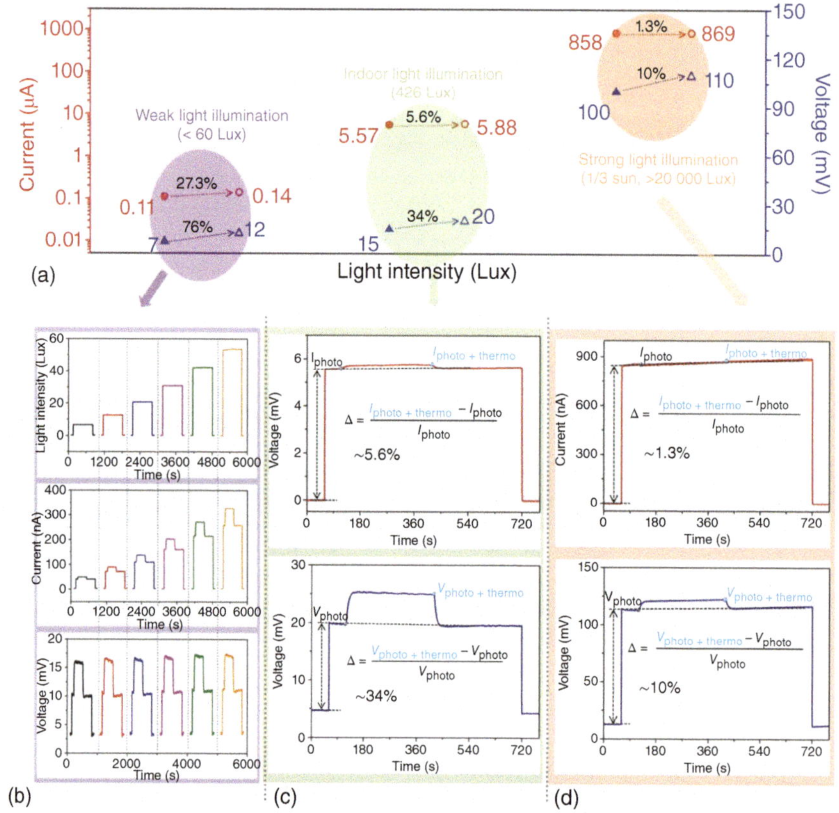

Figure 9.12 Output electric signals from thermo-photoelectric coupled effect nanogenerator with InP/ZnO structure. (a) Enhancement of output current and voltage under weak, indoor, and strong light illumination. (b–d) Output current and voltage under weak (b), moderate (c), and strong (d) light illumination. Source: Reproduced with permission from Zhang and Yang [61]. Copyright 2017, John Wiley and Sons.

thermoelectric-photoelectric current and voltage were named as $I_{photo+thermo}$ and $V_{photo+thermo}$. As illustrated in Figure 9.12a, thermoelectric effect could effectively increase the current as well as voltage upon irradiation by various lights. In a wide range of light intensities, the enhancement ratio increased with decreasing light intensity. In addition, the voltage enhancement ratio was much higher than the current enhancement ratio under the tested conditions. When the device was illuminated by weak light, with light intensity enhancing, I_{photo} and current enhancement ΔI exhibited a significant increase while V_{photo} and voltage enhancement ΔV showed a slight enhancement, as illustrated in Figure 9.12b. The current and voltage enhancement ratios were about 27.3% and 76%, which were almost constant as the light intensity changed within 7–50 Lux. When the device was irradiated by moderate light illumination, such as indoor light with illumination of 426 Lux, I_{photo} was 5.57 μA. When illumination and cooling processes were applied on this nanogenerator simultaneously, current signals were increased to 5.88 μA because of the thermo-photoelectric effect and a similar voltage enhancement could be obtained as well. The corresponding current and voltage enhancement ratios were about 5.6% and 34%, respectively, as shown in Figure 9.12c. To evaluate the output performance of the coupled nanogenerator under strong light intensity, 1/3 sunlight with the intensity of about 33 W/m^2 was utilized to irradiate this device. Figure 9.12d presents the recorded electric signals. As can be seen, upon individual illumination, I_{photo} and V_{photo} were about 0.86 mA and 0.1 V, respectively. When the illumination–cooling process was applied on the nanogenerator, voltage signals were significantly promoted with a voltage enhancement ratio of about 10%; however, no obvious current enhancement was observed. The temperature of the device could be significantly increased due to the strong light. The observed phenomena were ascribed to the coupling effects of temperature and temperature gradient on the output performance of the device.

Figure 9.13 illustrates the output performance of an ITO/BFO/Ag thermo-photoelectric effect nanogenerator [54]. Electric signals from this coupled nanogenerator were systematically measured under irradiation, irradiation + heating, as well as irradiation + cooling conditions. Figure 9.13a,b presents typical current–time curves together with voltage–time curves under irradiation as well as irradiation + heating conditions. If this device was irradiated by 365 nm light for a while, stable photocurrent and photovoltage were observed. After heating was employed on the coupled device, the steady photocurrent started to increase and finally reached a higher value; however, the steady photovoltage began to decrease and ultimately reach a lower value. Such phenomena were attributed to the thermo-photoelectric effect, which could be observed even if the light intensity changed within a wide range of 3.0–271.4 mW/cm^2. On the other hand, under illumination + cooling condition, stable photocurrent reduced to a smaller value while the photovoltage enhanced to a large value due to the coupling of thermoelectric and photoelectric effects. To directly exhibit the influence of thermo-photoelectric effect on the generated electric signals, the relationships between electric signals and light intensity under illumination and illumination + heating/cooling conditions are summarized (Figure 9.13c,d). Obviously, the current and voltage increased with increasing light intensity under illumination and illumination + heating/cooling conditions. On one hand, by introducing the heating process, the current produced could be significantly improved. On the other hand, the output voltage could be effectively promoted by employing the cooling process. By marking the output current and voltage under illumination and illumination + heating/cooling conditions as I_{photo} and V_{photo}, and $I_{thermo+photo}$ and $V_{thermo+photo}$, the current and voltage increasing ratios were calculated and have been presented in Figure 9.13e,f. As is seen, the current increasing ratio under illumination + heating condition exhibited an increasing behavior as the light intensity enhanced, achieving a maximum of 60% by increasing the light intensity to 271.4 mW/cm^2. Voltage increasing ratio under illumination + cooling condition increased with increasing light intensity, reaching the highest value of 75% when the light intensity was increased to 271.4 mW/cm^2.

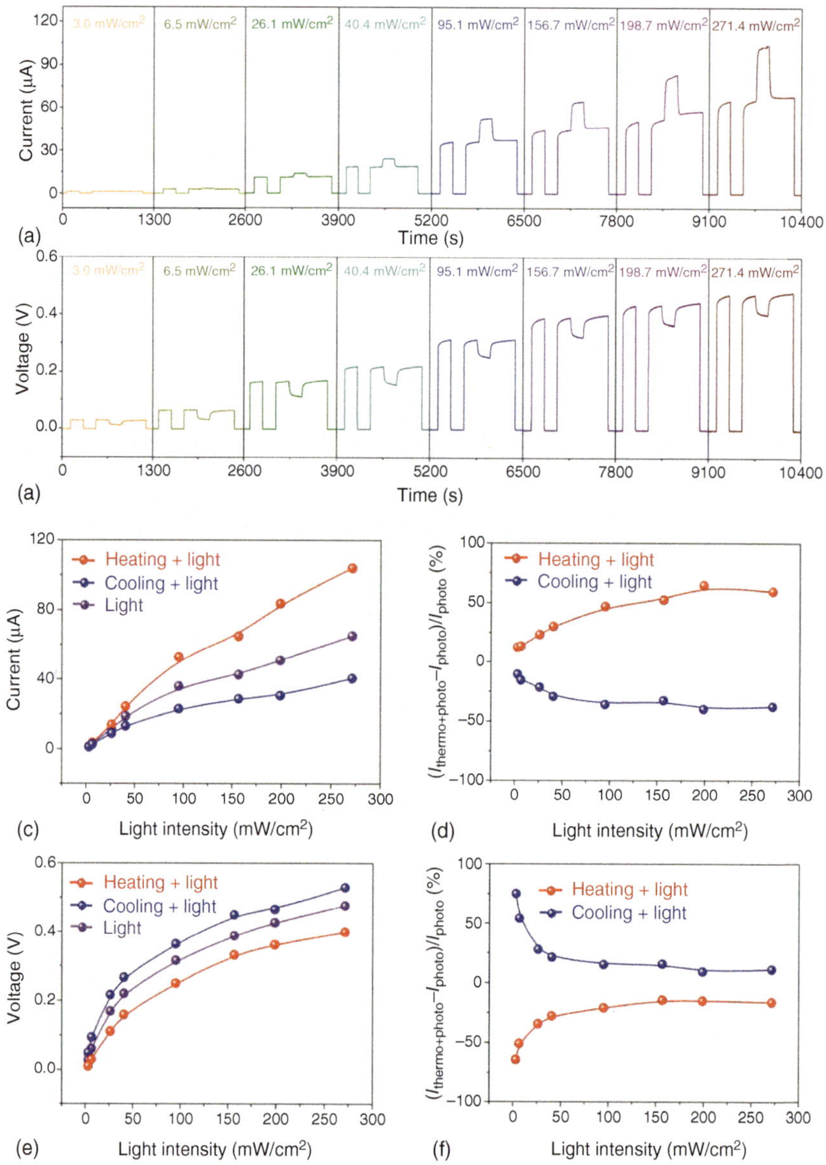

Figure 9.13 Output electric signals induced by thermo-photoelectric effect in ITO/BFO/Ag nanogenerator. (a,b) Output current (a) and voltage (b) of the device under light and heating + light conditions. (c–f) Dependence of short-circuit current (c) and open-circuit voltage (d) on light intensity, and the corresponding enhancement ratio of short-circuit current (e) and open-circuit voltage (f). Source: Reproduced with permission from Qi et al. [54]. Copyright 2018, American Chemical Society.

9.4.2 Output Performance for Scavenging Thermal and Mechanical Energies

The output characteristics of a pyro-piezoelectric effect nanogenerator with Ag/BTO/Ag structure were investigated in detail by Song et al. [55]. The coupled nanogenerator could individually/simultaneously scavenge thermal and mechanical energies. When individual heating process was applied on the device, pyroelectric signals were generated. By connecting a 100 MΩ resistance, the peak voltage generated from the device linearly enhanced from 0.075 to 0.638 V as the temperature increased from 2.03 to 13.57 °C, as illustrated in Figure 9.14a,b. Under pressing condition, the device generated AC piezoelectric voltage. When increasing the applied pressure from 15.4 to 27.6 kPa, the peak output voltage exhibited a linear increasing behavior from 0.059 to 0.58 V with an increasing rate of 0.044 V/kPa, as shown in Figure 9.14c. To obtain the pyro-piezoelectric performance of this nanogenerator, voltage signals upon heating + pressing (ΔT was 13.57 °C and the pressure was ranging from 15.4 to 27.6 Pa) were investigated, as presented in Figure 9.14d. Obviously, when pressing was applied on this nanogenerator, the peak voltage increased on account of the combined pyro-piezoelectric effect. In addition, the enhancement effect became remarkable as the applied pressure was further increased. Furthermore, Lee et al. [63] investigated the output performance of a PVDF based pyro-piezoelectric effect nanogenerator. The coupled nanogenerator could generate a peak piezoelectric voltage of 0.7 V/1.0 V when it was stretched/compressed, respectively. Upon heating or cooling, a peak pyroelectric voltage of 0.4 V could be generated by the device. Under heating-stretching/releasing condition, the pyroelectric and piezoelectric voltage produced by the device were perfectly integrated to generate a peak pyro-piezoelectric voltage of 1.1 V/1.4 V.

9.4.3 Output Characteristics for Harvesting Light and Mechanical Energies

Output electric signals of a piezo-photoelectric effect nanogenerator with perovskite/ZnO heterojunction configuration were reported [62]. Without strain, when the device was irradiated via a solar simulator, the obtained current, voltage, as well as power conversion efficiency were 2.07 nA, 0.59 V, and 0.0216%, which could be increased to 15.1 nA, 0.74 V, and 0.298% by applying a compressive strain of 0.8% because of the piezo-photoelectric effect. Zhu et al. [65] studied the output characteristics of a SnS/ZnO heterojunction piezo-photoelectric coupled nanogenerator when vertical or bending stress was applied on this coupled nanogenerator. It was found that current density and voltage increased from 4.58 to 4.69 mA/cm^2 and 0.54 to 0.75 V by employing 320 kPa vertical pressure due to the piezo-photoelectric effect. In addition, when the coupled device was working under the bending mode, the current density, voltage, and conversion efficiency could be significantly enhanced to 4.71 mA/cm^2, 0.58 V, and 1.3% by utilizing −0.88% compressive strain on the coupled device.

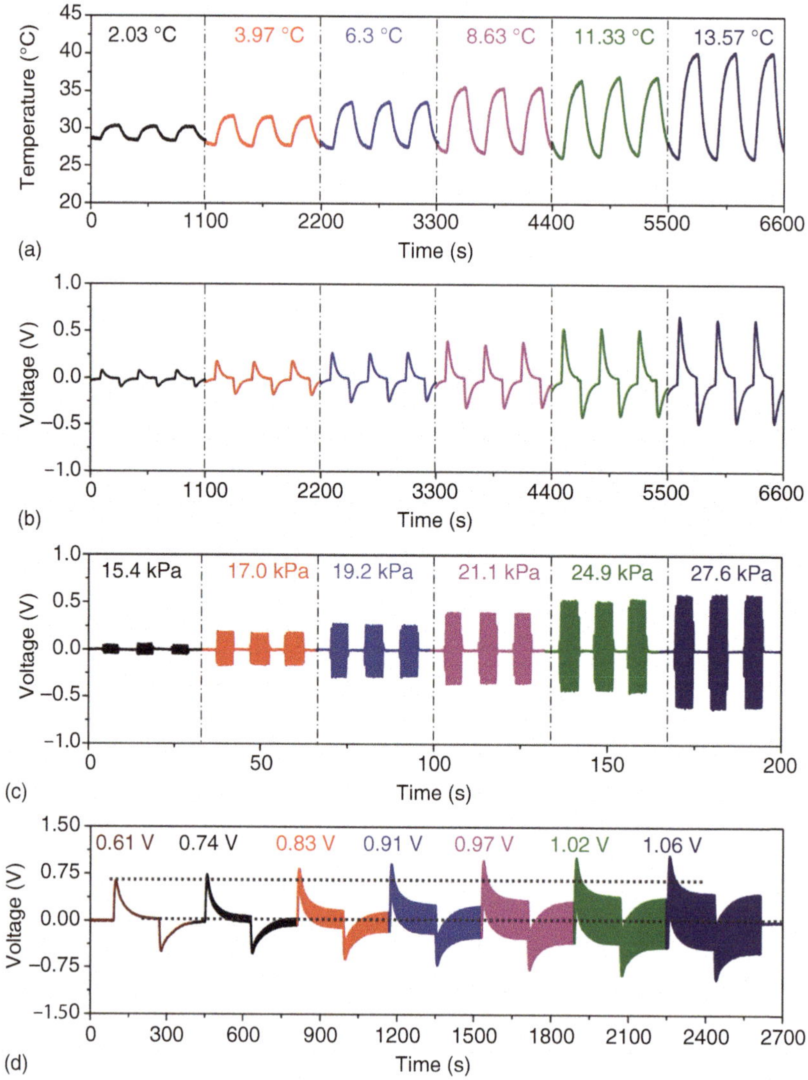

Figure 9.14 Output electric signals induced by pyro-piezoelectric effect in ITO/BTO/Ag nanogenerator. (a,b) Temperature variation (a) and the corresponding output voltage (b) of the device under heating conditions. (c) Output voltage of the device under pressing conditions. (d) Output voltage of the device under heating + pressing conditions. Source: Reproduced with permission from Song et al. [55]. Copyright 2019, John Wiley and Sons.

9.4.4 Output Characteristics for Harvesting Thermal, Light, and Mechanical Energies

Piezo-pyro-photoelectric and piezo-tribo-pyro-photoelectric effect nanogenerators have attracted immense attention on account of their outstanding ability to individually/simultaneously scavenge light, thermal, and mechanical energies. Piezo-pyro-photoelectric effect nanogenerators have the ability of

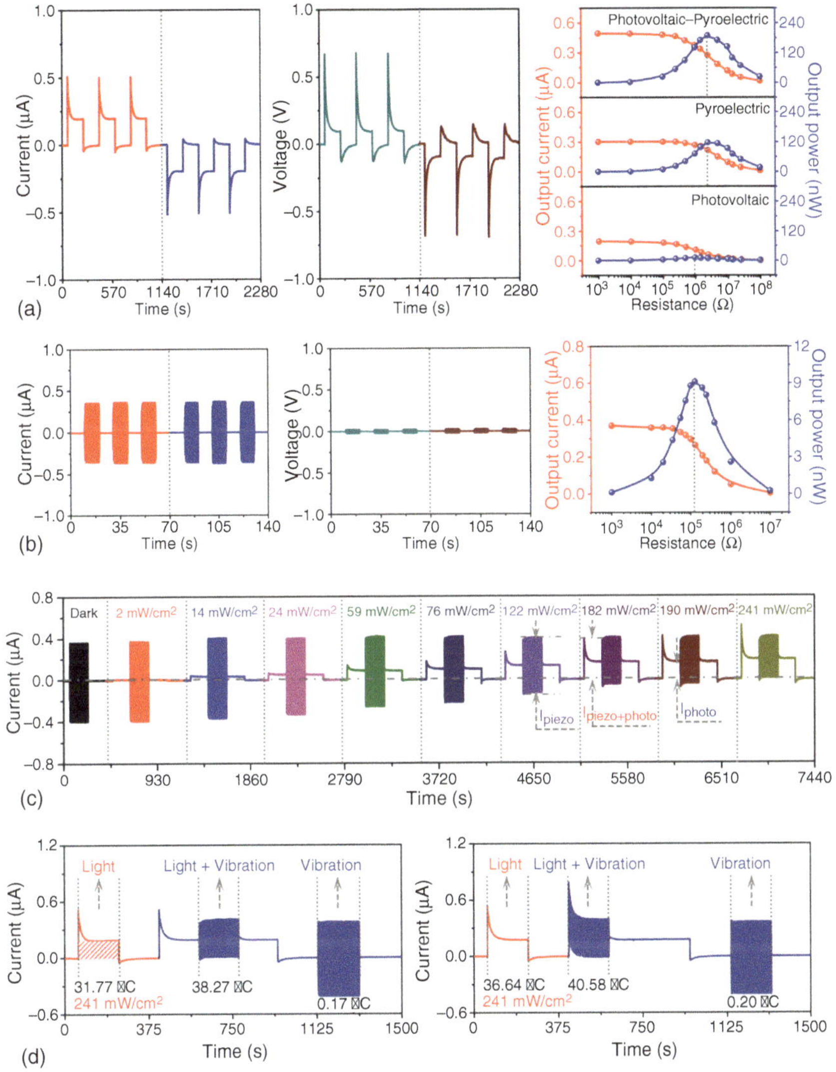

Figure 9.15 Output electric signals from piezo-pyro-photoelectric effect nanogenerator with ITO/BTO/Ag structure. (a) Output current, voltage, and power induced by pyro-photoelectric (a) and piezoelectric effect (b). (c) Output current resulted from piezo-pyro-photoelectric effect under light + vibration conditions with different light intensities. (d) Comparisons of output current induced by pyro-photoelectric, piezo-pyro-photoelectric, and piezoelectric effect. Source: Reproduced with permission from Ji et al. [41]. Copyright 2019, Royal Society of Chemistry.

converting light, thermal, and/or mechanical energies into electricity. Ji et al. [41] systematically investigated the output performance of a piezo-pyro-photoelectric coupled nanogenerator with ITO/BTO/Ag configuration. Figure 9.15a illustrates the current, voltage, and power from this nanogenerator upon 405 nm light irradiation. Upon illumination, current–time patterns together with voltage–time patterns exhibited similar two-stage behavior, where the peak output was related to pyro-photoelectric effect and the plateau output was relevant to the photoelectric effect. As can be seen, the pyro-photoelectric current/voltage and photoelectric current/voltage were 0.55 μA/0.67 V and 0.19 μA/0.1 V, respectively. It is reported that output electric signals decreased on reducing the light intensity. By defining the difference between the pyro-photoelectric current and the photoelectric current as the pyroelectric current, the resistance-dependent current as well as power induced by pyro-photoelectric, pyroelectric, and photoelectric effects were analyzed. Obviously, with increased load resistance, all the output currents reduced; however, all the output powers increased first and then dropped. The optimized output power related to pyro-photoelectric effect, pyroelectric effect, and photoelectric effect were 189, 117, and 11 nW, which were obtained at the external resistances of 2.5, 1.0, and 2.5 MΩ, respectively. Under vibration condition, the coupled nanogenerator could generate AC pizoelectric signals. By analyzing piezoelectric current under various vibration frequencies (5–40 Hz), the inherent frequency of the device was determined to be about 30 Hz. Figure 9.15b illustrates pizoelectric signals from this device upon vibration (30 Hz), indicating that the peak current/voltage could reach up to 0.384 μA/0.03 V, and the maximum power of 117 nW was obtained by loading a 0.13 MΩ resistance. Figure 9.15c presents the output performance induced by piezo-pyro-photoelectric effect under light–vibration coupled condition, showing that the piezoelectric current I_{Piezo} reduced as the light intensity increased. By monitoring the temperature-dependent piezoelectric coefficient d_{33} of the device, it was found that the piezoelectric coefficient d_{33} decreased as the temperature increased. Because of the light-induced thermal effect, the temperature of the device could reach a higher value with increasing light intensity, which would significantly reduce the piezoelectric coefficient d_{33}. Consequently, the piezoelectric current I_{Piezo} dropped as the light intensity enhanced. In addition, the increasing temperature under illumination could be suppressed when the device was vibrated due to the cooling effect induced by air flow, resulting in the improvement of the photoelectric current because of the interaction of the pyroelectric and photovoltaic effects. As a result, the transferred charges (calculated by integrating the output current within an entire working period) of the device under light–vibration coupling conditions were greatly promoted when compared to that under light or vibration conditions. A similar phenomenon could be observed even if the order of light as well as vibration was changed (Figure 9.15d). Such phenomenon became more remarkable when the light intensity was further increased. The enhancement of transferred charges was induced by the piezo-pyro-photoelectric effect.

The output characteristics of a piezo-tribo-pyro-photoelectric effect nanogenerator with PZT as the main active component have been systematically introduced when heat, light, and wind were applied [42]. This device is an integration of PyENG, PVC, and TPiENG. Figure 9.16a presents the current signal from individual PyENG while the temperature of this nanogenerator cyclically changed from 23 to 38 °C. The current–time curve indicates that the peak pyroelectric current was 480 nA. Figure 9.16b illustrates the current signals from individual PVC illuminated by a full spectrum solar simulator, where

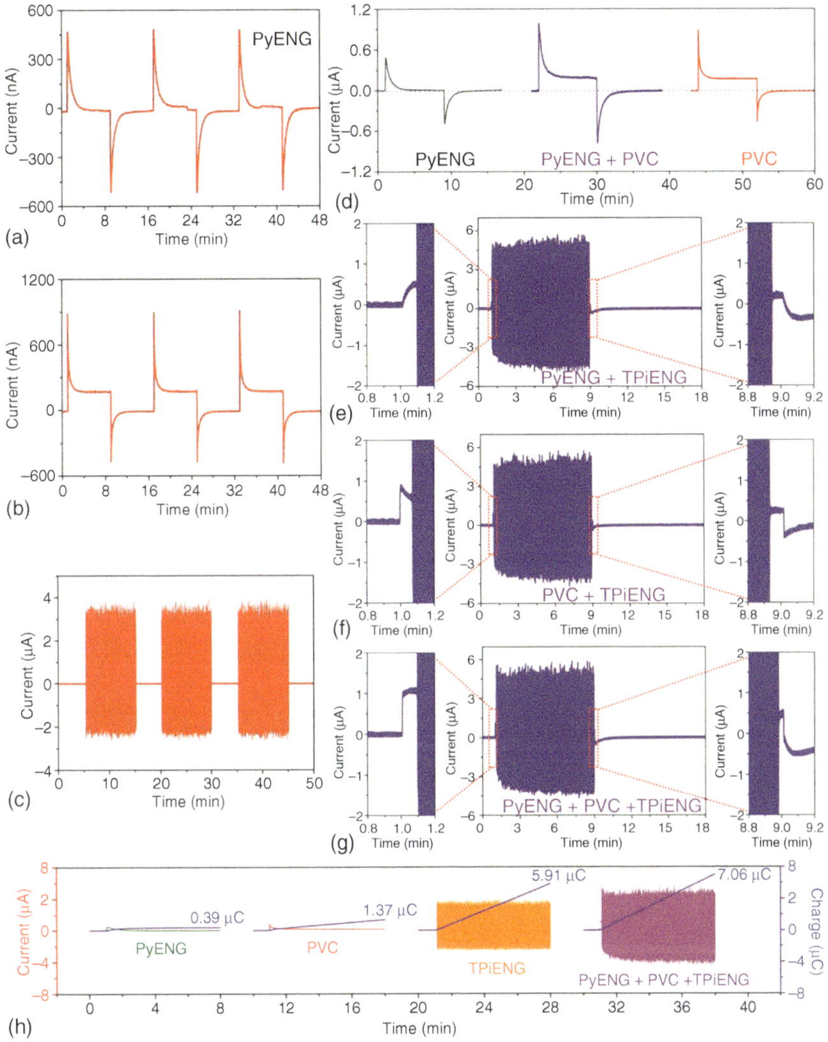

Figure 9.16 Output current of piezo-tribo-pyro-photoelectric effect nanogenerator. (a–c) Output current induced by pyroelectric effect (a) photovoltaic effect (b) and piezo-triboelectric effect (c). (d–g) Output current induced by pyro-photoelectric effect (d), piezo-tribo-pyroelectric effect (e), piezo-tribo-photoelectric-electric effect (f) and piezo-tribo-pyro-photoelectric effect (g). (h) Transferred charges induced by pyroelectric effect, photovoltaic effect, tribo-piezoelectric effect, and piezo-tribo-pyro-photoelectric effect. Source: Reproduced with permission from Zhang et al. [42]. Copyright 2017, John Wiley and Sons.

the time-dependent current curve showed typical two-stage behavior due to pyroelectric–photoelectric effect. The peak and plateau current signals were 890 and 170 nA, respectively. Current signals from individual TPiENG were recorded when the wind speed was ~15 m/s. Figure 9.16c depicts testing results, indicating that the AC piezo-triboelectric output current has a peak value of about 3.8 μA.

By loading various resistances, the maximum output powers were computed, which were 13 μW for PyENG, 7 μW for PVC, and 44 nW for TPiENG. The internal resistance of TPiENG was 8, which was much smaller than that of PyENG (0.2) as well as PVC (0.1). The output electricity induced by multiple effects was further studied by combining various individual units (Figure 9.16d–g). As displayed in Figure 9.16d, if heating + light was applied on this nanogenerator, the peak and plateau current values of PyENG + PVC were about 1 μA and 200 nA, respectively, which were larger than the current from individual PyENG or PVC. Such results indicated that PyENG and PVC could be efficiently combined to synchronously collect thermal energy together with light energy. Figure 9.16e–g presents current signals of this device under various coupled conditions, indicating that the output current peak was increased to more than 5 μA if PyENG, PVC, and TPiENG worked together. The coupled effect could be clearly observed by enlarging the current–time curves at the starting and ending of wind flow. Transferred charges of individual as well as coupled units were calculated and are summarized in Figure 9.16h. As is seen, compared to individual unit, the PyENG + PVC + TPiENG could deliver more charges (~7 μC). By integrating the PyENG, PVC, and TPiENG, the piezo-tribo-pyro-photoelectric effect nanogenerator could synchronously harvest thermal energy, solar energy, as well as mechanical energy to deliver a peak current, a peak voltage, and a plateau voltage of about 5 μA, 80 V, and 50 V, respectively.

9.5 Applications

As a powerful approach to convert various energies into electricity, multi-effects coupled nanogenerators can be widely utilized in the area of energy-storage device charging, photodetection, multifunctional sensing, and image sensing.

9.5.1 Energy-Storage Device Charging

Multi-effects coupled nanogenerators exhibit great potential to provide electricity for diverse electronics. To improve the continuous working time of electronics, energy-storage devices are usually required. The charging performance of multi-effects coupled nanogenerators for energy-storage devices have drawn considerable attention.

Capacitors and Li-ion batteries are commonly used as energy-storage devices in electronics. Ji et al. [41] studied the charging characteristics relevant to a BTO-based piezo-pyro-photoelectric effect nanogenerator. The coupled nanogenerator had a cantilever structure, which could scavenge light and vibration energies. Figure 9.17a presents the current produced by this nanogenerator under diverse working conditions. Although AC-type piezoelectric current was generated by the device under vibration and wind condition, DC-type outputs were finally obtained on account of piezo-pyro-photoelectric effect, making it possible for the device to directly charge capacitors without rectifiers. Charging measurements of this nanogenerator under various

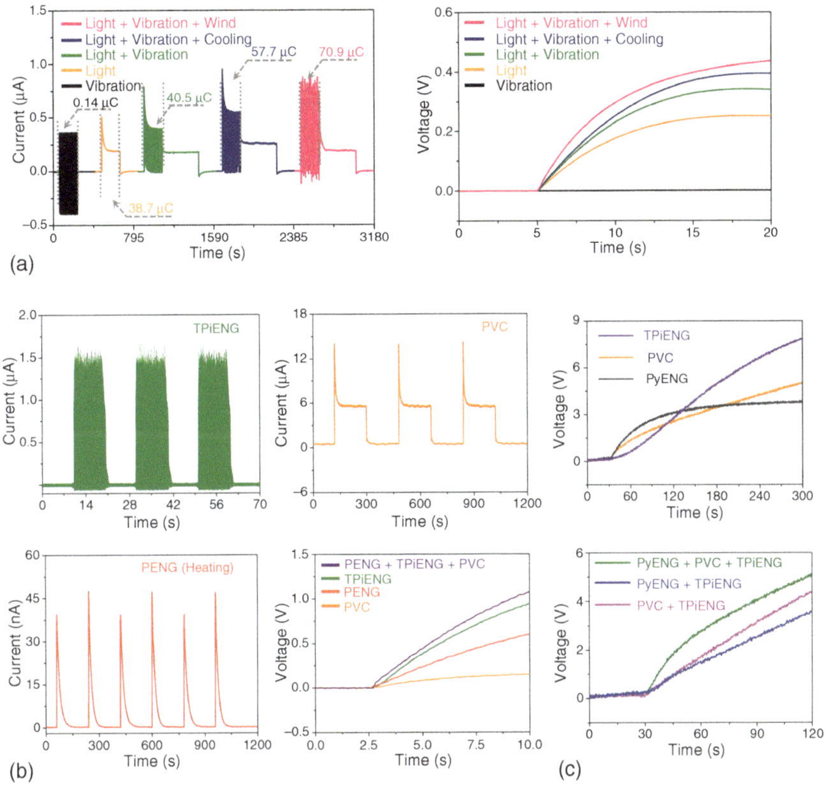

Figure 9.17 Charging performance of multi-effects coupled nanogenerators. (a) Output current and charging performance of cantilever-structured ITO/BTO/Ag multi-effects coupled nanogenerators under different working conditions. Source: Reproduced with permission from Ji et al. [41]. Copyright 2019, Royal Society of Chemistry. (b) Rectified output current and charging performance of BTO-based piezo-tribo-pyro-photoelectric couple effect nanogenerator. Source: Reproduced with permission from Ji et al. [64]. Copyright 2018, John Wiley and Sons. (c) Charging performance of PZT-based piezo-tribo-pyro-photoelectric couple effect nanogenerator under different working conditions. Source: Reproduced with permission from Zhang et al. [42]. Copyright 2017, John Wiley and Sons.

working conditions indicated that the charging speed is of the following order: "vibration" < "light" < "light + vibration" < "light + vibration + cooling" < "light + vibration + wind," which was consistent with the order of transferred charges. The voltage of a capacitor (4.7 μF) was rapidly improved to 0.44 V within only 15 seconds. It is obvious that the coupling of multi-effects can significantly enhance the charging performance of nanogenerators. Furthermore, charging performance of a BTO-based piezo-tribo-pyro-photoelectric effect nanogenerator was reported as well [64]. To charge a capacitor, a rectifier was employed to convent AC electric signals into DC ones. Rectified peak output currents of individual PENG, PVC, and TPiENG were about 40, 14, and 1.48 μA, respectively. Rectified peak outputs from PENG + TPiENG + PVC were 1.5 μA and 7 V, respectively. Figure 9.17b depicts the charging performance of the

PVC, PENG, TPiENG and the PENG + TPiENG + PVC. Obviously, the charging speed of PENG + TPiENG + PVC was much faster than that of the individual PVC, PENG, and TPiENG, and after a 10 seconds charging process, the voltage of the 0.33 μF capacitor could be rapidly enhanced to 1.1 V. Zhang et al. [42] investigated the charging performance of a piezo-tribo-pyro-photoelectric effect nanogenerator. Figure 9.17c presents the charging measurements for a capacitor (10 μF). For an individual unit, PyENG showed the fastest charging speed in the beginning, while TPiENG exhibited the best long-term charging performance. The fastest charging speed was obtained only if PyENG, PVC, and TPiENG worked together; meanwhile, the voltage of the capacitor could be increased to about 5.1 V within only 90 seconds.

9.5.2 Electrics Powering

Rapid advancements of micro/nano materials facilitate the progress of portable and wearable electronic devices. Harvesting clean and renewable energy for the electronics from the ambient circumstances has become a crucial issue. Zhang and Yang [61] successfully powered a temperature sensor by utilizing a thermo-photoelectric effect nanogenerator with InP/ZnO heterojunction. To obtain a proper output voltage value, two coupled nanogenerators (SC-1 and SC-2) with matched output current were connected in series. Figure 9.18a indicates that SC-1, SC-2, and SC-1 + SC-2 nearly had the same short-circuit currents, while the open-circuit voltage produced by SC-1 + SC-2 was larger. Considering that SC-1 + SC-2 had a relatively large output current, an electricity management element was employed to further improve the photovoltage. Finally, the photovoltage from SC-1 + SC-2 was boosted from 0.7 to 1.6 V. Figure 9.18b exhibits the working temperature sensor powered by the system of SC-1 + SC-2 and the power management circuit under a solar simulator. As can be seen, a temperature value of 29.6 °C was clearly displayed, indicating great potential of the thermo-photoelectric effect nanogenerator for powering intelligent electronics.

9.5.3 General Sensing

A self-powered 365 nm photodetector based upon an ITO/porous ZnO/Ag thermo-photoelectric effect nanogenerator was reported [43]. To evaluate the performance of the photodetector, electric signals upon irradiation at different temperatures were systematically investigated. The temperature gradient between the top and bottom sides induced by the thermoelectric module was recorded and marked as ΔT. Figure 9.19a presents the output current under illumination at room temperature, showing that stable output current signals with good on–off repetition were produced when the device was illuminated. Upon illumination (0.408 W/cm^2), the response time τ_{res} was around 5.12 seconds, and the recovery time τ_{rec} was about 6.89 seconds. The dependence of the measured current on light intensities is summarized in Figure 9.19b. As is seen, the output current linearly increased with the light intensity arising under any temperature gradient ΔT, indicating the excellent performance of this photodetector for

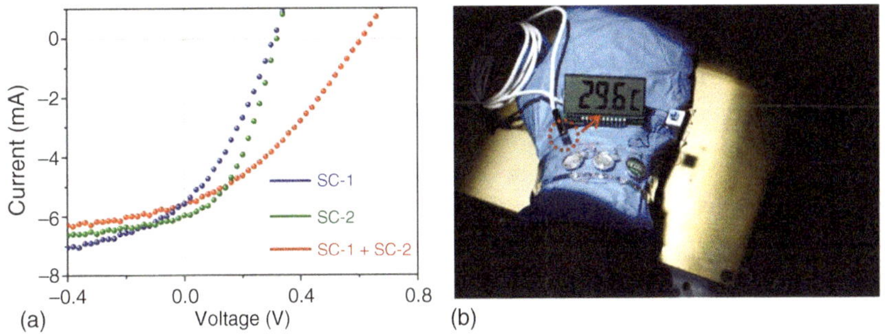

Figure 9.18 Performance of thermo-phototronic effect based nanogenerators for electronics powering. (a) Current–voltage curves of InP/ZnO based multi-effects coupled nanogenerators in series under illumination. (b) Photographic image of temperature sensor powered by the device. Source: Reproduced with permission from Zhang and Yang [61]. Copyright 2017, John Wiley and Sons.

365 nm photodetection. In addition, under the same light intensity, the output current increased with the temperature gradient ΔT reducing. By fitting the light-intensity-dependent current curves, sensitivity and its enhancement ratio of the photodetector were obtained, as summarized in Figure 9.19c. It is obvious that the sensitivity and its enhancement ratio reduced as the temperature gradient ΔT increased. When the temperature gradient ΔT was −2.4 K, the sensitivity attained its maximum value, which was 13.70 μA/(W cm^2). Meanwhile, a maximum sensitivity enhancement ratio of 21.78% was achieved. The enhanced performance of this device resulted from the coupling of thermoelectric and photoelectric effects.

9.5.4 Multifunctional Sensing

A self-powered multifunctional sensor based upon a pyro-piezoelectric nanogenerator was introduced by Song et al. [55]. This device could be employed to detect temperature and pressure variations. The coupled nanogenerator was composed of three layers, where a circular BTO ceramic disk was sandwiched between the Ag electrodes. To make the device flexible, the device was coated with PDMS. Figure 9.20a presents typical output voltages of the device under heating, pressing, and heating–pressing conditions. By applying the heating process with a temperature gradient ΔT of 8.63 °C on the device, a peak pyroelectric voltage of 0.34 V could be observed. Upon pressing with a pressure of 17 kPa, the device could generate a peak piezoelectric voltage of 0.15 V. When the heating and pressing processes were applied on the device simultaneously, the device could deliver a peak pyro-piezoelectric voltage of 0.49 V, which was the same as the sum of the peak pyroelectric voltage and peak piezoelectric voltage. Such a phenomenon indicated that pyro-piezoelectric effect could remarkably promote the performance of this detector. In addition, the rise times of this device under individual heating condition was 8.6 seconds, which could be reduced to 7.8 seconds by introducing the pressing process. The temperature sensing performance of this detector under diverse pressures was systematically investigated

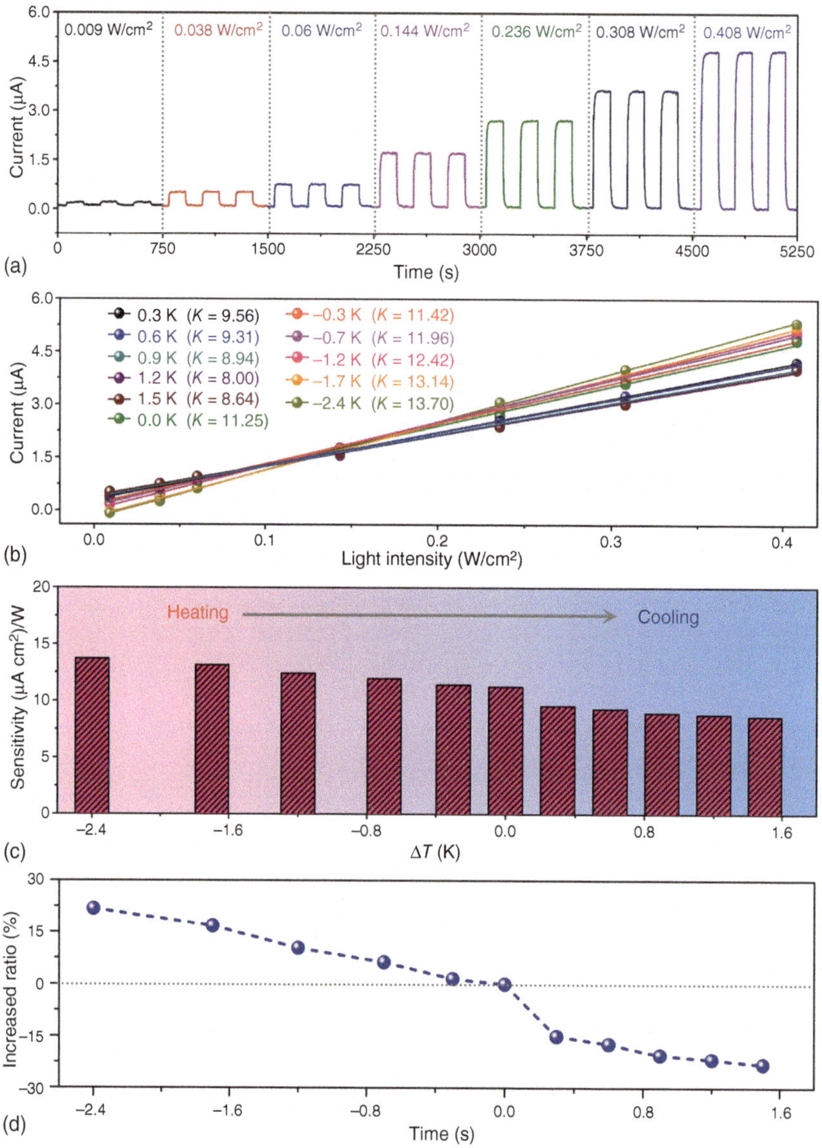

Figure 9.19 Performance of multi-effects coupled nanogenerator with ITO/porous ZnO/Ag structure for 365 nm photodetection. (a) Output current of the device under illumination with different light intensities at room temperature. (b) Light-intensity-dependent output current and the corresponding sensitivity. (c) Temperature dependence of sensitivity. (d) Temperature-dependent enhancement ratio of sensitivity. Source: Reproduced with permission from Gao et al. [43]. Copyright 2019, John Wiley and Sons.

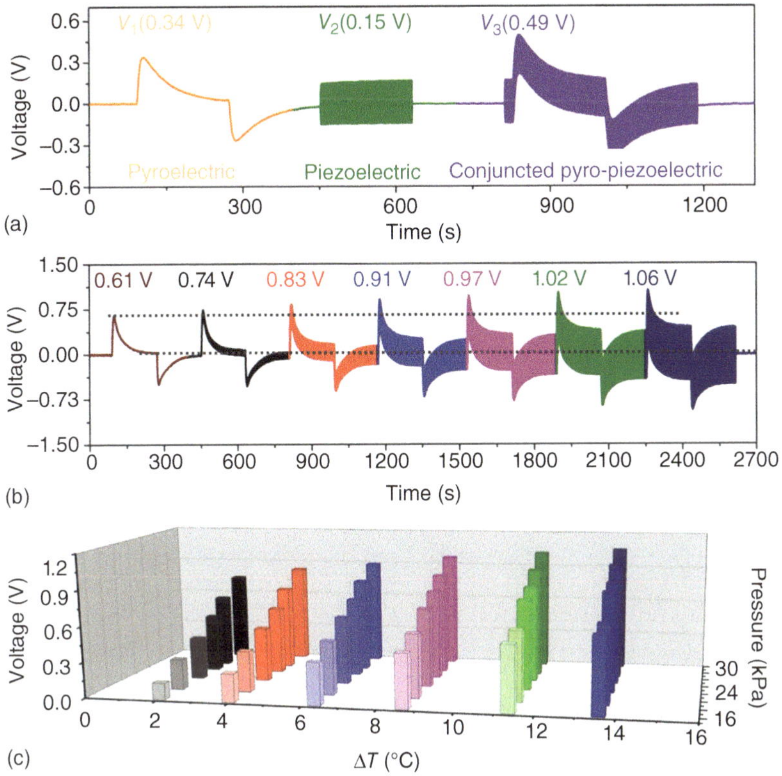

Figure 9.20 Multi-effects coupled ITO/BTO/Ag nanogenerator for temperature and vibration frequency sensing. (a) Comparisons of output voltages under heating ($\Delta T = 8.63\,°C$), pressing (pressure of 17 kPa), and heating–pressing combined conditions. (b) Under ΔT of 13.57 °C, output voltage of the device under different pressure gradients. (c) Temperature-dependent output voltage under different pressures. Source: Reproduced with permission from Song et al. [55]. Copyright 2019, John Wiley and Sons.

(Figure 9.20b). When individual heating process with temperature gradient ΔT of 13.57 °C was applied on this device, a peak pyroelectric voltage of 0.61 V was observed, which could be promoted to 1.06 V by increasing the pressure on the device to 27.6 kPa because of the pyro-piezoelectric effect. Figure 9.20c illustrates the output voltage of this detector under diverse heating–pressing conditions. On one hand, under a constant temperature gradient ΔT, voltage signals increased linearly as pressure increased. On the other hand, under a constant pressure, the output voltage exhibited a linear increasing behavior as the temperature gradient ΔT increased. In addition, similar enhancement phenomena were observed under various

cooling–pressing conditions, indicating the excellent sensing performance of the device for detecting temperature and pressure variations.

Based upon the photovoltaic–pyroelectric–piezoelectric effect nanogenerator, a BTO-based multifunctional detector for sensing stimulation from light, thermal, as well as vibration energies was prepared [53]. The coupled nanogenerator with cantilever structure was composed of interdigitated ITO electrodes, a ferroelectric BTO ceramic tablet, and a Ag film on the bottom of the BTO tablet. To effectively monitor the vibration frequency, masses with the weight of 0.6 g were fastened at the free end of this sensor. Upon individual irradiation (365 nm), the output current from this sensor showed linear increasing behavior as the light intensity increased, indicating the feasibility of using this sensor to monitor the light intensity. The sensitivity related to the photovoltaic–pyroelectric effect of this device was evaluated to be 0.17 nA/(mW cm^2). The maximum power of the device was 45.9 nW by loading a 100 MΩ resistance. For temperature sensing, the pyroelectric current from the multifunctional sensor linearly increased as the temperature gradient ΔT increased, showing that this device was suitable for temperature sensing. The sensitivity values of the device for temperature monitoring under cooling and heating were 4.80 and 1.75 nA/K, respectively. Meanwhile, by loading an optimal resistance of 750 MΩ, the maximum output power of 49.7 nW was achieved. When individual vibration process was applied on the multifunctional sensor, the peak–peak piezoelectric current of the sensor exhibited a perfectly linear increase, with the vibration frequency increasing from 11 to 17 Hz and 17 to 26 Hz, suggesting that the sensor could be utilized to monitor variation of vibration frequency. The corresponding frequency sensitivity values of the sensor were 21.02 and −14.07 nA/Hz for monitoring low-frequency (≤17 Hz) and high-frequency (≥17 Hz) vibration, respectively. The optimal loading resistance for the multifunctional sensor working under vibration condition was 1 MΩ, and correspondingly, a maximum output power of 7.5 nW was obtained. Figure 9.21a presents a representative output current curve when the sensor was working under illumination–cooling condition. Figure 9.21b exhibits the relevant sensitivity. For any light intensity, current signals showed linear increasing behavior as ΔT increased, indicating that the sensor was suitable for temperature sensing under illumination–cooling condition. In addition, the performance of the device for more complicated sensing was investigated when illumination, heating (ΔT was 12.9 K), and vibration processes were sequentially applied on the device, as seen in Figure 9.21c,d. Obviously, the output current of the sensor showed characteristics of piezoelectric signals when the device was working under the illumination–heating–vibration condition. The peak–peak output current showed good linear increasing behavior as the frequency enhanced from 11 to 17 Hz and 17 to 26 Hz as well, indicating that the multifunctional sensor could be successfully utilized for coupled sensing of stimulation from light, thermal, as well as vibration energies by employing the coupling effect based upon photovoltaic, pyroelectric, and piezoelectric effects.

9.5.5 Image Sensing

A 4 × 4 array self-powered image photodetector to monitor 365 nm light intensity distribution was successfully prepared by Qi et al. [54]. Every photodetection unit was a thermo-photoelectric effect nanogenerator with ITO/BFO/Ag structure. Optical images of the multichannel photodetector before/after being covered by an L-shaped mask under illumination were presented in Figure 9.22a. The 16 channels of the photodectector were connected to a

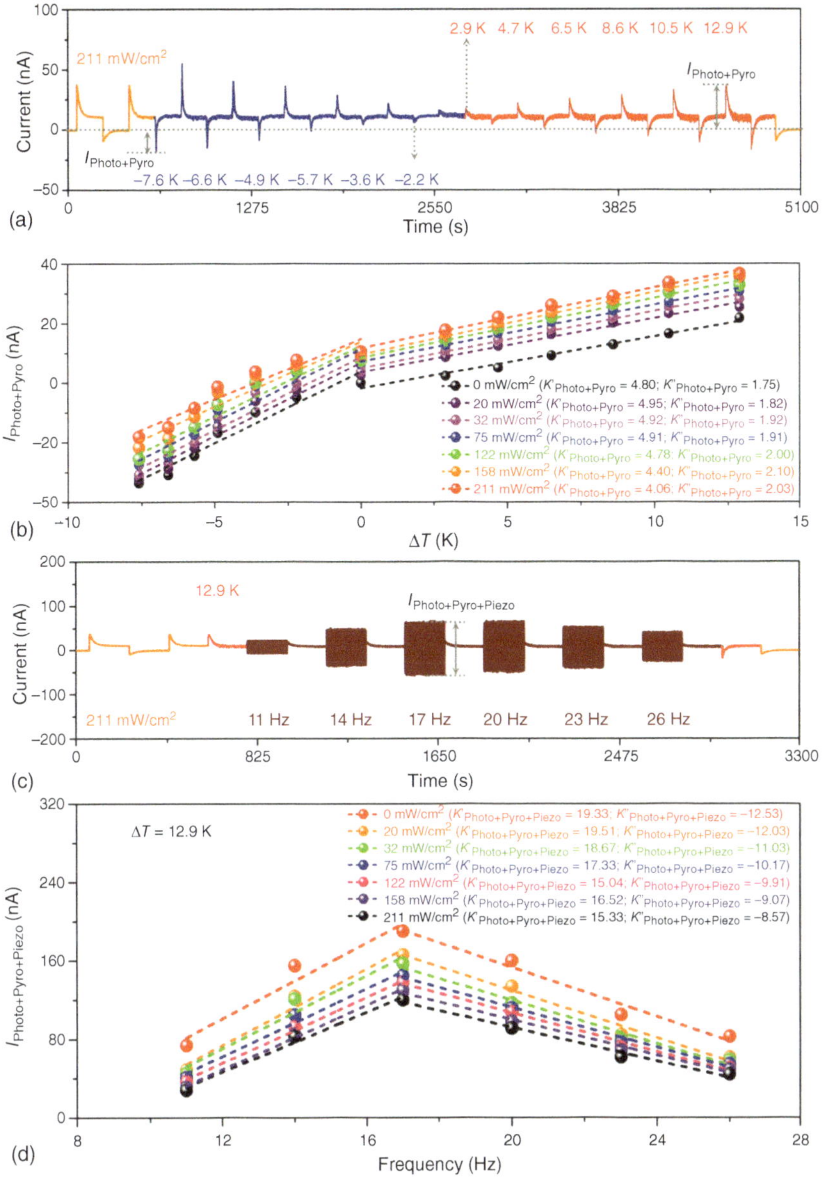

Figure 9.21 Multi-effects coupled ITO/BTO/Ag nanogenerator for multifunctional sensing. (a,b) Upon illumination, output current (a) and the corresponding sensitivity (b) of the device for temperature sensing. (c,d) Output current (a) and the corresponding sensitivity (b) of the device for monitoring vibration frequency under light + heating conditions. Source: Reproduced with permission from Ji et al. [53]. Copyright 2019, John Wiley and Sons.

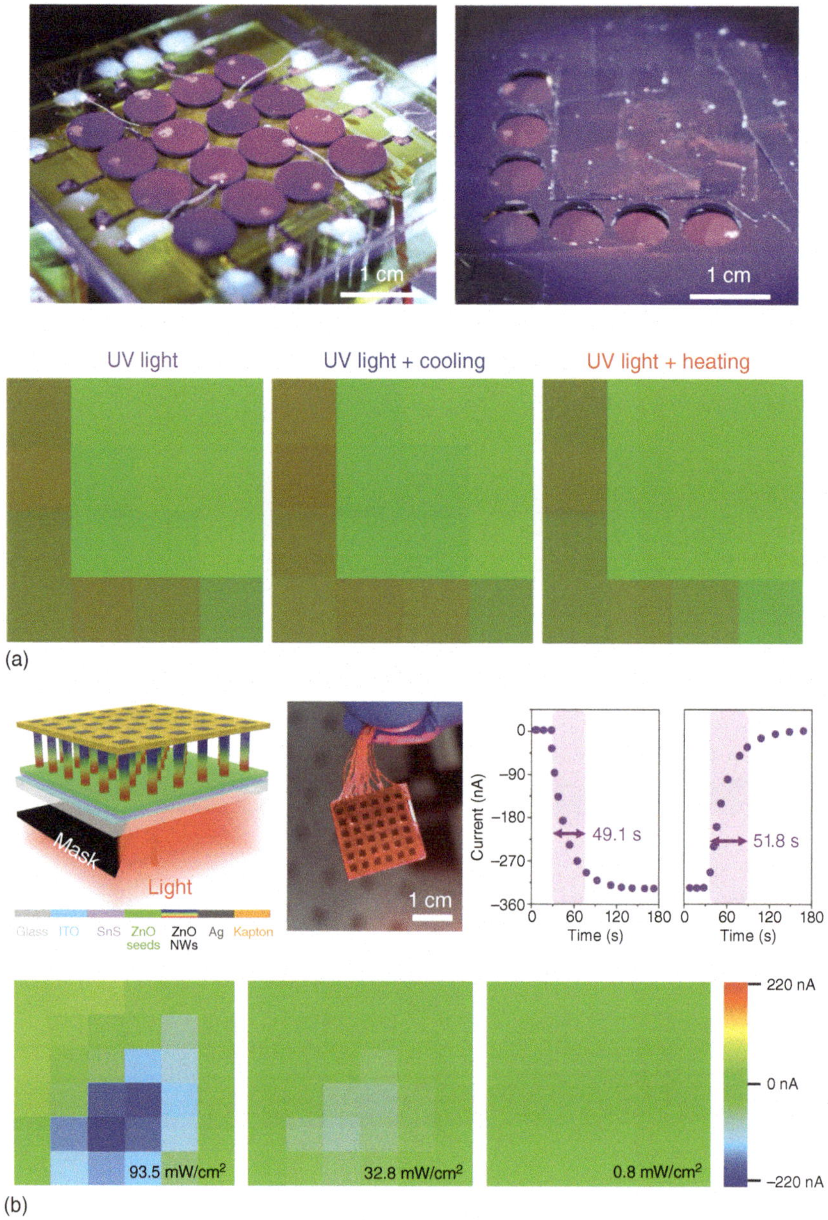

Figure 9.22 Multi-effects coupled nanogenerators for image photodetection. (a) Photographs and performance of image photodetector based on ITO/BFO/Ag multi-effects coupled nanogenerators. Source: Reproduced with permission from Qi et al. [54]. Copyright 2018, American Chemical Society. (b) Schematic diagram, photographic image, and performance of image photodetector based on SnS/ZnO nanowires multi-effects coupled nanogenerator. Source: Reproduced with permission from Ouyang et al. [44]. Copyright 2018, Elsevier.

multichannel data acquisition system, so that the real-time output voltage value from each photodetection unit could be recorded simultaneously and exhibited in an image figure with corresponding colors. Similar colors stood for similar output voltage values. The fabricated photodetector array exhibited excellent uniformity under dark and illumination conditions. Image figures induced by L-shaped light illumination under room temperature, low temperature, and high temperature were investigated and the obtained image figure under cooling condition exhibits darker colors, indicating that the performance of the photodetector array could be significantly improved by thermo-photoelectric effect. In addition, the photodectector could also be utilized to detect light with more complicated shapes, such as square-annular shape, diagonal-line shape, and so on. Furthermore, Ouyang et al. [44] fabricated a self-powered 6×6 photodetector array based upon thermo-photoelectric effect nanogenerators (Figure 9.22b). This multichannel photodetector was composed of a transparent ITO electrode, a layer of SnS/ZnO heterojunction, and a 6×6 matrix of Ag electrode. Photographic image of the device indicates that the size of every Ag unit was about $1.5 \times 1.5\,mm^2$. Each thermo-photoelectric effect nanogenerator worked as a photodetection element and could be utilized to monitor 365 nm light. The real-time output current of each unit was recorded so that image figures that carried the information of light intensity and illuminated position could be obtained according to the measured output current. The response and recovery times of this sensor were 49.1 and 59.8 seconds upon irradiation ($93.5\,mW/cm^2$). When the bottom-right part of this multichannel photodetector was illuminated, the color of the bottom-right part of the image figure turned to blue and became lighter with the light intensity decreasing, indicating excellent performance of the photodetector array.

Figure 9.23 illustrates a 4×4 array self-powered flexible sensor based on the pyro-piezoelectric effect for sensing variations of temperature and pressure [55]. The sensor array consisted of 16 Ag/BFO/Ag pyro-piezoelectric effect nanogenerators working as response units. All the units were integrated in a PDMS layer. Photographs of the device illustrated in Figure 9.23a,b indicate that the device could be bent at large angles. Every nanogenerator was connected to a multichannel data acquisition system. Real-time output voltage could be obtained simultaneously and displayed as image figures of different shapes and colors according to the measured voltage. Figure 9.23b–d exhibits the output voltage of each element and the corresponding image figures of various shapes under heating, pressing, and heating–pressing conditions, where "1," "3," and "5" stood for pyroelectric voltage, and "2," "4," "6" were representative of pyro-piezoelectric voltage. Obviously, by applying F-shaped, inverted T-shaped, and X-shaped stimulation with heating and heating–pressing on the device, F-shape, inverted T-shape, and X-shape could be clearly observed in the image figures. In addition, output voltage values induced by pyro-piezoelectric effect were much higher than that caused by individual pyroelectric or piezoelectric effect; the stimulated elements exhibited much darker color under heating–pressing conditions than under heating or pressing conditions, indicating that the sensing ability of this sensor was promoted through pyro-piezoelectric effect. Moreover, such sensor

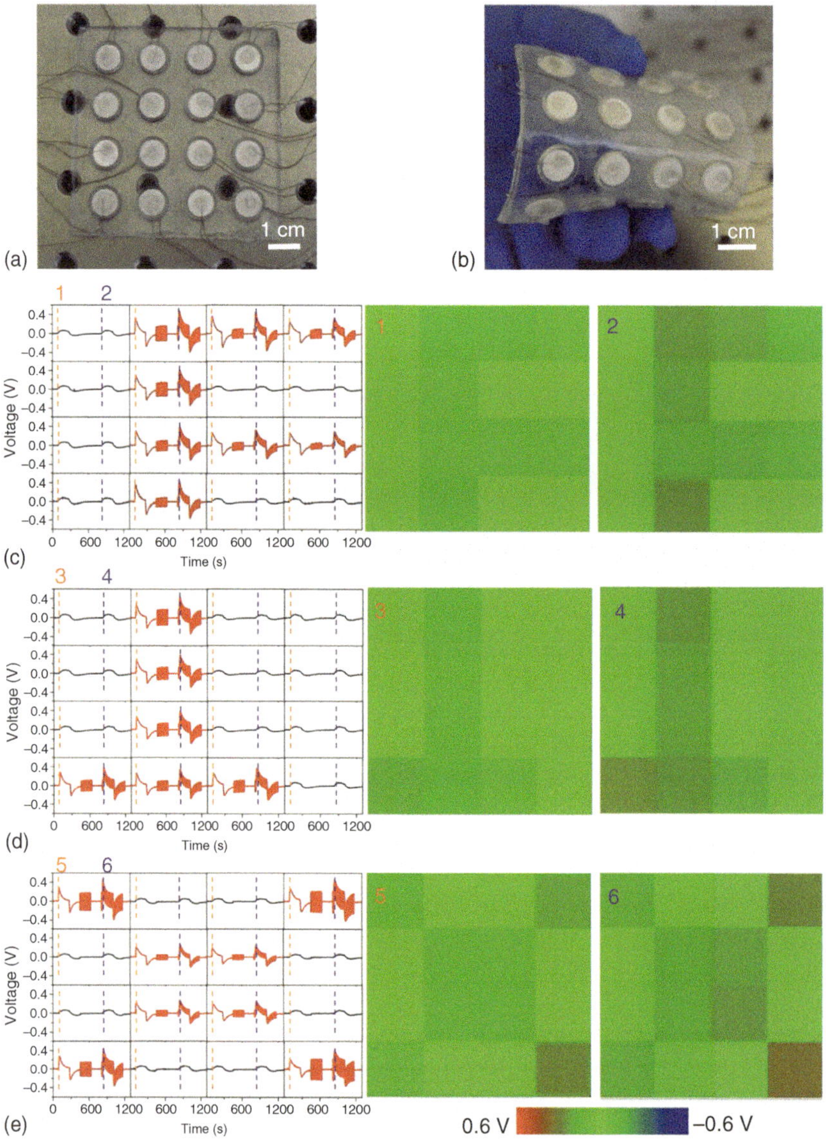

Figure 9.23 ITO/BTO/Ag multi-effects coupled nanogenerator for simultaneous temperature and pressure image sensing. (a,b) Photographs of the device before (a) and after bending (b). (c–e) Output current and the corresponding image figures when "F"-shaped (c), inverted "T"-shaped (d), and "X"-shaped (e) heating and heating + pressing combined effect were applied on the device. Source: Reproduced with permission from Song et al. [55]. Copyright 2019, John Wiley and Sons.

arrays exhibited excellent responses to temperature and pressure simultaneously generated from people's typing motions, which could be designed as a keyboard password.

9.6 Conclusions and Prospects

In conclusion, recent progress in the field of multi-effects coupled nanogenerators have been systematically described. Multi-effects coupled nanogenerators are constructed based on the coupling of pyroelectricity, piezoelectricity, triboelectricity, thermoelectricity, and photovoltaics of multifunctional materials in one structure. The highly integrated multi-effects coupled nanogenerators can generate electric power by individually/simultaneously harvesting multiple kinds of energies, including thermal, light, as well as mechanical energies. Multiple effects can interact with each other to improve the behavior of carriers by modulating the built-in electric field, energy band, and depolarized electric field, consequently promoting the output electric power of multi-effects coupled nanogenerators. The coupled effect nanogenerators can not only be employed to efficiently power electrics but also be utilized as self-powered sensors to motor stimulation from various energies. Although plenty of advancements in the area of multi-effects coupled nanogenerators have been achieved, there are still some challenges. Crucial factors to be considered to improve the power generation of multi-effects coupled nanogenerators are proper materials and novel device configurations. Investigations in terms of materials modification and configuration optimization are still required to further promote the electricity produced by multi-effects coupled nanogenerators. Notably, eco-friendliness and biocompatibility should be taken into consideration when selecting multifunctional materials for these coupled nanogenerators. The working mechanism as well as lifetime of these coupled nanogenerators needs deeper investigations for better practical applications. Furthermore, applying the laboratory-scale achievement into commercial range still remains a challenge. Nevertheless, in view of the growing achievements in this area and more attention received from the researchers, multi-effects coupled nanogenerators are believed to be available for real application before long.

References

1 Wang, J., Wu, C.S., Dai, Y.J. et al. (2017). Achieving ultrahigh triboelectric charge density for efficient energy harvesting. *Nat. Commun.* 8: 88.
2 Chen, B., Yang, Y., and Wang, Z.L. (2018). Scavenging wind energy by triboelectric nanogenerators. *Adv. Energy Mater.* 8: 1702649.
3 Kim, J., Lee, J.H., Ryu, H. et al. (2017). High-performance piezoelectric, pyroelectric, and triboelectric nanogenerators based on P(VDF-TrFE) with controlled crystallinity and dipole alignment. *Adv. Funct. Mater.* 27: 1700702.

4 Jiang, Q., Chen, B., Zhang, K.W., and Yang, Y. (2017). Ag nanoparticle-based triboelectric nanogenerator to scavenge wind energy for a self-charging power unit. *ACS Appl. Mater. Interfaces* 9: 43716–43723.
5 Zi, Y.L., Lin, L., Wang, J. et al. (2015). Triboelectric–pyroelectricpiezoelectric hybrid cell for high efficiency energy-harvesting and self-powered sensing. *Adv. Mater.* 27: 2340–2347.
6 Yang, Y., Zhang, H.L., Zhu, G. et al. (2013). Flexible hybrid energy cell for simultaneously harvesting thermal, mechanical, and solar energies. *ACS Nano* 7: 785–790.
7 Wang, X., Wang, Z.L., and Yang, Y. (2016). Hybridized nanogenerator for simultaneously scavenging mechanical and thermal energies by electromagnetic–triboelectric–thermoelectric effects. *Nano Energy* 26: 164–171.
8 Lee, S., Bae, S.-H., Lin, L. et al. (2013). Flexible hybrid cell for simultaneously harvesting thermal and mechanical energies. *Nano Energy* 2: 817–825.
9 Zheng, Q., Shi, B.J., Li, Z., and Wang, Z.L. (2017). Review recent progress on piezoelectric and triboelectric energy harvesters in biomedical systems. *Adv. Sci.* 4: 1700029.
10 Ji, Y., Gao, T.T., Wang, Z.L., and Yang, Y. (2019). Configuration design of $BiFeO_3$ photovoltaic devices for self-powered electronic watch. *Nano Energy* 64: 103909.
11 Xu, C., Pan, C., Liu, Y., and Wang, Z.L. (2012). Hybrid cells for simultaneously harvesting multi-type energies for self-powered micro/nanosystems. *Nano Energy* 1: 259–272.
12 Alluri, N.R., Saravanakumar, B., and Kim, S.-J. (2015). Flexible, hybrid piezoelectric film ($BaTi_{(1-x)}Zr_xO_3$)/PVDF nanogenerator as a self-powered fluid velocity sensor. *ACS Appl. Mater. Interfaces* 7: 9831–9840.
13 Klochko, N.P., Klepikova, K.S., Kopach, V.R. et al. (2019). Development of semi-transparent ZnO/FTO solar thermoelectric nanogenerator for energy efficient glazing. *Solar Energy* 184: 230–239.
14 Xue, H., Yang, Q., Wang, D.Y. et al. (2017). A wearable pyroelectric nanogenerator and self-powered breathing sensor. *Nano Energy* 38: 147–154.
15 Yin, X., Liu, D., Zhou, L.L. et al. (2019). Structure and dimension effects on the performance of layered triboelectric nanogenerators in contact-separation mode. *ACS Nano* 13 (1): 698–705.
16 Niu, L., Liu, F.C., Zeng, Q.S. et al. (2019). Controlled synthesis and room-temperature pyroelectricity of $CuInP_2S_6$ ultrathin flakes. *Nano Energy* 58: 596–603.
17 Bowen, C.R., Taylor, J., LeBoulbar, E. et al. (2014). Pyroelectric materials and devices for energy harvesting applications. *Energy Environ. Sci.* 7: 3836–3856.
18 Wang, X.Q., Tan, C.F., Chan, K.H. et al. (2017). Nanophotonic-engineered photothermal harnessing for waste heat management and pyroelectric generation. *ACS Nano* 11: 10568–10574.
19 Wang, X.Q., Tan, C.F., Chan, K.H. et al. (2018). In-built thmo-mechanical cooperative feedback mechanism for self-propelled multimodal locomotion and electricity generation. *Nat. Commun.* 9: 3438.

20 Andrei, V., Bethke, K., and Rademann, K. (2016). Thermoelectricity in the context of renewable energy sources: joining forces instead of competing. *Energy Environ. Sci.* 9: 1528–1532.

21 He, M.H., Lin, Y.-J., Chiu, C.-M. et al. (2018). A flexible photo-thermoelectric nanogenerator based on MoS_2/PU photothermal layer for infrared light harvesting. *Nano Energy* 49: 588–595.

22 Zhang, G.J., Liao, Q.L., Zhang, Z. et al. (2016). Novel piezoelectric paper-based flexible nanogenerators composed of $BaTiO_3$ nanoparticles and bacterial cellulose. *Adv. Sci.* 3: 1500257.

23 Gao, M., Li, L.H., Li, W.B. et al. (2016). Direct writing of patterned, lead-free nanowire aligned flexible piezoelectric device. *Adv. Sci.* 1600120.

24 Zhang, Q., Zhang, Z., Liang, Q.J. et al. (2019). Green hybrid power system based on triboelectric nanogenerator for wearable/portable electronics. *Nano Energy* 55: 151–163.

25 Wang, Z.L. (2017). On Maxwell's displacement current for energy and sensors: the origin of nanogenerators. *Mater. Today* 20: 74–82.

26 Yi, F., Zhang, Z., Kang, Z. et al. (2019). Recent advances in triboelectric nanogenerator-based health monitoring. *Adv. Funct. Mater.* 1808849.

27 Khandelwal, G., Chandrasekhar, A., Alluri, N.R. et al. (2018). Trash to energy: a facile, robust and cheap approach for mitigating environment pollutant using household triboelectric nanogenerator. *Appl. Energy* 219: 338–349.

28 Cao, D.W., Wang, C.Y., Zheng, F.G. et al. (2012). High-efficiency ferroelectric-film solar cells with an n-type Cu_2O cathode buffer layer. *Nano Lett.* 12 (6): 2803–2809.

29 Li, L., Lu, H., Yang, Z.Y. et al. (2013). Bandgap-graded CdS_xSe_{1-x} nanowires for high performance field-effect transistors and solar cells. *Adv. Mater.* 25: 1109–1113.

30 Bie, Y.Q., Liao, Z.M., Zhang, H.Z. et al. (2011). Self-powered, ultrafast, visible-blind UV detection and optical logical operation based on ZnO/GaN nanoscale p–n junctions. *Adv. Mater.* 23: 649–653.

31 Ma, N., Zhang, K.W., and Yang, Y. (2017). Photovoltaic–pyroelectric coupled effect induced electricity for self-powered photodetector system. *Adv. Mater.* 29: 1703694.

32 Qi, J., Ma, N., and Yang, Y. (2018). Photovoltaic–pyroelectric coupled effect nanogenerators for self-powered photodetector system. *Adv. Mater. Interfaces* 5: 1701189.

33 Lee, K.Y., Gupta, M.K., and Kim, S.-W. (2015). Transparent flexible stretchable piezoelectric and triboelectric nanogenerators for powering portable electronics. *Nano Energy* 14: 139–160.

34 Guo, Y.B., Zhang, X.S., Wang, Y. et al. (2018). All-fiber hybrid piezoelectric-enhanced triboelectric nanogenerator for wearable gesture monitoring. *Nano Energy* 48: 152–160.

35 Ma, N. and Yang, Y. (2018). Boosted photocurrent in ferroelectric $BaTiO_3$ materials via two dimensional planar-structured contact configurations. *Nano Energy* 50: 417–424.

36 Shin, S.-H., Kim, Y.-H., Lee, M.H. et al. (2014). Hemispherically aggregated $BaTiO_3$ nanoparticle composite thin film for high-performance flexible piezoelectric nanogenerator. *ACS Nano* 8 (3): 2766–2773.

37 Sahatiya, P., Kannan, S., and Badhulika, S. (2018). Few layer MoS_2 and in situ poled PVDF nanofibers on low cost paper substrate as high performance piezo-triboelectric hybrid nanogenerator: Energy harvesting from handwriting and human touch. *Appl. Mater. Today* 13: 91–99.

38 You, M.H., Wang, X.X., Yan, X. et al. (2018). Self-powered flexible hybrid piezoelectric-pyroelectric nanogenerator based on non-woven nanofiber membranes. *J. Mater. Chem. A* 6: 3500–3509.

39 Zhang, K.W., Ouyang, B.S., Wang, Y.H. et al. (2019). Coupling enhancement of photo-thermoelectric conversion in a lateral ZnO nanowire array. *ACS Appl. Energy Mater.* 2: 7647–7654.

40 Yang, Q., Guo, X., Wang, W.H. et al. (2010). Enhancing sensitivity of a single ZnO micro-/nanowire photodetector by piezo-phototronic effect. *ACS Nano* 4: 6285–6291.

41 Ji, Y., Zhang, K.W., Wang, Z.L., and Yang, Y. (2019). Piezo-pyro-photoelectric effects induced coupling enhancement of charge quantity in $BaTiO_3$ materials for simultaneously scavenging light and vibration energies. *Energy Environ. Sci.* 12: 1231–1240.

42 Zhang, K.W., Wang, S.H., and Yang, Y. (2017). A one-structure-based piezo-tribo-pyro-photoelectric effects coupled nanogenerator for simultaneously scavenging mechanical, thermal, and solar energies. *Adv. Energy Mater.* 7: 1601852.

43 Gao, T.T., Ji, Y., and Yang, Y. (2019). Thermo-phototronic effect enhanced photodetectors based on porous ZnO materials. *Adv. Electron. Mater.* 5: 1900776.

44 Ouyang, B.S., Zhang, K.W., and Yang, Y. (2018). Photocurrent polarity controlled by light wavelength in self-powered ZnO nanowires/SnS photodetector system. *Science* 1: 16–23.

45 Ouyang, B.S., Chang, C., Zhao, L.D. et al. (2019). Thermo-photoelectric coupled effect induced electricity in N-type SnSe:Br single crystals for enhanced self-powered photodetectors. *Nano Energy* 66: 104111.

46 Jood, P., Mehta, R.J., Zhang, Y.L. et al. (2011). Al-doped zinc oxide nanocomposites with enhanced thermoelectric properties. *Nano Lett.* 11: 4337–4342.

47 Yang, Y., Pradel, K.C., Jing, Q.S. et al. (2012). Thermoelectric nanogenerators based on single Sb-doped ZnO micro/nanobelts. *ACS Nano* 6: 6984–6989.

48 Han, X., Du, W.M., Yu, R.M. et al. (2015). Piezo-phototronic enhanced UV sensing based on a nanowire photodetector array. *Adv. Mater.* 27: 7963–7969.

49 Song, W.D., Wang, X.F., Xia, C. et al. (2017). Improved photoresponse of a-axis GaN microwire/p-polymer hybrid photosensor by the piezo-phototronic effect. *Nano Energy* 33: 272–279.

50 Mitra, M., Drayton, J., Cooray, M.L.C. et al. (2007). Piezo-photovoltaic coupling in CdS-based thin-film photovoltaics. *J. Appl. Phys.* 102: 034505.

51 Kumar, M., Patel, M., Kim, J., and Lim, D. (2017). Enhanced broadband photoresponse of a self-powered photodetector based on vertically grown SnS layers via the pyro-phototronic effect. *Nanoscale* 9: 19201–19208.

52 Ko, Y.J., Kim, D.Y., Won, S.S. et al. (2016). Flexible $Pb(Zr_{0.52}Ti_{0.48})O_3$ films for a hybrid piezoelectric-pyroelectric nanogenerator under harsh environments. *ACS Appl. MaterInterfaces* 8: 6504–6511.

53 Ji, Y., Wang, Y.H., and Yang, Y. (2019). Photovoltaic–pyroelectric–piezoelectric coupled effect induced electricity for self-powered coupled sensing. *Adv. Electron. Mater.* 5: 1900195.

54 Qi, J., Ma, N., Ma, X.C. et al. (2018). Enhanced photocurrent in $BiFeO_3$ materials by coupling temperature and thermo-phototronic effects for self-powered ultraviolet photodetector system. *ACS Appl. Mater. Interfaces* 10: 13712–13719.

55 Song, K., Zhao, R.D., Wang, Z.L., and Yang, Y. (2019). Conjuncted pyro-piezoelectric effect for self-powered simultaneous temperature and pressure sensing. *Adv. Mater.* 31: 1902831.

56 Shi, B.J., Zheng, Q., Jiang, W. et al. (2016). A packaged self-powered system with universal connectors based on hybridized nanogenerators. *Adv. Mater.* 28: 846–852.

57 Tamang, A., Ghosh, S.K., Garain, S. et al. (2015). DNA-assisted β-phase nucleation and alignment of molecular dipoles in PVDF film: a realization of self-poled bioinspired flexible polymer nanogenerator for portable electronic devices. *ACS Appl. Mater. Interfaces* 7: 16143–16147.

58 Sultana, A., Ghosh, S.K., Alam, M.M. et al. (2019). Methylammonium lead iodide incorporated poly(vinylidene fluoride) nanofibers for flexible piezoelectri-pyroelectric Nanogenerator. *ACS Appl. Mater. Interfaces* 11: 27279–27287.

59 Wang, S.H., Wang, Z.L., and Yang, Y. (2016). A one-structure-based hybridized nanogenerator for scavenging mechanical and thermal energies by triboelectric-piezoelectric-pyroelectric effects. *Adv. Mater.* 28: 2881–2887.

60 Lee, J.H., Ryu, H.J., Kim, T.Y. et al. (2015). Thermally induced strain-coupled highly stretchable and sensitive pyroelectric nanogenerators. *Adv. Energy Mater.* 5: 1500704.

61 Zhang, K.W. and Yang, Y. (2017). Thermo-phototronic effect enhanced InP/ZnO nanorod heterojunction solar cells for self-powered wearable electronics. *Adv. Funct. Mater.* 27: 1703331.

62 Hu, G.F., Guo, W.X., Yu, R.M. et al. (2016). Enhanced performances of flexible ZnO/perovskite solar cells by piezo-phototronic effect. *Nano Energy* 27-33.

63 Lee, J.-H., Lee, K.Y., Gupta, M.K. et al. (2014). Highly stretchable piezoelectric-pyroelectric hybrid nanogenerator. *Adv. Mater.* 26: 765–769.

64 Ji, Y., Zhang, K.W., and Yang, Y. (2018). A one-structure-based multieffects coupled nanogenerator for simultaneously scavenging thermal, solar, and mechanical energies. *Adv. Sci.* 5: 1700622.

65 Zhu, L.P., Wang, L.F., Xue, F. et al. (2016). Piezo-phototronic effect enhanced flexible solar cells based on n-ZnO/p-SnS core-shell nanowire array. *Adv. Sci.* 4: 1600185.

10

Coupled Nanogenerators for New Physical Effects

10.1 Introduction

Coupled nanogenerators have been extensively reported due to possible applications in multiple energies scavenging and sensor systems [1–5]. Coupled nanogenerators are based on several physical effects such as piezoelectric, triboelectric, pyroelectric, thermoelectric, and photovoltaic effects. Piezoelectric nanogenerators are based on the mechanical strain-induced flow of electrons, requiring specific piezoelectric materials such as $Pb(Zr,Ti)O_3$ [6, 7] and $BaTiO_3$ [8, 9]. The working of triboelectric nanogenerators is based on the periodic motions between two materials, which can induce the flow of electrons in the circuits due to electrostatic induction [10–15]. Pyroelectric nanogenerators are based on the spontaneous polarization variations with temperature, driving the electrons to flow between two electrodes [16, 17]. Thermoelectric nanogenerators are based on the temperature difference in the device, driving electrons or holes to flow in the thermoelectric materials [18, 19]. Photovoltaic nanogenerators are based on the spontaneous polarization induced inner electric fields, which can separate the charges to flow in the ferroelectric materials [20]. By comparing the different nanogenerators, it can be clearly seen that the electrons can flow in the external circuits for the piezoelectric/triboelectric/pyroelectric nanogenerators, while the electrons and holes can flow inside of the materials or devices for the thermoelectric and photovoltaic nanogenerators.

The coupling processes between the different physical effects are very interesting such as the coupling enhancement between the piezoelectric and photovoltaic effects [1]. The vibration-induced cooling is the result of coupling enhancement between the two effects. The corresponding mechanisms of coupling enhancements between these effects need to be understood, where some new physical effects can be seen in the coupling processes. The pyro-phototronic effect is focused on effective coupling of pyroelectric and photoelectric effects in the semiconducting materials [21]. The ferro-pyro-phototronic effect is focused on effective coupling of pyroelectric and photovoltaic effects in the ferroelectric materials [22]. The thermo-phototronic effect is focused on effective coupling of thermoelectric and photoelectric effects in semiconducting materials [23].

*Corresponding author: yayang@binn.cas.cn

Hybridized and Coupled Nanogenerators: Design, Performance, and Applications,
First Edition. Ya Yang.

10.2 Pyro-Phototronic Effect

10.2.1 Introduction

Some semiconducting materials have been extensively reported to have pyroelectric effect [24]. The pyroelectric current can be given by

$$I = pA\frac{\partial T}{\partial t} \tag{10.1}$$

where T is the temperature, A is the device area, and p is the pyroelectric constant [25]. The measured pyroelectric coefficient for ZnO is about $-9.4\,\mu Cm^{-2}/K$, and CdS has a measured pyroelectric coefficient of about $-4.0\,\mu Cm^{-2}/K$ [26]. Zhang et al. first reported the pyro-phototronic effect in 2016 by using the coupling of semiconductor, pyroelectricity, and photoexcitation [21], as illustrated in Figure 10.1. The temperature variation-induced pyro-potential can be utilized to modulate the transport process of photoinduced electrons and holes in semiconducting materials. The contributions of the pyro-phototronic effect are mainly due to both the pyroelectric nanogenerator-induced electricity and the corresponding energy band bending variations at the metal–semiconductor interfaces or p–n junctions.

10.2.2 Possible Semiconducting Materials

The pyro-phototronic effect is focused on semiconducting materials, which have pyroelectric characteristics. ZnO nanomaterials have been reported to have pyro-phototronic effect [21]. Moreover, other semiconducting materials

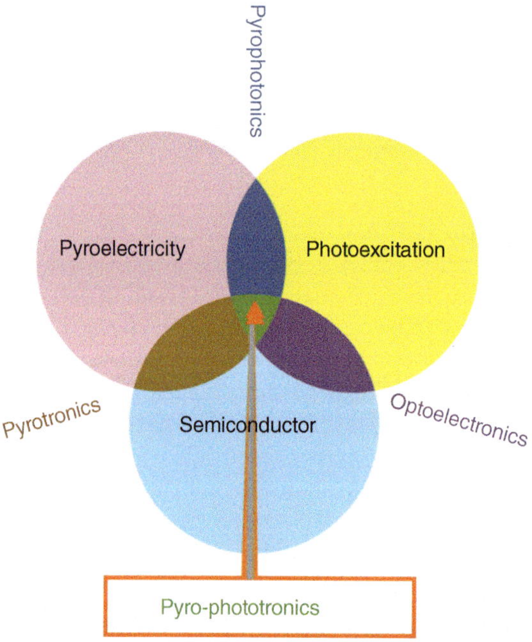

Figure 10.1 Schematic diagram of pyro-phototronics. Source: Reproduced with permission from Zhang et al. [21]. Copyright 2016, American Chemical Society.

are also good candidates such as GaN and AlN. Owing to the wide bandgaps of these materials, they have much better applications in photodetectors instead of solar cells, where the pyro-phototronic effect was used to increase the output performance of various photodetectors. The chosen materials require high pyroelectric constants although they are semiconducting materials, so that the single crystalline semiconducting nanomaterials may have better performance than bulk or film materials. The thickness of the materials also needs to be considered since other effects such as thermoelectric effect will appear if the used materials are too thick. Usually, it is difficult to obtain a stable temperature difference across several tens of nanometers in the nanomaterials. If the thickness becomes several tens of micrometers, the thermoelectric effect will be obvious.

10.2.3 Applications

The photoelectric process in the P3HT/ZnO nanowires p–n heterojunction photovoltaic cell can be modulated by using pyro-phototronic effect [21]. The corresponding device consists of an indium tin oxide (ITO) glass, 3 μm ZnO nanowires, the P3HT and PEDOT:PSS, and Ag nanowires. As depicted in Figure 10.2a, the temperature variations of the device can be recorded by a temperature sensor. Figure 10.2b,c displays the output current signals under different time and temperature conditions. The corresponding output current signals include a pyroelectric current signal and a photovoltaic current signal. Figure 10.2b shows that the corresponding output current can be increased to about 105 nA under indoor light illumination, reaching a sharp peak value of about 124 nA under the cooling condition. The corresponding output current can be increased by 18%, and the output voltage can be enhanced by 152%. The photovoltaic performances of P3HT/ZnO can be obviously modulated by this effect.

To understand the corresponding contributions of the pyroelectric nanogenerator and photovoltaic cell in the device, the different circuits of the two parts are marked, as illustrated in Figure 10.2d–f. Initially, the photovoltaic effect can drive the free carriers to flow from Ag to ITO electrodes under light illumination condition, as displayed in Figure 10.2d. At this time, no current signals were found to flow in the corresponding circuit due to the constant polarization at constant temperature. When the device is under cooling condition, the temperature in the ZnO nanowires is decreased, leading to a polarized distribution of charges in the ZnO, where the positive charges were found at the P3HT/ZnO interface, as displayed in Figure 10.2e. These positive charges at the interface dramatically lower the barrier height at the interface, inducing the increase of the photovoltaic current. Moreover, the pyroelectric nanogenerator-induced current has the same flow direction as the photovoltaic current. When the cooling temperature and light illumination were fixed at constant values, the stable output current plateau exhibits a larger value than that before cooling. The enhancement of this is associated with the band bending of the heterojunction interface, which can obviously promote the photovoltaic performance of the device. When the cooled temperature was increased to room temperature, the electrons can flow along the opposite direction for the pyroelectric nanogenerator, as illustrated in Figure 10.2f.

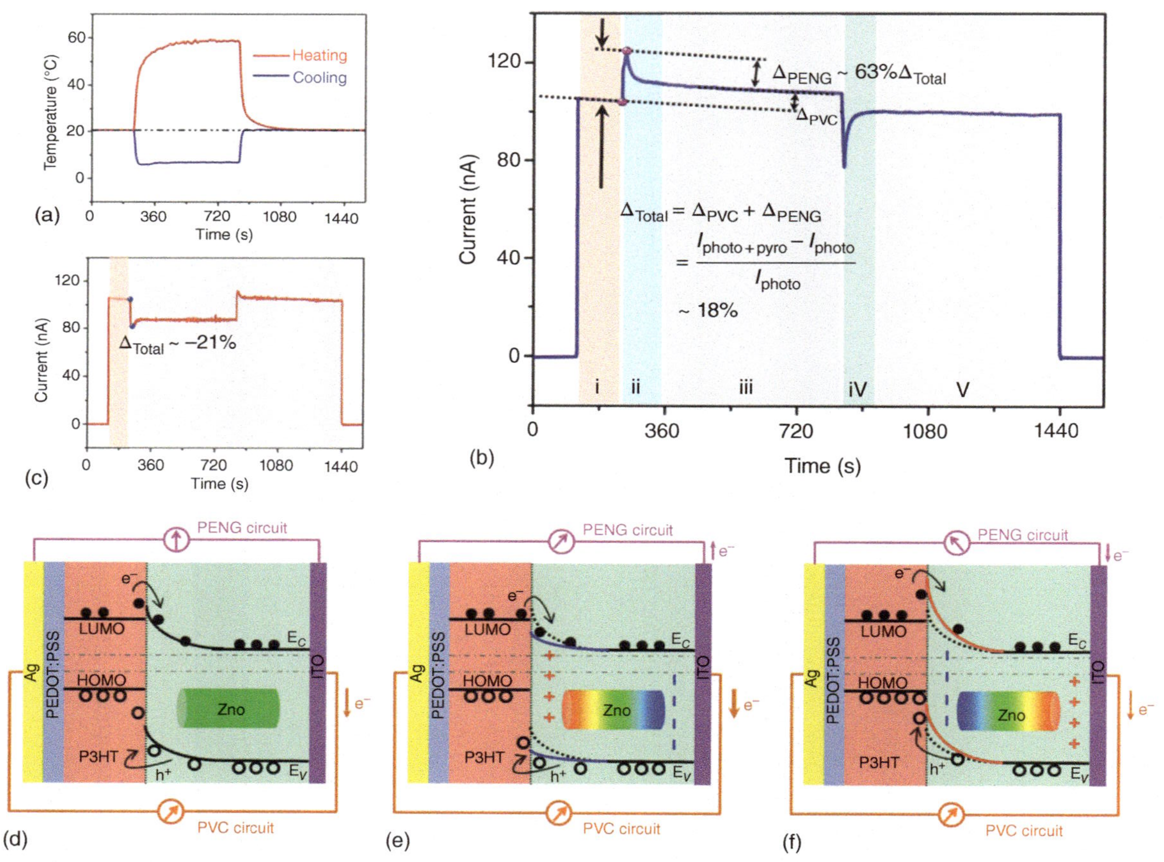

Figure 10.2 Working mechanism of the pyro-phototronic effect enhanced solar cell under indoor light illumination. (a) The temperature changing curves under heating and cooling conditions. (b) *I*–*t* curve of the pyro-phototronic effect enhanced solar cell under cooling condition. (c) *I*–*t* curve of the pyro-phototronic effect enhanced solar cell under heating condition. (d–f) Schematic energy band diagrams of the devices without temperature change (d), with temperature cooling condition (e), and with temperature heating condition (f). Source: Reproduced with permission from Zhang et al. [21]. Copyright 2016, American Chemical Society.

The pyro-phototronic effect has potential applications in enhancing the output performances of solar cells, photodetectors, light-emitting diodes, and so on.

10.3 Ferro-Pyro-Phototronic Effect

10.3.1 Introduction

Professor Ya Yang and coworkers first invented the ferro-pyro-phototronic effect in 2018 [22], which is focused on the coupling among pyroelectricity, ferroelectric materials, and photoexcitation. Pyro-phototronic effect is focused on the effective coupling among pyroelectricity, semiconductor, and photoexcitation for enhancing the related performances of solar cells and photodetectors [21]. However, it is difficult to realize a large performance modulation by using ZnO nanowires due to the low pyroelectric constant of about 1.5 nC/cm^2 K. Other high-performance pyroelectric materials are needed to realize high-performance pyroelectric–photovoltaic coupled energy scavenging devices or sensors. Some ferroelectric materials such as $Pb(Zr,Ti)O_3$ [27–29], $BaTiO_3$ [30], and $LiNbO_3$ [31, 32] have been found to exhibit much better pyroelectric performances, as compared with the semiconducting ZnO nanomaterials. Photovoltaic performances can also be observed in these ferroelectric materials under light illumination with short wavelengths [33]. The effective separation of charges under light illumination was realized due to the built-in electric field induced by the pyroelectric effect in these ferroelectric materials instead of semiconducting materials. The ferro-pyro-phototronic effect has been displayed in Figure 10.3. Pyrophotonic effect is the effective coupling between pyroelectricity and photoexcitation. It is crucial for confirming the relationships between pyroelectricity and the related output performances in these ferroelectric materials and understanding the corresponding coupling mechanisms under different temperature variation conditions.

10.3.2 Possible Ferroelectric Materials

The ferro-pyro-phototronic effect is focused on ferroelectric materials, which have good pyroelectric characteristics. Pyroelectric effect has primary and secondary coefficients, where the primary pyroelectric coefficient has been defined as the measured pyroelectric coefficient under clamped condition. Secondary pyroelectric coefficient has been defined as the pyroelectric coefficient due to thermal expansion. The ferroelectric $BaTiO_3$ has a primary coefficient of about −260 μCm^{-2}/K and a secondary coefficient of about 60 μCm^{-2}/K, so that the measured pyroelectric coefficient is about −200 μCm^{-2}/K [34]. The ferroelectric $Pb(Zr_{0.95}Ti_{0.05})O_3$ has a reported pyroelectric coefficient of about −268 μCm^{-2}/K and the ferroelectric $Sr_{0.5}Ba_{0.5}Nb_2O_6$ material has a reported pyroelectric coefficient of about −550 μCm^{-2}/K [34]. The ferroelectric $LiNbO_3$ material has a reported pyroelectric coefficient of about −83 μCm^{-2}/K, and $LiTaO_3$ has a reported pyroelectric coefficient of about −176 μCm^{-2}/K [34]. Moreover, some 2D ferroelectric materials have been reported such as 4 nm

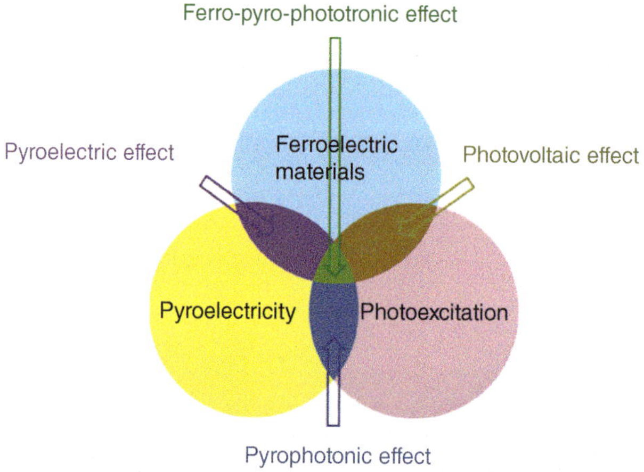

Figure 10.3 Schematic diagram of ferro-pyro-phototronic effect. Source: Reproduced with permission from Zhao et al. [22]. Copyright 2018, Elsevier.

$CuInP_2S_6$ [35], 3 nm In_2Se_3 [36–40], and 1.4 nm WTe_2 [41]. Some theoretical predictions about the ferroelectric characteristics in monolayer 2D materials were also reported [42–45]. Although these 2D materials exhibit ferroelectricity, it is still difficult to obtain high pyroelectric coefficients from these materials.

10.3.3 Applications

The ferro-pyro-phototronic effect was reported to change the photocurrent in radically polarized $BaTiO_3$ film for increasing the performances of photodetectors [22]. Figure 10.4a displays the measured output current signals under individual "light," "heating," and "light + heating" conditions, indicating that a coupling weakness between the "light" and "heating" conditions can be seen. As shown in Figure 10.4b, the coupling effect induced enhancement ("1 + 1 > 2") of the measured photocurrent between the "light" and "heating" conditions can be clearly seen under the "light + cooling" condition. The corresponding output current plateau under the "light + cooling" condition is obviously larger by comparing it with that under the individual "light" condition, suggesting that temperature condition is crucial for enhancing the output performances in ferroelectric $BaTiO_3$ materials.

To understand the mechanism of the enhancement of the photocurrent plateau, the change of energy band structures for the $BaTiO_3$ device has been shown in Figure 10.4c–f. Figure 10.4c presents the corresponding valence and conduction energies of ITO electrodes and $BaTiO_3$, where the $BaTiO_3$ has the Schottky contact with the two ITO electrodes, resulting in two Schottky barriers at the interfaces between ITO electrodes and $BaTiO_3$. The different polarized charges can be produced by polarizing the $BaTiO_3$ materials, so that the Schottky barrier height (SBH) can be increased by the negative polarized charges, but the SBH can be decreased by the positive polarized charges produced. As depicted in Figure 10.4d, the electrons and holes produced under UV light illumination

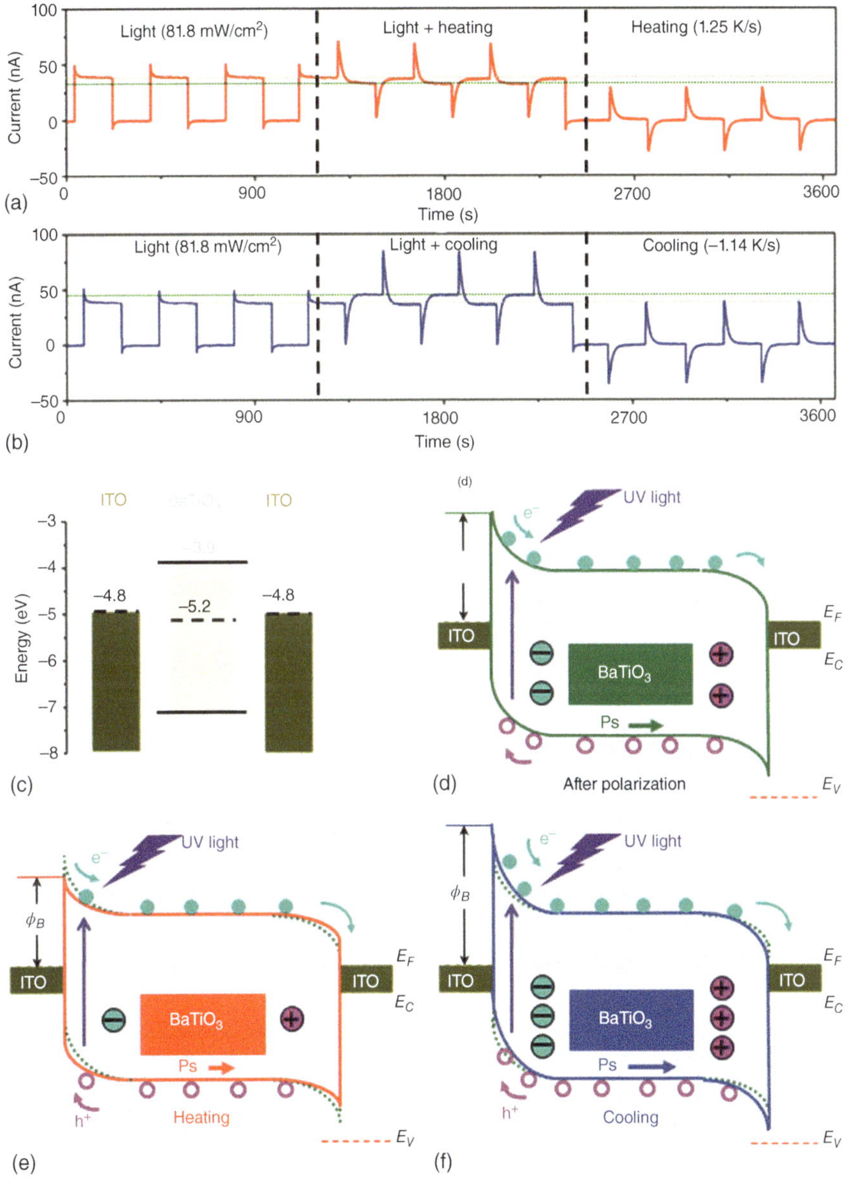

Figure 10.4 The mechanism of the ferro-pyro-phototronic effect induced enhancement of photocurrent. (a,b) I–t curves of the device under light + heating (a) and light + cooling (b) conditions. (c) Energy-level diagram of the device. (d,e,f) Schematic diagrams of the energy band structures of the device without temperature change (d), with heating condition (e), and with cooling condition (f). Source: Reproduced with permission from Zhao et al. [22]. Copyright 2018, Elsevier.

were dramatically separated to deliver the measured signals. When the $BaTiO_3$ material is under heating condition, the decreased polarized charges can induce opposite variations for the left and right Schottky barriers, as presented in Figure 10.4e. As a result, the effective separation of the charges induced by light illumination can be weakened, decreasing the corresponding output current signals, as depicted in Figure 10.4a. Moreover, the SBH on the left side was increased but the SBH on the right side was decreased due to the cooling temperature-induced enhancement of polarized charges in the ferroelectric $BaTiO_3$ material, as displayed in Figure 10.4f. As a result, the separation of charges under light illumination becomes faster than that in Figure 10.4d,e, resulting in the observed coupling enhancement for the photocurrent plateau in Figure 10.4b. The results indicate that coupling enhancement can be found in ferroelectric $BaTiO_3$ material under cooling conditions.

Cooling temperature can enhance the photocurrent in $BaTiO_3$ material by using ferro-pyro-phototronic effect. It is necessary to investigate the photocurrent in $BaTiO_3$ material under low temperature conditions [46]. Figure 10.5a presents the photocurrent values of the $BaTiO_3$ material under different temperatures ranging from 80 to 360 K. A maximum value of about 393 nA can be seen at 220 K, and the photocurrent exhibits an obvious decrease with increasing temperatures from 220 to 360 K. An enhanced ratio of over 120% can be found at low temperatures ranging from 80 to 240 K. To explain the relationship between the measured photocurrent and the corresponding temperatures, the remnant polarization Pr values under different temperature conditions can be obtained from *P–E* hysteresis loops, as depicted in Figure 10.5b. The remnant polarization values in both dark and light illumination can be increased by increasing the temperature, obtaining the largest value at a temperature of about 160 K, but the value decreases when the temperature was increased further.

To understand the temperature dependence of photocurrent signals in the $BaTiO_3$ material, the energy band diagrams have been depicted in Figure 10.5c–e. The work functions of $BaTiO_3$ and ITO electrode are 5.3 and 4.5 eV, respectively. By using polarized $BaTiO_3$ materials, the SBH on the left side can be increased by the negative polarized charges, but the SBH on the right side can be decreased by the polarized charges. Usually, the photocurrent can be determined by two processes including light-induced carrier generation and separation due to the inner electric field. Under low temperature conditions, the enhanced photocurrent signals are due to the ferro-pyro-phototronic effect. First, the electrons induced by light illumination can be increased due to both the band–band excitation and the activation from shallow trap levels, as illustrated in Figure 10.5c. Second, the polarization-induced built-in electric field in the $BaTiO_3$ can effectively separate the electrons and holes produced. Third, the pyroelectric effect induced by light illumination can also have a contribution to the increased output current signals. Under room temperature, the contribution from the shallow trap levels has decreased, as displayed in Figure 10.5d. The polarization electric field in the $BaTiO_3$ can also decrease with increasing temperature, resulting in the decrease of photocurrent. When the device is under a temperature of about 360 K, the polarization electric field becomes much weaker, resulting in the large recombination between electrons and holes since the electrons and holes

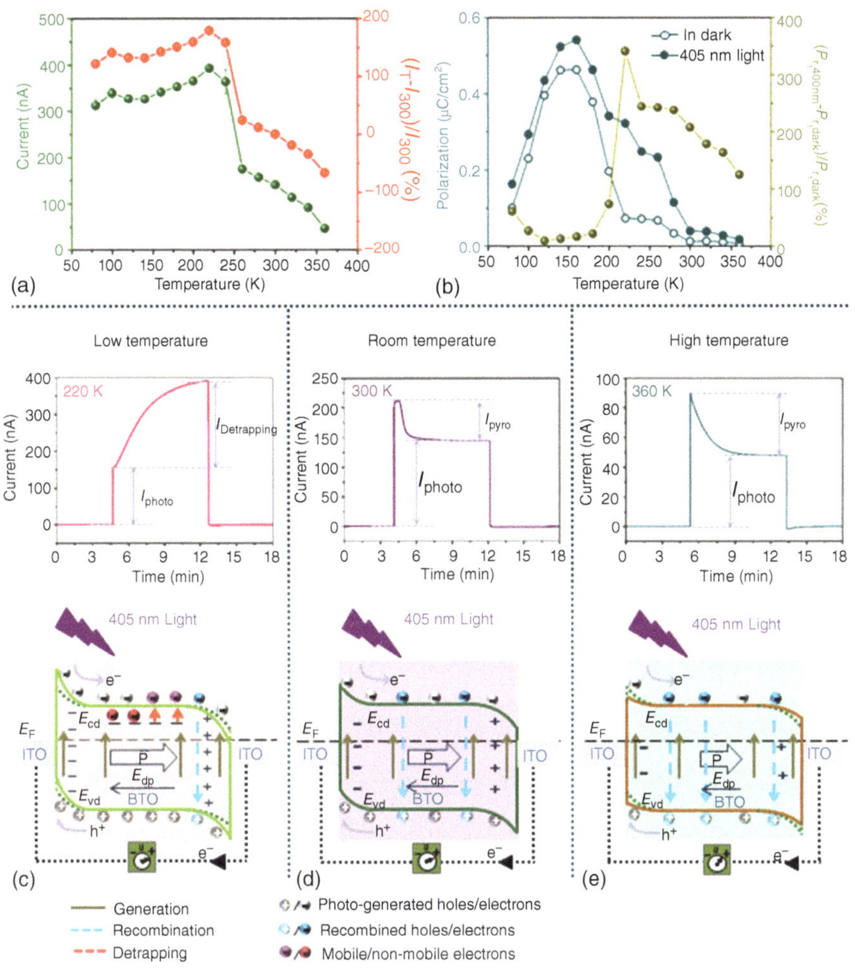

Figure 10.5 The ferro-pyro-phototronic effect induced enhancement of photocurrent under low temperatures. (a) Measured photocurrent signals under different temperatures and the corresponding enhancement ratios. (b) The remnant polarization values of the device under different temperatures and the corresponding enhancement ratios. (c–e) The working mechanisms of the ferro-pyro-phototronic effect induced enhancement of photocurrent signals under low temperature (c), room temperature (d), and high temperature (e). Source: Reproduced with permission from Ma and Yang [46]. Copyright 2019, Elsevier.

cannot be effectively separated, as presented in Figure 10.5e. As a result, the photocurrent can be obviously decreased under high temperature conditions. Although the ferroelectric $BaTiO_3$ materials exhibit high photocurrent signals under the cooling temperature condition, it is not the same for all the ferroelectric materials for this phenomenon, where some ferroelectric materials may have much larger photocurrent signals under high temperature conditions. However, the ferro-pyro-phototronic effect can be used to effectively modulate the photocurrent signals in these ferroelectric materials.

10.4 Thermo-Phototronic Effect

10.4.1 Introduction

Professor Ya Yang and coworkers first invented the thermo-phototronic effect in 2017 [23], which is focused on the three-way coupling among thermoelectricity, semiconductor, and photoexcitation. Two dominant techniques of scavenging solar energy are based on solar–photovoltaic and solar–thermal devices [47–49]. It is important to couple the two approaches to realize the maximum harvesting of solar energy, where the one-circuit-based coupling systems are desirable for realizing this functionality. The intrinsic effect of thermoelectricity on the photovoltaic performances of the devices needs to be confirmed. The thermo-phototronic effect is focused on the interaction between the thermoelectric effect and the photoexcitation. A thermoelectric potential induced by the temperature difference across the device can be utilized to induce electrons or holes to flow, separate the charges, and change the energy band barrier height at the interfaces. Moreover, the energy band of the thermoelectric material can be tilted due to the thermoelectric effect-induced built-in electric field in the devices. The interface barriers between the thermoelectric material and the metal electrodes can be increased and decreased by the accumulated negative and positive charges, respectively.

As illustrated in Figure 10.6, the thermophotonic effect is focused on the effective coupling of thermoelectricity and photoexcitation, and the thermo-phototronic effect is focused on the effective coupling among thermoelectricity, semiconductor, and photoexcitation. The thermo-phototronic effect has potential applications in solar cells, photodetectors, and so on. The thermo-phototronic effect-based related devices require both high thermoelectric and high photoelectric characteristics. Some materials have good thermoelectric performance, but the corresponding photoelectric performance of these materials is not good. Appropriate carrier concentration of thermoelectric materials is a key factor to look for in good thermo-phototronic materials.

10.4.2 Possible Thermoelectric Materials

The thermo-phototronic effect is based on thermoelectric materials, requiring good thermoelectric performances. Inorganic semiconductors and alloys including Bi_2Te_3, PbTe, and Sb_2Te_3 are the most widely reported thermoelectric materials [50, 51]. The thermoelectric figure of merit (ZT) has been given as

$$ZT = \frac{S^2\sigma}{k}T \tag{10.2}$$

where S is the Seebeck coefficient, σ is the electrical conductivity, and k is the thermal conductivity [52]. These three material properties are interrelated, resulting in the challenge of optimizing materials for good thermoelectric energy scavenging. The molecular-beam epitaxy method has been used to obtain artificial superlattices such as PbTe [53, 54], Bi_2Te_3 [55], and SiGe [56], where an ultrahigh ZT value of larger than 2.0 was reported in the Bi_2Te_3/Sb_2Te_3 superlattice

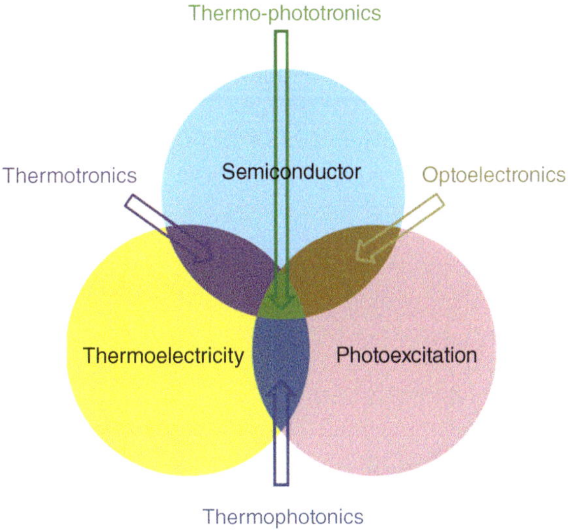

Figure 10.6 Schematic diagram of thermo-phototronics. Source: Reproduced with permission from Zhang and Yang [23]. Copyright 2017, John Wiley and Sons.

in 2001. Moreover, the bulk SnSe crystal was reported to have a high ZT value of larger than 2.0 due to the strongly anisotropic transport properties [57]. SnSe, which has room-temperature electric resistivity values from 10^1 to $10^5\,\Omega\,\text{cm}$, is also a good optoelectronic material.

Since thermoelectric-induced potential has the important effect of modulating the photovoltaic performance of a device due to thermo-phototronic effect, the Seebeck coefficient S is a key factor in this process, which can be expressed as

$$S = \frac{\Delta V}{\Delta T} \tag{10.3}$$

where V is the thermoelectric voltage and T is the thermoelectric temperature. The high Seebeck coefficient α can induce a high voltage difference across the device, which can lead to good modulations of the output performance. The single Sb-doped ZnO microwire has a Seebeck coefficient of about $-350\,\mu\text{V/K}$ [58]. Ionic liquid electrolytes have Seebeck coefficients of larger than 1.5 mV/K [59]. The strong ionic Soret effect in polymeric electrolytes can induce a large Seebeck coefficient of over 10 mV/K [60].

10.4.3 Applications

The thermo-phototronic effect has been first used to modulate the photocurrent in InP/ZnO nanowires heterojunction for enhancing the performance of solar cells in 2017 [23]. The device consists of the bottom Ag electrode, p-type InP material, ZnO nanowire array, and the Ag nanowire electrode. Under a fixed temperature difference of about 3.5 °C in the device, the corresponding output voltage and current signals were increased by 27.3% and 76% under a light illumination of 20 Lux, respectively. The enhancement is associated with the inner thermoelectric electric field-induced charge transfer in the InP material due to thermo-phototronic effect. Under light illumination conditions, the light-induced holes can flow into the bottom Ag electrode, but the electrons induced by the light can flow into the top Ag electrode due to the separation of charges by the p–n junction created by the inner electric field. When the InP side is under a cooling temperature, an inner thermoelectric potential can be created by the temperature difference across the InP material, which can enhance the transfer of holes in the InP. The conduction and valence bands of InP can be lifted at the interface due to the accumulated thermoelectric charges. As a result, the output performances of the device can be increased under the cooling temperature and light illumination conditions. The modulation of thermo-phototronic effect is associated with the thermoelectric-induced built-in electric field to induce charges to flow in the material and the variations of energy band barrier at the interfaces due to the thermoelectric-induced accumulated charges.

The modulation of p–n junction photocurrent by thermo-phototronic effect has been also reported in the ZnO nanowires/SnS heterojunction as a photodetector array to distinguish the detected light wavelength in the ultraviolet and visible regions [61]. The switching of the interface modulation was realized by using the pyro-phototronic effect inside ZnO nanowires and the thermo-phototronic effect inside SnS. By using the thermo-phototronic effect-induced thermoelectric potential, the responsivity of about −155 μA/W can be obtained. The strategy of controlling the photocurrent polarity can be utilized to improve the resolution of dynamic light detection. The dual-polarity response can also be reported in the ZnO nanowires/Sb_2Se_3 film heterojunction photodetector array for optical communication [62]. As shown in Figure 10.7a, the fabricated device includes a transparent ITO electrode, a p-type Sb_2Se_3 film, an n-type ZnO nanowires layer, and a Ag electrode, where each pixel unit has an area of about 3 × 3 mm. Figure 10.7b displays the thickness of each layer in the fabricated device. The corresponding output current of the device can flow from ITO to Ag electrodes under light illumination with the long wavelength region from 760 to 880 nm, as presented in Figure 10.7c. The flow of the current is due to the p–n junction-induced inner electric field, as presented in Figure 10.7d. However, under light illumination with the short wavelength region, the corresponding output current of the device can flow from Ag to ITO electrodes in Figure 10.7e. The light-induced thermoelectric electric filed can drive the holes to flow from ITO electrode to ZnO in the device due to the thermo-phototronic effect of the Sb_2Se_3 thin film, as displayed in Figure 10.7f. The dual polarity of photocurrent signals is related to the competition of both the photoelectric effect and the thermo-phototronic effect. Moreover, the corresponding current polarity switching wavelength could be changed by choosing the thickness of the Sb_2Se_3 thin film or the absorption coefficient, Seebeck coefficient of the material. More improvements of these current dual-polarity photodetectors can be realized by using the other semiconducting materials to increase the responsivity.

Instead of p–n junction, the modulation of photocurrent by using thermo-phototronic effect has been reported in single crystal SnSe materials [63]. The current coupling enhancement is attributed to the same current directions of thermoelectric and photoelectric signals, where the thermo-phototronic effect-induced thermoelectric electric field can accelerate the flow of the electrons induced by light illumination. As illustrated in Figure 10.8, the

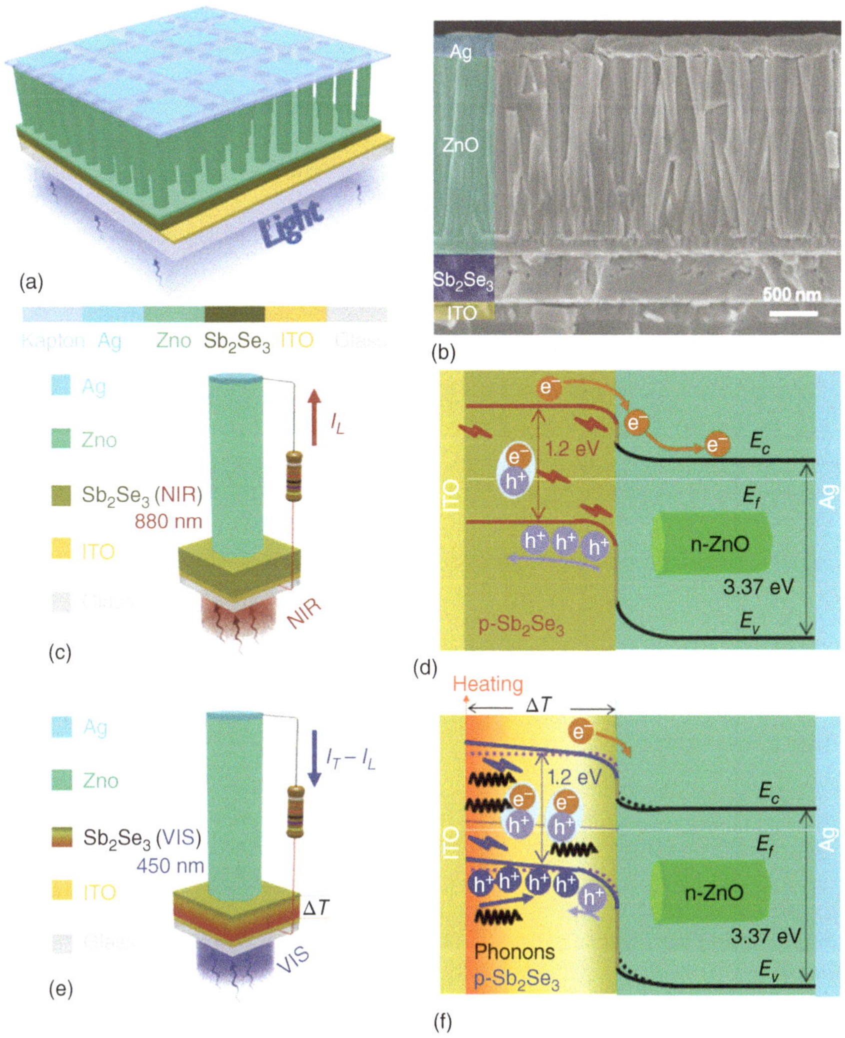

Figure 10.7 The thermo-phototronic effect induced photocurrent. (a) Schematic diagram of the device array. (b) Cross-sectional scanning electron microscopic (SEM) image of the fabricated device. (c,d) Schematic diagrams of the working state (c) and the energy band diagram (d) of the device under 880 nm light illumination. (e,f) Schematic diagrams of the working state (e) and the energy band diagram (f) of the device under 450 nm light illumination. Source: Reproduced with permission from Ouyang et al. [62]. Copyright 2019, Elsevier.

thermo-phototronic effect has been used to explain the coupling enhancement of output current signals in the SnSe materials. The four different operation states of the device have been displayed in Figure 10.8a,c,e,g. The energy band diagrams have been illustrated in Figure 10.8b,d,f,h. Under individual light illumination with a wavelength of about 760 nm, the photocurrent can flow between the two electrodes in Figure 10.8a. The corresponding flow of the electrons and holes in the SnSe can be clearly observed in Figure 10.8b. Under the individual cooling condition, a temperature difference exists across the SnSe, as presented in Figure 10.8c. Figure 10.8d shows the flow of electrons in the n-type SnSe. A thermoelectric inner electric field can be created between the two electrodes, where the energy band can be tilted up between the two electrodes.

Both the photocurrent and the thermoelectric current can flow between the two electrodes, as presented in Figure 10.8e. The acceleration of electrons by the thermoelectric effect-induced electric field can effectively decrease the recombination of electron–hole pairs, as displayed in Figure 10.8f. The photocurrent can be enhanced by the thermo-phototronic effect. Although the incline of energy band is not good for driving the electrons to flow in the SnSe, the thermo-phototronic effect-induced thermoelectric inner electric field dominates the flow of electrons induced by both the thermoelectric effect and photoelectric effect. Under the simulations of light illumination and heating condition, the thermoelectric effect-induced current has been changed to flow between the two electrodes, as displayed in Figure 10.8g. The energy band of SnSe can be tilted due to the thermoelectric effect, as depicted in Figure 10.8h, where the flow directions of the electrons induced by the two effects are opposite. The current coupling enhancement is attributed to the same current directions for the thermoelectric effect and photoelectric effect-induced current signals under the temperature variation process, where the electrons induced by photoelectric effect can be accelerated by the thermo-phototronic effect. This method can be also used for the other thermoelectric materials to enhance the output performances, exhibiting potential applications in photodetectors, solar cells, and so on.

10.5 Conclusions and Prospects

Pyro-phototronic effect is focused on the effective coupling among pyroelectricity, semiconductor, and photoexcitation. The ferro-pyro-phototronic effect is focused on the effective coupling among ferroelectric materials, pyroelectricity, and photoexcitation. The thermo-phototronic effect is focused on the effective coupling among thermoelectricity, semiconductor, and photoexcitation. All the three new physical effects are based on coupled nanogenerators for simultaneously or individually scavenging different energies from the living environment. Both the pyro-phototronic effect and thermo-phototronic effect are based on using semiconducting materials, while the ferro-pyro-phototronic effect is focused on using ferroelectric materials. All the three new physical effects will have potential applications in energy scavenging such as solar cells, and self-powered sensor systems such as photodetectors. The coupling enhancement mechanisms of these new physical effects will be investigated and more high-performance related materials will be chosen for research in future.

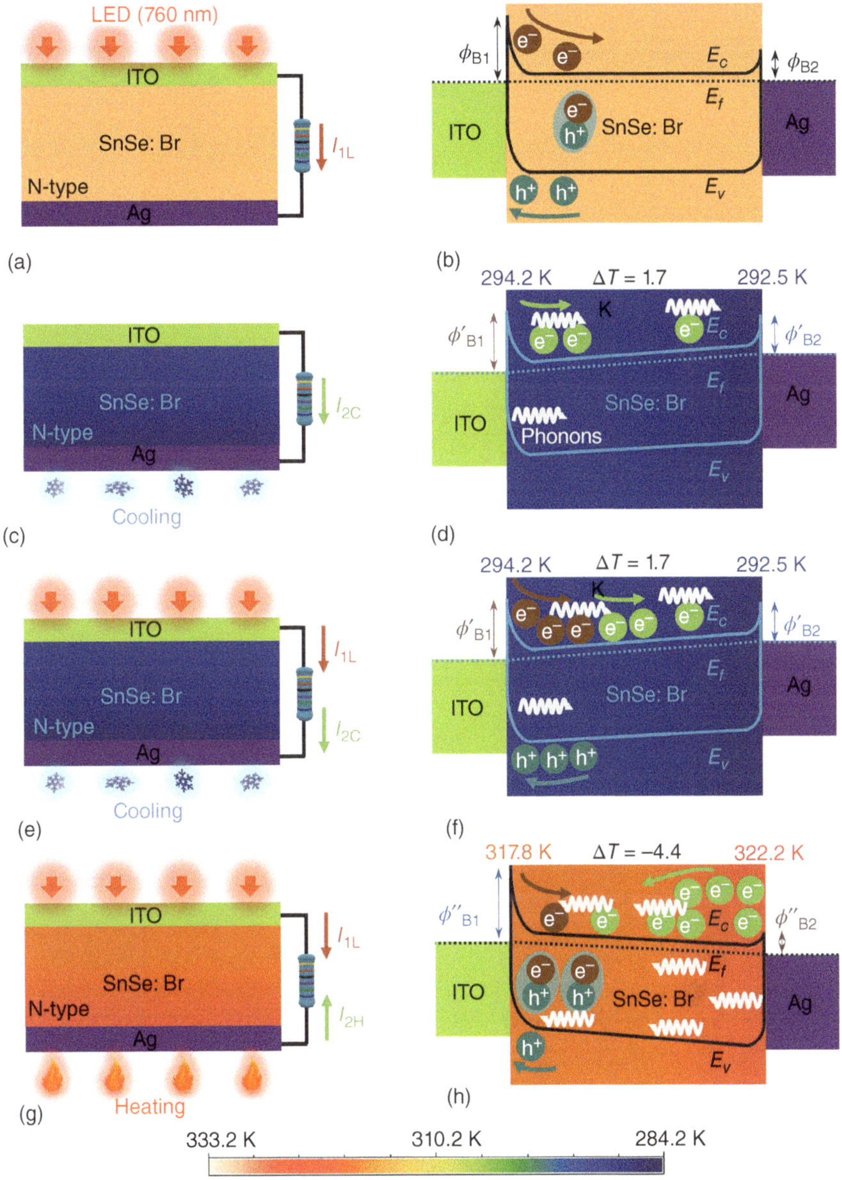

Figure 10.8 Schematic diagrams of the device and the corresponding energy band diagrams under different conditions. (a,b) Schematic working state (a) and energy band diagram (b) of the device under light illumination. (c,d) Schematic working state (c) and energy band diagram (d) of the device under cooling condition. (e,f) Schematic working state (c) and energy band diagram (d) of the device under light + cooling condition. (g,h) Schematic working state (g) and energy band diagram (h) of the device under light + heating condition. Source: Reproduced with permission from Ouyang et al. [63]. Copyright 2019, Elsevier.

References

1 Ji, Y., Zhang, K., Wang, Z.L., and Yang, Y. (2019). Piezo–pyro-photoelectric effects induced coupling enhancement of charge quantity in $BaTiO_3$ materials for simultaneously scavenging light and vibration energies. *Energy Environ. Sci.* 12: 1231.

2 Zhang, K., Wang, S., and Yang, Y. (2017). A one-structure-based piezo-tribo-pyro-photoelectric effects coupled nanogenerator for simultaneously scavenging mechanical, thermal, and solar energies. *Adv. Energy Mater.* 7: 1601852.

3 Ji, Y., Wang, Y., and Yang, Y. (2019). Photovoltaic–pyroelectric–piezoelectric coupled effect induced electricity for self-powered coupled sensing. *Adv. Electron. Mater.* 5: 1900195.

4 Ma, N., Zhang, K., and Yang, Y. (2017). Photovoltaic–pyroelectric coupled effect induced electricity for self-powered photodetector system. *Adv. Mater.* 29: 1703694.

5 Ji, Y., Zhang, K., and Yang, Y. (2017). A one-structure-based multi-effects coupled nanogenerator for simultaneously scavenging thermal, solar, and mechanical energies. *Adv. Sci.* 5: 1700622.

6 Lee, E.J., Kim, T.Y., Kim, S.W. et al. (2018). High-performance piezoelectric nanogenerators based on chemically-reinforced composites. *Energy Environ. Sci.* 11: 1425.

7 Xu, S., Hansen, B.J., and Wang, Z.L. (2010). Piezoelectric-nanowire-enabled power source for driving wireless microelectronics. *Nat. Commun.* 1: 93.

8 Shi, K., Sun, B., Huang, X., and Jiang, P. (2018). Synergistic effect of graphene nanosheet and $BaTiO_3$ nanoparticles on performance enhancement of electrospun PVDF nanofiber mat for flexible piezoelectric nanogenerators. *Nano Energy* 52: 153.

9 Shin, S.H., Choi, S.Y., Lee, M.H., and Nah, J. (2017). High-performance piezoelectric nanogenerators via imprinted sol–gel $BaTiO_3$ nanopillar array. *ACS Appl. Mater. Interfaces* 9: 47.

10 Fan, F., Tian, Z., and Wang, Z.L. (2012). Flexible triboelectric generator! *Nano Energy* 1: 328.

11 Yang, Y., Zhang, H., and Wang, Z.L. (2014). Direct-current triboelectric generator. *Adv. Funct. Mater.* 24: 3745.

12 Yang, Y., Zhang, H., Lin, Z.H. et al. (2013). Human skin based triboelectric nanogenerators for harvesting biomechanical energy and as self-powered active tactile sensor system. *ACS Nano* 7: 9213.

13 Wang, S., Mu, X., Yang, Y. et al. (2015). Flow-driven triboelectric generator for directly powering a wireless sensor node. *Adv. Mater.* 27: 240.

14 Wang, S., Mu, X., Wang, X. et al. (2015). Elasto-aerodynamics-driven triboelectric nanogenerator for scavenging air-flow energy. *ACS Nano* 9: 9554.

15 Chen, B., Yang, Y., and Wang, Z.L. (2017). Scavenging wind energy by triboelectric nanogenerators. *Adv. Energy Mater.* 8: 1702649.

16 Yang, Y., Guo, W., Pradel, K.C. et al. (2012). Pyroelectric nanogenerators for harvesting thermoelectric energy. *Nano Lett.* 12: 2833.

17 Zhang, K., Wang, Y., Wang, Z.L., and Yang, Y. (2019). Standard and figure-of-merit for quantifying the performance of pyroelectric nanogenerators. *Nano Energy* 55: 534.

18 Zhang, D., Wang, Y., and Yang, Y. (2019). Design, performance and application of thermoelectric nanogenerators. *Small* 15: 1805241.

19 Zhang, D., Zhang, K., Wang, Y. et al. (2019). Thermoelectric effect induced electricity in stretchable graphene-polymer nanocomposites for ultrasensitive self-powered strain sensor system. *Nano Energy* 56: 25.

20 Ji, Y., Gao, T., Wang, Z.L., and Yang, Y. (2019). Configuration design of single $BiFeO_3$ photovoltaic device for self-powered electronic watch. *Nano Energy* 64: 103909.

21 Zhang, K., Wang, Z.L., and Yang, Y. (2016). Enhanced P3HT/ZnO nanowire heterojunction solar cells by pyro-phototronic effect. *ACS Nano* 10: 10331.

22 Zhao, K., Ouyang, B., and Yang, Y. (2018). Photocurrent enhancement of radially polarized ferroelectric $BaTiO_3$ materials by ferro-pyro-phototronic effect. *iScience* 3: 208.

23 Zhang, K. and Yang, Y. (2017). Thermo-phototronic effect enhanced InP/ZnO nanorod heterostructure solar cells for self-powered wearable electronics. *Adv. Funct. Mater.* 27: 1703331.

24 Lines, M.E. and Glass, A.M. (1977). *Principles and Applications of Ferroelectrics and Related Materials*. Oxford: Clarendon Press.

25 Lang, S.B. (2005). Pyroelectricity: from ancient curiosity to modern imaging tool. *Phys. Today* 58: 31.

26 Lee, M.H., Guo, R., and Bhalla, A.S. (1998). Pyroelectric sensors. *J. Electroceram.* 2: 229.

27 Vats, G., Kumar, A., Ortega, N. et al. (2016). Giant pyroelectric energy harvesting and a negative electrocaloric effect in multilayered nanostructures. *Energy Environ. Sci.* 9: 1335.

28 Guo, Y., Akai, D., Swada, K., and Ishida, M. (2007). Ferroelectric and pyroelectric properties of highly (110)-oriented $Pb(Zr_{0.40}Ti_{0.60})O_3Pb(Zr_{0.40}Ti_{0.60})O_3$ thin films grown on $Pt/LaNiO_3/SiO_2/SiPt/LaNiO_3/SiO_2/Si$ substrates. *Appl. Phys. Lett.* 90: 232908.

29 Ko, Y.J., Kim, D.Y., Won, S.S. et al. (2016). Flexible $Pb(Zr_{0.52}Ti_{0.48})O_3$ films for a hybrid piezoelectric–pyroelectric nanogenerator under harsh environments. *ACS Appl. Mater. Interfaces* 8: 6504.

30 Ban, Z.G. and Alpay, S.P. (2003). Dependence of the pyroelectric response on internal stresses in ferroelectric thin films. *Appl. Phys. Lett.* 82: 3499.

31 Savage, A. (1966). Pyroelectricity and spontaneous polarization in $LiNbO_3$. *J. Appl. Phys.* 37: 3071.

32 Weis, R.S. and Gaylord, T.K. (1985). Lithium niobate: summary of physical properties and crystal structure. *Appl. Phys. A* 37: 191.

33 Dubourdieu, C., Bruley, J., Arruda, T.M. et al. (2013). Switching of ferroelectric polarization in epitaxial $BaTiO_3$ films on silicon without a conducting bottom electrode. *Nat. Nanotechnol.* 8: 748.

34 Li, X., Lu, S.G., Chen, X.Z. et al. (2013). Pyroelectric and electrocaloric materials. *J. Mater. Chem. C.* 1: 23.

35 Liu, F., You, L., Seyler, K.L. et al. (2016). Room-temperature ferroelectricity in $CuInP_2S_6$ ultrathin flakes. *Nat. Commun.* 7: 12357.
36 Zhou, Y., Wu, D., Zhu, Y. et al. (2017). Out-of-plane piezoelectricity and ferroelectricity in layered α-In_2Se_3 nanoflakes. *Nano Lett.* 17: 5508.
37 Cui, C., Hu, W.J., Yan, X. et al. (2018). Intercorrelated in-plane and out-of-plane ferroelectricity in ultrathin two-dimensional layered semiconductor In_2Se_3. *Nano Lett.* 18: 1253.
38 Xiao, J., Zhu, H., Wang, Y. et al. (2018). Intrinsic two-dimensional ferroelectricity with dipole locking. *Phys. Rev. Lett.* 120: 227601.
39 Poh, P.M., Tan, S.J.R., Wang, H. et al. (2018). Molecular-beam epitaxy of two-dimensional In_2Se_3 and its giant electroresistance switching in ferroresistive memory junction. *Nano Lett.* 18: 6340.
40 Zheng, C., Yu, L., Zhu, L. et al. (2018). Room temperature in-plane ferroelectricity in van der Waals In_2Se_3. *Sci. Adv.* 4: eaar7720.
41 Fei, Z., Zhao, W., Palomaki, T.A. et al. (2018). Ferroelectric switching of a two-dimensional metal. *Nature* 560: 336.
42 Shirodkar, S.N. and Waghmare, U.V. (2014). Emergence of ferroeletricity at a metal- semiconductor transition in a 1T monolayer of MoS2. *Phys. Rev. Lett.* 112: 157601.
43 Fei, R., Kang, W., and Yang, L. (2016). Ferroelectricity and phase transitions in monolayer group-IV monochalcogenides. *Phys. Rev. Lett.* 117: 097601.
44 Ding, W., Zhu, J., Wang, Z. et al. (2017). Prediction of intrinsic two-dimensional ferroelectrics in In_2Se_3 and other II_2-VI_3 van der Waals materials. *Nat. Commun.* 8: 14956.
45 Bruyer, E., Sante, D.D., Barone, P. et al. (2016). Possibility of combining ferroelectricity and Rashba-like spin splitting in monolayers of the 1T-type transition-metal dichalcogenides MX_2 (M= Mo, W; X= S, Se, Te). *Phys. Rev. B* 94: 195402.
46 Ma, N. and Yang, Y. (2019). Boosted photocurrent via cooling ferroelectric $BaTiO_3$ materials for self-powered 405 nm light detection. *Nano Energy* 60: 95.
47 Lenert, A., Bierman, D.M., Nam, Y. et al. (2014). A nanophotonic solar thermophotovoltaic device. *Nat. Nanotechnol.* 9: 126.
48 Kraemer, D., Jie, Q., McEnaney, K. et al. (2016). Concentrating solar thermoelectric generators with a peak efficiency of 7.4%. *Nat. Energy* 1: 16153.
49 Sundarraj, P., Maity, D., Roy, S.S., and Taylor, R.A. (2014). Recent advances in thermoelectric materials and solar thermoelectric generators-a critical review. *RSC Adv.* 4: 46860.
50 Snyder, G.J. and Toberer, E.S. (2008). Complex thermoelectric materials. *Nat. Mater.* 7: 105.
51 Heremans, J.P., Dresselhaus, M.S., Bell, L.E., and Morelli, D.T. (2013). When thermoelectrics reached the nanoscale. *Nat. Nanotechnol.* 8: 471.
52 Blackburn, J.L., Ferguson, A., Cho, C., and Grunlan, J.C. (2018). Carbon-nanotube-based thermoelectric materials and devices. *Adv. Mater.* 30: 1704386.
53 Harman, T.C., Walsh, M.P., Laforge, B.E., and Turner, G.W. (2005). Nanostructured thermoelectric materials. *J. Electron. Mater.* 34: L19.

54 Harman, T.C., Taylor, P.J., Spears, D.L., and Walsh, M.P. (2000). Thermoelectric quantum-dot superlattices with high ZT. *J. Electron. Mater.* 29: L1.
55 Venkatasubramanian, R., Sivola, E., Colpitts, T., and O'Quinn, B. (2001). Thin-film thermoelectric devices with high room-temperature figures of merit. *Nature* 413: 597.
56 Koga, T., Cronin, S.B., Dresselhaus, M.S. et al. (2000). Experimental proof-of-principle investigation of enhanced in (001) oriented Si/Ge superlattices. *Appl. Phys. Lett.* 77: 1490.
57 Zhao, L.-D., Lo, S.-H., Zhang, Y. et al. (2014). Ultralow thermal conductivity and high thermoelectric figure of merit in SnSe crystals. *Nature* 508: 373.
58 Yang, Y., Pradel, K.C., Jing, Q. et al. (2012). Thermoelectric nanogenerators based on single Sb-doped ZnO micro/nanobelts. *ACS Nano* 6: 6984.
59 Abraham, T.J., MacFarlane, D.R., and Pringle, J.M. (2013). High Seebeck coefficient redox ionic liquid electrolytes for thermal energy harvesting. *Energy Environ. Sci.* 6: 2639.
60 Zhao, D., Wang, H., Khan, Z.U. et al. (2016). Ionic thermoelectric supercapacitors. *Energy Environ. Sci.* 9: 1450.
61 Ouyang, B., Zhang, K., and Yang, Y. (2018). Photocurrent polarity controlled by light wavelength in self-powered ZnO nanowires/SnS photodetector system. *iScience* 1: 16.
62 Ouyang, B., Zhao, H., Wang, Z.L., and Yang, Y. (2019). Dual-polarity response in self-powered ZnO NWs/Sb_2Se_3 film heterojunction photodetector array for optical communication. *Nano Energy* 68: 104312. https://doi.org/10.1016/j.nanoen.2019.104312.
63 Ouyang, B., Chang, C., Zhao, L.-D. et al. (2019). Thermo-photoelectric coupled effect induced electricity in n-type SnSe:Br single crystals for enhanced self-powered photodetectors. *Nano Energy* 66: 104111.

Index

Hybridized and Coupled Nanogenerators: Design, Performance, and Applications,
First Edition. Ya Yang.